Multiple Electron Resonance Spectroscopy

Multiple Electron Resonance Spectroscopy

Edited by
Martin M. Dorio
Diamond Shamrock Corporation
Painesville, Ohio

and

Jack H. Freed
Cornell University
Ithaca, New York

Plenum Press · New York and London

Library of Congress Cataloging in Publication Data

Main entry under title:

Multiple electron resonance spectroscopy.

Includes index.
1. Electron paramagnetic resonance spectroscopy — Addresses, essays, lectures.
I. Dorio, Martin M. II. Freed, Jack H.
QD96.E4M84   543'.08   78-27381
ISBN 0-306-40123-1

© 1979 Plenum Press, New York
A Division of Plenum Publishing Corporation
227 West 17th Street, New York, N.Y. 10011

Printed in the United States of America

Contributors

Neil M. Atherton, Department of Chemistry, The University of Sheffield, Sheffield, England

Reinhard Biehl, Institut für Molekülphysik, Freie Universität Berlin, Boltzmannstrasse 20, Berlin 33, Germany

Harold C. Box, Biophysics Department, Roswell Park Memorial Institute, Buffalo, New York 14263

Larry R. Dalton, Departments of Chemistry and Biochemistry, Vanderbilt University, Nashville, Tennessee 37235, and Department of Chemistry, State University of New York at Stony Brook, Stony Brook, New York 11794

Lauraine A. Dalton, Department of Molecular Biology, Vanderbilt University, Nashville, Tennessee 37235. *Present Address:* Department of Chemistry, State University of New York at Stony Brook, Stony Brook, New York 11794

Martin M. Dorio, Diamond Shamrock Corporation, T. R. Evans Research Center, Painesville, Ohio 44077

Jack H. Freed, Department of Chemistry, Cornell University, Ithaca, New York 14853

Marvin D. Kemple, National Bureau of Standards, Washington, D.C. 20234. *Present Address:* Department of Physics, Indiana University–Purdue University, Indianapolis, Indiana 46205

Larry Kevan, Department of Chemistry, Wayne State University, Detroit, Michigan 48202

Lowell D. Kispert, Department of Chemistry, The University of Alabama, Tuscaloosa, Alabama 35486

Daniel S. Leniart, Instrument Division, Varian Associates, 611 Hansen Way, Palo Alto, California 94303

Haim Levanon, Radiation Laboratory and Department of Chemistry, University of Notre Dame, Notre Dame, Indiana 46556. *Permanent Address:* Department of Physical Chemistry, The Hebrew University, Jerusalem, Israel

Klaus Möbius, Institut für Molekülphysik, Freie Universität Berlin, Boltzmannstrasse 20, Berlin 33, Germany

P. A. Narayana, Department of Chemistry, Wayne State University, Detroit, Michigan 48202

Richard H. Sands, Biophysics Research Division, Institute of Science and Technology, and Department of Physics, University of Michigan, Ann Arbor, Michigan 48109

Charles P. Scholes, Department of Physics, State University of New York at Albany, Albany, New York 12222

Contents

9. ENDOR and ELDOR on Iron–Sulfur Proteins 331
Richard H. Sands

10. Radiation Biophysics 375
Harold C. Box

11. Polymer Studies 393
Martin M. Dorio

Multiple Electron Resonance Spectroscopy: An Introduction

Martin M. Dorio

Since its discovery,[1] electron paramagnetic resonance (EPR) has flourished to the point where several hundred papers per year are now published. This has resulted in the concomitant increase in review articles, which now appear regularly.[2–4] "It is doubtful that ... any chemist would have been so bold as to predict the great diversity of systems which have proved amenable to study by EPR spectroscopy."[5] EPR has, of course, proven itself useful because of its high sensitivity in the identification and study of a variety of paramagnetic species. Systems studied range from single crystals containing unpaired electrons to complex biological and polymeric matrices. However, the limitations of the spectral interpretations have become increasingly apparent as the user population has become more sophisticated in the types of problems to which EPR was applied. EPR linewidth changes, for example, may occur over the range of rotational correlation times from approximately 1×10^{-9} to 1×10^{-7} sec. Hence, EPR is less applicable at regions of much slower or faster molecular motions. Further, a scientist is frequently confronted with significantly overlapped EPR spectra due to the presence of multiple radical species. Again, interpretations of the spectral displays are difficult. The study of molecular energy transfer and relaxation processes was also quite difficult from EPR spectra alone.

Multiple electron resonance spectroscopy (MERS) had its birth 11 years after Zavoisky's work with George Feher's early experiments on electron-nuclear double resonance[6] (ENDOR). Since then, an ever increasing number of experimental and theoretical papers have appeared. These have, by the

Martin M. Dorio • Diamond Shamrock Corporation, T. R. Evans Research Center, Painesville, Ohio

addition of the nuclear radiofrequency field, broadened the capabilities of scientists to study increasingly complex problems. With the advent of commercial instrumentation, ENDOR has now become routinely available in a great variety of fields.

Four years after Feher's now classic ENDOR work, the first application of dual microwave frequencies was published by P. Sorokin, G. Lasher, and I. Gelles[7] as a test of the ideas of Bloembergen *et al.*[8] on cross-relaxation. In that work[7] on nitrogen centers in diamond, four simultaneous nitrogen spin flips were found to account for the cross-relaxation. The extraction of individual nitrogen cross-relaxation times was carried out. So, from the very first, electron–electron double resonance (ELDOR) had shown its capabilities to permit the study and resolution of molecular relaxation questions.

Subsequently, Unruh and Culvahouse[9] performed pulsed ELDOR on Co^{2+} in lanthanide crystals. Again, relaxation information was obtained. The temperature dependence of the relaxation rates was obtained in the region of 1.18–4.2°K. Unruh and Culvahouse have been the only authors to date to apply ELDOR to transition metals. Presumably this is due to the difficulties of working at liquid helium temperatures.

Moran[10] introduced continuous pumping into ELDOR technology, providing in his paper a significant amount of information regarding the theory and experimental technique. His spectrometer permitted double resonance investigations to well within 100 kHz of the observing frequency, which he had positioned at the center of the absorption signal. His ELDOR response was found to be symmetrical about the observing point and increased in magnitude as the pump power increased. Moran's intention in his work was to test Portis' model[11] of static spin packets distributed within an inhomogeneous envelope (the $T_1 - T_2$ model). The saturation behavior for the system had previously indicated that the absorption signal exhibited behavior intermediate between the purely homogeneous and inhomogeneous cases. Only with weak pump power were the theoretical and experimental curves the same. When the power increased, forbidden transitions manifested themselves as structure on the ELDOR spectra. These were the first examples of forbidden structure, and the ELDOR magnitudes were larger and persisted for greater frequency separation than the $T_1 - T_2$ model predicted.

Roughly 8 years after the experiments began, the first solution application of ELDOR was explored both in James Hyde's laboratory[12] at Varian and in Russia.[13] Since then, the advent of commercial instrumentation has, as in ENDOR, broadened the problems that can be attacked. It should be noted that few if any of the latest ELDOR instruments allow the close 100 kHz frequency separation the Moran instrument did.

It is our purpose in this book to present an up-to-date review of the details of these new multiple resonance techniques. We have spent considerable space on detail, using as our basis the texts by Kevan and Kispert,[14]

Abragam and Bleaney,[15] and others.[16-19]

Basically, this volume deals with those spectroscopic techniques that use EPR as a detection method. Chapters 2 through 5 cover the experimental and theoretical aspects of multiple resonance spectroscopy. Chapters 6 through 12 are systems-oriented and deal with the multiple resonance techniques applied to crystals, glasses, bioproteins, polymers, and triplets. The concepts of the first five chapters illustrate the strength of MERS to solve a broad range of problems. Chapters 13 and 14 are rather detailed introductions to two of the latest new applications: TRIPLE resonance and optical perturbations in EPR. The latter is to be distinguished from the OMDR (*optical magnetic double resonance*) techniques, in which the optical system is the detection method.

References

1. E. J. Zavoisky, *J. Phys. U.S.S.R.* **9**, 211 (1945).
2. J. S. Hyde, *Ann. Rev. Phys. Chem.* **25**, 407 (1974).
3. N. M. Atherton, *Electron Spin Resonance* (Specialist Periodical Reports, The Chemical Society, London) **1**, 32 (1972); **2**, 35 (1974).
4. J. H. Freed, *Ann. Rev. Phys. Chem.* **23**, 265 (1972).
5. J. E. Wertz and J. R. Bolton, *Electron Spin Resonance, Elementary Theory and Practical Applications*, p. xii, McGraw-Hill Book Co., New York (1972).
6. G. Feher, *Phys. Rev.* **103**, 834 (1956).
7. P. P. Sorokin, G. J. Lasher, and I. L. Gelles, *Phys. Rev.* **118**, 939 (1960).
8. N. Bloembergen, S. Shapiro, P. S. Pershan, and J. O. Artman, *Phys. Rev.* **114**, 445 (1959).
9. W. P. Unruh and J. W. Culvahouse, *Phys. Rev.* **129**, 2441 (1963).
10. P. Moran, *Phys. Rev.* **135**, A247 (1964).
11. A. M. Portis, *Phys. Rev.* **91**, 1071 (1953).
12. J. S. Hyde, J. C. W. Chien, and J. H. Freed, *J. Chem. Phys.* **48**, 4211 (1968).
13. V. A. Benderskii, L. A. Blumenfeld, P. A. Stunzhas, and E. A. Sokolov, *Nature* **220**, 365 (1968).
14. L. Kevan and L. D. Kispert, *Electron Spin Double Resonance Spectroscopy*, John Wiley and Sons, New York (1976).
15. A. Abragam and B. Bleaney, *Electron Paramagnetic Resonance of Transition Ions*, Oxford Press, New York (1970).
16. N. M. Atherton, *Electron Spin Resonance*, John Wiley and Sons, New York (1973).
17. H. M. Swartz, J. R. Bolton, and D. C. Borg, *Biological Applications of Electron Spin Resonance*, Wiley-Interscience, New York (1972).
18. J. E. Wertz and J. R. Bolton, *Electron Spin Resonance, Elementary Theory and Practical Applications*, McGraw-Hill Book Co., New York (1972).
19. P. F. Knowles, D. Marsh, and H. W. E. Rattle, *Magnetic Resonance of Biomolecules*, John Wiley and Sons, New York (1976).

2

Instrumentation and Experimental Methods in Double Resonance

Daniel S. Leniart

1. ENDOR

1.1. The Basic ENDOR Spectrometer

1.1.1. General Remarks

The ENDOR technique was introduced by Feher[1] for the investigation of single crystals in 1956 and continued through its embryonic stages to be used as a tool for the investigation of solid state phenomena. In 1963, the technique was extended to systems in the liquid phase by Cederquist[2] as applied to metal–ammonia solutions. In 1964, Maki chose the organic radical galvanoxyl as a candidate for ENDOR in an organic solvent, *n*-heptane, and collaborating with Hyde at Varian Associates, a prototype instrument was constructed that produced the first ENDOR spectrum of an organic free radical in solution.[3]

This marked the beginning of what is now termed " high-power " ENDOR and led to the development of the first commercially available system delivered in the fall of 1967. Although low-power ENDOR will be discussed, the major emphasis of this chapter is devoted to high-power ENDOR instrumentation and experimental variables. Currently, almost all high-power ENDOR spectrometers are capable of performing low-power ENDOR experiments, the key piece of hardware being a microwave cavity supporting an ENDOR coil, usually a single-turn loop, that conforms optimally to the

---

Daniel S. Leniart • Instrument Division, Varian Associates, Palo Alto, California

sample shape so as to produce the maximum rf field when used in conjunction with a 5- or 10-W broad-band low-power amplifier. This section gives a review of the basic ENDOR spectrometer system as a whole and, in particular, the components that comprise the essential ENDOR subsystems.

Since roughly 1967 a great body of ENDOR literature has appeared which simplifies the interpretation of complex hyperfine splitting patterns observed in ESR spectra of organic free radical in nonviscous solvents by enabling the investigator to accurately measure the magnitude of the various hyperfine splittings present. In essence, this is possible because equivalent nuclei of spin I give rise to $2NI + 1$ ESR hyperfine lines, while they exhibit only a single ENDOR line.

The same resolution increase has been studied with inhomogeneously broadened single ESR lines or ESR powder patterns. Here again, the ENDOR spectra give rise to lines occurring at specific rf frequencies that allow for spectral interpretations unable to be made with just the ESR data. Examples of the measurement of hyperfine splittings of a variety of samples are given elsewhere in this volume.

However, the ENDOR experiment allows one to measure a variety of parameters other than the value of the hyperfine splitting. The following section describes linewidth and percent enhancement ENDOR studies that allow one to study relaxation mechanisms, molecular dynamics, Heisenberg exchange effects, and correlation times. Methodology for measuring important experimental parameters needed for interpretation, such as rf field at the sample, are given.

Finally, Section 1.3 deals with the experimental parameters under operator control and the effects that variation of these parameters have on the ENDOR spectrum. These include field modulation amplitude effects, microwave power effects, coherence effects, pulse rate effects, and temperature effects.

1.1.2. *ENDOR Instrumentation*

The ENDOR spectrometer is comprised of those components in Figure 1 and includes an rf oscillator and servo-controlled rf power amplifier. This rf amplifier chain subsystem is responsible for generating rf fields at the sample capable of saturating nuclear transitions having spin–lattice relaxation times, T_{1n}, ranging from seconds to microseconds. The servo system tunes an LC circuit by varying the amplifier capacitance with respect to the rf frequency being swept. Probably the single most significant component is the resonant cavity structure that supports an rf field that is mutually orthogonal to both the microwave magnetic field and the main dc magnetic field. This subsystem is described below.

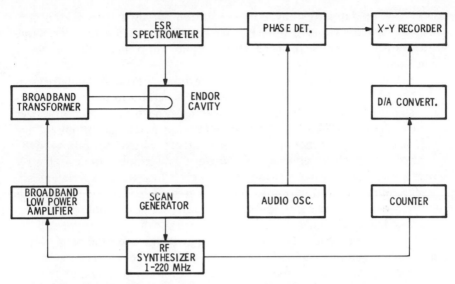

Figure 1. Block diagram of a typical low-power broad-band ENDOR spectrometer.

After positioning the dc magnetic field on the ESR line, lines, or spectral region of interest and then locking the microwave frequency to the magnetic field (i.e., locking to a particular resonance condition), the ENDOR experiment is performed by sweeping the radiofrequency over the range of allowable NMR transitions. Any perturbation of the ESR signal amplitude or line shape caused by inducing an NMR transition at a particular radiofrequency results in a net dc change from the phase-sensitive detector that serves as the Y-axis drive for the recorder. If the rf output serves as the X axis of the X–Y recorder, then a signature appears relating the changes in ESR signal intensity at particular rf frequencies. This is the basic ENDOR experiment.

1.1.2.1. *Models of ENDOR Spectrometers—Low Power Instrumentation*[4]

a. *Scan generator programmer.* The scan generator is generally of modern digital design and is based on a digital clock. The clock pulses are transformed to an analog voltage via accumulation by binary counters—the output of which is fed to a D/A converter. One chooses a circuit of such design that the linearity and accuracy of the scan sequence design specification complies with the requirements necessary to do the experiment. Several scan sequences are provided by the scan generator. The operator has the option of selecting rf sweeps of increasing or decreasing frequency as well as single scan in both directions (triangular waveform). A hold position is convenient for stopping the scan at a particular frequency by holding the ramp voltage constant at any position along the scan. A selection switch capable of adjusting the minimum output voltage as well as the range enables the operator to study both broad and narrow ENDOR lines at any position between 1 and 220 MHz. In order to provide a first-derivative ENDOR display, the scan generator output ramp may be frequency modulated.

b. *Audio oscillator and frequency synthesizer.* The audio oscillator provides the FM modulation signal for the scan generator as well as the reference signal for the phase-sensitive detector. The frequency synthesizer produces the 1–220 MHz nuclear resonance frequencies. The output voltage of the scan generator generally is used to control the frequency output of the rf source. Although a frequency synthesizer has a number of desirable requirements, those most important to a successful ENDOR experiment are linearity, stability, and reproducibility. Increased performance in these specifications is, in general, directly proportional to selling price. The frequency linearity should be such that identical changes in ramp voltage produce the same frequency differences at any position on the band, while the frequency stability should be governed solely by the scan generator and not the synthesizer. The reproducibility specification is governed by the fact that, each time the scan generator clock produces a specific digitally controlled analog output voltage, the frequency is the same. It is our experience that frequency synthesizers offered by Hewlett-Packard, General Radio, and Programmed Test Sources, among others, meet the general requirements for the ENDOR experiment.

c. *Broad-band low-power amplifiers.* The general broad-band power amplifier used with a standard low-power ENDOR system provides a gain of roughly 1000 over the standard frequency range of 1–220 MHz. The amplification should be linear over these frequencies with no frequency discrimination. Although the output power depends on the requirements of the experiment, 10 W continuous wave is common in many ENDOR systems, and the magnitude of H_{rf} at the sample is controlled by the rf coupling loop surrounding the sample of interest. Electronic Navigation Industries (ENI), Amplifier Research Corporation, and RF Communications are among the many commercial brands of low rf power amplifier lines commonly used in low-power ENDOR systems.

d. *Broad-band transformer.* The matching of the amplifier output to the rf loop housed in the microwave cavity is a function of the broad-band transformer and the quintessence of a low-power ENDOR spectrometer. Since the impedance (reactance) will vary directly as the frequency (e.g., a factor of 220 from 1 to 220 MHz), the current in the rf loop, and thus H_{rf}, will vary in a like manner. However, by matching the amplifier output impedance to that of the ENDOR coil at all frequencies, H_{rf} at the sample should be constant. Gunnel[5] has proposed very practical broad-band impedance-matching networks suitable for both low and high frequencies in the form of transmission-line transformers. Transmission-line transformers are wound

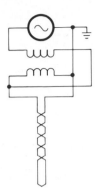

Figure 2. An example of the rf transmission line used in low-power
ENDOR.

so that the interwinding capacity, which resonates with the leakage inductance in conventional transformers producing a loss peak, is a component of
the characteristic impedance of the line. Hence, no resonances occur to limit
the bandwidth. The high-frequency response is maintained by use of a short
length of transmission line, usually in the form of a twisted pair, while the
low-frequency response of these transformers is determined by the primary
inductance and requires high-permeability ferrite toroids (Figure 2).
Ruthroff[6] describes a typical ENDOR impedance transformer.

   e. *ENDOR cavity.* The ENDOR cavity structure used for low-power
work is a fairly straightforward yet extremely critical component of the
spectrometer system. The Q of the microwave portion of the cavity must be
degraded only slightly upon insertion of the rf loop. Typical low-power
ENDOR cavities[7-9] must be capable of being matched with samples of varying dielectric constant. One mechanism of critically coupling the cavity is the
insertion of dielectric material at various depths and positions within the
waveguide. The approach, called a slide-screw tuner, enables one to completely immerse the cavity for low-temperature studies. A standard iris coupling mechanism can be used with flowing-gas temperature control systems.

   f. *Rf loop.* The exact shape and size of a low-power rf coil is determined
many times by the size and shape of the sample. It should conform as closely
as possible to the sample geometry, e.g., an rf coil for a small single crystal
study might be quite different than that used for a nonlossy solvent study. In
general, the coil is a single-turn cylindrical loop and positioned along the
axis of a cylindrical microwave cavity so that, under normal conditions, the
oscillating rf magnetic field produced by the loop is orthogonal to both
the microwave and dc magnetic fields. In general, a 0.5–1 G rf field in
the rotating frame is produced when the loop is coupled to a 5–10 W rf
power amplifier. This is usually sufficient to saturate nuclear transitions
having T_{1n} of 10^{-3} to 10^{-4} sec.

   *1.1.2.2. Models of ENDOR Spectrometers—High-Power Instrumenta-
tion.* There are three types of commercially available high-power ENDOR

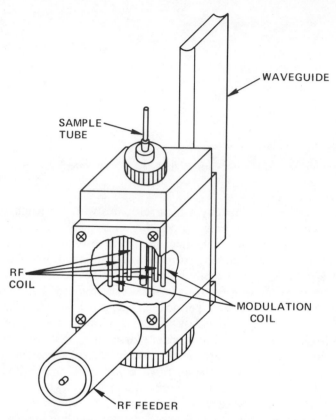

Figure 3. An example of a TE_{011} cylindrical cavity with a set of internal rf coils and internal modulation coils.

spectrometers. JEOL and BRUKER offer a continuous wave (CW) model[10-15] that employs a wideband power amplifier and enables experiments to be carried out with the radiofrequency being continuously swept and detected. The Instrument Division of Varian Associates offers a pulsed ENDOR system that pulses the rf field and simultaneously gates the detection system as the nuclear radiofrequency is swept.[16]

The model ES-EDX1 JEOL ENDOR spectrometer is compatible with the JES-PE3X ESR spectrometer system and consists of a TE_{011} cylindrical cavity with a set of internal rf coils and modulation coils both placed inside the cavity structure as shown in Figure 3. The rf coil is configured in a two-loop arrangement and is made from four copper rods having a 2-mm outside diameter. The loaded cavity Q is estimated to be ~ 6000. The rf source is Hewlett-Packard sweep generator model 8601A (0.1–100 MHz), which is frequency-modulated at a frequency of 6.5 kHz. The output is amplified by a distributed amplifier to about 20 W, and this amplified output is used to drive a high-power amplifier up to 200 W. The output circuit of the rf power amplifier is a parallel resonant circuit derived from the inductance of the rf coil in the ENDOR cavity and a variable vacuum capacitor. The latter is used for tuning the LC circuit as a function of rf frequency and provides a high-power frequency range of 3–50 MHz in three separate bands. A standard double-modulation-frequency technique is employed.[16] The ESR signal is field-modulated at 80 Hz. Upon passing through a nuclear resonance, the enhanced ESR signal has both a 6.5-kHz component from the frequency-modulated radiofrequency and an 80-Hz component from the field modulation. Phase-sensitive detection first at 6.5 kHz suppresses the 6.5-kHz carrier and passes the ± 80-Hz sidebands. If no ENDOR information is present, no signal passes through the 6.5-kHz-tuned amplifier and no 80-Hz ESR (or ENDOR) signal appears on the recorder. If an ENDOR resonance is traversed, a 6.5-kHz signal appears with sidebands at ± 80 Hz, as well as an 80-Hz ESR signal. The latter is suppressed along with the carrier, and the sidebands containing the ENDOR information are passed by the tuned stage at the higher frequency and then are phase-sensitive detected at the field modulation frequency (80 Hz). The dc output of the second audio phase detector is displayed on the recorder.

The commercial BRUKER ER-420 ENDOR spectrometer is compatible with ER-420 ESR spectrometer and employs a microwave helix instead of a cavity structure to support the microwave field.[10] The helix diameter is 3 mm and is mechanically stabilized by a Teflon mounting structure that houses the slow-wave structure. Wrapped around the outside of the helix is an rf loop having a diameter of 12 mm and constructed from a pair of coils, each having 13 turns of silver-plated copper wire, connected in parallel. The small distance between coils allows for rf fields up to 14 G in the rotating frame to be achieved over a typical frequency range of 8–38 MHz with a single coil. Other coil configurations allow sweeps between 2–10 and 35–60 MHz. A 20–100 W broad-band power amplifier is employed. Because of the diameter of the microwave helix, the sample size is limited to sample tubes that have an outside diameter of less than 3 mm. The ENDOR signals are obtained by frequency-modulating the rf at 9.3 kHz and field-modulating at some lower frequency. The ENDOR signal, as in the JEOL spectrometer, is phase-sensitive detected first at the higher frequency and then at the lower frequency.

The Varian E-1700 ENDOR spectrometer system is compatible with all complete Varian E-Line and E-Line Century Series X-band ESR spectrometers having a Varian 9-in.-or-larger magnet system. The design, first conceived in 1965 and later modified in 1973, differs from other models in that the rf power generally is amplituded modulated (i.e., pulsed) at a frequency of 1 kHz. In order to reduce sample and cavity heating arising from the intense rf fields produced by a standard 1-kW "high-power" amplifier, a duty cycle with steps of 5, 10, 25, 50, and 100% is employed to allow for operator control of the average power dissipated in the cavity structure. With the standard E-1735 large-access ENDOR cavity, typical rf fields of 10–15 G in the rotating frame are attained. The four rods composing the rf loop in this cavity are configured in a square and separated by 1.250 in., allowing for insertion of large volume samples—typically powders or low-loss organic solvents. With the E-1737 typical-access ENDOR cavity (i.e., 11-mm stack diameter), rf fields of 40–55 G may be achieved. The rf rods are configured in a rectangular pattern and separated by ~ 0.7 in. on the long side and ~ 0.4 in. on the shorter side. The smaller rf dimensions are optimized for smaller sample volumes, such as high-dielectric or lossy samples, single crystals, and sample-limited experiments. The detection system is gated simultaneously with the rf pulses and is on only during the time of the rf pulse to amplify the ESR or ENDOR signal present and is off otherwise, preventing the amplification of noise when the rf field is not present. In addition to amplitude modulation of the rf field, the magnetic field is modulated at some lower frequency (e.g., 35 Hz), and the ENDOR signal is ultimately phase-sensitive detected at 35 Hz and presented for display. A typical sweep range of the radiofrequency is 6–44 MHz at 1 kW of rf power.

The transmitter section is designed for other magnetic resonance experiments besides high-power ENDOR. The transmitter section, with the final 1-kW amplifier omitted, is excellent for this purpose. In this mode, the frequency range extends from 220 kHz to 220 MHz. Likewise, this configuration will also be useful as a radiofrequency source for wideline NMR, nuclear quadrupole resonance, and dynamic nuclear polarization studies.

Another important property of the ENDOR experiment is evident when one studies organic free radicals having narrow (less than a gauss) linewidths and/or well-separated hyperfine lines. Under these experimental conditions, ESR resonant condition must be held very constant in order to insure that the ENDOR information always is obtained at the ESR extremum. Hence, a normal ESR spectrometer employs an AFC loop to lock the klystron frequency to the cavity resonance and a Hall probe to separately control the magnetic field. Increased stability necessary for many ENDOR studies is provided by a field/frequency lock system quite common to NMR. Microwaves used to irradiate the sample under investigation are also coupled to a second DPPH sample located in a microwave helical structure and situated in the magnet gap. Any drift in the microwave source (unlikely) results in a corrective change in the magnetic field; similarly, any drift in the magnetic field (more likely) is corrected by an applied voltage to the magnet power supply. Finally, any drift in the cavity resonance frequency (quite likely, especially at high rf powers) is accompanied by a corresponding klystron frequency change via the AFC, and this again results in a corrective change in the magnetic field.

Since many commercial Varian ENDOR spectrometers are currently in use and the author is experienced with such systems, a more detailed description is given below to point out the general principles involved in ENDOR instrumentation.

1.1.2.3. *Varian E-1700 ENDOR Spectrometer.* A block diagram of the E-1700 ENDOR is given in Figure 4. Spectrometer operational control is derived from the E-1704 system control module. The unit permits selection of the various modes of operation including:

1. EPR using 1-kHz field modulation.
2. Second-derivative ESR using 1-kHz and 34.7-Hz field modulations.
3. ENDOR using 1-kHz radiofrequency modulation and 34.7-Hz field modulation.
4. ENDOR using 1-kHz field modulation and 34.7-Hz radiofrequency modulation.
5. ENDOR-enhanced ESR—the ESR spectrum observed when the radiofrequency is on and set to an ENDOR transition. It is useful for measuring absolute percent enhancements.
6. ENDOR-induced ESR—the difference between the ESR spectrum and the ENDOR-enhanced ESR spectrum.

Operational mode (3) is the most commonly used ENDOR configuration and is described in detail in Section 1.2. The radiofrequency is generated from the E-1707 frequency synthesizer operating between 400 kHz and 160 MHz. The spectral purity must be excellent and the harmonic content should be between 27 and 30 dB below the carrier level, while the output level variation should be less ± 0.3 dB from 2 to 160 MHz and ± 1 dB from 400 kHz to 2 MHz. The output of frequency synthesizer is fed to the E-1709 rf driver amplifier, a low-power broad-band amplifier that provides 10 W output over the entire frequency range of the synthesizer. This output is fed to the E-1703 high-power rf amplifier, which further amplifies the output of the 10-W amplifier to 1 kW peak power and is a servocontrolled, tracking, narrow-band amplifier tuned to resonate with the rf coil around the sample. The tuning range is 6–44 MHz in a single band. The output of the rf power amplifier is coupled directly to the E-1735 or E-1737 ENDOR cavity complete with associated hardware for variable-temperature capability—the latter being a very important parameter for generating ENDOR signals. The long-term resonance stability needed for ENDOR is provided by the E-272B field/frequency lock module. A long-term stability of 1 ppm per centigrade degree insures optimum ENDOR performance on narrow ESR lines.

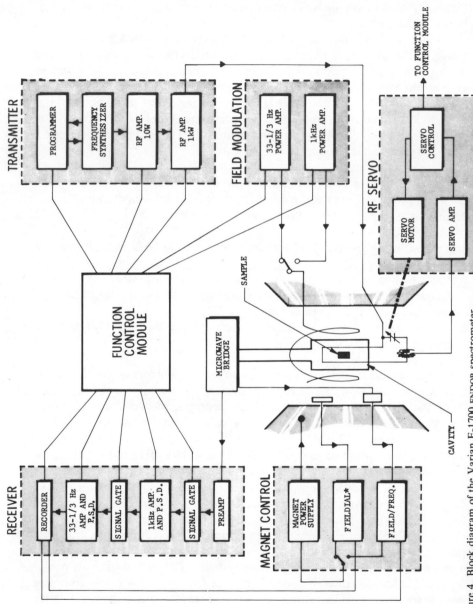

Figure 4. Block diagram of the Varian E-1700 ENDOR spectrometer.

The E-1708 ENDOR programmer acts as the scan generator and includes the following capabilities:

1. Scan width control from 20 kHz to 100 MHz in 12 steps.
2. Scan time control (the time required to scan the selected width) 0.5–256 min in 10 steps.
3. Radiofrequency step size is 10 Hz or [(scan width)/10,000], whichever is greater. The rf step size is much less than any ENDOR linewidths observed to date.
4. Recorder control. The synthesizer controls the X axis of the E-Line recorder. The ENDOR spectra are linear in frequency and are on calibrated charts.
5. Radiofrequency scan start control. The radiofrequency scan starts from this selected frequency and proceeds either up or down as desired at a rate determined by the scan width and scan time controls.
6. Digitally controlled frequency modulation with selectable deviations from ± 1 to ± 999 kHz in 1-kHz steps.
7. Selectable duty cycle (amplitude modulation) 5, 10, 25, and 50%.
8. Direct-reading LED display of the radiofrequency.

The E-1706 low-frequency module is a modified version of the standard ESR unit and provides the 1-kHz receiver and phase-sensitive detector. The E-1705 power supply provides the required power for all of the modules described above except the rf power amplifier. The latter necessitates a separate power source, the E-1702 power supply.

Certain of the ESR spectrometer components are critical to ENDOR performance, and proper modifications must be made to insure good performance. The preamplifier, located in the microwave bridge, must be reshielded and modified to greatly reduce the effects of radiofrequency interference (RFI). The normal low-frequency module supplied with the ESR spectrometer is modified to permit operation at 34.7 Hz rather than 35 Hz in order to eliminate RFI from higher harmonics.

1.2. *ENDOR Measurements*

1.2.1. *Initial Setup*

The first step in the experimental setup is to obtain an oscilloscope display of the normal first-derivative ESR spectrum. If the hyperfine lines are well separated, then one generally selects the ESR signal with the greatest intensity by shifting it to the center of the scope display and reducing the audio field modulation (i.e., the second modulating field regulating the oscilloscope X axis) until only a single extremum of the ESR hyperfine line fills the display. The resonance condition is locked with the field frequency unit.

However, if the spectrum has many hyperfine lines or is a powder, then the field modulation amplitude is adjusted to encompass a significant fraction of the spectral width to increase the observable ENDOR intensity. Again, the exact dc field is chosen and the resonance condition stabilized with the field frequency lock.

Next, the temperature of the sample is preset to some initial value, and the incident microwave power is varied until the ESR signal display is maximized; usually this will take place at the onset of saturation. Radiofrequency power is applied to the sample, and the entire cavity structure is allowed to thermally equilibrate. Finally, the radiofrequency is swept at some given rate using an appropriate time constant. If no ENDOR signal appears, the sample temperature should be readjusted, microwave saturation rechecked, and the rf scan reinitiated. Once the ENDOR signal is obtained, the microwave power and modulation amplitude (i.e., the ESR parameters) are readjusted to maximize the ENDOR signal-to-noise ratio. Finally, the duty cycle and rf power (ENDOR parameters) are selected that give the largest ENDOR signal or desired experimental observable. Section 1.3 describes some of the subtleties associated with the experimental variables mentioned above.

1.2.2. Signal Tracing

1.2.2.1. *ENDOR and EI-ESR at 1-kHz rf Pulsing and 34.7-Hz Field Modulation.* The standard 34.7-Hz field-modulated ENDOR and ENDOR-induced ESR signal paths are depicted in Figure 5. The frequency components of the signal-emanating microwave detector crystal are the 34.7-Hz ESR signal and the 1-kHz carrier with the ENDOR information contained in the sidebands at ± 34.7 Hz (i.e., 1 kHz $\pm$ 34.7 Hz and a 1-kHz signal generated by the rf pulsing but containing no desirable information). The latter signal, in fact, has unwanted effects as described in Section 1.3. This signal train is fed to the input gate through the input filter, removing both the dc level from microwave rectification and the low-frequency noise components. The 2-kHz gating frequency up-converts all of the local noise components near the 34.7-Hz ENDOR information to 2 kHz $\pm$ 34.7 Hz. All frequencies outside of the bandwidth of tuned stage of the 1-kHz amplifier are discriminated against allowing passage of the 1 kHz $\pm$ 34.7 Hz ENDOR information to be phase-sensitive detected at frequency 1 kHz. The components and waveforms of the signals passing through the gate, amplifier, and PSD (*phase-sensitive detector*) are shown in detail in Figure 6. The 1-kHz PSD down-converts the ENDOR information from 1 kHz $\pm$ 34.7 Hz to 34.7 Hz, and this signal is filtered (high-pass, knee < 20 Hz), amplified, and phase-sensitive detected at the audio frequency. The resulting dc ENDOR signal is displayed on the recorder.

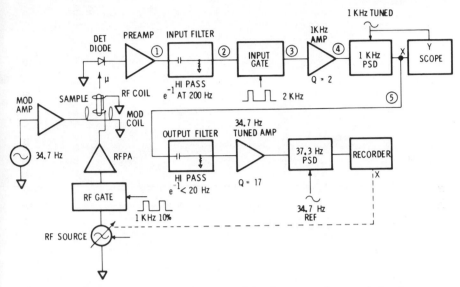

Figure 5. Signal paths for 34.7-Hz field-modulated ENDOR and ENDOR-induced ESR.

The signal emanating from the detector diode has the waveform shown in Figure 6 regardless of whether one sets the dc field to the positive or negative lobe of the ESR signal, the only effect on the waveform being a phase shift of 180°. Since the ESR signal at this point is an absorption, the rf information must always have the same sense, i.e., *all pulses must be above the waveform ac signal*, as long as the ESR signal is enhanced. This interpretation is different than that shown in Figure 2 of Leniart et al.[17] and is believed to be correct.

The purpose of the input filter is to eliminate noise at the audio frequencies where the ENDOR information will ultimately be detected. Also, the input gate, the 1-kHz tuned amplifier, and the audio output filter serve a similar purpose.

The waveform of the amplitude modulated carrier signal that provides the 1-kHz PSD input signal (Figure 6) depends to a certain extent on the field modulation amplitude. Employing large-modulation amplitudes of such magnitude that a portion of the field being scanned does not result in an ENDOR enhancement generates a type of suppressed carrier waveform at this point in the signal chain. For example, the waveform may be

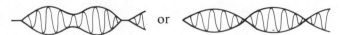 or

1.2.2.2. *ΔF Operation.* ΔF operation is especially convenient for emphasis of small changes in curvature that accompany normal first-derivative

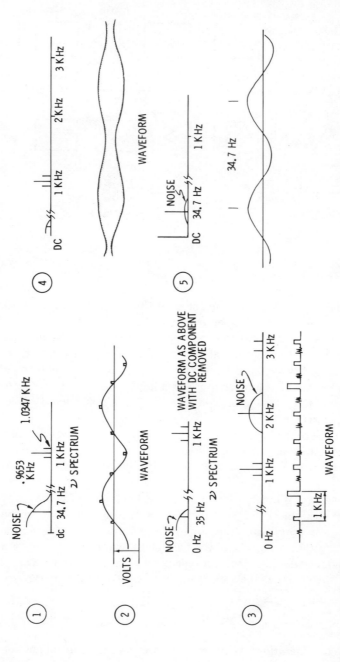

Figure 6. Frequency components and waveforms of ENDOR signal passing through the spectrometer system: (1) emanating from the detector diode, (2) emanating from the input filter, (3) from the input gate, (4) from the 1-kHz amplifier, (5) from the 1-kHz PSD.

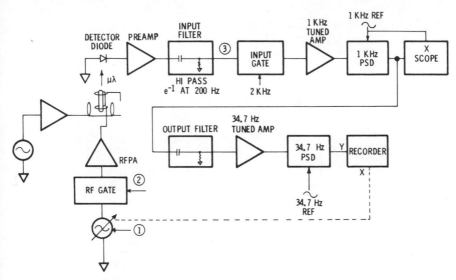

Figure 7. Typical ENDOR block diagram for modulating the magnetic field at 34.7 Hz and square wave modulating the rf at a 1-kHz rate with a selectable frequency deviation.

displays. Also, the mode tends to diminish any baseline drift problems attributed the RFI at high rf powers.

Figure 7 shows the typical block diagram for modulating the magnetic field at 34.7 Hz and square-wave frequency modulating the rf at a 1-kHz rate with a selectable frequency deviation. The frequencies can, of course, be reversed. The rf signal from the frequency synthesizer is square-wave frequency modulated at a 1-kHz rate and fed to the rf gate. The gate chops the frequency-modulated signal, but this time at a 2-kHz rate (see Figure 8). In this manner, alternate pulses are shifted in frequency above and below the value shown on the LED display by an amount equal to the magnitude of the frequency deviation.

For example, suppose the rf frequency is at $v_0 = 15.5$ MHz and the frequency deviation Δf is chosen to be 100 kHz. Two pulses of rf will appear, one at 15.6 MHz and the other at 15.4 MHz, i.e., at $v_0 \pm \Delta f$, with the two frequencies being interchanged at a 1 kHz rate. Thus, if v_0 is held constant, a second 15.6-MHz pulse will appear 1 msec after the preceding 15.6-MHz pulse. To perform the ENDOR experiment, the ENDOR frequency v_0 is swept in the standard manner. Note from Figure 8 that the phase of the 1-kHz AM frequency is the same at both 50% and 10% duty cycles. The fundamental FM frequency is 1 kHz, even though the rf pulses to the power amplifier are at a 2-kHz rate.

The signal that emerges from the detector diode is almost identical to that shown in Figure 6, except that the rf pulses containing the ENDOR

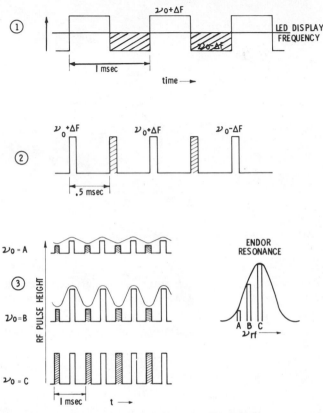

Figure 8. An example of square wave frequency modulation. (1) The input signal with a 50% duty cycle; (2) typical ENDOR operation with a 10% duty cycle; (3) ENDOR output from square frequency modulation while traversing an ENDOR resonance. Hatched areas indicate $v_0 - \Delta F$; unhatched areas indicate $v_0 + \Delta F$.

information occur twice as often. The input filter again acts to reduce noise at the ultimate detection frequency of 34.7 Hz. The input gate now passes the 2-kHz rf pulse train but again generating signals containing the ENDOR information at $2 \text{ kHz} \pm (1 \text{ kHz} \pm 34.7 \text{ Hz}) = 1 \text{ kHz} \pm 34.7 \text{ Hz}$ and $3 \text{ kH} \pm 34.7$ Hz. Other signals generated by the 34.7-Hz field modulation are filtered as described in Figure 7. The gate output is shown in Figure 8. The ENDOR signal again passes through the tuned stage of the 1-kHz amplifier as previously described, and the 1-kHz component is PSD and sent to the audio filter, amplified, and PSD for final presentation on the recorder. The 1-kHz component of the ENDOR signal emanating from the gate is proportional to the difference of the magnitude of the ENDOR enhancement at $v_0 \pm \Delta f$, i.e., to the difference of the rf pulse heights of the crosshatched $(v_0 - \Delta f)$ and open $(v_0 + \Delta f)$ pulses shown in Figure 8. The difference is small when v_0 is at the wings (A), largest where the slope is greatest (B), and zero right at the center (C). This generates the first-derivative line shape that is observed in the ΔF mode of operation.

1.2.3. *Other Studies Using ENDOR*

Initially, the major reason for performing double resonance experiments on free radicals in liquids was for the simplification of ESR spectra. Das et al.[18] have shown that the six classes of protons in vitamin K_3 (menadione) give rise to 128 ESR lines, but there are just 12 ENDOR lines, six above and six below the free-proton frequency. In addition, differences in splitting constants of the order of 50 mG could easily be obtained from the ENDOR spectra of menadione. Hyde[19] has shown that selective deuteration coupled with ENDOR allowed for a complete analysis of the complicated ESR spectrum (4050 lines) of the triphenylphenoxyl radical dissolved in benzene. Other work pertaining to the unraveling of complex ESR spectra by the use of ENDOR has been reported[20-24] and has led to straightforward interpretation of the proton hyperfine splittings. Hyde et al.[25] extended the double resonance method to systems that exhibit so-called powder ESR spectra. Among these systems are liquid crystal solvents, macromolecules that tumble slowly in a given solvent, and various paramagnetic species dispersed in rigid media such as glasses and powders. Thus, ENDOR has proven itself to be a useful probe for the study of complex radicals in different environments; *however, almost all of these studies to date have been of a qualitative nature, with the only quantitative parameter measured being the hyperfine splitting constant.*

The ENDOR experiment is capable of yielding additional parameters that can be quantitatively measured and analyzed in terms of the spin-relaxation processes that affect the ENDOR signal. This type of analysis allows for the evaluation of certain relaxation parameters related only to the double resonance experiment.[26,27]

In order to understand the principles of these studies, it is helpful to review the fundamental processes that lead to an ENDOR signal. Upon partially saturating a particular electronic transition with a microwave frequency, an ESR signal of a given amplitude is obtained. If a second field having an NMR radiofrequency is slowly altered, it is found that at values corresponding to the resonance of a given type of nuclei a change takes place in the spin populations of the electronic eigenlevels. This change affects the amplitude of the original ESR signal, and it is the magnitude of this "difference in amplitudes" that is observed as a function of the rf frequency that has come to be commonly known as ENDOR. In this technique, detection takes place at the microwave frequency, and the NMR radiofrequency is used as a probe for the electron–nuclear interaction.

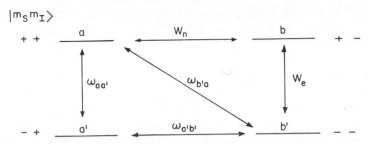

Figure 9. The induced transitions (ω), relaxation paths (W), and eigenlevels for double resonance in a radical with $S = \frac{1}{2}$, $I = \frac{1}{2}$. (From Leniart.[27])

In general, the population corresponding to the eigenlevel $|M_S = -\frac{1}{2}\rangle$ of the particular electronic transition being monitored is increased. Qualitatively, this increase stems from the fact that nuclear transitions induced by the rf field make indigenous relaxation paths available to the electron spins more effective. This is easily demonstrated by referring to Figure 9. Normally, the most effective relaxation path for an electron spin in state a is directly to a' via the spin–lattice relaxation $W_{aa'} = W_e$. The second path, $a \rightarrow b \rightarrow b' \rightarrow a'$, is usually not as effective, since the nuclear transition probability $W_{ab} = W_n$ (i.e., nuclear spin–lattice relaxation) is much less than W_e. If one thinks of the W as conductances (of spins), then the effective conductance of path 2 is

$$W_{\text{eff}} = \left(\frac{1}{W_{n_1}} + \frac{1}{W_e} + \frac{1}{W_{n_2}} \right)^{-1} \qquad (1)$$

where W_{n_1}, W_e, and W_{n_2} are W_{ab}, W_{bb}, and $W_{b'a'}$, respectively. When the rf field is introduced at a frequency corresponding to the resonance of the nuclear transition $b' \rightarrow a'$, the nuclear spin-flip rate tends to infinity (i.e., states a and b are "shorted together") and the conductance of path 2 is increased:

$$W_{\text{eff}} = \left(\frac{1}{W_{n_1}} + \frac{1}{W_e} \right)^{-1} \qquad (2)$$

This increase of effectiveness by path 2 in relaxing the electron spin from $a \rightarrow a'$ is the basis of the ENDOR phenomenon.

The amplitude of the double resonance signal depends on the magnitude of the effectiveness of spin relaxation via path 2. If $W_n < W_e$, then the bottleneck of the relaxation path is the nuclear spin–lattice transition probability. The predominant mechanism contributing to W_n is the anisotropic dipolar interaction of the unpaired electron spin with the magnetic nuclei composing the molecule (intramolecular dipolar interaction). The strength of this interaction is proportional to the ratio of the viscosity η to the temperature T; hence, experiments carried out at reduced temperature should remove the bottleneck caused by W_n. If $W_n > W_e$, as is the case in certain fluorine-containing compounds, then W_e becomes the bottleneck. Since W_e is proportional to temperature divided by viscosity (T/η), then working at elevated temperatures should increase ENDOR signals. The maximum signal occurs at that temperature where $W_e = W_n$. Thus, a quantitative measurement of the ENDOR amplitude (i.e., the difference in heights of the ESR signals when the rf field is on and off nuclear resonance) can be related to the relaxation properties of the spin system under observation. This experiment has been coined a "percent enhancement study."[27]

The ENDOR linewidths are an important spectral characteristic because they can be partly attributed to a broadening of the eigenlevels between which transitions are induced rather than to the incoherence of the rf and microwave frequencies or to the resolving power of the spectrometer. Hence, dynamic processes that influence these eigenlevels are reflected in the linewidths. In addition, other processes, such as Heisenberg spin exchange, readily affect the ENDOR spectra, implying that this technique may be used as a springboard for investigation of exchange phenomena.[26]

Thus, by quantitatively monitoring the effective ENDOR linewidth, i.e., the absorption half-width at half-height $\Delta_{n_{1/2},\,1/2}$, relaxation parameters may be obtained that directly reflect the molecular dynamics. These experiments have been termed "ENDOR linewidth studies." Both of these experiments are described in detail.

1.2.3.1. *Linewidth Studies.* ENDOR linewidth studies may be carried out in the modulation sequence of 35-Hz field modulation and 1-kHz rf amplitude modulation or the reverse sequence. The fastest modulating frequency employed in the experiment must be less than the inverse of the electron and/or nuclear spin–lattice relaxation times. Otherwise, rather involved theoretical corrections must be made to the experimental data to extract meaningful results. After letting the sample come to thermal equilibrium at a given temperature, the ENDOR spectra of a particular line are obtained for a series of rf powers. At any particular rf power, the signal should be traversed about four times, sweeping the radiofrequency in the direction that provides best tracking of the rf frequency by the servo system to maintain the tank circuit resonance of the RFPA (radiofrequency power amplifier) (optimally twice in either direction if the servo loop shape is perfect). The field modulation amplitude that gives the largest obtainable ENDOR signal amplitude often results in overmodulation of the ESR spectrum. Care must be taken to insure that the modulation amplitude is low enough so as not to affect the ENDOR line shape (however, it does affect the percent enhancement study, as shown in the next section). Suppose, for example, the narrowest proton ENDOR linewidth $\Delta_{1/2,\,1/2} = 35$ kHz, or (since $\gamma_H = 4.2$ kHz/G) a linewidth of about 8 G. A typical modulation amplitude for the initial experiments

Table 1

Comparison of ENDOR and Lorentzian Line Shapes for PBSQ[a]

Fraction of absorption amplitude	$\Delta/\Delta_{1/2,\,1/2}$ (Lorentzian)	$\Delta/\Delta_{1/2,\,1/2}$ (ENDOR)[b]	Deviation from Lorentzian (%)[c]	$\Delta/\Delta_{1/2,\,1/2}$ (ENDOR)[d]	Deviation from Lorentzian (%)[c]
0.9	0.337	0.379	12.5	0.324	3.9
0.8	0.507	0.526	3.7	0.543	7.1
0.7	0.667	0.695	4.2	0.667	0.0
0.6	0.832	0.853	2.5	0.857	3.0
0.5	1.000	1.000	0.0	1.000	0.0
0.4	1.248	1.189	4.7	1.257	0.7
0.3	1.525	1.453	4.7	1.495	2.0
0.2	1.960	1.853	5.5	1.952	0.4
0.1	2.871	2.695	6.1	2.838	1.1

[a] $\Delta_{1/2,\,1/2}$ is the half-width at half-height; Δ is the half-width at a particular fraction of the absorption amplitude.
[b] The absorption mode of configuration 1 was used (15-kHz field modulation and 40-Hz pulsing).
[c] Percent error $= [(R_L - R_{expt})/R_L] \times 100$, where R is the reduced width $(\Delta/\Delta_{1/2,\,1/2})$ of a true Lorentzian (L) or experimental (expt) line shape.
[d] The absorption mode of configuration 2 was used (40-Hz field modulation and 6-kHz pulsing).

might be 1 G. Second, several runs should be made using modulation amplitudes just above and below this value—say, at 5 and 2 G. If the line shape is unaffected, then 1 G of modulation amplitude is a good choice for the ENDOR linewidth studies. Next, the temperature is changed to a new value and the study is repeated until data for a series of rf powers have been obtained at four or five different temperatures. Finally, in order to analyze the data in terms of a Lorentzian line-shape function, the line shape of the experimental ENDOR signal must be compared to that of a true Lorentzian. Table 1 gives such a comparison for the *p*-benzosemiquinone (PBSQ) molecule dissolved in ethanol.

After verifying the line shape (noting, however, the deviation from Lorentzian character becomes more pronounced at high rf powers[19]) an attempt can be made to relate the ENDOR half-width at half-height to the experimentally measured rf power in the following manner:

$$\Delta_{1/2,\,1/2} = (S/T_{2n})^{1/2} \tag{3}$$

where S is the nuclear saturation factor given by

$$S = 1 + (\Delta_n T_{2n})^2 + d_n^2(\Omega_n - \Omega_{e,\,n}^2\delta)T_{2n} \tag{4}$$

and

$$\Delta_n = (W_n - W_0) \tag{5a}$$

$$d_n = \tfrac{1}{2}\gamma_n B_n J \tag{5b}$$

$$\delta^{-1} = \Omega_e + 1.33(T_{2e}d_e^2)^{-1} \tag{5c}$$

while $\Omega_n - \Omega_{e,n}^2\delta$ are the saturation parameters that represent the "effective" nuclear T_{1n}. See Chapter 3 for a more detailed description of these parameters. Thus, when the nuclear resonance condition $\Delta_n = 0$ is fulfilled, equation (3) can be rewritten in a manner similar to that used to analyze progressive saturation data[17]:

$$\Delta_{1/2,\,1/2}^2 = \frac{1}{T_{2n}^2} + \frac{\Omega_n'}{T_{2n}}d_n^2 \tag{6a}$$

where

$$\Omega_n' = \Omega_n - \Omega_{e,n}^2\delta \tag{6b}$$

Experimentally, one measures $\Delta_{1/2,\,1/2}$ for a series of d_n leading to a plot of $\Delta_{1/2,\,1/2}^2$ versus d_n^2 having an intercept at zero rf power $(B_n = 0)$ proportional to the true ENDOR linewidth and a slope proportional to the effective nuclear spin–lattice relaxation time. The generalized form of equation (6a) is

$$\Delta_{1/2,\,1/2}^2 = i + mB_n^2 \tag{7}$$

A typical example of the experimental plot of equation (7) is shown in Figure 10, while Figure 11 shows the temperature dependence of the ENDOR linewidths for three different compounds in two solvent systems.

1.2.3.2. *Percent Enhancement Studies.* There is a variety of instrumental configurations that may be employed to measure the magnitude of the percent enhancement of an ESR signal. The measurement, in theory, is a simple one—record the amplitude of the ESR signal in the standard manner with the rf frequency first on then off a nuclear resonance frequency. The percent enhancement (PE) is simply the difference in amplitudes multiplied by 100 and divided by the off-nuclear resonance (i.e., the standard ESR signal) amplitude.

Unfortunately, the experimental counterpart is not quite so simple. For a variety of reasons, double modulation techniques optimize the ENDOR sensitivity, but this scheme generally demands a specific order of phase sensitive detection, namely, first the high frequency then the low frequency. As shown below, for PE measurements, this is fine for high-frequency field modulation and low-frequency rf modulation but untenable when the frequencies are reversed. Nevertheless, all ENDOR spectrometers should be capable of measuring this parameter in one mode of operation.

a. *Absolute method.* The method of measuring the absolute percent enhancement when field modulating at 1 kHz and radiofrequency pulsing at 34.7 Hz will be described by following the course of the signal from the

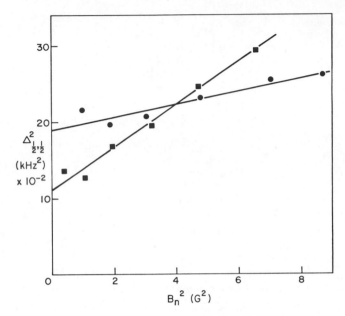

Figure 10. Plots of $\Delta^2_{1/2,\,1/2}$ as a function of B^2_n for PBSQ ($\bullet$) and durosemiquinone (DSQ) ($\blacksquare$) in ethanol at $-30°C$. (From Leniart *et al.*[17])

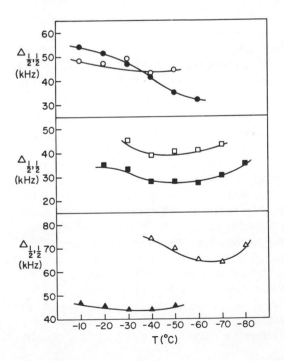

Figure 11. Unsaturated ENDOR linewidth as a function of temperature for dilute solutions of semiquinones. $\triangle$ = PBSQ in dimethoxyethane (DME); $\blacktriangle$ = PBSQ in ethanol; $\square$ = DSQ in DME; $\blacksquare$ = DSQ in ethanol; $\bigcirc$ = 2,5-DMPBSQ in ethanol ring proton; $\bullet$ = 2,5-DMPBSQ in ethanol methyl proton. (From Leniart *et al.*[17])

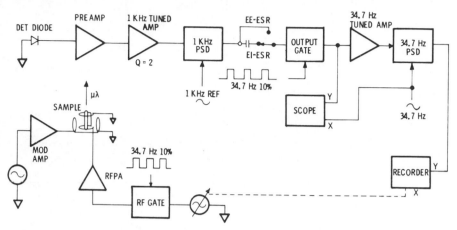

Figure 12. Typical ENDOR block diagram for modulating the magnetic field at 1 kHz (34.7-Hz rf pulsing) and observing 1-kHz ENDOR, ENDOR-induced ESR, and ENDOR-enhanced ESR.

microwave detector diode. A description of the ENDOR experiment is given in Figure 12.

When one is doing an ENDOR experiment, the dc magnetic field H_0 is set at one of the maxima of the first-derivative ESR signal and the nuclear rf frequency is swept. If the rf frequency sweep is held at the point of maximum ENDOR amplitude (center of nuclear resonance), the frequency components of the crystal signal are a 1-kHz ESR signal with amplitude B and 1 kHz $\pm$ 34.7 Hz (i.e., 965.3-Hz and 1034.7-Hz) ENDOR signals of amplitude A (see Figures 13 and 14). In addition, a 34.7-Hz signal generated from the rf pulses is passed by the detector crystal but rejected by the narrow-band filter of the 1-kHz amplifier. The amplification of the 1-kHz PSD is represented by X. The composite signal enters the PSD and is down-converted in frequency to two components, one a dc voltage of amplitude XB and the other a 34.7-Hz signal of amplitude XA.

The output of the 1-kHz phase detector goes to the function selector switch, which ac couples the signal train to the output gate. The dc ESR signal is blocked at the function selector switch, and only the 34.7-Hz ENDOR signal passes into the gate. Since the gate is open only when the nuclear rf is on, the rf-induced signal is passed to the audio phase detector, where it is amplified, detected, and displayed on the recorder. If the gain of the audio phase detector is represented by Y, the final signal amplitude is XYA. This scheme may now be compared to that of the ENDOR-enhanced ESR technique used to measure directly the absolute PE.

If the function selector switch is changed to the ENDOR-enhanced ESR position, the 1-kHz phase detector output is dc coupled to the output gate, and both the dc ESR signal of amplitude XB and the 34.7-Hz ENDOR signal of

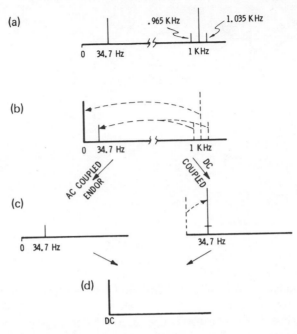

Figure 13. Frequency components for 1-kHz field-modulated ENDOR experiments as the three types of signals (1-kH ENDOR, EI-ESR, EE-ESR) pass through the spectrometer system—output of (a) microwave detector diode, (b) 1-kHz PSD, (c) gate for ENDOR, (c′) gate for ENDOR-enhanced ESR, (d) 34.7-Hz PSD.

amplitude XA are incident on the gate. As described above, the 34.7-Hz ENDOR signal passes through the gate to the audio phase detector. In this configuration, the dc ESR signal is chopped by the output gate. The signal incident on the 34.7-Hz phase detector has two components—34.7-Hz ENDOR of amplitude XA and 34.7-Hz chopped dc ESR of amplitude XB. The output of the audio detector unit in this case would be $XY(A + B)$. If the nuclear rf frequency were set beyond the observable wings of the ENDOR line, the output of the audio unit would contain no ENDOR component and would have amplitude XYB.

In the experimental measurement, the dc magnetic field is varied. The audio phase detector output is then either an enhanced or unenhanced ESR spectrum, depending upon the value of the rf frequency. Since the amplitude of the on-nuclear resonance ESR (ENESR) is given by $XY(A + B)$ and the off-nuclear resonance value (UNENESR) is XYB, the PE is obtained as follows:

$$PE = \frac{(\text{ENESR}) - (\text{UNENESR})}{(\text{UNENESR})} (100) \qquad (8)$$

$$PE = \frac{XY(A + B) - XYB}{XYB}(100) = \frac{A}{B}(100) \qquad (9)$$

Care must be taken so that the rf frequency chosen when recording the off nuclear resonance ESR signal does not affect the PE.

The absolute PE measured at any single rf power level B_n^2 can be compared directly with the ENDOR signal amplitude obtained at the same B_n^2, and a calibration curve can be plotted for ENDOR signal amplitude versus

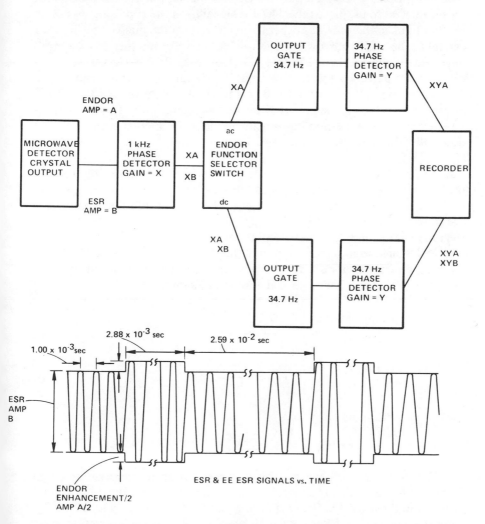

Figure 14. A block diagram of the ENDOR apparatus used to determine the absolute percent enhancement while pulsing the rf field at 34.7 Hz and modulating the dc magnetic field at 1 kHz.

PE. This calibration curve is easily confirmed by simply measuring the
ENDOR signal amplitude and corresponding PE at a series of different B_n^2.
The entire series of experiments is then repeated at four or five different
temperatures.

b. *Relative method.*[17] In the ENDOR configuration using 34.7-Hz field
modulation and 1-kHz rf pulsing, it is not possible to do the absolute PE
experiment described above. The 34.7-Hz ESR signal and the 1 kHz $\pm$
34.7 Hz ENDOR signal and a pure 1-kHz signal component from the pulsing
(of no interest to us, just as the 34.7-Hz signal from the rf pulsing was of no
interest to us in the previous experiment) pass from the microwave detector
crystal to the 1-kHz narrow-band amplifier and PSD and then to the output
gate and finally to the 34.7-Hz PSD. However, the 35-Hz ESR signal is
completely filtered by the narrow-band 1-kHz amplifier, thus rendering it
impossible to see a dc ESR signal in this mode of operation (see Figure 15).
However, since this is the preferred ENDOR configuration in terms of sensitivity and stability, a method for measuring the percent enhancement is quite
desirable. This procedure is outlined below.

A sample whose PE has been previously measured with the 34.7-Hz rf
pulsing configuration is used as a standard reference sample. The relative
method involves the measurement of the 1-kHz rf pulsed ENDOR amplitudes
and the 34.7-Hz ESR amplitudes of both the standard and the unknown. If
the gains of the 1-kHz and 34.7-Hz PSD units are again denoted by X and Y,
then the ENDOR and ESR signal amplitudes are XYA and XYB, respectively.
The 35-Hz field-modulation amplitude and microwave power settings used
in these measurements are very important and will be discussed in detail
later. The PE of the unknown is determined as follows:

$$\left.\begin{array}{l} \text{ENDOR } (XYA) \\ \text{ESR } (YB) \end{array}\right\} \text{ Data for unknown}$$

$$\left.\begin{array}{l} \text{ENDOR } (XYM) \\ \text{ESR } (YN) \end{array}\right\} \text{ Data for standard} \tag{10}$$

PE known

Unknown PE $= (XYA)(YN)(\text{Std. PE})/(YB)(XYM)$

The PE of the unknown may be determined over the temperature range
of interest for one value of B_n^2 by the relative method. This value of B_n^2 can be
included in those used in the ENDOR linewidth–PE variation with B_n^2 study.
Since the value of the PE is known for one data point in each variation of B_n^2
experiment, it is possible to determine the value of the PE at other values of
B_n^2 from the ratio of the ENDOR intensities.

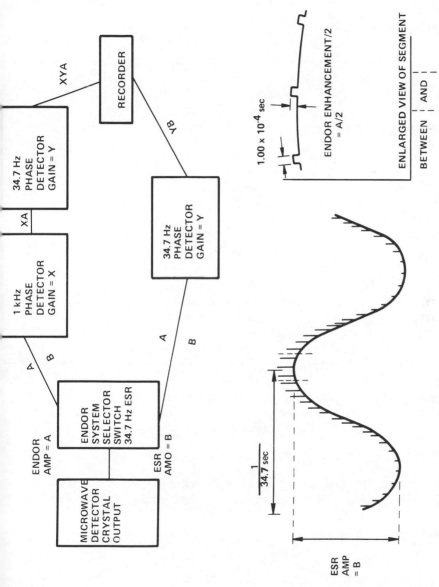

Figure 15. A block diagram of the ENDOR apparatus used to determine the relative percent enhancement while pulsing the rf field at 1 kHz and modulating the dc magnetic field at 34.7 Hz.

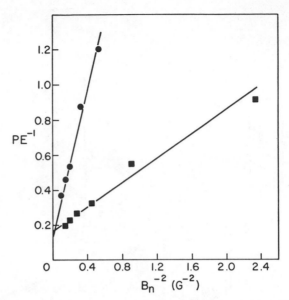

Figure 16. Plot of PE^{-1} as a function of B_n^{-2} for PBSQ (●) and DSQ (■) in ethanol at $-30°C$. (From Leniart et al.[17])

c. *Data analysis of percent enhancement studies.* Freed has shown that the analytical expression for the present enhancement of a four-level $(S = \frac{1}{2}, I = \frac{1}{2})$ ENDOR system, neglecting coherence effects, can be written as

$$2\left(\frac{PE}{100}\right)^{-1} = -1 + \frac{\Omega_n \Omega_e}{\Omega_{e,n}^2} + \left(\frac{\Omega_e}{\Omega_{e,n}^2 T_{2n}}\right) d_n^{-2} \qquad (11)$$

where the Ω are the saturation parameters for the electron (e), nuclear (n), and cross (e, n) relaxation paths and T_{2n}^{-1} is the ENDOR linewidth extrapolated to zero rf power. For a multispin system, the data are plotted as

$$(PE)^{-1} = i + mB_n^{-2} \qquad (12)$$

The interpretation of the slope and intercept for multispin systems is given in references 17, 27, and 31. Experimental data showing a plot of equation (12) is given in Figure 16, while Figure 17 shows an example of the temperature dependence of the PE for a variety of compounds and solvents.

1.3. *Experimental Effects in ENDOR Studies*

There is a variety of conditions and parameters under which ENDOR experiments are attempted that produce a series of effects that, if properly

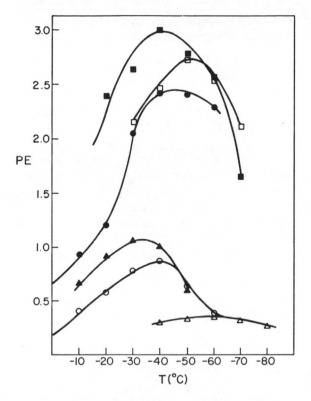

Figure 17. ENDOR PE as a function of temperature for solutions of semiquinones. $\triangle$ = PBSQ in DME; $\blacktriangle$ = PBSQ in ethanol; $\square$ = DSQ in DME; $\blacksquare$ = DSQ in ethanol; $\bigcirc$ = 2,5-DMPBSQ in ethanol ring proton; $\bullet$ = 2,5-DMPBSQ in ethanol methyl protons. (From Leniart *et al.*[17])

understood, can be either quite useful or an experimental " no-man's-land." In this section, an attempt is made to review the common effects uncovered to date in order to provide the reader with an awareness of the consequences of the experimental conditions chosen to carry out an ENDOR experiment.

1.3.1. *Modulation Amplitude Effects*

1.3.1.1. *Effects on the Percent Enhancement*
a. *Theory.* A theoretical analysis of the effects of the modulation amplitude on the percent enhancement has been given in Connor[26] and Leniart.[27] Using Wahlquist's[28] study of modulation broadening of Lorentzian spectra and modifying it to include microwave saturation (used in the ENDOR experiment), one obtains an expression for the first-derivative

amplitude of a saturated Lorentzian detected with sinusoidal modulation of amplitude H_{w_m} that can be expressed as

$$v' = \pm 6T_2^2 Z^{-3/2}[(U_P - 2)/U_P(2U_P - 3)]^{1/2} \tag{13a}$$

where

$$U_p = 2 + (\tfrac{4}{3})\beta^2 + (r/3)\beta(\beta^2 + \tfrac{3}{4})^{1/2} \tag{13b}$$

$$Z = (1 + \gamma^2 T_1 T_2) \tag{13c}$$

$$\beta = \tfrac{1}{2}(T_2^{-1} Z^{1/2}/H_{w_m}) \tag{13d}$$

and T_2^{-1} is the unsaturated ESR full width at half-height. The ENDOR effect can be treated as a perturbation in T_2^{-1} or Z induced by the nuclear rf field. The presence of the rf field (at nuclear resonance) decreases the degree of microwave saturation of the ESR line, thereby causing Z to decrease. This decrease in effective saturation causes an increase in signal amplitude, which is the PE observed and measured in ENDOR. If one redefines β as

$$\beta = \beta' + \varepsilon\beta' \gg \varepsilon \tag{14}$$

where ε is the rf-induced ESR linewidth change, then the decrease in effective $T_2^{-1} Z^{-1/2}$ can be examined by rewriting equation (13d) as

$$T_2^{-1} Z^{1/2} = 2(\beta' + \varepsilon)H_{w_m} \tag{15}$$

CASE 1: $H_{w_m} \ll T_2^{-1}$; therefore $\beta' + \varepsilon \gg 1$. Equation (13b) becomes

$$U_p \cong 2.5 + \tfrac{8}{3}(\beta' + \varepsilon)^2 \tag{16}$$

and equation (13a) becomes

$$(v)' \cong \frac{3\sqrt{3}\, T_2^{-1}}{16 H_{w_m}^3 \beta'^4}\left(1 - \frac{4\varepsilon}{\beta'}\right) \tag{17}$$

Hence, for modulation amplitudes much smaller than the ESR linewidth, PE $= 4\varepsilon/\beta'$ or 4% if we choose $\varepsilon/\beta' = 0.01$. Likewise, the results for three other cases are summarized.

CASE 2: $H_{w_m} = \tfrac{1}{2}T_2^{-1}$, i.e., modulation amplitude = ESR linewidth. To include the effect of the rf field, let

$$\beta = 1 + \varepsilon \tag{18a}$$

Equation (13b) becomes

$$U_P = 5.09 + 5.43\varepsilon \tag{18b}$$

and equation (13a) becomes

$$(v)' \cong \frac{0.22 T_2^{-1}}{H_{w_m}^{-3}}(1 - 3.4\beta\varepsilon) \tag{19}$$

or the PE decreases to $3.4\varepsilon/\beta'$ when one modulates the saturated ESR line with an amplitude equal to $\frac{1}{2}T_2^{-1}$, i.e., the unsaturated absorption half-width at half-height.

CASE 3: $H_{w_m} = T_2^{-1}$, i.e., the modulation amplitude that maximizes the ESR signal. To include the effect of the rf field, let

$$\beta = \tfrac{1}{2}(1 + \varepsilon') \tag{20a}$$

where $\varepsilon' = \varepsilon/2$, giving

$$U_p = 3.0 + \tfrac{3}{2}\varepsilon' \tag{20b}$$

and

$$(v)' \cong \frac{2T_2^{-1}}{H_{w_m}^3}(1 - 3\varepsilon') \tag{20c}$$

which reduces the PE to $3\varepsilon/\beta'$ or 3% for modulation amplitudes that maximize the ESR signal height.

CASE 4: $H_{w_m} \gg T_2^{-1}$, i.e., strong overmodulation of the ESR signal. In this case, equation (13d) shows that

$$\beta \ll 2 + 1.15\beta \qquad \text{and} \qquad U_p \sim 2 + 1.15\beta, \tag{21}$$

giving

$$(v)' = \tfrac{3}{4}^{3/4}(H_{w_m}^3\beta^{5/2})^{-1}(1 - \tfrac{5}{2}\varepsilon) \tag{22}$$

or the PE is reduced to $2.5\varepsilon/\beta$ or 2.5% when strongly overmodulating the ESR signal.

b. *Experiment.* Figure 18 shows the magnitude of the percent enhancement plotted as a function of the modulation amplitude for a dilute sample of durosemiquinone in ethanol. It is important to note that the effect of modulation amplitude is independent of the rf pulsing configuration or the frequency of the field modulation. In addition, under the conditions of obtaining the data shown in Figure 18, the ENDOR linewidth was found to be independent of modulation amplitude.

Experimentally, when measuring percent enhancements, one prefers to do the experiments using the very low modulation amplitudes described in Case 1, where the PE is essentially invariant and a maximum. Unfortunately, however, the ESR signal itself is greatly reduced when the modulation amplitude is $\ll T_{2e}^{-1}$, and this may have a disastrous effect on the ENDOR signal-to-noise ratio. In some cases, ENDOR amplitudes may be reduced to 10% of their maximum values—a price that is many times too high to pay and still obtain accurate experimental measurements.

Thus, for practical purposes, it is necessary to perform such experiments under conditions of optimum ENDOR sensitivity, i.e., the maximized product

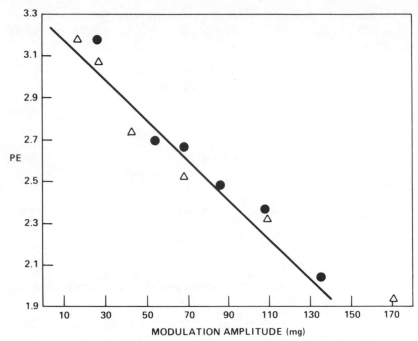

Figure 18. Percent enhancement of a dilute sample of DSQ in ethanol at $-40°C$ plotted as a function of modulation amplitude. $\bullet$ = 6-kHz rf pulsing, $\triangle$ = 35-Hz rf pulsing, $B_n^2 = 4.4 \ G^2$. (From Section 2-III in Leniart *et al.*[17])

of PE times ESR signal amplitude. After a few trials, it is reasonable to establish that for well-separated ESR hyperfine lines, the ENDOR signal is optimized under the experimental conditions described in Case 3. Under these conditions, the measured percent enhancement is predicted to be reduced to 75% of its maximum value. This correction factor may be verified experimentally by choosing a " strong ENDOR " sample having well-separated hyperfine lines as a standard. Using identical experimental conditions as for all other ENDOR studies, except reducing the modulation amplitude to < 10% of the ESR linewidth, gives an experimental correction factor that can be compared with theory. An example of such a study is given in Table 2. Such a procedure can be used to verify the theoretical predictions in the remaining two cases (Case 2 and 4) described in the previous section.

The agreement between theory and experiment is reasonable enough to encourage the use of such a simple correction factor. Better theoretical estimates can be made by taking into account the variation of modulation amplitude along the sample for ENDOR cavities having fairly large active regions.

1.3.1.2. *Effects on the ENDOR Linewidth.* The amplitude of the Zeeman field modulation described in the previous section was large enough

Table 2

Determinations of Overmodulation Correction Factor Using a Dilute Solution of DSQ in Ethanol

Temperature (°C)	Pulse rate	Correction factor[a]
−40	35 Hz	1.67 ± 0.09
−50	35 Hz	1.72 ± 0.15
−60	35 Hz	1.57 ± 0.06
−40	6 kHz	1.63 ± 0.07
−50	6 kHz	1.66 ± 0.05
−60	6 kHz	1.56 ± 0.09

[a] Factor by which the PE of an overmodulated ESR line is decreased from the maximum PE obtained when modulation amplitude is less than 10% of the ESR linewidth.

to affect the magnitude of the percent enhancement but still small enough to leave the ENDOR line shape unaffected. However, for a complex, unresolved ESR spectrum or a powder pattern, the amplitude of modulation might cover a significant fraction of the spectral width (this is analogous to Case 4). Under these experimental conditions, the amplitude of modulation may definitely reflect itself in the ENDOR line shape. In general, the difference in the gyromagnetic ratios of the electron and, let's say, the proton is large enough, $\gamma_e/\gamma_H = 658$, so that a modulation amplitude of 1 G for ESR lines having $T_{2e}^{-1} = 1$ G $= 2.8$ MHz is extremely small for comparable ENDOR linewidths of 2.8 MHz or 658 G. However, this condition is not met if, e.g., the ENDOR linewidth is 50 kHz or ~ 12 G and the corresponding complex ESR spectrum of the organic free radical in solution is ~ 25 G. Optimum ENDOR signal amplitudes are often obtained when $H_{w_m} \sim \frac{1}{2}$ spectral width or ~ 12 G. Now modulation of the NMR resonance becomes important and, as previously described, PSD at the audio detector produces a second component that is superimposed on the standard ENDOR signal. This may be either a first-derivative signal superimposed on the ENDOR absorption or a second-derivative signal superimposed on the (ΔF) ENDOR first-derivative spectrum. This distortion of ENDOR line shape is evident in the ENDOR spectra of dihydropleiadenes[29] given in Figure 19.

1.3.2. Microwave Power Effects

1.3.2.1. *Effects on Percent Enhancement.* Theoretical ENDOR analysis[30] has shown that for one nucleus of spin $I = \frac{1}{2}$ or for "average ENDOR,"[31] the PE is related to the microwave power B_e^2 as

$$(PE)^{-1} = r + sB_e^{-2} \tag{23}$$

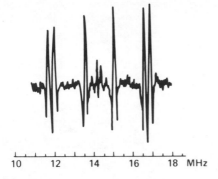

Figure 19. The central portion of the ENDOR spectrum of 7,12-dihydropleiadene. (From Allendoerfer *et al.*[29])

where r and s are dependent upon the relaxation properties, NMR frequency, and B_n^2, provided one holds the ESR frequency and B_0 constant. Computer simulations of more general cases for PBSQ have also predicted such a linear dependence on B_e^{-2}. A typical experimental result that demonstrates this relationship is given in Figure 20, while in Figure 21 the asymptotic behavior of PE with B_e^2 for large B_e^2 is illustrated.

Ideally, studies of the variation of the PE as a function of nuclear rf power should have been carried out in the region where the approximation $d_e^2 \to \infty$ could be utilized. In order to reach this limit, one would have to use microwave powers that would produce microwave coherence broadening of the ENDOR lines (see next section). However, since the ENDOR lineshape should be preserved when making accurate PE measurements, this region should be avoided. Reproducible PE experiments can be carried out at microwave powers that maximize the ESR derivative amplitude at each temperature being studied.

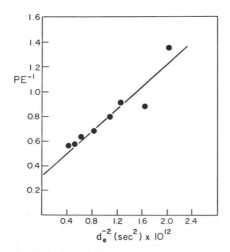

Figure 20. $(PE)^{-1}$ plotted as a function of d_e^{-2} for a dilute solution of PBSQ in ethanol at $-40°C$; $B_n^2 = 4.0 \ G^2$. (From Section 2-III in Leniart *et al.*[17])

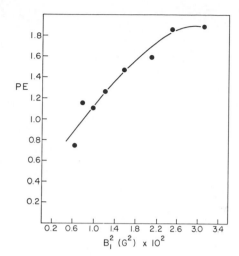

Figure 21. PE plotted as a function of B_1^2 for a dilute solution of PBSQ in ethanol at $-40°C$; $B_n^2 = 4.0\ G^2$.

1.3.2.2. Microwave Coherence Effects. Figure 22 shows an ENDOR line of the single methide proton in Coppinger's radical as a function of incident microwave power. The ENDOR line is observed to broaden and split as the incident microwave power is increased by roughly 20 dB. The effect of a splitting of the NMR resonance Z_n'' by the microwave field d_e^2 is well understood theoretically[32] and has been reproduced by computer simulation. Generally, the microwave power that gives the most intense and narrowest ENDOR line is free from any line-broadening effects attributed to the microwave field.

1.3.3. Nuclear Radiofrequency Coherence Effects

1.3.3.1. Effect on PE and Linewidth Studies. In experiments with narrow-line ESR samples, it is easy to distort the ENDOR line shape from that of a simple Lorentzian when the nuclear rf power is high. Typical examples are given in Figure 23. One is able to predict such line shapes from the general computer solution. It is instructive, however, to study the analytical results for a single $I = \frac{1}{2}$ case. The unsaturated ESR absorption signal amplitude at exact resonance is given by

$$Z_e'' = \frac{q\omega_e d_e T_{2e}(T_{2x}^2 \Delta_n^2 + 1 + T_{2e} T_{2x} d_n^2)}{T_{2e}^2 \Delta_n^2 + (1 + T_{2e} T_{2x} d_n^2)^2} \qquad (24)$$

Simplifying this expression by letting $T_{2e} \cong T_{2x}$ and using rf powers that are of such magnitude that

$$d_n^2 T_{2e}^2 \ll 1 \qquad (24a)$$

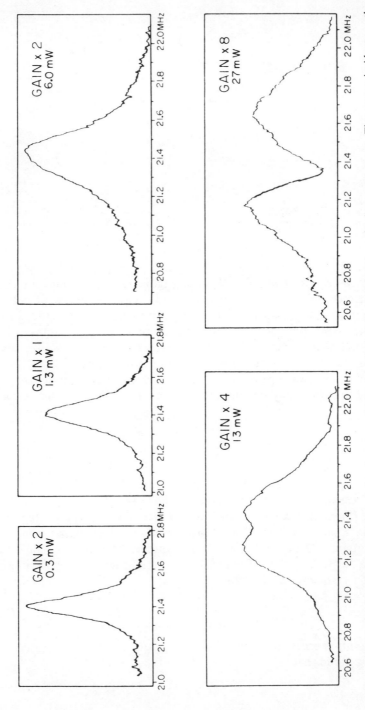

Figure 22. The ENDOR line from the methylenyl proton of Coppinger's radical as a function of microwave power. The power incident on the cavity is given in milliwatts, and the relative recorder gain is indicated. (From Freed *et al.*[31])

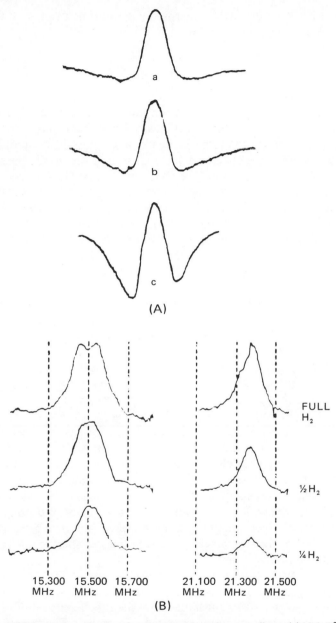

Figure 23. (A) Distorted ENDOR line shapes caused by coherence effects: (a) high B_1^2, long T_{2e}, medium B_n^2; (b) high B_1^2, long T_{2e}, high B_n^2; (c) low B_1^2, long T_{2e}, high B_n^2. (From Section 2-III in Leniart *et al.*[17]) (B) Effect of rf field (H_2) on the ring proton (left) and methylenyl proton (right) ENDOR lines from Coppinger's radical (3×10^{-4} M) in heptane at $-85°C$. The splitting of the ring proton line at 15.5 MHz illustrates a nuclear–nuclear coherence effect. (From Freed *et al.*[32])

(which is satisfied for T_{2e} of ~ 1 microsecond and rf fields of ~ 10 G in the rotating frame) gives

$$Z_e'' = q\omega_e d_e T_{2e} \, 1 - \frac{T_{2e}^2 \, d_n^2}{1 + T_{2e}^2 \, \Delta_n^2} \qquad (25)$$

Equation (25) indicates that the ESR amplitude decreases as d_n^2 increases. Also note that this rf-induced effect has a Lorentzian line shape with a half-width at half-height the same as the ESR line. Since the ESR amplitude is reduced, the phase of the coherence line will be opposite to the phase of the usual ENDOR-enhanced line. Also, the coherence line will be broader than the ENDOR line, since $T_{2e} \ll T_{2n}$ (typical values are ENDOR $\Delta_{1/2,\,1/2} = 112$ kHz). The sum of two such lines gives line shapes like those in Figure 23A.

The broad coherence line increases in amplitude as the nuclear rf power increases. The best method of observation of this line is achieved by using low microwave powers and high nuclear rf powers. The coherence reduction signal is independent of microwave power, but the normal ENDOR signal is weaker.

Similarly, one can use a high value of B_1^2 in the ENDOR experiment to help minimize the effect of the coherence line on the ENDOR line shape. Using a high-duty cycle and comparatively low nuclear rf power also minimizes the distortion of the ENDOR line shape.[32]

The effect described above is based on the nuclear rf field shifting the ESR resonant frequencies and it can result from the ENDOR of a single proton as described in Freed et al.[32] However, this coherence effect is not the same as that resulting from the nuclear rf field shifting the NMR resonant (ENDOR) frequencies. The latter requires an ENDOR transition between two or more equivalent nuclei. An example of this effect is shown in Figure 23B.

In general, coherence effects interfere with the relaxation studies, because they distort the line shapes, destroying any simple dependences the ENDOR line shapes may have on rf power. Consequently, linewidth and PE experiments must be performed at sufficiently low values of B_n to guarantee that the coherence effects are not present. However, they can be used advantageously to aid in the assignment of ENDOR lines to specific nuclei. This is described in the next section.

1.3.3.2. *ENDOR Assignments Based on Coherence Effects.* The rf coherence shift of the ENDOR frequencies depends mainly on the rf field strength and hence is, in theory, distinguishable from the rf coherence shift of the ESR frequencies, the latter being microwave-field dependent. The former effect is important, since multiple quantum transitions can be seen only for spin systems having a total nuclear spin $I > \frac{1}{2}$. When dealing with proton ENDOR, this means that this type of coherence effect will be seen for ENDOR lines that correspond to two or more equivalent protons in the radical. Thus, coher-

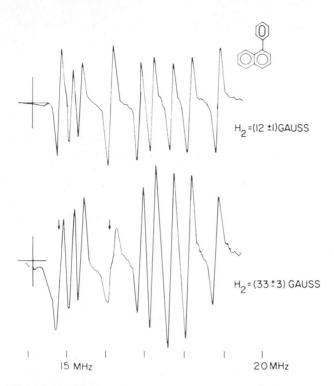

$H_2 = (12 \pm 1)$ GAUSS

$H_2 = (33 \pm 3)$ GAUSS

I5 MHz   2 0 MHz

Figure 24. Effect of rf field (H_2) on the ENDOR spectra of 1-phenylnaphthalene anion in DME. The spectrum above the free proton frequency is shown except for a line at 22.2 MHz. The ENDOR lines showing broadening due to a nuclear–nuclear coherence effect are marked by arrows and correspond to degenerate hyperfine couplings. (From Dinse *et al.*[33])

ence effects can be used to pinpoint those ENDOR lines that correspond to two or more equivalent protons.

An interesting example of the application of this coherence effect is shown by the ENDOR spectrum of 1-phenylnaphthalene anion.[33] Figure 24 shows the ENDOR spectra at two different rf field strengths. A maximum of 12 hyperfine constants corresponding to 12 sets of nonequivalent protons is possible for this radical, but the ENDOR spectrum at the lower rf field strength only shows 10 ENDOR lines above the free proton frequency (the line from the largest splitting is not shown). Consequently, one or two of these lines must correspond to more than one equivalent proton. It is seen that two of these lines do in fact broaden upon application of a high rf field, indicating that these two lines must correspond to two equivalent protons each.

However, the experimenter must be extremely careful when making these assignments solely via coherence effects. He would be wise to base his final decision on the overall assignment on additional grounds, such as ENDOR linewidths, intensities, or deuteration.

1.3.4. *Effect of the Radiofrequency Pulse Rate*

Care must be taken to insure that the ESR and nuclear spin systems can respond quickly to the amplitude-modulated square-wave rf pulses. Freed[30] has given a qualitative discussion concerning transient effects in pulsed rf ENDOR experiments by noting that for pulse time $t \sim W_e^{-1}$, the ESR signal could be enhanced because of (1) power absorption from the microwave field at a variable rate during the application of an rf pulse and (2) a net absorption of energy as radical populations with $m_s = +\frac{1}{2}$ build up when the rf field is on, i.e., a possible "heating" of the electron spins. If the rf pulse duration of a 1-kHz amplitude-modulated system at a 10% duty cycle is 10^{-4} sec, then one must take care when the nuclear relaxation parameters T_{2n} and Ω_n and the electronic relaxation parameter Ω_e have comparable times. These conditions most often may be met when ENDOR experiments are carried out at liquid-helium temperatures.

To insure that a steady-state condition is being observed and that transient phenomena are not present, one may reduce the duty cycle to a lower value, thereby making the pulse width even shorter, and monitor the ENDOR signal for an increase in the enhancement from a "heating" of the spin

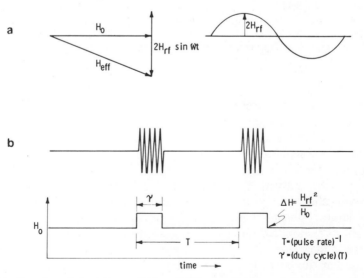

Figure 25. The effect of rf pulsing—a field modulation produced at the sample at a frequency equal to the pulse rate. (a) The effective field generated by a linearly oscillating rf field. (b) A shift in the magnitude of the dc magnetic field as the rf field is pulsed.

system. Additionally, one may switch the pulse rate from 1 kHz to 34.7 Hz and repeat the ENDOR linewidth and PE experiments to verify that the ENDOR relaxation parameters remain unchanged.

A second effect of the rf pulsing is to superimpose a field modulation on the sample at a frequency equal to the pulse rate. The effect has not been quantitatively described in the literature but is rather straightforward to analyze. The rf field is applied at right angles to the main dc applied magnetic field as shown in Figure 25a. The effective field H_{eff} is given by

$$H_{\text{eff}}^2 = H_0^2 + 4H_{\text{rf}}^2 \sin^2 \omega t \tag{26}$$

solving for H_{eff}, assuming $(H_{\text{rf}}/H_0)^2 \ll 1$, equation (14) becomes

$$H_{\text{eff}} = H_0 + \frac{H_{\text{rf}}^2}{H_0} - \frac{H_{\text{rf}}^2}{H_0} \cos 2\omega t \tag{27}$$

Thus, each time the rf field is pulsed on or off, the total dc field is shifted, as shown in Figure 25b. Fourier analysis of the rf pulse shape shown in Figure 25b gives

$$f(t) = \frac{1}{T} \sum_{-\infty}^{\infty} c_n \exp(j\omega_n t) \tag{28a}$$

$$c_n = \int_{-T/2}^{T/2} f(t) \exp(-j\omega_n t) \tag{28b}$$

giving

$$C_0 = \tau \frac{H_{\text{rf}}^2}{H_0} \tag{28c}$$

$$C_{\pm 1} = \frac{T}{\pi} \sin \frac{\tau}{T\pi} \tag{28d}$$

Therefore

$$f(t) = \frac{\tau}{T} \frac{H_{\text{rf}}^2}{H_0} + \frac{2H_{\text{rf}}^2}{\pi H_0} \sin \frac{\tau}{T\pi} \cos \omega_{n=1} t \tag{29}$$

where

$$\omega_n = \frac{2\pi n}{T}$$

Inserting the values for a standard X-band ENDOR experiment gives $H_0 = 3333$ G, $H_{\text{rf}} = 10$ G, $\tau/T = 0.1$ (10% duty cycle), a dc field shift from equation (29) of 3 mG each time the rf field is pulsed, and a peak-to-peak field modulation at frequency $\omega_{n=1}$ of

$$2\left(\frac{2H_{\text{rf}}^2}{\pi H_0} \sin \frac{\tau}{T}\right) = \frac{400}{\pi(3333)} \sin 0.1\pi = 12 \text{ mG}$$

This may cause an unwanted ESR signal arising from an effective field modulation of 12 mG at the rf pulsing frequency. One may monitor this effect by eliminating any field modulation to the cavity and doing a standard ESR experiment pulsing only the rf frequency. If no ESR signal is observed, the effect is transparent. If an effect is present, one might use it to calibrate the rf field at high rf powers.

2. ELDOR

2.1. General Remarks

In an ELDOR experiment, the intensity change of an ESR signal is recorded as a function of irradiating the sample with a second microwave field. In general, the observing microwave frequency that monitors the change in ESR intensity is fixed and set at some nonsaturating power level, while the pumping microwave frequency is variable and set at a power level that strongly saturates the ESR transition(s) being irradiated. The magnitude of the change in ESR intensity is governed by the relaxation mechanism(s) that couple the spin systems being pumped and observed—the more effective the coupling, the greater the ELDOR response. Thus, the ELDOR intensity provides a quantitative means for studying a variety of dynamic processes that govern relaxation between the pumped and observed spin systems. In samples displaying hyperfine structure, the absolute difference in the two microwave frequencies at ELDOR resonance affords the experimenter with a quantitative measure of the hyperfine splitting constant and may be useful in separating overlapping spectra from a mixture of different radicals.

A varietvariety of investigators, using the principles described above, have studied numerous problems leading to advances in the understanding of relaxation phenomena, exchange processes, molecular structure, and solid state dynamics. If the reader were to review the instrumental details of the ELDOR literature cited in this book, it would be clear that the equipment necessary to perform this double resonance experiment falls into two general categories.

One needs a basic spectrometer system and an ELDOR cavity. Although differences appear in the various spectrometer systems employed, all generally use a magnet, two microwave sources, and standard lock-in detection. Any differentiation usually arises in the particular method of display for the type of study under investigation. The primary experimental challenge is the design of a satisfactory microwave probe that will house the sample and is capable of (1) simultaneously supporting two microwave fields that are independent and isolated from one another and (2) providing a means to vary the temperature at the sample. This device is *the* critical component of an ELDOR spectrometer and has been designed in three or four basic configurations. These may be described as (1) single-helix microwave probes, (2) helix–helix microwave probes, (3) cavity–helix microwave structures; and (4) cavity–cavity microwave structures. In the following section a general discussion of the basic ELDOR spectrometer and cavity configurations will be presented with an adjunct to specific equipment where appropriate.

2.2. Models of ELDOR Spectrometers

The only known commercial ELDOR spectrometer currently being used is the Varian E-800 frequency-swept ELDOR system designed as an accessory to the standard ESR spectrometer.[37,42] Several home-made ELDOR spectrometers, differing largely in the design of the microwave probe being used, have been reported.[34–36,38–41,43–45]

2.2.1. The Basic ELDOR Spectrometer

A simplified block diagram of the Varian E-800 ELDOR spectrometer is shown in Figure 26. The microwave source of the standard bridge (usually a klystron) resonates at some fixed frequency v_{obs} and is most commonly used as the observing frequency. A second microwave source, generally a klystron but possibly a traveling wave tube (TWT), a backward wave oscillator (BWO), or some other microwave device, used either as an oscillator or amplifier and capable of providing of the order of one or more watts of microwave power, is housed in a second bridge. This microwave device is referred to as the pump source v_p and can be continuously swept over a range of about 600 MHz. That is, $v_p \simeq v_{obs} \pm 300$ MHz, assuming that the resonant frequency of the observing mode is located at the center of the pump frequency range. The pump microwave bridge contains a second AFC system that locks the pump klystron to the instantaneous resonance of the appropriate mode in a cavity–cavity type bimodal ELDOR probe. This lock is accomplished by incorporating a continuous AFC-controlled servomotor that is capable of tuning the pump frequency. The pump AFC generates an error signal in the standard manner, and that signal is relayed to both the klystron reflector tracking mechanism and to a servomotor that controls the klystron tuning shaft. The servomotor continuously changes the absolute value of the pump frequency so that it remains within the capture range of the AFC system. The AFC then slightly perturbs the klystron frequency via electrical control of the reflector voltage to achieve a lock with the pump cavity resonance. Of course, when a broad-band helix is used to support the pump mode, no AFC system is necessary.

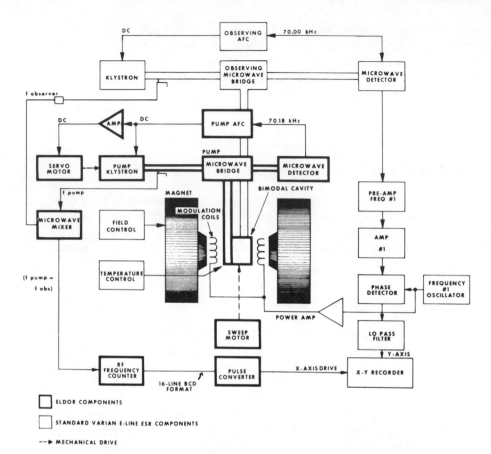

Figure 26. Block diagram of the Varian E-800 ELDOR spectrometer. The heavily ruled lines outline the components added to the standard ESR spectrometer. The modulation frequency can be either 100, 10, or 1 kHz or 270 or 35 Hz.

In order to perform an ELDOR experiment, the isolation between the pumping and observing modes of the microwave probe should be better than the 40 dB over the typical range of frequency deviation, $v_{obs} - v_{pump} = \pm 300$ MHz. This is a difficult requisite to achieve, especially in a cavity–cavity-type structure, and nearly all designers have chosen to use additional filtering of the pump power immediately preceding the microwave detection crystal. The general solution is to utilize a transmission microwave cavity tuned to the observing frequency. Its purpose is twofold. First, it insures that the bias on the observing detector diode remain constant as the pump frequency is swept. The transmission device achieves this by greatly reducing the pump power that may impinge on the detector crystal from frequency-dependent isolation changes of the microwave probe. Second, it prohibits any ESR signals arising from the pump frequency to be superimposed with those from the observing frequency and ultimately be microwave-rectified by the detector crystal. The ELDOR signal, after passing the detector crystal, is then phase detected in the conventional manner, as described in Section 1.2.2, and presented to the Y axis of the recorder for display. A variety of X axis displays is possible (see Section 1.1.2), the most common being that employed during a swept frequency experiment. In this mode of operation, the frequency difference $v_{obs} - v_{pump} = \Delta v$ is displayed on a counter while the digital output of counter is converted to an analog signal that drives the X axis of the ESR recorder. Thus, e.g., one may choose to begin the ELDOR sweep with the pump frequency 2.8 MHz above the observing frequency (left-hand side of the chart paper) and conclude with the pump frequency 56 MHz above the observing frequency (right-hand side of the chart display), the resulting spectrum constituting the ELDOR response to fixing the observing frequency (i.e., sitting at some magnetic field H_0) and continuously sweeping the pump frequency through the ESR spectrum at fields of 1 to 20 G above H_0 (i.e., 2.8 MHz = 1 G).

All other components of an ELDOR spectrometer system closely resemble a standard ESR spectrometer. The second microwave bridge obtains its power from either the ESR console power supply or a separate stand-alone unit dedicated to the ELDOR bridge itself. To achieve stability of the observing resonance condition a standard field/frequency lock (see Section 1.1.2) may be added to the existing ESR spectrometer.

The basic ELDOR spectrometer has been modified to some degree by various workers throughout the world. The next few sections describe these modifications and the discussion is centered around the significant instrumental change, namely, the microwave probe.

2.2.2. *Benderskii and Blumenfeld—Soviet Union*

In December of 1968, two Soviet scientists from the Institute of Chemical Physics in Moscow submitted an article describing an ELDOR spectrometer using a spiral resonator. This paper, describing the first attempt at constructing such an instrument in the Soviet Union, was published in April of 1969.[38] In this article, Sokolov and Benderskii describe essentially the ELDOR spectrometer system shown in Figure 27. This figure shows microwave power from both the pump and observing klystrons passing through circulator number one and impinging on a single helical resonator surrounded by a set of field modulation coils operating at a frequency of 1 MHz. The single broad-band helix is capable of supporting both microwave frequencies simultaneously. In this ELDOR system, the two frequencies are fixed at some predetermined frequency interval Δv, and the magnetic field is swept. The intensity of the observed ESR line is recorded and the

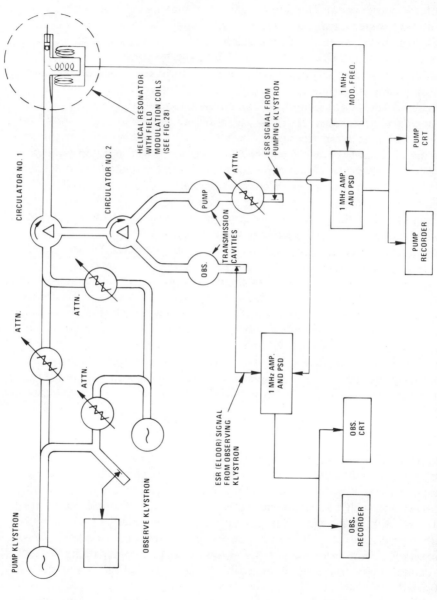

Figure 27. Block diagram of the Soviet ELDOR spectrometer employing a single helical resonator. (From Sokolov and Benderskii.[38])

frequency interval is changed. The entire process is repeated until enough points are gathered to plot the change in ESR intensity at the observing field as a function of the magnitude of Δv.

Upon traversing an ELDOR resonance, a signal is reflected from the helix, through circulators one and two, into the pump transmission cavity. This cavity is tuned in such a manner as to pass all ESR information generated at the pump frequency and to reflect all ESR information occurring at any other frequency, including the observing frequency. The reflected signal is then redirected through circulator number two into the observing transmission cavity. This filter passes the observed ESR (ELDOR) signal for microwave rectification by the observing detector crystal. The 1-MHz component is amplified, phase sensitive detected, and presented for display in the conventional manner. Likewise, a display of the ESR information resulting from the pump frequency is available for convenient display in much the same manner as described.

The single-helix ELDOR probe is shown in Figure 28. The top of the spiral is connected to a cylinder, and both are inserted in a Dewar vessel enabling temperature variation of the sample—a very important ELDOR parameter. Both microwave frequencies pass through a section of reducing waveguide, called a coordinator, that acts to match the impedance of the transmission line and microwave helix over a broad frequency range. The fine tuning of this impedance match is achieved by adjusting what is effectively a sliding short at the end of the waveguide. Optimum coupling of the microwave power occurs in the matched condition where the bandpass of the reflecting resonator exceeds 1 GHz. In the unmatched condition, the reflection coefficient of the helix varies periodically with frequency, and the bandpass is reduced to the order of 30–300 MHz. Analysis of the operation of the single resonator shows that both the inside and outside helix diameters determine the corresponding coil separation, i.e., pitch of the helix, necessary for optimum performance. In addition, the presence of a dielectric material surrounding the helical structure, such as a Dewar or a sample, introduces a significant variation in microwave field homogeneity along the spiral axis, leading to a sensitivity reduction. In theory, the limiting sensitivity of this slow-wave structure of 1-mm diameter having a filling factor of unity is 9×10^{10} (spins/ΔH), where ΔH is the ESR linewidth in gauss. In practice, the ESR sensitivity of the observing channel of this spectrometer system is 10^{12} (spins/ΔH). The ELDOR sensitivity is obtained simply by multiplying the ESR sensitivity by the ELDOR reduction factor $R(0 < R < 1)$.

In 1970, a modification to this ELDOR spectrometer is described[40] whereby a transmission helix is employed. The primary advantage to be gained is that this device is easier to match over a broad frequency band than is a reflection helix. In addition, the ELDOR signal was detected by

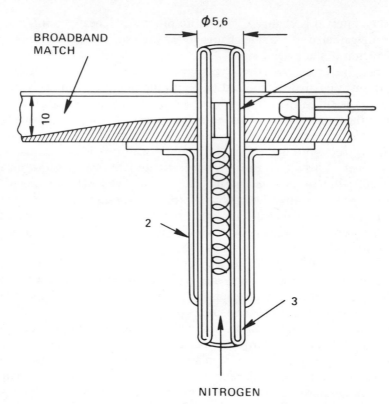

Figure 28. A spiral resonator from the Soviet ELDOR spectrometer with coordinator. The spiral diameter is 1.5 mm with a winding spacing of 0.8 mm and a wire thickness of 0.2 mm. The total length is 40 mm. (1) Connection antenna; (2) high-frequency modulation loop; (3) Dewar vessel. (From Sokolov and Benderskii.[38])

chopping the pump power at a 30-Hz rate and, subsequently, narrow-band amplifying and synchronously detecting at this frequency in the observing channel.

2.2.3. *Veith and Hausser—Germany*

In 1971, Vieth, Brunner, and Hausser[45] describe an ELDOR spectrometer with a microwave probe comprised of a slow-wave helix inside a microwave cavity; the former is used to support the pump mode and the latter to support the observe mode. A simplified block diagram of the basic spectrometer system is shown in Figure 29. The basic spectrometer system employs a double-modulation technique similar to that used in ENDOR and described in Section 1.1. The microwave cavity is designed with a set of internal modulation windings operating at 100 kHz, while the output of the

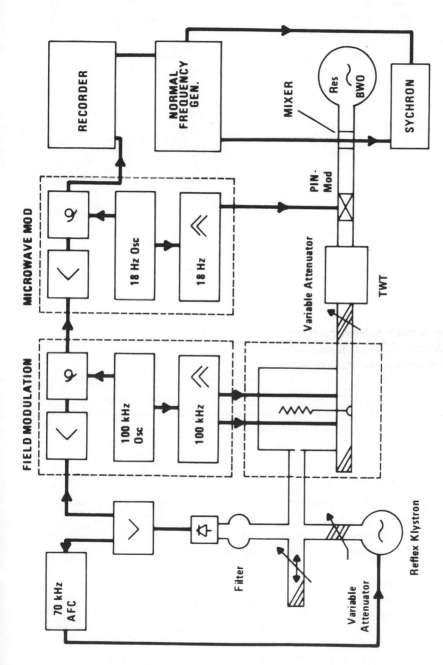

Figure 29. Block diagram of the German ELDOR spectrometer. (From Vieth *et al.*[45])

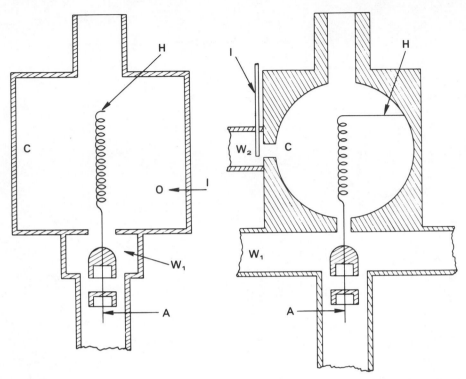

Figure 30. Two views of a cylindrically resonant cavity (C), having a helix (H) at the cavity center and connected at one end to an antenna (A) coupling microwave power from the pump waveguide (W_1). The top of the helix is connected to the cavity wall. Power from (W_2), the observing waveguide, is coupled to the cavity via the iris (I). (U.S. patent no. 3,798,532.)

pump klystron is PIN (*p*ositive *i*ntrinsic *n*egative) (amplitude) modulated at a rate of 18 Hz. The signal emanating from the microwave probe passes through a microwave transmission filter and is subsequently amplified and phase-sensitive detected first at 100 kHz and then at 18 Hz.

The original cavity–helix design described in Vieth's thesis[46] made no provision for temperature variation of the sample; however, this technical problem has been overcome, and recent versions of the device incorporate this important feature. Figure 30 depicts a simplified diagram of this structure showing two views of a cylindrically resonant cavity (C) having a helix (H) traversing the center of the cavity along the cylindrical axis and connected at one end to an antenna (A), the latter providing an **E** to **H** coupling to the pump microwave field in the waveguide (W_1). The top end of the helix is connected to the wall of C. A second waveguide (W_2) supplies the observ-

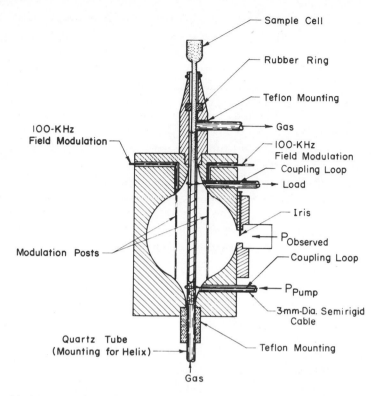

Figure 31. A cross-sectional view of the Vieth and Hausser ELDOR cavity using a microwave helix.

ing microwave field and is coupled to C via the iris (I), generating a micro-wave field at the helix that is perpendicular to the cavity (and helical) axis. The ELDOR sample is then located inside the helix. In order to insure orthogonal-ity of the microwave fields associated with the cavity and helix, the structure was designed so that the helix could be tilted and/or laterally shifted both along and perpendicular to the cavity axis. In addition, mode isola-tion may be achieved with insertion of dielectric material at various cavity locations.

A version of the cavity–helix structure with a Dewar is shown in Figure 31. The helix is formed by gluing silver tape to the inner Dewar wall, and sufficient isolation is obtained by inserting the Dewar so that the helical windings are orthogonal to the electric field lines of a TE_{112} cavity mode. Unwanted pump microwave field generated outside the helix is minimized by reducing the helical diameter relative to the microwave wavelength until the constraint of limited sample volume seriously degrades the ELDOR sensitivity. Hausser *et al.* have found that at 8.6 GHz, a helical diameter of 3.8 mm and a pitch of 3 mm optimizes the radial homogeneity of the microwave field. Once again, insertion of a sample affects the axial field homogeneity, and care must be taken to couple the helix (match helix to waveguide impedance) as near to critical as possible. The helical dimensions limit nonpolar samples to tubes having a length of 40 mm and an outside diameter slightly less than 3.5 mm, while, for polar samples, the tube diameter is more nearly 0.5 to 1 mm. The sensitivity of such a system is very nearly equal to that of the Russian ELDOR system.

Recently, the Heidelberg group has computerized the operation of the ELDOR spectrometer[47] as shown in Figure 32 and modified the probe by symmetrically coupling the pump power into and out of the helix. The advantage of such a modification is a substantial improvement in the axial homogeneity of the pumping microwave magnetic field.

2.2.4. *Other Cavity Designs*

A variety of cavity–cavity-type microwave probes has been designed for ESR induction experiments and forbidden transition studies as well as ELDOR investigations. Design requirements of bimodal cavities for these experiments vary in that the first requires that the two orthogonal modes resonate at the same fixed frequency, the second necessitates that the two modes resonate at different fixed frequencies, while the third requires independent tuning of the different fixed resonant frequencies of each of the two modes with respect to each other. Figure 33 shows the prototype bimodal cavity used in the first liquid-phase ELDOR experiment,[37] where the sample experiences the fields of two orthogonal modes, but in addition, where there exists a region of space for each mode where the field of the other mode is unsupported. By the introduction of dielectric material into either of the unshared regions, the corresponding frequency of one mode may be altered without affecting the frequency of the second mode. Since the irises are in the unshared portion of the cavity, a high degree of isolation (30 dB) between the modes is inherently assured. Further isolation may be achieved by the insertion of resistive and reactive elements into the cavity at the positions depicted in the figure. Glassware in the form of a Dewar and/or sample asymmetrically shifts the resonant frequency of the two modes, since the electric field along the length of the insert is constant for one mode and varies as a cosine for the other mode. This type of problem is compensated for by adjustment of the tuning screws located in the unshared regions of the cavity.

A second bimodal cavity design used in the initial ELDOR studies at

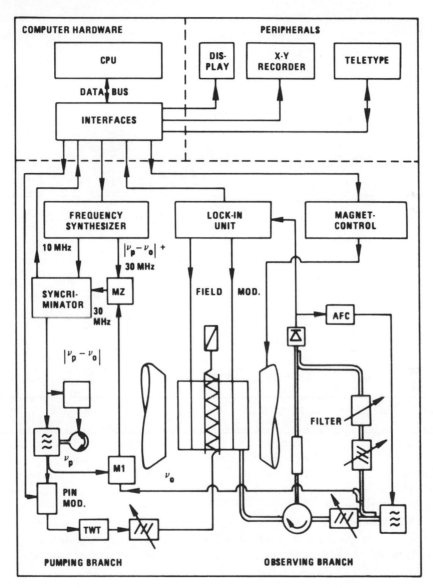

Figure 32. Block diagram of a computer-controlled ELDOR spectrometer. (From Stetter *et al.*[47])

Varian Associates is shown in Figure 34. The cavity incorporates a TE_{102} fully shared observing mode and a TE_{103} partially shared pump mode. The principles of operation are exactly the same as described above with the exception being that only the pump mode frequency is tunable.

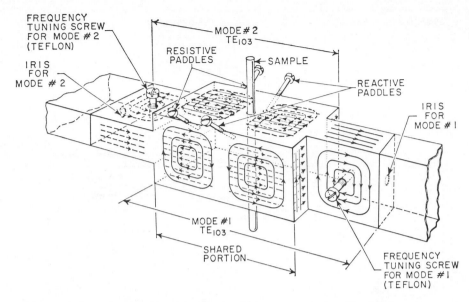

Figure 33. The prototype bimodal cavity in which two rectangular TE_{103} modes are crossed and have two half-wavelengths common. The lines of rf magnetic flux of mode 1 are indicated by solid lines and of mode 2 by dashed lines. (From Hyde *et al.*[37])

The group of J. Smidt *et al.* at the Delft Technische Hogeschool in Holland have developed a bimodal ELDOR reflection cavity from a ceramic material called Wonderstone. Albeit information is somewhat sketchy, the cavity is easily tunable and fairly compact, although a criticism of the design is that no provision is made to vary or control the temperature. In addition, provision is made to display both the ELDOR spectrum and the ESR spectrum simultaneously by sweeping both $H_0 - v_p$ and H_0. One of the advantages of the $H_0 - v_p$ sweep is the relative ease of signal interpretation in the case of relaxation studies.

The Italian group of Chiarini *et al.*[48] have constructed an ELDOR probe employing very high pump powers to observe incipient NMR spin-decoupling effects in an ELDOR experiment. Bimodal cavity development by the groups of Conciauro[49] at the University of Pavia and Franconi[50] at the University of Venice has produced a series of induction structures that may be extended to encompass the ELDOR technique.

The final version of the probe used in the commercial ELDOR spectrometer made by Varian Associates is a bimodal cavity structure that incorporates a variable temperature capability and a variable pump frequency. Cavity frequency tuning is accomplished by insertion of a conducting element, with respect to the rf electric field, into a region of the pumping mode where the electric field of the observing mode is unperturbed. Since

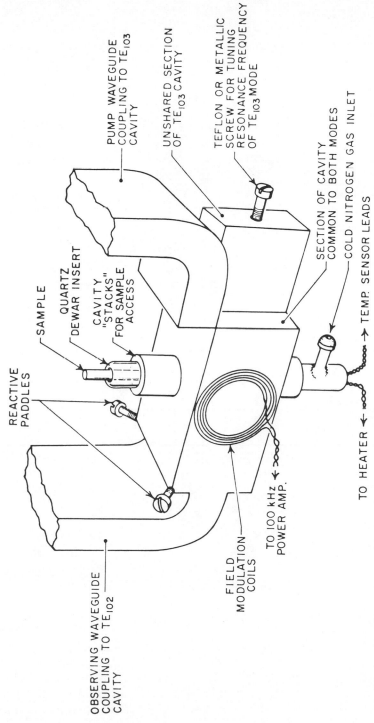

REACTIVE PADDLES

SAMPLE

QUARTZ DEWAR INSERT

CAVITY "STACKS" FOR SAMPLE ACCESS

PUMP WAVEGUIDE COUPLING TO TE$_{103}$ CAVITY

UNSHARED SECTION OF TE$_{103}$ CAVITY

TEFLON OR METALLIC SCREW FOR TUNING RESONANCE FREQUENCY OF TE$_{103}$ MODE

SECTION OF CAVITY COMMON TO BOTH MODES

COLD NITROGEN GAS INLET

TO HEATER ← → TEMP. SENSOR LEADS

TO 100 kHz POWER AMP.

FIELD MODULATION COILS

OBSERVING WAVEGUIDE COUPLING TO TE$_{102}$ CAVITY

Figure 34. The bimodal cavity in which a rectangular TE$_{103}$ mode supports the pumping microwave field and is crossed with a rectangular TE$_{102}$ mode by which double resonance signals are observed. (From Hyde et al.[37])

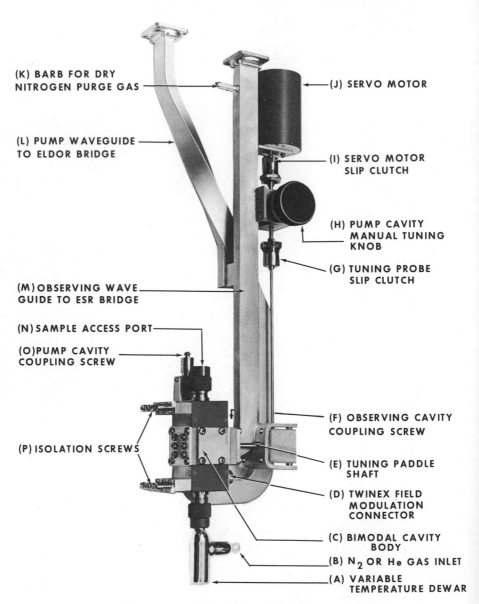

(K) BARB FOR DRY NITROGEN PURGE GAS

(J) SERVO MOTOR

(L) PUMP WAVEGUIDE TO ELDOR BRIDGE

(I) SERVO MOTOR SLIP CLUTCH

(H) PUMP CAVITY MANUAL TUNING KNOB

(G) TUNING PROBE SLIP CLUTCH

(M) OBSERVING WAVE GUIDE TO ESR BRIDGE

(N) SAMPLE ACCESS PORT

(O) PUMP CAVITY COUPLING SCREW

(F) OBSERVING CAVITY COUPLING SCREW

(P) ISOLATION SCREWS

(E) TUNING PADDLE SHAFT

(D) TWINEX FIELD MODULATION CONNECTOR

(C) BIMODAL CAVITY BODY

(B) N₂ OR He GAS INLET

(A) VARIABLE TEMPERATURE DEWAR

Figure 35. An external view of the E-802 Varian ELDOR cavity.

this cavity has been used by a variety of groups throughout the world, it will be discussed in detail in the following section.

2.3. *The Varian ELDOR Cavity*

Frequency-swept ELDOR is defined as the change (reduction) in an ESR signal—as monitored by a constant-frequency observing microwave source—that is caused by a frequency-swept pumping microwave source. The resulting ELDOR display consists of the change in ESR signal height of the observed ESR line as a function of the difference frequency of the two microwave sources. The Varian E-802 ELDOR cavity is designed to provide such a display.

An external view of this rectangular, crossed TE_{102} bimodal cavity is given in Figure 35. The microwave structure (C) contains a pair of standard modulation coils (D) and houses a cylindrical Dewar insert (A) for direct control of the sample temperature. Emerging from a standard commercial microwave bridge is the observing waveguide (M) that provides the means for transmission of microwaves at some constant frequency to the observing cavity portion of the bimodal ELDOR probe. The ELDOR microwave bridge, sitting directly atop the standard bridge, connects the tunable pumping klystron to the pumping resonator portion of the cavity via the pump waveguide (L).

2.3.1. *Microwave Modes*

The observing and pumping waveguides are located on each of two intersecting perpendicular cavity walls, and the irises of each guide are also actually perpendicular. Both the pump and observing cavities are rectangular and generate two orthogonal TE_{102} microwave modes as shown in Figure 36. The microwave magnetic field lines of flux in the observing mode are the vertical lines (Y) coupled from the observing waveguide to the cavity via the vertical iris (F), whereas the magnetic field lines of the pumping mode are shown by the horizontal solid lines (Z) coupled from the pumping waveguide to the cavity via the horizontally positioned iris (O). The microwave magnetic field maximum for the entire bimodal structure is a line formed by the two intersecting stacks of modal planes and occurring at the exact center of the cavity. From this figure, it can be seen that the pump microwave magnetic field along a line sample is constant, while the observing microwave field varies as a cosine.

A careful study of Figure 36 shows that there are four unshared regions that exist inside the cavity structure. The observing mode possesses an

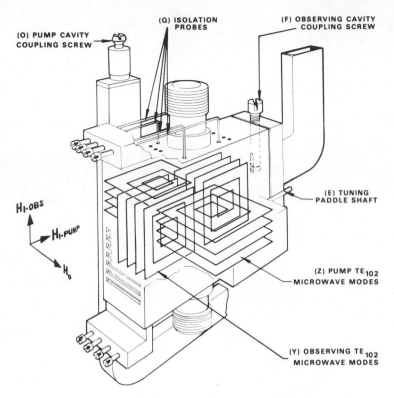

Figure 36. The microwave modes of the E-802 Varian ELDOR cavity.

unshared region at the coupling screw (F) located at the rear of the cavity and another unshared region at the front of the cavity adjacent to the optical irradiation slots. The pump mode reflects analogous symmetry with unshared regions at the pump coupling screw (O) on the left side of the cavity and a similar configuration at the right side of the cavity. The property of generating a cavity structure that is not completely spatially degenerate is accomplished by the insertion of conducting rods at positions (U) and (W) as depicted in Figure 37. These internal septa provide a conductive pathway along which each mode can be supported and constrained. In the area of the ELDOR cavity where the observing and pumping modes overlap, the walls that constrain the observing mode from further overlapping the pump mode are formed by the horizontal conducting rods (U), whereas the walls that constrain the pump mode from further overlapping the observing mode are formed by vertical rods at position (W). This design feature allows unshared regions within the cavity that provide for indepen-

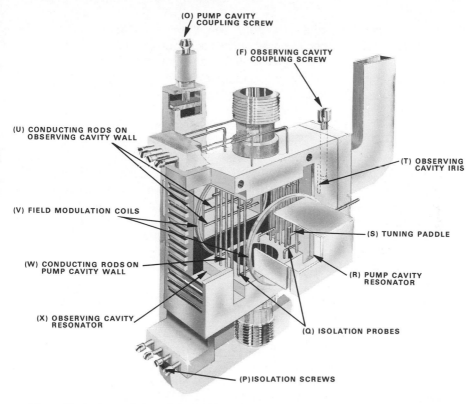

Figure 37. An internal view of the E-802 Varian ELDOR cavity. (U.S. patent no. 3,609,520.)

dent coupling, tuning, and mode isolation. Each of these features is discussed below.

2.3.2. Cavity Coupling

The TE_{102} pump mode is coupled into the ELDOR cavity via the pumping cavity iris located below the pump cavity coupling screw (position O), while the observe mode is coupled into the cavity at position T by the observing cavity iris (F). The observing frequency coupling screw (F) consists of a small disk of conducting metal attached to a threaded Teflon rod that passes parallel to the long axis of the observing iris (T). The pump-coupling mechanism is a sliding short—a piece of metal that moves back and forth in the waveguide as the pump coupling screw (O) is rotated. The purpose of each coupler is to transmit all of the incident microwave power into its cavity mode without reflection to either source and to bilaterally couple ESR signal from the cavity to the observing detector in an optimum manner. In practice, as the pump frequency is swept, the pump cavity coupling may change to such a degree that the sweep must be halted at regular intervals to adjust the pump cavity coupling screw. Such an experimental operation done properly does not affect the ELDOR spectrum.

2.3.3. Cavity Tuning

The pump cavity frequency is varied by simultaneous insertion of two tuning paddles, constructed in the shape of a T (S in Figure 37), located inside the pump cavity. The tuning paddles are conductive elements inserted parallel to the electric field in the two unshared regions of the pumping mode, thus selectively perturbing the pump resonant frequency without disturbing the resonant frequency or the coupling of the observing cavity mode. These two paddles are attached to the tuning shaft (E) located on either side of the observing waveguide (M) of Figure 35. The degree of paddle insertion is automatically controlled by a slave dc motor (J) so that a continuous variation of the pump resonant frequency may be attained. The dc motor receives its input from the ELDOR microwave bridge, where the operator preselects the upper and lower limits of the pump frequency sweep via a set of thumbwheel switches calibrated in megahertz. The magnitude and rate of the pump sweep determines the subsequent depth and velocity of paddle insertion. To track the changing resonant frequency of the pump cavity, the pump klystron frequency is varied by applying a correction voltage generated by the pump AFC circuit to a servomotor located in the ELDOR bridge. This servomotor controls the pump klystron tuning shaft and mechanically restores the pump microwave frequency to the exact instantaneous cavity resonance in the standard manner. In this way, the tuning paddle–AFC combination generates a continuously resonant pumping frequency.

In addition, the pump frequency can be varied manually by turning the tuning knob H shown in Figure 35. A slip clutch (I) protects the dc motor during manual operation, and a second slip clutch (G) protects the motor when the gear train travels against the upper or lower mechanical stop.

Variable temperature quartz Dewars, needed to obtain ELDOR spectra as a function of temperature, have an effect on the cavity tuning range. The more quartz (dielectric material) inserted into the cavity, the smaller the frequency difference range Δv attainable between the two modes. Thus, commercial ELDOR Dewars employ a special thin quartz wall (~ 0.5 mm) in order to provide the maximum ELDOR frequency range possible. Typically, the pump frequency range is 200 MHz $< v_0 <$ 350 MHz. If one were to use a standard Varian variable-temperature Dewar insert in the ELDOR cavity frequency, then a decrease in the frequency range to 100 MHz $< v_0 < 200$ MHz would be typical.

Alternative mechanisms for tuning the pump cavity are either to rotate the conductive elements in the unshared regions of the pump modes or to replace the T paddles with another set of elements differing in both geometry and material, e.g., a set of Rexolite rods, that generate a continuous shift in the pump cavity resonant frequency.

2.3.4. *Cavity Mode Isolation*

To conduct an ELDOR experiment, the pumping and observing micro-wave modes must be mutually isolated from each other for the reasons discussed in Section 2.1. Construction of the E-802 bimodal structure with 40 dB of (power) isolation (see below) between the two microwave modes in the presence of both a Dewar and a cylindrical sample is possible over the entire range of pumping frequency sweep. Any asymmetry present inside the ELDOR cavity has a potential effect on the mutual geometry of the pump and observe modes resulting in a loss of isolation. A slight nonorthogonality between modes may be corrected by the insertion of conducting isolation screws (P) that appear in the front of the cavity (Figures 35, 36, or 37). Rotation of any of the eight isolation screws (P) introduces a perturbation to the appropriate mode(s) in each of the unshared regions at the eight corners of the cavity (Q). In theory, the isolation screws effect the geometry of one mode relative to the other to produce a high degree of isolation under a variety of external conditions. In practice, the isolation is affected to a significant degree by at most two of the eight screws; however, this is sufficient to achieve the degree of isolation necessary to perform the ELDOR experiment.

Since the isolation screws are designed to compensate for small asymmetries of the pump and observe modes, Dewars and samples themselves may play a major role in cavity isolation if they induce nonorthogonalities beyond the compensating ability of the isolation screws. The quartz Dewar, if not made of two uniform, closely concentric tubes, can create isolation problems. If large (4-mm-diameter) single crystals of noncylindrical symmetry are inserted into the cavity structure and rotated, the anisotropic dielectric loss may cause a distortion of the orthogonality of the two microwave modes. Normally, isolation screws are used to correct for this. However, much time will be saved by employing smaller crystals (< 2 mm) to minimize the anisotropic variation of cavity isolation.

To measure the magnitude of the isolation between the pumping and observing modes, a power meter is attached to the observing waveguide immediately preceding the transmission cavity filter. The following conditions are present prior to the test:

1. $v_0 \equiv v_p$; i.e., the resonant frequencies of both the pump and observe modes are made coincident.
2. Both cavities are critically coupled.
3. The Dewar and, preferably, the sample are in place.

Microwave power is applied to the pump mode, and any leakage to the observe mode is detected via the power meter attached to the observing waveguide. This method involves a power measurement and is reflected in dB by 10 log (P_{pump}/P_{obs}). An alternative method involves measuring an induced signal from the observe mode. The latter is a *voltage* measurement and, expressed in dB of isolation, is twice that of the corresponding power measurement. The literature freely mixes terminology of isolation measurements, and the reader must be aware of the proper implication. Isolation measurements made when $v_0 \neq v_p$ are less demanding on the orthogonality of the two modes, since the observing cavity now acts as a filter that diminishes the amount of pump power at frequency v_p that passes through the observing resonance at frequency v_0. Thus, the most stringent isolation test for a cavity is made when both frequencies are coincident.

2.3.5. ELDOR Spectral Presentation

In general, the three parameters that determine the specific ELDOR display to be recorded are the pumping frequency, the observing frequency, and the magnetic field. The particular selection of which parameters will be varied and which remain fixed determine the nature of the ELDOR display.

2.3.5.1. *Various Options.* The first option, for example, would be to fix all three parameters to some particular ELDOR resonant condition and observe the spectral response to an external perturbation (e.g., temperature, light) as a function of time. This type of display may, e.g., be useful in the investigation of optical effects on nuclear relaxation, electron polarization, or other such phenomena that are coupled to the ELDOR experiment.

The second option is to fix the value of two parameters while varying the third. This is the particular display that is currently employed in the great majority of ELDOR work. There are two modes in which spectra are generally recorded—field swept and frequency swept.

a. *Field-swept ELDOR.* Of the two options it is far easier experimentally, although certainly less convenient from the viewpoint of spectral analysis, to perform an ELDOR experiment by presetting the two microwave frequencies to some fixed separation Δv and sweeping the magnetic field. If the pump and observe modes are uncoupled, an ordinary first-derivative ESR line emerges from the observing cavity and is displayed on the recorder. If, however, the two modes are coupled, the resulting spectrum appears identical in nature to the previously recorded ESR signal, except that the intensity is then recorded for a variety of frequency intervals Δv.

b. *Frequency-swept ELDOR.* In addition, commercial Varian ELDOR equipment permits a sweep of the pump frequency while both the observing frequency and the magnetic field remain constant, providing a display that strictly adheres to the basic definition of ELDOR given in Section 1. If the spectral width to be investigated occurs over a magnetic field, the pumping frequency v_p is then preset to scan a frequency range that is greater than the observing frequency v_{obs}. Likewise, if the position of the ESR spectrum to be probed is at a magnetic field that is greater than that being observed, the magnitude of the pumping frequency is less than that of the observing frequency. The frequency difference of the two klystrons is displayed on a counter having a range of 500 MHz and may be stepped though the desired scan range with a resolution of 10 or 100 kHz. As previously discussed (Section 2.1), this frequency difference is translated to an analog drive for an ordinary XY recorder. The recorder Y-axis plots the change in ESR amplitude as the pumping frequency is swept resulting in a true ELDOR presentation.

The third option is to fix one parameter while simultaneously changing the other two. For example, fixing one microwave frequency and sweeping both the second microwave frequency and the magnetic field is possible. Such an experiment is conceivable where the pumping frequency and the magnetic field are fixed at some preset interval, perhaps equal to a hyperfine splitting, and both are swept while the observing frequency is fixed. Theoretical considerations show the possibility of obtaining spin relaxation data directly from the resulting ELDOR line shapes, enabling a relatively easy evaluation of T_{1e} and T_{2e}.[47,51,52] In addition, a sweep of $(H_0 - v_p)$ as well as H_0 allows for simultaneous registration of both the ELDOR and ESR spectra, and the Delft group has filed patent applications in this regard.[52]

Other possibilities involve sweeping both microwave frequencies while fixing the magnet field. One experiment would be to fix $v_p - v_0 = \Delta v$ and to sweep both frequencies through the ESR spectrum of interest displaying the spectral response to Δv. By repeating the experiment for a variety of Δv, the spectral response that is recorded should be analogous to that of a spectrum analyzer. Similar results would be expected by continuously varying Δv about some fixed field position, i.e., $-50 < \Delta v < 50$ MHz. However, the experimental design of a microwave structure capable of such an experiment would be terribly difficult and demanding. For the sake of completeness, one may conceive of varying both frequencies as well as the magnetic field simultaneously, albeit this is only a concept at this time.

2.3.5.2. *Other Factors Affecting the ELDOR Display*

a. *Field modulation.* As shown in the chapters by Freed and Dalton, the ELDOR effect is dependent upon the relative magnitudes of the relaxation parameters W_i ($i = n, e, X_1, X_2$), and under certain experimental conditions, the relaxation parameters themselves become functions of rapid applied oscillations such as the field modulation. This dependence increases so that, as molecular motion slows, relaxation parameters (T_1) become longer and more intertwined with instrumental parameters such as the field modulation amplitude and frequency. Consider a slow-tumbling ESR spectrum composed of a Gaussian distribution of spin packets each of which is Lorentzian in nature. When the modulation cycle time is comparable to or faster than T_i ($i = e, n, X_1, X_2$), there is insufficient time for individual spin packet relaxation during each magnetic field modulation cycle. Thus, before a spin packet can completely relax, it is brought through resonance again giving rise to a pseudorelaxation time dependent on modulation frequency. In this manner the magnitude and line shape of the ELDOR spectrum are related to the field modulation and any parameters obtained from such data must be corrected for such effects.

There are two ways to avoid the problem altogether. Modulation is employed in magnetic resonance instrumentation for a number of reasons, the primary one being sensitivity. If this technique is inherent to all designs then one may (a) reduce the field modulation frequency and/or amplitude to an acceptable value, or (b) amplitude modulate the microwave frequency (see Section 2.2.4) in the standard manner, eliminating field modulation altogether. Each of these techniques will allow ELDOR parameters to be extracted from the data without the necessity of incorporating modulation frequency corrections.

b. *Adiabatic rapid passage.* On the other hand, it is possible to utilize the interplay between field modulation, molecular motion, and relaxation phenomena in an advantageous way. If the magnetization of the sample under investigation dutifully follows the effective magnetic field H_{eff} (the resultant of H_0 and H_1) upon passage through resonance, the system is said to be governed by the conditions of adiabatic rapid passage. This implies that the magnetization passes through resonance at a rate fast enough so that any relaxation back to H_0 in this time period is prohibited, but not so fast so as to prohibit the magnetization from following (i.e., maintaining its initial angle with respect to H_{eff}) the effective field exactly. Instrumental parameters such as field modulation frequency and amplitude may be utilized making the experimental conditions optimal for nonadiabatic rapid-passage studies.[53] This situation has the inherent property of generating ELDOR signals that are observed by setting the reference phase to the PSD 90° out of phase with respect to the modulation frequency. Since the rate at which individual spin packets pass through resonance depends also on the magnitude of the rotational diffusion of the paramagnetic molecules themselves, it can be seen that by monitoring the ELDOR signal under various conditions of nonadiabatic rapid passage (i.e., 90° out of phase), displays will appear that not only permit a determination of T_{1e} and T_{2e} but also enable elucidation of the precise dependence of these parameters upon the molecular diffusion rates and intramolecular dynamics.[54,55] In addition, increased ELDOR signal amplitudes and sensitivity are quite common for solid and slow-tumbling liquid samples upon adjustment of the reference phase to the PSD to be 90° from that of the modulation frequency at the sample.

2.4. Summary

The primary advantage of scanning the pump microwave frequency in fast-tumbling liquids is that a single scan through the ESR spectrum renders the relaxation-induced couplings of one particular observed ESR line with all components of the spectrum. Likewise, in solids or slow-tumbling liquids, this spectral presentation is a measure of the interaction of the rest of the spectrum, with one particular portion corresponding to the observing resonance position. In the former display, the ELDOR spectrum is composed of discreet lines at particular frequencies v_p, whereas the latter display may be more continuous in nature with varying ELDOR intensity as a function of the pump frequency. A comparison of sweeping v_p and $(H_0 - v_p)$ shows primarily advantages of convenience in the latter sweep, although in certain cases, the increase in sensitivity may be significant as shown in Figure 38.

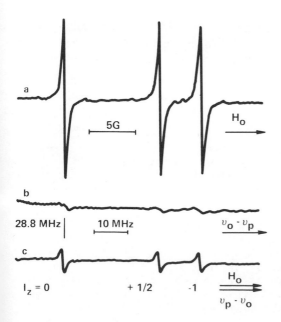

Figure 38. (a) High-field portion of the ESR spectrum of perdeuterated Tanone (33 % ^{15}N) in ethylbenzene, concentration 10^{-3} moles/liter; $T = 290°$K. $\tau_{rot} = 1.1 \times 10^{-11}$ sec. (b) Frequency-swept ELDOR spectrum of the same sample, observed line $I_z = -\frac{1}{2}$. Note that the pumping frequency is below the observing frequency. (c) Field- and frequency-swept ELDOR spectrum taken simultaneously with spectrum (a), pumped line $I_z = -\frac{1}{2}$.

This comparison shows:

1. When sweeping v_p, the ELDOR spectrum may be saturation broadened, rendering less resolution and sensitivity as compared with a sweep of $(H_0 - v_p)$.
2. The dependence of the ELDOR parameters on modulation frequency, although not eliminated, may be minimized by sweeping $(H_0 - v_p)$, since v_p is always resonant, i.e., $\Delta_{\omega p} T_{2p} \equiv 0$, simplifying the analysis somewhat.
3. Quantitative measurement of the ELDOR effect must be related to the normal ESR spectrum. A sweep of $(H_p - v_p)$ permits simultaneous presentation of both the ESR and ELDOR signals, whereas a v_p sweep gives the ELDOR only.

When modulation effects on ELDOR spectra are undesirable, PIN modulation of the pump frequency is advantageous. In the opposite case, when differing modulation frequencies are used as a probe for the molecular dynamics, oftentimes ELDOR data obtained with the reference phase of the modulation frequency shifted by 90° produces an increase in the ELDOR sensitivity as well as a more direct display of the lattice dynamics.

In summation, of the many spectral presentations discussed in this section, it seems reasonable to conclude that the optimum display to be employed depends on the specific nature of the parameters that are to be studied.

ACKNOWLEDGMENT

The author wishes to express his thanks to Messrs. James H. Jacobsen, Robert G. MacNaughton, and Robert C. Sneed of the ESR Group at Varian for a series of most helpful, interesting, and valuable discussions.

References

1. G. Feher, *Phys. Rev.* **103**, 834 (1956).
2. A. Cederquist, Ph.D. thesis, Washington University, St. Louis, Missouri (1963).
3. J. S. Hyde and A. H. Maki, *J. Chem. Phys.* **40**, 3117 (1964).
4. See, e.g., *American Laboratory* (1974).
5. H. Gunnel, *Brown-Boveri Rev.* **31**, 327 (1944).
6. P. Ruthroff, *Proc. IRE* **47**, 337 (1959).
7. K. H. Hausser and F. Reinhold, *Z. Naturforsch.* **16A**, 1114 (1961); *Phys. Lett.* **2** 53 (1962); R. B. Clarkson, Ph.D. thesis, Princeton University, Princeton, New Jersey (1969).
8. A. L. Terhune, J. Lambe, L. Kikuchi, and J. Baker, *Phys. Rev.* **123**, 1265 (1961).
9. G. D. Watkins and J. W. Corbett, *Phys. Rev.* **134A**, 1359 (1964).
10. D. Schmalbein, A. Witte, R. Roder, and G. Laukien, *Rev. Sci. Instr.* **43**, 1664 (1972).
11. A. H. Maki, R. D. Allendoerfer, J. C. Danner, and R. T. Keys, *J. Amer. Chem. Soc.* **90**, 4225 (1968); R. D. Allendoerfer and D. J. Eustace, *J. Phys. Chem.* **75**, 2765 (1971).

12. N. S. Dalal, D. E. Kennedy, and C. A. McDowell, *J. Chem. Phys.* **59**, 3403 (1973).
13. K. P. Dinse, K. Möbius, and R. Biehl, *Z. Naturforsch.* **28A**, 1069 (1973).
14. I. Mayagawa, R. B. Davidson, H. A. Helms, Jr., and B. A. Wilkinson, Jr., *J. Mag. Res.* **10**, 156 (1973).
15. T. Yamamoto, M. Kono, K. Sato, T. Miyamae, K. Mukai, and K. Ishizu, *JEOL News* **10a**, 6 (1972).
16. J. S. Hyde, *J. Chem. Phys.* **43**, 1806 (1965); *Varian Instrument Division Bull., E-700 High Power ENDOR System* (1971); J. S. Hyde, T. Astlind, L. E. G. Eriksson, and A. Ehrenberg, *Rev. Sci. Instr.* **41**, 1598 (1970).
17. D. S. Leniart, H. D. Connor, and J. H. Freed, *J. Chem. Phys.* **63**, 165 (1975).
18. M. R. Das, H. D. Connor, D. S. Leniart, and J. H. Freed, *J. Amer. Chem. Soc.* **90**, 4354 (1968).
19. J. S. Hyde, *J. Phys. Chem.* **71**, 68 (1967).
20. A. H. Maki, R. Allendoerfer, J. C. Danner, and R. T. Keyes, *J. Amer. Chem. Soc.* **90**, 4225 (1968).
21. N. M. Atherton and A. J. Blackhurst, *J. Chem. Soc. Faraday Trans. II*, **68**, 470 (1972).
22. K. Möbius, H. van Willigen, and A. H. Maki, *Mol. Phys.* **20**, 289 (1971).
23. N. S. Dalal, D. E. Kennedy, and C. A. McDowell, *J. Chem. Phys.* **59**, 3403 (1973).
24. F. Gerson, J. Jachimowiez, K. Möbius, R. Biehl, J. S. Hyde, and D. S. Leniart, *J. Mag. Res.* **18**, 471 (1975).
25. J. S. Hyde, G. H. Rist, and L. E. G. Eriksson, *J. Phys. Chem.* **72**, 4269 (1968).
26. H. D. Connor, Ph.D. thesis, Cornell University, Ithaca, New York (1972).
27. D. S. Leniart, Ph.D. thesis, Cornell University, Ithaca, New York (1971).
28. H. Wahlquist, *J. Chem. Phys.* **35**, 1709 (1961).
29. R. D. Allendoerfer, P. E. Gallagher, and P. T. Landsburg, *J. Amer. Chem. Soc.* **94**, 7702 (1972).
30. J. H. Freed, *J. Chem. Phys.* **43**, 2312 (1965).
31. J. H. Freed, D. S. Leniart, and H. D. Connor, *J. Chem. Phys.* **58**, 3089 (1973).
32. J. H. Freed, D. S. Leniart, and J. S. Hyde, *J. Chem. Phys.* **47**, 2762 (1967).
33. K. P. Dinse, R. Biehl, K. Möbius, and M. Plato, *J. Mag. Res.* **6**, 444 (1972).
34. P. P. Sorokin, G. J. Lasher, and I. L. Gelles, *Phys. Rev.* **118**, 939 (1960).
35. W. P. Unruh, and J. W. Culvahouse, *Phys. Rev.* **129**, 2441 (1963).
36. P. R. Moran, *Phys. Rev.* **135A**, 247 (1964).
37. J. S. Hyde, J. C. W. Chien, and J. H. Freed, *J. Chem. Phys.* **48**, 4211 (1968).
38. E. A. Sokolov and V. A. Benderskii, *Prib. Tekh. Eksper.* **2**, 232 (1969). [English Translation: *Instrum. Exper. Technol.* **2**, 526–527.]
39. P. A. Stunkas, V. A. Benderskii, L. A. Blumenfeld, and E. A. Sokolov, *Opt. Spektros.* **28**, 278 (1970).
40. P. A. Stunkas, V. A. Benderskii, and A. A. Sokolov, *Opt. Spektros.* **28**, 487 (1970).
41. V. A. Benderskii, L. A. Blumfeld, P. A. Stunkas, and E. A. Sokolov, *Nature* **220**, 365 (1968).
42. J. S. Hyde, R. C. Sneed, and G. H. Rist, *J. Chem. Phys.* **51**, 1404 (1969).
43. T. S. Kuau, D. S. Tinti, and M. A. El-Sayed, *Chem. Phys. Lett.* **4**, 507 (1970).
44. M. Leung, and M. A. El-Sayed, *Chem. Phys. Lett.* **16**, 454 (1972).
45. H. M. Vieth, H. Brunner, and K. H. Hausser, *Z. Naturforsch* **26A**, 167 (1971).
46. H. M. Vieth, *Elektron–Elektron doppel resonanz von nitraxid-Radikalen in Lösung*, Ph.D. thesis, University of Heidelberg, Heidelberg, West Germany (1973).
47. E. Stetter, H. M. Vieth, and K. H. Hausser, *J. Mag. Res.* **23**, 493 (1976).
48. F. Chiarini, M. Martinelli, and P. A. Rolla, Multiple quantum transitions in ELDOR spectroscopy, *Lett. Nuovo Cimento Soc. Ital. Fis.* [2] **5**(2), 197–201 (1972) (in English).
49. G. Conciauro, and E. Randazzo, *Rev. Sci. Instr.* **44**, 1087 (1973).
50. G. Conciauro, M. Puglisi, C. Franconi, P. Galuppi, and E. Randazzo, *J. Mag. Res.* **9**, 363 (1973).

51. J. Smidt, and A. Mehlkopf, private communication.
52. E. Van der Drift, A. F. Mehlkopf, and J. Smidt, *Chem. Phys. Lett.* **36**, 385 (1975).
53. M. M. Dorio and J. C. W. Chien, *Macromolecules* **8**, 734 (1975); M. M. Dorio, and J. C. W. Chien, *J. Chem. Phys.* **62**, 3963 (1975); M. M. Dorio, Ph.D. thesis, University of Massachusetts, Amherst (1975).
54. M. Smigel, L. Dalton, J. Hyde, and L. Dalton, *Proc. Natl. Acad. Sci.* **71**, 1925 (1974).
55. J. S. Hyde, M. Smigel, L. Dalton, and L. Dalton, *J. Chem. Phys.* **62**, 1655 (1975).

3

Theory of Multiple Resonance and ESR Saturation in Liquids and Related Media

Jack H. Freed

1. Introduction to Saturation

1.1. General Considerations

The well-known result from the steady-state solution of the Bloch equations is that the absorption is given by the y component of magnetization $\tilde{M}_y$ in the rotating frame[1]:

$$\tilde{M}_y = \frac{\gamma H_1 T_2}{1 + (T_2\,\Delta\omega)^2 + \gamma_e^2 \beta_1^2 T_1 T_2}\, M_0 \tag{1}$$

with M_0 the equilibrium magnetization, β_1 the strength of the rf magnetic field, and γ_e the electron-spin gyromagnetic ratio. When we switch to a quantum mechanical description, we can calculate

$$M_\pm = M_x \pm iM_y = (\tilde{M}_x \pm i\tilde{M}_y)e^{\pm i\omega t} \tag{2}$$

statistically from its associated quantum mechanical operator

$$\mathscr{M}_\pm = \mathscr{N} \hbar\gamma_e S_\pm \tag{3}$$

where $\mathscr{N}$ is the concentration of electron spins, by taking a trace of the spin density matrix $\sigma(t)$ with the spin operator $S_\pm$:

$$M_\pm(t) = \mathscr{N} \hbar\gamma_e \operatorname{Tr} \sigma(t)S_\pm \tag{4}$$

The trace is invariant to a choice of zero-order basis states.

Jack H. Freed • Department of Chemistry, Cornell University, Ithaca, New York

The basis of our analysis here will be the equation of motion for $\sigma(t)$ appropriate in the motional narrowing region, which has been derived in several places.[1-3] That is, we write in operator form

$$\dot{\sigma} = -i[\mathscr{H}_0 + \varepsilon(t), \sigma] + R(\sigma - \sigma_{eq}) \tag{5a}$$

which may be rewritten in terms of matrix elements as

$$\dot{\sigma}_{\alpha\alpha'}(t) = -i\omega_{\alpha\alpha}\sigma_{\alpha\alpha'} - i[\varepsilon(t), \sigma]_{\alpha\alpha'} + \sum_{\beta\beta'} R_{\alpha\alpha'\beta\beta'}(\sigma_{\beta\beta'} - \sigma_{eq\beta\beta'}) \tag{5b}$$

where $\mathscr{H}_0$ is the zero-order spin Hamiltonian leading to the resonant transition frequencies $\omega_{\alpha\alpha'}$ between eigenstates α and α', $\varepsilon(t)$ includes the interaction of the spins with the various oscillating rf and microwave fields, which induce the resonances as well as any high-frequency modulation of the dc field, while R is the relaxation matrix with elements $R_{\alpha\alpha'\beta\beta'}$. The relaxation matrix yields the linewidths and the transition probabilities for relaxation from nonequilibrium population distributions. It is clear, from the form of equations (5) that in order to get stable exponential relaxation, we must require that the real part of $R_{\alpha\alpha'\beta\beta'}$ be negative (i.e., Re $R < 0$). We will give simple examples of the use of the relaxation matrix below. The inclusion of the equilibrium density matrix σ_{eq} in equations (5) is part of a high-temperature approximation for which we write

$$\sigma_{eq} = \frac{\exp(-\hbar\mathscr{H}_0/kT)}{\text{Tr}[\exp(-\hbar\mathscr{H}_0/kT)]} \approx \frac{1}{A}\left(1 - \frac{\hbar\mathscr{H}_0}{kT}\right) \tag{6}$$

where A is the total number of spin eigenstates and k is Boltzmann's constant.

1.2. A Simple Line: Two-Level System

We have from equation (5b) that when $E(t) = 0$ and $R = 0$, then

$$\sigma_{\alpha\alpha'}(t) = \exp(-i\omega_{\alpha\alpha'}t)\sigma_{\alpha\alpha'}(0) \tag{7}$$

Thus, if $\sigma_{\alpha\alpha'}(0) \neq 0$, then $\sigma_{\alpha\alpha'}(t)$ will be oscillatory. Now suppose we have only a simple line with $\omega_0 = \omega_{ab}$, where a and b are the $M_s = \frac{1}{2}$ and $-\frac{1}{2}$ levels, and there are no other spin levels. Then

$$\langle b|S_-|a\rangle = \langle a|S_+|b\rangle = 1 \tag{8}$$

and

$$\text{Tr}[\sigma(t)S_+] = \sigma(t)_{ba}S_{+ab} = \sigma(t)_{ba} \tag{9}$$

with

$$\sigma_{ba}(t) = \exp[(-i\omega_{ba} + R_{ba,ba})t]\sigma_{ba}(0) \tag{10}$$

We have in equation (10) included only the "diagonal" element of this four-indexed variable. In this form, we see that $-R_{ba, ba} = T_2^{-1}$ for this simple line whose relaxation is uncoupled to any other lines. Since $\text{Re } R$ is negative, $\sigma_{ba} \to 0$ for $t \gg |\text{Re } R|^{-1}$. Thus, there will be no steady-state absorption unless we include effects of the rf field. So we add to the Hamiltonian

$$\hbar\varepsilon(t) = \tfrac{1}{2}\hbar\gamma_e B_1[S_+ e^{-i\omega t} + S_- e^{+i\omega t}] \tag{11}$$

which is the interaction of the spin with a rotating field $\mathbf{B}_1 = B_1(\cos\omega t\mathbf{i} + \sin\omega t\mathbf{j})$. Then for our simple line the $\langle b| - |a\rangle$ matrix element of equation (5a) is

$$\dot{\sigma}_{ba} = (i\omega_0 + R_{ba, ba})\sigma_{ba} - id(\sigma_{bb} - \sigma_{aa})e^{i\omega t} \tag{12}$$

where

$$d = \tfrac{1}{2}\gamma_e B_1\langle b|S_-|a\rangle = \tfrac{1}{2}\gamma_e B_1 \tag{13}$$

Now the power absorbed from the rotating field is just[1]:

$$P = \omega B_1 \tilde{M}_y = \frac{-\omega B_1 i}{2}(M_+ e^{-i\omega t} - M_- e^{i\omega t}) \tag{14}$$

where from equation (4) $M_\pm \propto \text{Tr}_\sigma[\sigma(t)S_\pm]$ and S_{+ab} requires $\sigma(t)_{ba}$ in the trace. Thus, only the component of $\sigma(t)_{ba}$ oscillating as $e^{i\omega t}$ will give a net time-averaged power absorption. So, let

$$\sigma_{ba} = Ze^{i\omega t} \tag{15}$$

and assume Z is time independent to achieve the steady-state solution. [More rigorously, the steady state σ_{ba} may be expanded in Fourier components $n\omega$, where $n = 0, 1, 2, \ldots$, but only the component oscillating at frequency ω need be considered (see Freed *et al.*[5]).] Thus we have

$$(\Delta\omega + iR_{ba, ba})Z = d(\sigma_{bb} - \sigma_{aa}) \tag{16}$$

where $(\sigma_{bb} - \sigma_{aa})$ is the population difference in the two states. Now note that σ is Hermitian, so $\sigma_{ab} = \sigma_{ba}^*$ and

$$\sigma_{ab}^* = Z^* e^{-i\omega t} \tag{17}$$

Thus,

$$P \propto \text{Im } Z \equiv Z'' \tag{18}$$

We may begin to suspect that Z plays the role of $\tilde{M}_+$ (while Z^* is $\tilde{M}_-$). Also, we have already noted that

$$R_{ba, ba} = -(1/T_2)_{ba} = -(1/T_2)_{ab}$$

We now need the diagonal spin-density matrix elements σ_{bb} and σ_{aa}, which in steady state are not oscillating in time. We get from equation (5b),

$$R_{aa,\,aa}\sigma_{aa} + R_{aa,\,bb}\sigma_{bb} = di(Z - Z^*) = -2d\,\text{Im}\,Z \qquad (19a)$$

$$R_{bb,\,aa}\sigma_{aa} + R_{bb,\,bb}\sigma_{bb} = 2d\,\text{Im}\,Z \qquad (19b)$$

Here we see that $R_{aa,\,bb}$ and $R_{bb,\,aa}$ play the role of transition probabilities. Thus, we may write

$$R_{aa,\,bb} = R_{bb,\,aa} = W_{ab} = W_{ba} \qquad (20a)$$

while

$$R_{aa,\,aa} = -\sum_{\gamma \neq a} W_{a\gamma} \qquad (20b)$$

where W_{ab} is the transition probability from state b to state a, which leads to spin relaxation. Note that in the high-temperature approximation, $W_{ab} = W_{ba}$; i.e., the matrix formed from the transition probabilities is symmetric.

For simplicity, let $\gamma = b$ only (i.e., our simple line). Then we have

$$W_{ab}(\chi_a - \chi_b) = 2dZ'' \qquad (21)$$

where

$$\chi_a \equiv \sigma_{aa} - \sigma_{eqaa} \qquad (22a)$$

$$\chi_b \equiv \sigma_{bb} - \sigma_{eqbb} \qquad (22b)$$

so that the effect of the W_{ab} etc., is to lead to thermal equilibrium [cf. equation (6)] in the absence of $\varepsilon(t)$. Now equation (16) is rewritten as

$$(\Delta\omega - iT_2^{-1})Z + d(\chi_a - \chi_b) = q\omega_0 d \qquad (23)$$

where the high-temperature approximation [cf. equation (6)]

$$\sigma_{eqaa} - \sigma_{eqbb} \cong \frac{\exp(-E_a/kT) - \exp(-E_b/kT)}{\sum_\alpha \exp(-E_\alpha/kT)} \cong \frac{-\hbar\omega_{ab}}{kTA} \equiv -q\omega_0 \qquad (24)$$

has been used. Here, A, the number of spin states, is 2 in our example. We now need to solve the coupled equation

$$\begin{pmatrix} \Delta\omega & T_2^{-1} & d \\ -T_2^{-1} & \Delta\omega & 0 \\ 0 & -2d & W_{ab} \end{pmatrix} \begin{pmatrix} Z' \\ Z'' \\ \chi_a - \chi_b \end{pmatrix} = \begin{pmatrix} q\omega_0 d \\ 0 \\ 0 \end{pmatrix} \qquad (25)$$

This gives:

$$Z' = \Delta\omega\, T_2 Z'' \qquad (26a)$$

$$Z'' = \frac{q d\omega_0\, T_2}{1 + \Delta\omega^2\, T_2^2 + 4d^2 T_2 T_1} \qquad (26b)$$

where $T_1 \equiv (2W_e)^{-1}$ and

$$(\chi_a - \chi_b) = q\omega_0 4d^2 \frac{T_2 T_1}{1 + \Delta\omega^2 T_2^2 + 4d^2 T_2 T_1} \tag{26c}$$

These results are very similar to steady-state solutions of the Bloch equations, and we can get correspondence if

$$2M_0 = q\omega_0 = \frac{\hbar\omega_0}{AkT} = \sigma_{eqaa} - \sigma_{eqbb} \tag{27a}$$

$$T_2 = (T_2)_{ab}, \qquad T_1 = (T_1)_{ab} \tag{27b}$$

$$\gamma\mathbf{B} = -\omega_0\mathbf{k} + 2|d|(\mathbf{i}\cos\omega t + \mathbf{j}\sin\omega t) \tag{27c}$$

$$Z' = \tilde{M}_x, \qquad Z'' = \tilde{M}_y, \qquad Z = \tilde{M}_+ \tag{27d}$$

$$\chi_a - \chi_b = 2(M_0 - M_z) \tag{27e}$$

The above treatment is based on the high-field approximation $\omega_0/\gamma_e \equiv |B_0| \gg |B_1|$, as well as the fast motional condition $|\mathcal{H}_1|\tau_c \ll 1$, where $\mathcal{H}_1(t)$ is the random perturbation leading to the relaxation matrix $R_{\alpha\alpha'\beta\beta'}$. τ_c is the relaxation time for the random process modulating $\mathcal{H}_1(t)$, and this second inequality is the basis of the motional narrowing expression equation (5) for σ.[1-5] It also requires that $|\gamma B_1|\tau_c \ll 1$ in order that the **R** not be significantly affected by the presence of the rf field.[1,5]

1.3. A Simple Line: Multilevel System

The next most complicated case is a simple line coupled by relaxation to other spin eigenstates (see Figure 1). We now have

$$\sum_{\alpha \neq a} W_{a\alpha}(\chi_a - \chi_\alpha) = 2d\,\mathrm{Im}\,Z \tag{28a}$$

$$\sum_{\alpha \neq b} W_{b\alpha}(\chi_b - \chi_\alpha) = -2d\,\mathrm{Im}\,Z \tag{28b}$$

And, for $\alpha \neq a, b$, we get $A - 2$ equations

$$\sum_{\alpha}' W_{\alpha\beta}(\chi_\alpha - \chi_\beta) = 0, \qquad \beta \neq \alpha \tag{28c}$$

In equation (28c), we have assumed all transitions other than $a \to b$ are too far off resonance to have any appreciable off-diagonal density matrix elements; i.e., they are not excited by the rf field. Note that we can rewrite equation (23) as

$$Z = \frac{d[\omega_0 q - (\chi_a - \chi_b)]}{\Delta\omega - iT_2^{-1}} \tag{23'}$$

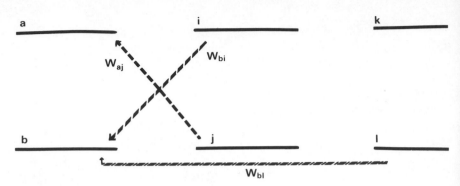

Figure 1. The pair of states corresponding to a simple line that are coupled by relaxation to other spin eigenstates.

When $\Delta\omega$ is very large, then $Z \to 0$ as $(d/\Delta\omega) \to 0$. In discussing this limit, we shall often use the convenient artifact of letting $d \to 0$ for that transition instead of more rigorously letting $(1/\Delta\omega) \to 0$.

The conservation of probability is

$$\text{Tr } \sigma = \text{Tr } \sigma_{eq} = 1 \qquad \text{or} \qquad \text{Tr } \chi = 0 \tag{29}$$

This is needed, because the above set of A equations are not all linearly independent. We can write these A equations in matrix notation as[6]:

$$\mathbf{W}\chi = \mathbf{U} \tag{30}$$

with $U_a = -U_b = 2d \text{ Im } Z$. When the rank of $\mathbf{W}$ is $A - 1$, then replacement of any one equation by equation (29) yields the matrix $\mathbf{W}^l$, which is now nonsingular, and we have

$$\chi = (\mathbf{W}^l)^{-1} \mathbf{U}^l \tag{30'}$$

Proper solutions of this $\mathbf{W}$ inversion are crucial in all saturation and double resonance analyses. It is possible to obtain general solutions by taking advantage of the general properties of $\mathbf{W}$ as developed in the Appendix.[6,7]

$$(\chi_a - \chi_b) = \Omega_{ba, ba} V_{ba}[q\omega_0 - (\chi_a - \chi_b)] \tag{31}$$

where from equations (16) and (22)

$$dZ'' = V_{ba}[q\omega_0 - (\chi_a - \chi_b)] \tag{32}$$

with

$$V_{ba} \equiv 2d^2 T_2/(1 + T_2^2 \Delta\omega^2) \tag{33}$$

and

$$\Omega_{ba, ba} \equiv 2C_{ba, ba}/C \tag{34}$$

where C is any cofactor of $\mathbf{W}$ (they are all equal, as may be shown from the properties of $\mathbf{W}^l$), and $C_{ba,\,ba}$ is the double cofactor of $\mathbf{W}$ obtained as the (signed) determinant resulting when the ath and bth rows and columns are deleted from $\mathbf{W}$ (see the Appendix).[6] More generally, we write

$$\Omega_{ij,\,kl} \equiv 2C_{ij,\,kl}/C \tag{35a}$$

where $C_{ij,\,kl}$ is the double cofactor of $\mathbf{W}$ obtained by deleting the ith and jth rows and the kth and lth columns of $\mathbf{W}$ and giving it the correct sign (cf. the Appendix).

The proof of equation (31) from equation (30') is as follows. First we note that equation (30') may be rewritten as

$$\chi_i = \sum_k (C_{ki}^l/|\mathbf{W}^l|)U_k^l \tag{36}$$

where C_{ki}^l is the kith cofactor of $\mathbf{W}^l$. Now

$$|\mathbf{W}^l| = \sum_k C_{lk} = AC \tag{37}$$

where the second equality follows from identity (1) in the Appendix [equation (A.22)]. Now, for this case of a single transition, we have

$$(\chi_a - \chi_b) = \frac{1}{AC} \sum_k (C_{ka}^l - C_{kb}^l)U_k^l$$

$$= \frac{1}{C} \sum_k C_{kl,\,ab}U_k^l = \sum_k \Omega_{kl,\,ab}U_k^l/2 \tag{38}$$

where identity (3) of the Appendix has been used. We now recognize that only $U_a = -U_b$ are nonzero, and since l is arbitrary, we let $l = b$ to yield equation (31).

The net result is to obtain our earlier results of equation (26), but now,

$$T_1 \to \tfrac{1}{4}\Omega_{ba,\,ba} \equiv \tfrac{1}{4}\Omega_{ba} \tag{39}$$

where Ω_{ba} is the saturation parameter for the $b \leftrightarrow a$ transition. It is not a simple T_1 nor decay time. In fact, there are as many as $(A - 1)$ different nonzero decay constants in the transient solution (which come from diagonalizing the $\mathbf{W}$ matrix). This Ω_{ba} may be regarded as a steady-state self-impedance representing the response of the $b \leftrightarrow a$ transition to the application of an rf field. That is, we rewrite equations (31) and (32) as

$$(\chi_a - \chi_b) = \Omega_{ba} \cdot (dZ'') \tag{40}$$

and make the electrical circuit analogy by letting $(\chi_a - \chi_b) = E$, $\Omega_{ba} = R$, and $dZ'' = I$. Thus, we see that inducing a resonant transition is formally equivalent to inducing a current flow, which causes a voltage drop $(\chi_a - \chi_b)$ proportional to the resistance Ω_{ba}.

2. *ELDOR*

Now we introduce a second ESR microwave field. Assume there are only two transitions of interest (see Figure 2). Now we have

$$\varepsilon(t) = \tfrac{1}{2}\gamma_e B_o[S_+ \exp(-i\omega_o t) + S_- \exp(+i\omega_o t)]$$
$$+ \tfrac{1}{2}\gamma_e B_p[S_+ \exp(-i\omega_p t) + S_- \exp(+i\omega_p t)] \tag{41}$$

where o and p refer to observing and pumping modes, respectively. We are looking to the applied fields to generate steady-state off-diagonal density matrix elements as a result of the resonance phenomena. We assume

$$|\gamma_e B_o|, \ |\gamma_e B_p|, \ |R| \ll |\omega_{aa'} - \omega_{bb'}| \sim |a| \tag{42}$$

so the hyperfine lines always remain well separated. Then we may have $\omega_{aa'} - \omega_o = -\Delta\omega_o \sim 0$, while $|\omega_{aa'} - \omega_p| \sim |a|$ and $\omega_{bb'} - \omega_p = -\Delta\omega_p \sim 0$, while $|\omega_{bb'} - \omega_o| \sim |a|$. Thus, the important elements are:

$$\sigma_{a'a} = \chi_{a'a} = Z_{a'a}\exp(i\omega_o t) \equiv Z_o\exp(i\omega_o t) \tag{43a}$$

$$\sigma_{b'b} = \chi_{b'b} = Z_{b'b}\exp(i\omega_p t) \equiv Z_p\exp(i\omega_p t) \tag{43b}$$

We obtain from equation (5)

$$(\Delta\omega_o - i/T_{2,\,o})Z_o + d_o(\chi_a - \chi_{a'}) = q\omega_{aa'}d_o \cong q\omega_e d_o \tag{44a}$$

$$(\Delta\omega_p - i/T_{2,\,p})Z_p + d_p(\chi_b - \chi_{b'}) = q\omega_{bb'}d_p \cong q\omega_e d_p \tag{44b}$$

Also, the analogs of equations (28) are now

$$\sum_{\alpha \neq a} W_{a\alpha}(\chi_a - \chi_\alpha) = 2d_o Z_o'' \tag{45a}$$

$$\sum_{\alpha \neq a'} W_{a'\alpha}(\chi_{a'} - \chi_\alpha) = -2d_o Z_o'' \tag{45b}$$

$$\sum_{\alpha \neq b} W_{b\alpha}(\chi_b - \chi_\alpha) = 2d_p Z_p'' \tag{45c}$$

$$\sum_{\alpha \neq b'} W_{b'\alpha}(\chi_{b'} - \chi_\alpha) = -2d_p Z_o'' \tag{45d}$$

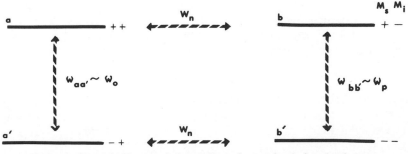

Figure 2. Example of ELDOR being performed on a simple four-level system.

These equations may be rewritten in matrix form as

$$(\mathbf{K} + i\mathbf{R})\mathbf{Z} = \mathbf{d}\chi + \mathbf{Q} \tag{46a}$$

$$(\mathbf{W}^j)(\chi) = -2\mathbf{d}^{\text{tr }j}\mathbf{Z}'' \tag{46b}$$

where

$$\mathbf{Q} \cong q\omega_e \begin{pmatrix} d_o \\ \\ d_p \end{pmatrix} \tag{47a}$$

$$\mathbf{K} = \begin{pmatrix} \Delta\omega_o & 0 \\ \\ 0 & \Delta\omega_p \end{pmatrix} \tag{47b}$$

$$-\mathbf{R} = \begin{pmatrix} T_{2,0}^{-1} & 0 \\ \\ 0 & T_{2,p}^{-1} \end{pmatrix} \tag{47c}$$

$$-\mathbf{d} = \begin{pmatrix} d_o & -d_o & 0 & 0 \\ \\ 0 & 0 & d_p & -d_p \end{pmatrix} \tag{47d}$$

$\mathbf{W}^j$ is a 4×4 transition probability matrix in the space of the four spin eigenstates with the jth row replaced by ones, and $\mathbf{d}^{\text{tr }j}$ is the transpose of $\mathbf{d}$ with the jth row replaced by zero. $\mathbf{Z}$ is a vector in the two-dimensional space of induced transitions. The formal solution is given by

$$\mathbf{Z}'' = \mathbf{M}^{-1}(-\mathbf{R}^{-1})\mathbf{Q} \tag{48a}$$

$$\mathbf{Z}' = (-\mathbf{R}^{-1})\mathbf{K}\mathbf{Z}'' \tag{48b}$$

$$\mathbf{d}\chi = -\mathbf{S}\mathbf{Z}'' \tag{48c}$$

where

$$\mathbf{M} = 1 + (\mathbf{R}^{-1}\mathbf{K})^2 + (-\mathbf{R}^{-1})\mathbf{S} \tag{49a}$$

and

$$\mathbf{S} = 2[\mathbf{D}(\mathbf{W}^j)^{-1}\mathbf{D}^{\text{tr }j}] \tag{49b}$$

Suppose $d_p = 0$. One recovers the single-line, simple saturation result, and by comparison, we find

$$S_{o,o} = d_o^2\Omega_{aa',aa'} \equiv d_o^2\Omega_{o,o} \tag{50}$$

One finds more generally (cf. Section 6.2.2 and the Appendix),

$$S_{\lambda,\eta} = d_\lambda d_\eta \Omega_{\lambda,\eta} \tag{51}$$

where $\Omega_{\lambda,\eta}$ is a cross-impedance (cross-saturation parameter), which is determined solely by the spin relaxation processes and represents the impedance at transition i from an external disturbance (e.g., a resonant rf field) on the transition j. [It is obtained by equation (35a) with $i \to j$ being the λth transition and $k \to l$ the ηth transition.]

Thus, equation (48c) is a generalization of equation (40) for the single resonance case. In fact, it gives

$$(\chi_a - \chi_{a'}) = d_o\Omega_{o,o}Z''_o + d_p\Omega_{o,p}Z''_p \tag{52a}$$

$$(\chi_b - \chi_{b'}) = d_p\Omega_{p,p}Z''_p + d_o\Omega_{p,o}Z''_o. \tag{52b}$$

with an electrical circuit analogy similar to that of equation (40). It follows from equations (47)–(51) that

$$\mathbf{M} = \begin{pmatrix} 1 + \Delta\omega_o^2 T_{2,o}^2 + d_o^2 T_o\Omega_o & d_o d_p\Omega_{o,p} T_{2,o} \\ d_p d_o\Omega_{p,o} T_{2,p} & 1 + \Delta\omega_p^2 T_{2,p}^2 + d_p^2 T_{2,p}\Omega_p \end{pmatrix} \tag{53}$$

where we have let $\Omega_{o,o} = \Omega_o$ and $\Omega_{p,p} = \Omega_p$. Then, from equation (48a),

$$Z''_o = q\omega_e T_{2,o} d_o \frac{1 - \xi_o/\Omega_{p,o}}{1 + \Delta\omega_o^2 T_{2,o}^2 + d_o^2 T_{2,o}(\Omega_o - \xi_o)} \tag{54}$$

with

$$\xi_o = d_p^2 T_{2,p}\Omega_{o,p}\Omega_{p,o}/(1 + \Delta\omega_p^2 T_{2,p}^2 + d_p^2 T_{2,p}\Omega_p) \tag{54'}$$

Now consider some special cases. Let us have $\Delta\omega_p = 0$ (represented by a superscript r) and very strong saturation of the pump mode:

$$d_p^2 T_p\Omega_p \gg 1 \tag{55}$$

(where we are now dropping the subscript 2 on the various T_2). Then

$$\xi_o^r(d_p^2 \to \infty) = \Omega_{o,p}\Omega_{p,o}/\Omega_p \tag{56}$$

which is just relaxation determined. We now let $T_{2,p} = T_p$, etc. Then

$$Z''_o = q\omega_e T_o d_o \frac{(\Omega_p - \Omega_{o,p})/\Omega_p}{1 + T_o^2 \Delta\omega_o^2 + d_o^2 T_o(\Omega_o\Omega_p - \Omega_{o,p}\Omega_{p,o})/\Omega_p} \tag{57}$$

If we also introduce the generalized no-saturation condition for the observing mode,

$$d_o^2 T_o[(\Omega_o\Omega_p - \Omega_{o,p}\Omega_{p,o})/\Omega_p] \ll 1 \tag{58}$$

one has the simple result that

$$Z''_o = \frac{T_o q\omega_e d_o}{1 + \Delta\omega_o^2 T_o^2} \left[1 - \frac{\Omega_{o,p}}{\Omega_p}\right] \tag{59}$$

Since Ω_p is always positive,[1] it follows from equation (59) that for $\Omega_{o,\,p} > 0$, the signal is reduced by the presence of the resonant pump field, while for $\Omega_{o,\,p} < 0$, the signal is amplified. The limiting (but not realistic) case for equation (59) occurs when W_n is very strong and W_e is negligible. (Here, W_e and W_n are, respectively, the lattice-induced electron spin-flip and nuclear spin-flip rates.) Then the case for the energy levels shown in Figure (2) is easily understood. Let P_i be the population of the ith state. Then saturation by ω_p causes $P_b = P_{b'}$; a strong W_n causes $P_a = P_b$ and $P_{a'} = P_{b'}$, leading to a reduction in intensity of the observed signal. This extreme will be seen to be equivalent to $\Omega_{o,\,p} = \Omega_o = \Omega_p$.

There are actually two effects that can be seen in ELDOR:

EFFECT 1. The no-saturation effect discussed above is a polarization effect (not unlike an Overhauser effect in NMR), but the two transitions involved have no level in common, and this places special requirements on the relaxation processes in order to obtain significant effects.

EFFECT 2. This effect is important *only* when Z_o'' is being saturated. It reflects the fact that the induced absorption mode Z_p'' acts as an induced transition, which, in conjunction with lattice-induced transitions, can facilitate the rate of energy transferred from the observing radiation field to the lattice via the spin system.

Effect 1 is the main effect in ELDOR, while the analog to Effect 2 is the dominant one in ENDOR. Further details are given in Hyde *et al.*[8] It should, however, be noted that typical ELDOR experiments yield derivative signals for low-enough field-modulation amplitude and frequency.[8] Complicating effects arising from high-enough field-modulation frequency (and amplitude) are discussed in Chapter 5.

3. ENDOR

3.1. General Considerations

We again consider our four-level system, but now,

$$\varepsilon(t) = \tfrac{1}{2}\gamma_e B_e[S_+ \exp(-i\omega_e t) + S_- \exp(+i\omega_e t)]$$
$$+ \tfrac{1}{2}\gamma_n B_n[I_+ \exp(-i\omega_n t) + I_- \exp(+i\omega_n t)]$$
$$+ \tfrac{1}{2}\gamma_e B_n[S_+ \exp(-i\omega_n t) + S_- \exp(i\omega_n t)]$$
$$+ \tfrac{1}{2}\gamma_n B_e[I_+ \exp(-i\omega_e t) + I_- \exp(i\omega_e t)] \tag{60}$$

In equation (60), the microwave field at frequency ω_e is to induce electron spin flips, while the rf field at frequency ω_n is to induce nuclear spin flips.

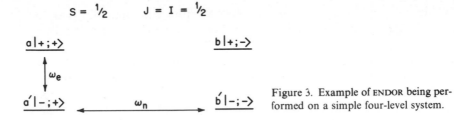

Figure 3. Example of ENDOR being performed on a simple four-level system.

Thus, the last term in equation (60) can be neglected as being too far off resonance to affect the nuclear spins. The third term in equation (60) does have a nontrivial effect on the effective transition moment of the nuclear spins.[6]

This arises from the correction to the high-field wave functions to first order in the off-diagonal hyperfine term: $a_n S_\pm I_\mp$ (e.g., for the four-level system discussed below, the states a' and b are more correctly $(|-+\rangle + \alpha|+-\rangle)$ and $(|+-\rangle - \alpha|-+\rangle)$ where the small mixing coefficient $\alpha = a_n/2B_0$. The effective transition moment is then found to be

$$d_n = d_{n0}(1 \pm (\gamma_e/\gamma_n)\bar{a}_n/2B_0) \tag{61}$$

where d_{n0} is the nuclear transition moment in the absence of this correction and the $\pm$ signs correspond to $M_s = \pm$).

Let us assume the four-level system shown in Figure 3. Let

$$\Delta_e \equiv \omega_e - \omega_{aa'} \approx 0 \tag{62a}$$

$$\Delta_n \equiv \omega_n - \omega_{a'b'} \approx 0 \tag{62b}$$

Then, for assumptions similar to those used for the ELDOR case, we expect important steady-state off-diagonal density matrix elements:

$$\chi_{a'a} = Z_{a'a} e^{i\omega_e t} \equiv Z_e e^{i\omega_e t} \tag{63a}$$

$$\chi_{b'a'} = Z_{b'a'} e^{i\omega_n t} \equiv Z_n e^{i\omega_n t} \tag{63b}$$

We obtain the series of equations:

$$(\Delta_e - i/T_e)Z_e + d_e(\chi_a - \chi_{a'}) + d_n Z_x = q\omega_e d_e \tag{64a}$$

$$(\Delta_n - i/T_n)Z_n + d_n(\chi_{a'} - \chi_{b'}) - d_e Z_x = q\omega_n d_n \tag{64b}$$

$$[\Delta_e + \Delta_n - i/T_x]Z_x - d_e Z_n + d_n Z_e = 0 \tag{64c}$$

where T_e, T_n, and T_x are the T_2 for the ESR, NMR, and cross-transitions, respectively, and all other terms we defined by analogy with previous definitions. Note the appearance of

$$\chi_{b'a} = Z_{b'a} \exp[i(\omega_e + \omega_n)t] \equiv Z_x \exp[i(\omega_e + \omega_n)t] \tag{65}$$

This is an overtone term—a two-quantum effect. Also,

$$\sum_{\alpha \neq a} W_{a\alpha}(\chi_a - \chi_\alpha) = 2d_e Z_e'' \tag{66a}$$

$$\sum_{\alpha \neq a'} W_{a'\alpha}(\chi_{a'} - \chi_\alpha) = -2d_e Z_e'' + 2d_n Z_n'' \tag{66b}$$

$$\sum_{\alpha \neq b} W_{b\alpha}(\chi_b - \chi_\alpha) = 0 \tag{66c}$$

$$\sum_{\alpha \neq b'} W_{b'\alpha}(\chi_{b'} - \chi_\alpha) = -2d_n Z_n'' \tag{66d}$$

Again we may write these equations in the matrix form given by equations (46) with the formal solution given by equations (48) and (49). Note that the **K**, or coherence matrix, is

$$\mathbf{K} = \begin{pmatrix} \Delta_e & 0 & d_n \\ 0 & \Delta_n & -d_e \\ d_n & -d_e & \Delta_e + \Delta_n \end{pmatrix} \tag{67}$$

and is no longer diagonal. Also, intensities are proportional to

$$\mathbf{Q} = q \begin{pmatrix} \omega_e d_e \\ \omega_n d_n \\ 0 \end{pmatrix} \tag{68}$$

but because $\omega_e/\omega_n \sim 660$ for protons, we may usually set

$$\mathbf{Q} \cong q \begin{pmatrix} \omega_e d_e \\ 0 \\ 0 \end{pmatrix} \tag{68'}$$

which amounts to neglecting the analog of Effect 1 in the ENDOR case.

3.2. Neglect of Coherence Effects

The coherence effects arise from the off-diagonal elements in the **K** matrix or, in other words, the contribution from Z_x. Consider the case of exact resonance, when $\Delta_e = \Delta_n = 0$, since this is the condition under which double resonance effects will be maximized. Equations (64)–(66) and (48)–(49) then yield

$$Z_e'' = Z_n'' = Z_x''' = 0 \tag{69a}$$

$$Z_e''' = \frac{q\omega_e d_e T_e}{1 + d_e^2(\Omega_e - \xi_e^r)T_e + d_n^2 T_x T_e} \tag{69b}$$

where

$$\xi_e^r = \frac{T_n d_n^2(T_x + |\Omega_{e,n}|)^2}{1 + d_n^2\Omega_n T_n + d_e^2 T_x T_n} \tag{69c}$$

Thus, from equation (69b), when

$$1 + d_e^2\Omega_e T_e \gg d_n^2 T_x T_e \qquad \text{(and } \xi_e^r \not\approx \Omega_e) \tag{70}$$

the coherence effect on Z_e''' may be neglected. ξ_e^r leads to an enhancement of a saturated ESR signal, since it effectively reduces the saturation parameter Ω_e. Now, when

$$1 + d_n^2\Omega_n T_n \gg d_e^2 T_x T_e \tag{71}$$

it follows from (69c) that the ratio ξ_e/Ω_e will not be affected by d_e, and further, if

$$|\Omega_{e,n}| \gg T_x \tag{72}$$

we may completely neglect the coherence effects.

   If there is appreciable saturation and

$$d_e^2 \sim d_n^2 \tag{73}$$

then we can replace equations (11)–(13) with the simpler set of conditions

$$\Omega_e, \Omega_n, |\Omega_{e,n}| \gg T_x \tag{74}$$

for the neglect of coherence effects. The inequalities of equation (74) are fulfilled if the T_1 or saturation parameters are much larger than the T_2 or inverse linewidths.

   Now our solutions for $\Delta_e, \Delta_n \approx 0$ are

$$Z_e'' = \frac{q\omega_e d_e T_e}{1 + (\Delta_e T_e)^2 + (\Omega_e - \xi_e)T_e d_e^2} \tag{75a}$$

$$\xi_e = \frac{d_n^2(\Omega_{e,n})^2 T_n}{1 + (\Delta_n T_n)^2 + d_n^2 T_n \Omega_n} \tag{75b}$$

If the ENDOR spectrum is monitored after subtraction of the ESR signal, then for $\Delta_e = 0$ and $\Omega_e T_e d_e^2 \gg 1$, we have

$$Z_{ENDOR}''' - Z_{ESR}''' = q\omega_e d_e\left(\frac{\Omega_{e,n}^2}{\Omega_e^2}\right)\frac{d_n^2 T_n}{1 + (\Delta_n T_n)^2 + [1 - (\Omega_{e,n}^2/\Omega_e\Omega_n)]T_n\Omega_n d_n^2} \tag{76}$$

Thus, the signal strength is proportional to $(\Omega_{e,n}/\Omega_e)^2$, and the shape is a Lorentzian of width T_n^{-1} and (modified) saturation parameter

$$\Omega_n\left(1 - \frac{\Omega_{e,n}^2}{\Omega_n\Omega_e}\right) \tag{77}$$

The enhancement of an ESR line due to ENDOR is then, from equation (76),

$$E \equiv \frac{Z''^r_{\text{ENDOR}} - Z''^r_{\text{ESR}}}{Z''_{\text{ENDOR}}} = \frac{\xi_e}{\Omega_e - \xi_e} \xrightarrow[\Delta_n \to 0]{d_n^2 \to \infty} \left(\frac{\Omega_n\Omega_e}{\Omega_{e,n}^2} - 1\right)^{-1} \tag{78}$$

where $d_n^2 \to \infty$ implies

$$d_n^2 T_n \Omega_n \gg 1 \tag{78'}$$

Further details may be found in Ref. 6. Note, however, that typical experimental arrangements yield the derivatives of Z''_{ESR} and Z''_{ENDOR} (see Chapter 2).

3.3. Coherence Effects in ENDOR

We now return to equations (62)–(67) and Figure 3. One obtains an **M** matrix [(see equation 49)]:

M =

$$\begin{pmatrix} 1 + T_e^2\Delta_e^2 + T_e(T_x d_n^2 + \Omega_e d_e^2) & T_e d_e d_n(\Omega_{e,n} - T_x) & T_e d_n[T_e\Delta_e + T_x(\Delta_e + \Delta_n)] \\ T_n d_e d_n(\Omega_{e,n} - T_x) & 1 + T_n^2\Delta_n^2 + T_n(T_x d_e^2 + \Omega_n d_n^2) & -T_n d_e[T_n\Delta_n + T_x(\Delta_e + \Delta_n)] \\ T_x d_n[T_e\Delta_e + T_x(\Delta_e + \Delta_n)] & -T_x d_e[T_n\Delta_n + T_x(\Delta_e + \Delta_n)] & 1 + T_x^2(\Delta_e + \Delta_n)^2 + T_x(T_e d_n^2 + T_x d_e^2) \end{pmatrix}$$

$$\begin{matrix} e & n & x \end{matrix}$$

$$\tag{79}$$

Thus we see that the general solution is quite complex, so we consider specific cases[9]:

3.3.1. Negative ENDOR—Weak Examining ESR Field

Here we may set all terms in equation (79) containing d_e equal to zero. Let us consider the ENDOR mode of sweeping through ω_n while $\Delta_e = 0$. Then

$$Z''_e = q\omega_e d_e T_e \frac{T_x^2\Delta_n^2 + 1 + T_e T_x d_n^2}{T_x^2\Delta_n^2 + (1 + T_e T_x d_n^2)^2} \tag{80}$$

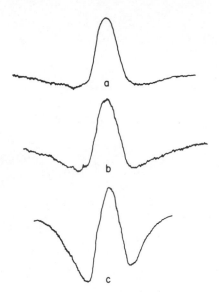

Figure 4. Distorted ENDOR line shapes caused by coherence effects. (a) High B_1^2, long T_{2e}, medium B_n^2. (b) High B_1^2, long T_{2e}, high B_n^2. (c) Low B_1^2, long T_{2e}, high B_n^2.

Thus, a finite d_n acts to reduce the resonant ESR signal, and a maximum reduction occurs for $\Delta_n = 0$. Thus, "negative ENDOR" can be used as a means to detect ENDOR signals.

Now, in order to get appreciable reductions, the quantity $T_e T_x d_n^2$ should not be too much smaller than unity. Usually, $T_x \sim T_e$ (see Ref. 9), so this condition may be restated as "$T_e^2 d_n^2$ should not be too small." Since it is often true that $T_e \ll \Omega_n$, it is not always possible to obtain sufficient rf power to make this method work. An example of negative ENDOR superimposed on a normal ENDOR signal is shown in Figure 4.

The cause of the reduction is readily seen by examining the expression for $\Delta_n = 0$ and ω_e swept. Then

$$Z_e'' = q\omega_e d_e \frac{T_e}{1 + T_e^2 \Delta_e^2 + T_e T_x d_n^2(1 - \xi_e)} \tag{81a}$$

with

$$\xi_e = \frac{(T_e + T_x)^2 \Delta_e^2}{1 + T_x^2 \Delta_e^2 + T_e T_x d_n^2} \tag{81b}$$

The denominator in equation (81a) can have more than one minimum. Thus, differentiating Z_e'' with respect to Δ_e and setting it equal to zero, we get extrema for

$$\Delta_e = 0 \tag{82a}$$

and

$$\Delta_e^2 = yT_x^{-2}[d_n(T_e + T_x)(T_x/T_e)^{1/2} - y] \tag{82b}$$

where

$$y = (1 + T_e T_x d_n^2)^{1/2} \tag{82c}$$

These expressions yield a critical value of d_n

$$d_n^{\text{crit}} = \pm T_x^{-1}(2 + T_x/T_e)^{-1/2} \xrightarrow{T_x \to T_e} \pm 1/\sqrt{3}\,T_e \qquad (83)$$

For $d_n > d_n^{\text{crit}}$, there are two peaks in the ESR experiment given by equation (82b), but for $d_n < d_n^{\text{crit}}$, there is only one peak at $\Delta_e = 0$. When $T_e T_x d_n^2 \gg 1$, one has $\Delta_e = \pm d_n$, and it is possible to use this as a method for measuring B_n.

3.3.2. Very Weak NMR Field

All terms in equation (79) containing d_n may be set equal to zero. Z_e'' becomes an ordinary saturated Lorentzian and is unaffected by d_n.

3.3.3. Strong ESR Field but Weaker NMR Field

Here we require that d_n be still strong enough to saturate the NMR, or

$$T_n \Omega_n d_n^2 \gtrsim 1 \qquad (84a)$$

so the induced relaxation effects of the NMR field on the ESR are *not* negligible. We further require that

$$d_e \gg d_n \qquad (84b)$$

such that the inequality of equation (70) holds but the inequality of equation (71) is *reversed*. This will lead to a coherence splitting of Z_n'' by d_e, but no splitting of Z_n'' by d_n. The coherence splitting on Z_n'' can then be observed via an ENDOR enhancement technique (see Figure 5), and it could be used to calibrate B_e (or d_e).

3.3.4. Splitting of ENDOR Lines by a Strong NMR Field

In this section, we start by assuming that we may neglect the electron–nuclear coherence effects of the type discussed in Sections 3.3.1 to 3.3.3 [i.e., conditions like equations (70) and (71) apply]. The simplest case of interest is then for $S = \frac{1}{2}$, $J = I = 1$ (see Figure 6). There is still a two-quantum transition Z_c involving coherence between the nuclear levels. We thus consider in $\mathbf{Z}$ the four transitions given by Z_e, $Z_\pm \equiv (1/2^{1/2})(Z_1 \pm Z_2)$, and Z_c, where

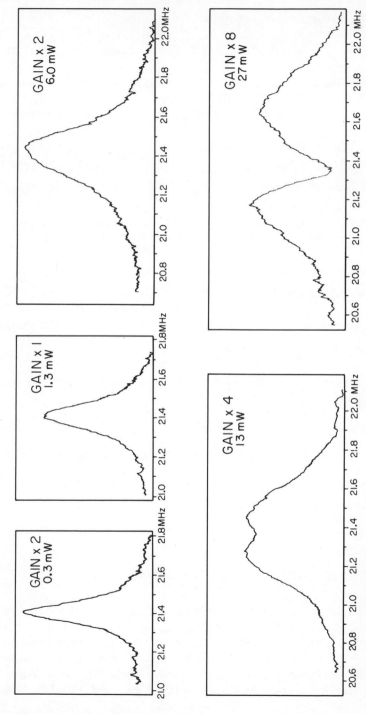

Figure 5. The ENDOR line from the methylenyl proton of Coppinger's radical as a function of microwave power. The power incident on the cavity is given in milliwatts, and the relative recorder gain is indicated. (From Freed *et al.*[9])

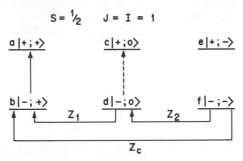

Figure 6. Transitions and eigenstates for double resonance in a radical with $S = \frac{1}{2}$ and a nuclear spin of $I = 1$ (or the $J = 1$ states of two equivalent nuclear spins). (From Freed *et al.*[9])

Z_e can be any of the three allowed ESR transitions in Figure 6. The **M** matrix in this case is

$$
\bar{\mathbf{M}} =
\begin{pmatrix}
1 + T_e^2 \Delta_e^2 + T_e S_e & (1/\sqrt{2}) T_e S_{e,+} & (1/\sqrt{2}) T_e S_{e,-} & 0 \\
(1/\sqrt{2}) T_+ S_{e,+} & 1 + T_+^2 \Delta_n^2 + T_+ S_+ & 0 & 0 \\
(1/\sqrt{2}) T_- S_{e,-} & 0 & 1 + T_-^2 \Delta_n^2 + T_- S_- + 2 d_n^2 T_e T_- & \sqrt{2}(T_- + 2T_e) T_- \Delta_n d_n \\
0 & 0 & \sqrt{2}\, T_e (T_- + 2T_e)\Delta_n d_n & 1 + 4 T_e^2 \Delta_n^2 + 2 T_e T_- d_n^2
\end{pmatrix}
$$

$$\tag{85}$$

where $S_{e,\pm} = S_{e,1} \pm S_{e,2}$ and $T_{\pm} = T_1 \pm |T_{1,2}|$ [where $T_1 = T_2 = -R_{1,1}(R_{1,1}^2 - R_{1,2}^2)^{-1}$ and $T_{1,2} = -R_{1,2}(R_{1,1}^2 - R_{1,2}^2)^{-1}$ (see Section 4.4.2)].

We see that the symmetric nuclear mode Z_+ does not couple to either Z_- or Z_c, but there is a coherence coupling between Z_- and Z_c. The mode Z''_+ represents the net rf induced absorption from state f to state b. Furthermore, it is easily shown that Z_+ is the only mode that is detected in a conventional NMR experiment, and this is consistent with the well-known result that the only effect of the coherence of the rf field on a simple nuclear resonance experiment appears as a saturation effect with no frequency shifts. The difference mode Z_- is not detected in NMR but does affect the ENDOR spectrum by means of its coupling to Z_e via a finite $S_{e,-}$. Since Z''_- represents the net rf induced absorption from states b and f to state d (i.e., it is proportional to $\sigma_f + \sigma_b - 2\sigma_d$), it does not correspond to any net energy absorption. The coupling of Z_c to Z_- and not Z_+ implies that it does not correspond to any net energy absorption. In fact, when $\Delta_n = 0$, one has $Z''_c = Z'_+ = Z'_- = 0$ and $Z'_c = 2^{1/2} d_n T_c Z''_-$, so at resonance there is only a dispersive mode at the frequency $2\omega_n$.

One may analyze the effects of the coupled nuclear transitions Z_- and Z_c in $\mathbf{M}$ given by equation (85) to show that for $T_c \sim T_- \equiv T$, there are extrema for

$$\Delta_n = 0 \qquad \text{and} \qquad \Delta_n^2 = \tfrac{1}{4}[3d_n(2X)^{1/2} - X] \tag{86}$$

where

$$X = T^{-2}(1 + T^2 d_n^2) \tag{86'}$$

Thus, for $d_n T \gg 1$, we have $\Delta_n = \pm d_n$ or two peaks separated by $2d_n = 2^{1/2}\gamma_n B_n$ (since $J = 1$). Also, there is a critical value $d_n^{\text{crit}} = (4T)^{-1}$ below which there is no splitting but only a single line centered at $\Delta_n = 0$. The effectiveness of the $Z_\pm''$ modes in contributing to ENDOR is determined by the magnitude of the $\Omega_{e,\pm}$. The results depend markedly upon whether the center ESR line or one of the outside ESR lines is saturated, and this may be understood in terms of simple symmetry considerations. We consider the center ESR line first. We note then that $Z_{db}'' = Z_1''$ and $Z_{df}'' = -Z_2''$ (where a minus corresponds to the transition arrow with a reversed direction) have identical effects on Z_{0e}''; i.e., they enhance symmetrically equivalent paths of relaxation (provided linear M_I-dependent relaxation effects are neglected). Furthermore, the Z_{0e}'' transition will have identical Overhauser type of effects on each of these nuclear spin transitions so that $Z_1'' = -Z_2''$ and $Z_+'' = 0$, thereby rendering this mode ineffective in any ENDOR enhancement of Z_{0e}''. [The way this negative relation between Z_1'' and Z_2'' shows up in the detailed analysis is from $\Omega_{0e,1} = -\Omega_{0e,2}$ (see the Appendix).] Therefore only Z_-'' can contribute to Z_{0e}'', and it will demonstrate a frequency splitting for large values of d_n. When either of the outside ESR lines is saturated, no such symmetry exists, so both nuclear spin modes may contribute, but in the region of appreciable ENDOR enhancements (i.e., $b \approx 1$), one finds that $\Omega_{\pm e,+} > \Omega_{\pm e,-}$. This means that the relative importance of the Z_-'' mode in the ENDOR enhancements is diminished and the coherence effects are much smaller for the outer ESR lines.

This basic treatment and analysis can be generalized to larger J values and to sets of equivalent nuclei. We show in Figure 7 typical computer simulations, demonstrating that the $M = 0$ line does indeed show the dominant effect. This coherence splitting has now been seen in a wide variety of cases. Further aspects of coherence effects are discussed in Ref. 9.

3.4. Triple Resonance[10]

We now suppose that the ENDOR experiment is expanded to include a second NMR exciting field that induces the $a \leftrightarrow b$ transition (see Figure 3), yielding a $Z_{ba} \equiv Z_{n'}$ added to equations (63). We shall, for simplicity, neglect any coherence effects, so that the **K** matrix is diagonal and three dimensional [see equation (8) but now also including $\Delta_{n'} \equiv \omega_{n'} - \omega_{ab} \approx 0$]. In this case the matrix **M** [cf. equation (49)] becomes (see Ref. 10).

$$
\mathbf{M} =
\begin{pmatrix}
1 + T_e^2 \Delta \omega_e^2 + T_e d_e^2 \Omega_e & d_e d_n \Omega_{e,\,n} T_e & d_e d_{n'} \Omega_{e,\,n'} T_e \\[2mm]
d_n d_e \Omega_{n,\,e} T_n & 1 + T_n^2 \Delta \omega_n^2 + T_n d_n^2 \Omega_n & d_n d_{n'} \Omega_{n,\,n'} T_n \\[2mm]
d_{n'} d_e \Omega_{n',\,e} T_{n'} & d_{n'} d_n \Omega_{n',\,n} T_{n'} & 1 + T_{n'}^2 \Delta \omega_{n'}^2 + T_{n'} d_{n'}^2 \Omega_{n'}
\end{pmatrix}
\tag{87}
$$

We again obtain equation (75a) as the solution where now

$$
\xi_e = \frac{d_n^2 \Omega_{e,\,n}^2 T_n (Y_{n'} + d_{n'}^2 T_{n'} \Omega_{n,\,n'}) + d_{n'}^2 \Omega_{e,\,n}^2 T_{n'} (Y_n + d_n^2 T_n \Omega_{n,\,n'})}{Y_n Y_{n'} - d_n^2 d_{n'}^2 \Omega_{n,\,n'}^2 T_n T_{n'}}
\tag{88}
$$

and

$$
Y_n = 1 + T_n^2 \Delta \omega_n^2 + T_n \Omega_n d_n^2
\tag{89}
$$

with an equivalent equation for $Y_{n'}$. Then for $\Delta_n \to 0$ and $d_n^2, d_{n'}^2 \to \infty$ we get

$$
\xi_e^r(d_n,\, d_{n'} \to \infty) = \frac{\Omega_{e,\,n}^2 (\Omega_{n'} + \Omega_{n,\,n'}) + \Omega_{e,\,n'}^2 (\Omega_n + \Omega_{n,\,n'})}{\Omega_n \Omega_{n'} - \Omega_{n,\,n'}^2}
\tag{90}
$$

compared to $\Omega_{e,\,n}^2 / \Omega_n$ for ENDOR. Then the expression for the limiting enhancement [equation (78)] becomes

$$
1 + E^{-1} \xrightarrow[\Delta \omega_n \to \infty]{d_n^2 \to \infty} \Omega_e / \xi_e^r(d_n,\, d_{n'} \to \infty)
\tag{91}
$$

Note that the limiting conditions on d_n^2 and $d_{n'}^2$ imply not only equation (78′), but also

$$
d_n^2 T_n (\Omega_n \Omega_{n'} - \Omega_{n,\,n'}^2) / \Omega_{n'} \gg 1
\tag{92}
$$

as well as the other pair of inequalities resulting from interchanging the indices n and n'. Equation (92) results from the coupled effects of both induced NMR absorptions.

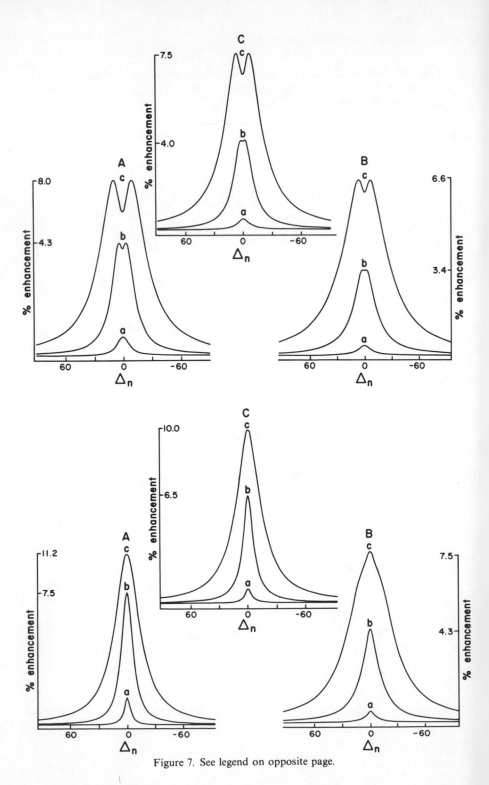

Figure 7. See legend on opposite page.

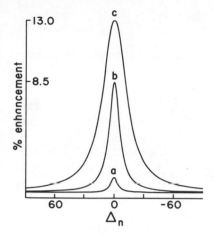

Figure 7. Top, opposite page: ENDOR line shape when the $M_J = 0$ ESR line is saturated, Bottom, opposite page: ENDOR line shape when the $M_J = 1$ ESR line is saturated. To the left: ENDOR line shape when the $M_J = \pm 2$, $J = 2$ ESR line is saturated. (A) $J = 1$, (B) $J = 2$, (C) the composite line from four equivalent spins of $I = \frac{1}{2}$ obtained from A and B properly weighted. All cases correspond to $b = 1$, $T_e^{-1} = 53.3$ units. In A, $d_e = 10.4$; in each, the line shapes a, b, and c correspond to $d_n = 1, 4,$ and 10 units, respectively. Each signal height is given relative to the magnitude of the original saturated ESR line. ($W_e = W_n = 1$ frequency unit.) (From Freed et al.[9])

4. Transition Probabilities

Consider now the general four-level system with all types of spin–lattice relaxation transitions as shown in Figure 8. We can solve for C_{ii} and $C_{ij, kl}$, the cofactors and double cofactors of the $\mathbf{W}$ matrix to obtain all the $\Omega_{ij, kl}$.

4.1. ELDOR—Generalized No-Saturation of Observing Mode

We have from equation (39) that the signal reduction is given by

$$R \equiv \frac{\% \text{ reduction}}{100} \tag{93}$$

$$= \frac{\Omega_{o, p}}{\Omega_p} = \frac{W_n^2 - W_{x_1} W_{x_2}}{W_e(2W_n + W_{x_1} + W_{x_2}) + (W_n + W_{x_1})(W_n + W_{x_2})}$$

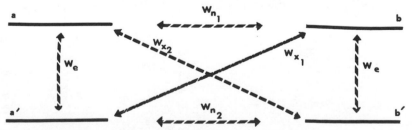

Figure 8. The general four-level system with all types of transitions due to spin–lattice relaxation.

where we have let $W_{n_1} = W_{n_2} = W_n$. Clearly, if $W_n^2 > W_{x_1} W_{x_2}$, one has a reduction in signal, while if $W_n^2 < W_{x_1} W_{x_2}$ there will be an enhancement.

a. Let $W_{x_1} = W_{x_2} = 0$ (i.e., only pseudosecular dipolar terms important). Then

$$R = \frac{W_n}{2W_e + W_n} = \frac{b}{2 + b} \tag{94}$$

where

$$b \equiv \frac{W_n}{W_e} \tag{94'}$$

or a reduction.

b. Let $W_{x_1} = W_n = 0$ (i.e., isotropic hyperfine modulation). Then $R = 0$ (i.e., no effect).

c. Let $W_{x_2} = 4W_n$, $W_{x_1} = \frac{2}{3}W_n$ (dipolar, extreme narrowing). Then

$$R = -\frac{b}{4 + 5b} \tag{95}$$

or an enhancement.

In the case of solids, one can also examine ELDOR enhancements for forbidden ESR transitions.[11] The limiting values for R are given in Table 1.

4.2. ENDOR—Limiting Enhancements

We have from equation (78) that the enhancement E is given by

$$1 + E^{-1} = \frac{\Omega_n \Omega_e}{\Omega_{e,n}^2}$$

$$= [W_n(2W_e + W_{x_1} + W_{x_2}) + (W_e + W_{x_1})(W_e + W_{x_2})]$$

$$\times [W_e(2W_n + W_{x_1} + W_{x_2}) + (W_n + W_{x_1})(W_n + W_{x_2})]$$

$$\times [W_{x_2}(W_e + W_n + W_{x_1}) + W_e W_n]^{-2} \tag{96}$$

a. Let $W_{x_1} = W_{x_2} = 0$ (i.e., only pseudosecular dipolar terms important). Then

$$E = \frac{1}{2[2 + b + b^{-1}]} \begin{cases} \xrightarrow{b \gg 1} \frac{1}{2}b^{-1} \\ \xrightarrow{b \ll 1} \frac{1}{2}b \\ \xrightarrow{b = 1} \frac{1}{8} \end{cases} \tag{97}$$

Table 1
ELDOR in the Four-Level System[a,b]

ELDOR experiment (o = observed, p = pumped)	ELDOR enhancement $\equiv -R$ [see equation (93)]	ELDOR frequency (first-order) $\lvert \omega_o - \omega_p \rvert = \Delta P$
(diagrams)	$\dfrac{2(W_{x_1} W_{x_2} - W_{n_1} W_{n_2})}{P}$	a
(diagrams)	$\dfrac{2(W_e^2 - W_{n_1} W_{n_2})}{q}$	$2\omega_n$
(diagram)	$\dfrac{(W_e + W_{n_1})(W_{x_2} + W_{n_2})(-2)}{P}$	$\left\lvert \omega_n - \dfrac{a}{2} \right\rvert$
(diagram)	$\dfrac{(W_e + W_{n_1})(W_{x_1} + W_{n_2})(-2)}{P}$	
(diagram)	$\dfrac{(W_e + W_{n_2})(W_{x_1} + W_{n_1})(-2)}{P}$	$\left\lvert \omega_n + \dfrac{a}{2} \right\rvert$
(diagram)	$\dfrac{(W_e + W_{n_2})(W_{x_2} + W_{n_1})(-2)}{P}$	
(diagram)	$\dfrac{(W_e + W_{n_1})(W_{x_2} + W_{n_2})(-2)}{q}$	$\left\lvert \omega_n - \dfrac{a}{2} \right\rvert$
(diagram)	$\dfrac{(W_e + W_{n_1})(W_{x_1} + W_{n_2})(-2)}{q}$	
(diagram)	$\dfrac{(W_e + W_{n_2})(W_{x_1} + W_{n_1})(-2)}{q}$	$\left\lvert \omega_n + \dfrac{a}{2} \right\rvert$
(diagram)	$\dfrac{(W_e + W_{n_2})(W_{x_2} + W_{n_1})(-2)}{q}$	

[a] From Rist and Freed.[11]

[b] $P = (W_{x_2} + W_{n_1})(2W_e + W_{x_1} + W_{n_2}) + (W_{x_1} + W_{n_2})(2W_e + W_{x_2} + W_{n_1})$;

$q = (W_e + W_{n_1})(W_e + W_{x_1} + W_{x_2} + W_{n_2}) + (W_e + W_{n_2})(W_e + W_{x_1} + W_{x_2} + W_{n_1})$.

b. Let $W_{x_1} = W_n = 0$ (i.e., isotropic hyperfine modulation). Then

$$E = W_{x_2}/W_e \tag{98}$$

This would theoretically be a most effective ENDOR mechanism if W_{x_2} were larger.

c. Case b but now the $a \leftrightarrow b$ ENDOR transition is saturated. Then $E = 0$.

d. Let $W_{x_2} = 4W_n$, $W_{x_1} = \frac{2}{3}W_n$ (dipolar, extreme narrowing).

$$E = \frac{b[22.5 + 60b + 40b^2]}{6 + b[25 + 34b + 15b^2]} \left\{ \begin{array}{l} \xrightarrow{\,b \gg 1\,} \frac{8}{3}b \\[2mm] \xrightarrow{\,b \ll 1\,} 3.75b \\[2mm] \xrightarrow{\,b = 1\,} 1.53 \end{array} \right. \tag{99}$$

This is also a very effective ENDOR mechanism if $b > 1$.

e. Case d but now the $a \leftrightarrow b$ ENDOR transition is saturated

$$E = \frac{b[2.5 + 10b + 10b^2]}{6 + b[45 + 84b + 45b^2]} \left\{ \begin{array}{l} \xrightarrow{\,b \gg 1\,} \frac{2}{9} \\[2mm] \xrightarrow{\,b \ll 1\,} \frac{5}{12}b \\[2mm] \xrightarrow{\,b = 1\,} \frac{1}{8} \end{array} \right. \tag{100}$$

4.3. Triple Resonance—Limiting Enhancements

We shall only consider case a, $W_{x_1} = W_{x_2} = 0$. Then one finds

$$E = \frac{W_e}{W_e + W_n} = \frac{1}{1 + b} \left\{ \begin{array}{l} \xrightarrow{\,b \gg 1\,} b^{-1} \\[2mm] \xrightarrow{\,b \ll 1\,} 1 \\[2mm] \xrightarrow{\,b = 1\,} \frac{1}{2} \end{array} \right. \tag{101}$$

Thus, for $b \lesssim 1$, one achieves much larger enhancements by triple resonance than by ENDOR, and for $b \ll 1$, one achieves the maximum possible signal enhancement of the saturated ESR of 100%. This is the case where both NMR transitions short out the weak W_n so that the ESR line is being relaxed via its own W_e process and equally well via a W_e process of the other hyperfine line. Thus, triple resonance is potentially a more powerful method than ENDOR.

However, for $b \lesssim 1$ the "effective" saturation of the NMR transitions is determined by satisfying the inequality (92). For the present case, equation (33) becomes

$$4d_n^2 T_n (2W_n + W_e)^{-1} \gg 1 \xrightarrow{\ b \ll 1\ } 4d_n^2 T_n / W_e \gg 1 \qquad (102)$$

while the simple ENDOR condition of equation (78′) is $d_n^2 T_n / W_n \gg 1$ (for $b \ll 1$), which is much easier to fulfill. Thus, the "effective" saturation of the NMR transitions for the triple resonance experiment can require substantially more rf field strengths than in ENDOR in order to realize its full potential.

4.4. Expressions for the Linewidths and the Transition Probabilities

In this section, we consider only the radical concentration-independent contributions to the spin relaxation. We assume a single set of completely equivalent nuclei with total nuclear spin quantum number J and total z component M. [We do not explitly indicate the distinction between degenerate states of the same values of J and M. Note that there will often be degenerate states for a given set of values of J and M. However, it is possible for dipolar terms (but not quadrupolar terms) to order the degenerate states according to a parameter κ or $J^{(\kappa)}$ such that the values of J *and* κ are preserved.[6] We do not explicitly indicate κ in the equations below.] We first consider the transition probabilities.[6]

4.4.1. Transition Probabilities

4.4.1.1. Nuclear Spin Transitions (or Pseudosecular Terms)
a. Dipolar

$$W_{(M_S, M) \to (M_S, M \pm 1)} = \tfrac{1}{2} j^D(0)[J(J+1) - M(M \pm 1)] \qquad (103)$$

where the electron–nuclear dipolar (END) spectral density $j^D(0)$ is

$$j^D(0) = \tfrac{1}{5} \gamma_e^2 \gamma_n^2 \hbar^2 \sum_m |D^{(m)}|^2 \tau_R \qquad (104)$$

with γ_e and γ_n the electronic and nuclear gyromagnetic ratios, respectively, $\hbar = h/2\pi$ with h as Planck's constant, and τ_R is the rotational correlation time, and it is assumed $|\omega_n \tau_R| \ll 1$. The dipolar coefficients are[12]

$$D^{(m)} = \left(\frac{6\pi}{5}\right)^{1/2} \langle \psi_e | r'^{-3} Y_{2,m}(\theta', \phi') | \psi_e \rangle \qquad (105)$$

where θ', ϕ', and r' are spherical polar coordinates that define the position of the unpaired electron with respect to a nucleus in the molecular coordinate frame.

b. *Quadrupolar.* (For this case only we consider a single nucleus of spin I):

$$W_{(M_S, M) \to (M_S, M \pm 1)} = 2j^Q(0)[I(I + 1) - M(M \pm 1)][2M \pm 1]^2 \quad (106a)$$

$$W_{(M_S, M) \to (M_S, M \pm 2)}$$

$$= 2j^Q(0)[I(I + 1) - M(M \pm 1)][I(I + 1) - (M + 1)(M \pm 2)]$$

$$(106b)$$

where

$$j^Q(0) = \frac{\tau_R}{80} \frac{e^2 Q^2}{\hbar^2 I^2 (2I - 1)^2} \sum_m |[\mathbf{V}\varepsilon]^{(m)}|^2 \quad (107)$$

with electric field gradient irreducible-tensor components[12]:

$$[\mathbf{V}\varepsilon]^{(0)} = -(\tfrac{3}{2})^{1/2} \langle \psi_e | V'_{zz} | \psi_e \rangle \quad (108a)$$

$$[\mathbf{V}\varepsilon]^{(\pm 1)} = \pm \langle \psi_e | V'_{xz} \pm i V'_{yz} | \psi_e \rangle \quad (108b)$$

$$[\mathbf{V}\varepsilon]^{(\pm 2)} = -\tfrac{1}{2} \langle \psi_e | V'_{xx} - V'_{yy} \pm 2i V'_{xy} | \psi_e \rangle \quad (108c)$$

Also, there will, in general, be cross-terms between the quadrupolar and dipolar interactions.[12]

4.4.1.2. *Electron Spin Transitions (or Nonsecular Terms)*

$$W_{(\mp, M) \to (\pm, M)} = 2j^D(0)M^2 + 4j^{(DG2)}(\omega_o)B_o M + 2j^{(G2)}(\omega_o)B_o^2 + W_e^{SR} \quad (109)$$

Here

$$j^D(\omega_o) = j^D(0)[1 + \omega_o^2 \tau_R^2]^{-1} \quad (110)$$

The g-tensor spectral density is

$$j^{(G2)}(\omega_o) = \frac{1}{20} \beta_e^2 \hbar^{-2} \left| \sum_{k=1}^{3} (g_k)^2 - 3g_s^2 \right| \frac{\tau_R}{1 + \omega_o^2 \tau_R^2} \quad (111)$$

where β_e is the Bohr magneton. The g-tensor dipolar cross-term spectral density is

$$j^{(DG2)}(\omega_0) = -\frac{1}{10} \gamma_e \beta_e \gamma_n \sum_m D^{(m)} g^{(m)} \frac{\tau_R}{1 + \omega_0^2 \tau_R^2} \quad (112)$$

with $g^{(0)} = 6^{-1/2}[2g'_z - (g'_x + g'_y)]$ and $g^{(\pm 2)} = \tfrac{1}{2}(g'_x - g'_y)$. The spin rotational contribution to W_e is in a semiclassic treatment:

$$W_e^{SR} = \frac{IkTC^2}{\hbar^2} \left(\frac{\tau_J}{1 + \omega_0^2 \tau_J^2} \right) \quad (113)$$

where I is the moment of inertia, C is the spin rotational constant of the radical (and we have assumed both to be isotropic), and τ_J is the correlation time for the angular momentum.[13] In liquids, usually $\tau_J \ll \tau_R$, ω_0^{-1}. One has for a Stokes–Einstein model

$$\tau_R = 4\pi\eta a^3/3kT \tag{114a}$$

$$\tau_J = [6IkT\tau_R]^{-1} \tag{114b}$$

More generally[13] $\mathbf{C} \cong -2\mathbf{A}\,\Delta\mathbf{g}$ where $\mathbf{A}$ is the inverse moment-of-inertia tensor and $\Delta\mathbf{g} = \mathbf{g} - 2.00231$. Then we have (for axially symmetric $\mathbf{A}$):

$$W_e^G \cong \sum_i (g_i - g_s)^2/40\tau_R \qquad \text{for } \omega_o^2\tau_R^2 \gg 1 \tag{115a}$$

$$W_e^{SR} \cong \sum_i (g_i - g_e)^2/18\tau_R \tag{115b}$$

If these are the dominant terms in W_e, then (see Ref. 7):

$$W_e \propto \tau_R^{-1} \tag{116a}$$

or

$$W_e \equiv AT/\eta \tag{116b}$$

Usually, $\tau_R > \omega_0^{-1}$ for free radicals in liquids below room temperature, since at X band $\omega_0^{-1} \cong 1.7 \times 10^{-11}$ sec. Then pseudosecular dipolar terms dominate in ELDOR or ENDOR. So, from equations (103) and (104),

$$W_n \propto \tau_R$$

or

$$W_n \equiv B\eta/T$$

Then

$$b = \left(\frac{B}{A}\right)\left(\frac{\eta}{T}\right)^2 \tag{117}$$

If we let

$$\eta \propto Te^{W/kT}, \qquad W > 0$$

we get b increasing significantly with decreasing T. This usually leads to better ELDOR and ENDOR signals at reduced temperatures.

The typical linear dependence of W_e on T/η, as well as the dependence of the linewidth parameters (from which W_n is determined) on η/T are shown in Figures 9 and 10.

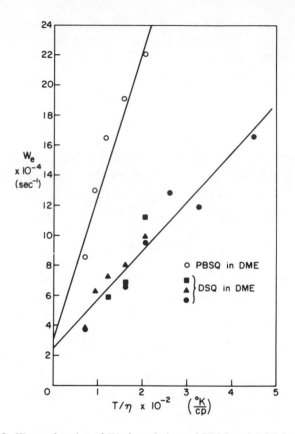

Figure 9. W_e as a function of T/η for solutions of PBSQ and DSQ in DME.

4.4.1.3. *Combined Electron Spin–Nuclear Spin Transitions (Cross-Relaxation from Nonsecular Terms)*

$$W_{M_s,\, M \to M_s \pm 1,\, M \mp 1} = \left[\tfrac{1}{3} j^{(D)}(\omega_0) + \tfrac{1}{2} j^I(\omega_0)\right][J(J+1) - M(M \mp 1)] \quad (118)$$

$$W_{M_s,\, M \to M_s \pm 1,\, M \pm 1} = 2 j^{(D)}(\omega_0)[J(J+1) - M(M \pm 1)] \quad (119)$$

The isotropic dipolar spectral density is

$$j^I(\omega) = \tfrac{1}{2}\gamma_e^2 \int_{-\infty}^{-\infty} [\langle a(t)a(t+\tau)\rangle - \bar{a}^2]e^{-i\omega\tau}\, d\tau \quad (120)$$

where $\bar{a}$ is the time-averaged hyperfine splitting. Note that for $j^D(\omega_0) \ll j^D(0)$ [and small $j^I(\omega_0)$], $W_x \ll W_n$ and pseudosecular terms dominate as noted above.

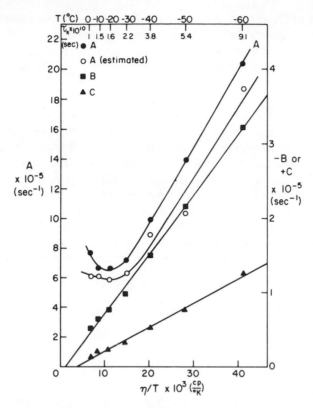

Figure 10. The ESR linewidth parameters A, B, and C for a dilute solution of PBSQ in ethanol as a function of η/T. W_n is usually determined from C.

4.4.2. Linewidths

One finds in the analysis of the relaxation matrix for the diagonal elements $R_{ab, ab}$ yielding the linewidths, that[1,2,6,12]

$$T_{2a, b}^{-1} = -\operatorname{Re} R_{ab, ab} = T'^{-1}_{2a, b} + \tfrac{1}{2}\left(\sum_{\gamma \neq a} W_{a\gamma} + \sum_{\gamma \neq b} W_{b\gamma} \right) \qquad (121)$$

where

$$T'^{-1}_{2a, b} = \int_0^\infty \langle \omega^{a, b}(t) \omega^{a, b}(t - \tau) \rangle \, d\tau \qquad (122)$$

and

$$\omega^{a, b}(t) \equiv [\mathscr{H}_1(t)_{aa} - \langle \mathscr{H}_1(t) \rangle_{aa}] - [\mathscr{H}_1(t)_{bb} - \langle \mathscr{H}_1(t) \rangle_{bb}] \qquad (123)$$

In equations (122) and (123), the angular brackets imply an ensemble average over the randomly fluctuating spin perturbation term $\mathscr{H}_1(t)$. The term $T'^{-1}_{2a,\,b}$ is seen to result from fluctuations in the eigenenergy difference between the two states a and b, which depends only on the diagonal elements of the perturbation $\mathscr{H}_1(t)$, so this contribution is known as the secular contribution to the linewidth. The terms in equation (131) of type $W_{a\gamma'}\,W_{b\gamma}$ give the mean of all the transitions away from states a and b. These are the nonsecular (as well as pseudosecular) terms yielding line broadening due to the Heisenberg uncertainty in lifetime effect. We also note that the imaginary parts of $R_{ab,\,ab}$ are not, in general, zero. One sees, from the form of equations (5), that these imaginary terms must cause frequency shifts and are therefore called "dynamic frequency shifts." They are of potential importance, but partly because they are not often studied, we will not discuss them further here. Their contributions in simple cases have been reviewed by Fraenkel.[14] We give below the expressions for the $T'^{-1}_{2a,\,b}$.[6, 12]

a.   States a and b involve an electron spin–flip transition.

$$T'^{-1}_{2a,\,b} = (1/4)[j^{(I)}(0) + (8/3)j^{(D)}(0)](M_a + M_b)^2$$
$$+ [j^{(IG_0)}(0) + (8/3)j^{(DG_2)}(0)]B_0(M_a + M_b)$$
$$+ [j^{(G_0)}(0) + (8/3)j^{(G_2)}(0)]B_0^2 \tag{124}$$

The new spectral density terms here are

$$j^{(G_0)}(0) = \beta_e^2\hbar^{-2}\tfrac{1}{2}\int_{-\infty}^{\infty}[\langle g_s(t)g_s(t+\tau)\rangle - \bar{g}_s^2]e^{-i\omega\tau}\,d\tau \tag{125}$$

where $\bar{g}_s$ is the time-averaged g value and

$$j^{(IG_0)}(0) = \tfrac{1}{2}\bar{g}_s\beta_e^2\hbar^{-2}\int_{-\infty}^{\infty}[\langle a(t)g_s(t+\tau)\rangle - \bar{a}\bar{g}_s]\,d\tau \tag{126}$$

Also, M_a and M_b are the z-component quantum numbers of the equivalent set of nuclear spins in states a and b, respectively. Note that an allowed ESR transition corresponds to $M_a = M_b$, while $M_a \neq M_b$ is a forbidden ESR transition.

b.   Linewidths for nuclear transitions (i.e., $\Delta M_s = 0$).

$$T'^{-1}_{2a,\,b} = (1/4)[j^{(I)}(0) + (8/3)j^{(D)}(0)][M_a - M_b]^2 \tag{127}$$

When $M_a - M_b = \pm 1$, we have a single-quantum NMR transition, while for $M_a - M_b = \pm n$, we have an n-tuple quantum NMR transition. The pseudosecular and nonsecular contributions to $T^{-1}_{2a,\,b}$ are obtained with the use of equation (121). However, there are also off-diagonal R-matrix elements that must be included for the nuclear transitions. They may be written as

$$-R^{\text{pseudosec}}_{M,\,M+1;\,M+1,\,M+2} = -\tfrac{1}{2}j^{(D)}(0)[f(J,\,-M)f(J,\,-(M+1))] \quad (128)$$

and

$$-R^{\text{pseudosec}}_{M-1,\,M;\,M,\,M+1} = -\tfrac{1}{2}j^{(D)}(0)[f(J,\,M)f(J,\,(M+1))] \quad (129)$$

where

$$f(J,\,\pm M) \equiv [(J \pm M)(J \mp M + 1)]^{1/2} \quad (130)$$

The subscripts on R indicate the values of M_J of the states being coupled. The quadrupolar terms will also make secular contributions as well as off-diagonal contributions to the matrix **B** for the nuclear transitions. Aspects of such terms are discussed elsewhere.[12]

In concluding this section, we note that more detailed analyses of these spectral densities in terms of realistic models both for isotropic as well as ordered fluids appear in several places.[12–23]

5. Heisenberg Spin Exchange and Chemical Exchange

Heisenberg spin exchange is a very important radical-concentration-dependent relaxation mechanism in normal liquids. It is probably the dominant one for $S = \tfrac{1}{2}$. It may be analyzed by a simple model, which also serves as a simple example of the stochastic Liouville approach (see Section 8). We assume radicals exist either as well-separated "monomers" or as interacting pairs or "dimers," each with mean lifetimes τ_2 and τ_1, respectively, and with density matrices σ and ρ, respectively. The equations of motion are then:

$$i\dot{\sigma} = \mathscr{H}_0^{(1)x}\sigma + i\frac{2}{\tau_2}\text{Tr}_s\rho - i\frac{2}{\tau_2}\sigma \quad (131)$$

$$i\dot{\rho} = (\mathscr{H}_0^{(1)x} + \mathscr{H}_0^{(2)x} + \mathscr{H}_J^x)\rho - i\tau_1^{-1}(\rho - \sigma \times \sigma) \quad (132)$$

where $\mathscr{H}_0^{(1)}$ is the spin Hamiltonian for radical 1, etc., and $\text{Tr}_s\rho = \tfrac{1}{2}(\text{Tr}_1\rho + \text{Tr}_2\rho)$ is a symmetrized trace over each of the two components of the interacting dimer, and we have used the superoperator form, e.g., $\mathscr{H}_0^x\sigma = [\mathscr{H}_0,\,\sigma]$. Also,

$$\mathscr{H}_J = JS^{(1)} \cdot S^{(2)} \quad (133)$$

where J is twice the exchange integral. One obtains a steady-state solution for σ in the rotating frame. It is then possible to show that when

$$|J|,\,\tau_1^{-1} \gg |a_i|,\,\gamma_e B_1 \quad (134)$$

Equation (131) is well approximated by

$$i\dot{\sigma} = \mathcal{H}_0^{(1)x}\sigma + i\omega_{HE}[\text{Tr}_s(\mathcal{P}\sigma \times \sigma\mathcal{P}) - \rho] \tag{135}$$

where

$$\omega_{HE} = \frac{1}{\tau_2}\frac{J^2\tau_1^2}{1 + J^2\tau_1^2} \tag{136}$$

is the Heisenberg exchange frequency. In equation (135), we have neglected a frequency shift term that is readily shown to be zero in the high-temperature approximation [i.e., $A\sigma \cong 1 + \sigma'$ with $|\sigma'| \ll 1$] (see equation 24). Here, $\mathcal{P}$ is the operator that permutes electron spins. The derivation of equation (135) is based on the fact that for spins $S = \frac{1}{2}$,

$$\mathcal{H}_j^x = \frac{1}{2}J\mathcal{P}^x \tag{137}$$

For simple Brownian diffusion of the radicals in solution, we have

$$\tau_2^{-1} = 4\pi Df\mathcal{N} \tag{138a}$$

$$\tau_1^{-1} = (6D/d^2)fe^u \tag{138b}$$

where $\mathcal{N}$ is the density of radicals, the diffusion coefficient is $D = kT/6\pi a\eta$, and d is the interaction distance for exchange. The factors f and fe^u are introduced for charged radicals to take account of Coulombic and ionic atmosphere effects.[24,25]

The result, equation (135), means that Heisenberg exchange appears as a simple exchange process analogous to chemical exchange processes for which the well-known Kaplan–Alexander[26] method applies. We let

$$\Phi_H(\chi) \equiv \omega_{HE}[\text{Tr}_s(\mathcal{P}\sigma \times \sigma\mathcal{P}) - \sigma] \tag{139}$$

and add this relaxation term to equation (5). One then finds that for well-separated hyperfine lines, the T_2 contributions are

$$T_{2,HE}^{-1}(\text{ESR}, \lambda) = \left(\frac{A - 2D(\lambda)}{A}\right)\omega_{HE} \tag{140}$$

$$T_{2,HE}^{-1}(\text{NMR}) = \frac{1}{2}\omega_{HE} \tag{141}$$

Here $D(\lambda)$ is the degeneracy of the λth transition, and the $T_2^{-1}(\text{NMR})$ is the width contribution to a well-resolved ENDOR line. The diagonal elements of equation (139) yield

$$[\Phi_H(\chi)]_{\alpha\pm\alpha\pm} = \frac{1}{2}\omega_{HE}[(\chi_{\alpha\mp} - \chi_{\alpha\pm}) \pm (\chi_+ - \chi_-)]$$

$$= \mp\frac{\omega_{HE}}{2}\left[\left(1 - \frac{2}{A}\right)\hat{\chi}_\alpha - \sum_{\gamma\neq\alpha}\frac{2}{A}\hat{\chi}_\gamma\right] \tag{142}$$

where

$$\chi_\pm = \frac{2}{A} \sum_\gamma \chi_{\gamma\pm} \tag{143a}$$

and

$$\chi_+ + \chi_- = 0 \tag{143b}$$

and

$$\hat{\chi}_\alpha \equiv \chi_{\alpha+} - \chi_{\alpha-} \tag{143c}$$

The notation $\alpha\pm$ in equations (142) and (143) refers to the αth nuclear spin configuration, and $M_s = \pm$.

The steady-state solution of equation (142) in equation (5) is

$$\chi_{\alpha+} - \chi_{\alpha-} = 2\chi_+ \tag{144}$$

i.e., differences in population between all pairs of levels differing only in M_s are equal. The unlinearized rate equations yield the steady-state result that all the ratios $\sigma_{\alpha-}/\sigma_{\alpha+}$ are equal.

If in chemical exchange (CE) (i.e., electron transfer), the predominant NMR relaxation of the diamagnetic radical precursors is the CE process, then CE appears to be just like HE in magnetic resonance experiments on the radicals and we indicate this by replacing ω_{HE} by the more general symbol ω_{EX} (see Ref. 7 for further details).

We now consider the $\mathbf{W}$ matrix including equation (142). Note first that equation (142) generates a matrix $\mathbf{W}^{HE}$, which is symmetric and which has the properties that the sums of all columns (rows) are all zero. Thus, all the theorems of the Appendix apply to the complete $\mathbf{W}$ matrix, including equation (142). Now, however, note that

$$[\Phi_{HE}(\chi)]_{\alpha+\alpha+} + [\Phi_{HE}(\chi)]_{\alpha-\alpha-} = 0 \tag{145}$$

so each pair of rows of $\mathbf{W}$ labeled $\alpha+$ and $\alpha-$ are linearly dependent. Thus, while $\mathbf{W}^{HE}$ is an $A \times A$ matrix, it is of rank $A/2$; i.e., HE does not act to change $(\chi_{\alpha+} + \chi_{\alpha-})$ but rather to equate all $(\chi_{\alpha+} - \chi_{\alpha-})$. One must add W_n or W_x terms to reduce this high-order singularity.

One may alternatively employ another method. Sufficient conditions for this method are:

Condition 1. All spin-flip relaxation transitions are of W_e, W_n, or ω_{HE} type (i.e., no W_x).

Condition 2

(a) $\quad W_{(+, M)\to(-, M)} = W_{(-, M)\to(+, M)}$ \hfill (146a)

(b) $\quad W_{(+, M)\to(+, M\pm 1)} = W_{(-, M)\to(-, M\pm 1)}$ \hfill (146b)

Then we may define a $\frac{1}{2}A$-dimensional square matrix $\hat{\mathbf{W}}$ (which is usually nonsingular or readily separated into nonsingular components) according to

$$[\hat{\mathbf{W}}\hat{\chi}]_\lambda \equiv [\mathbf{W}\chi]_{\lambda+} - [\mathbf{W}\chi]_{\lambda-} \tag{147}$$

This reduced eigenstate space is found to include only the $\hat{\chi}_\lambda \equiv \chi_{\lambda+} - \chi_{\lambda-}$, which are closely related to pure ESR transitions.

Then, if only ESR transitions are induced, we find

$$S_{\lambda,\,\eta} = 4d_\lambda d_\eta(\hat{\mathbf{W}}^{-1})_{\lambda,\,\eta} = d_\lambda d_\eta \Omega_{\lambda,\,\eta} \tag{148}$$

[This method can also be generalized for ENDOR (see Section 6.2).] Using this method, one can then prove[24]

$$\Omega_\lambda = \frac{2}{W_e D(\lambda)} \frac{1 + D(\lambda)b''}{1 + \frac{1}{2}Ab''} \tag{149}$$

$$b'' = \omega_{\mathrm{HE}}/AW_e \tag{149'}$$

with

$$T_{1,\,\lambda} = \tfrac{1}{4}[D(\lambda)\Omega_\lambda] \tag{150}$$

and

$$\Omega_{\lambda,\,\eta} = \frac{2}{W_e} \frac{b''}{1 + \frac{1}{2}Ab''} \quad \lambda \neq \eta \tag{151}$$

Equation (149) illustrates the "shorting-out" effect spin exchange has in coupling the different hyperfine lines [see equation (144)] without directly leading to electron spin-flips. It follows from equations (54) and (149)–(151) that

$$R \equiv \frac{Z''_{\mathrm{ESR}} - Z''_{\mathrm{ELDOR}}}{Z''_{\mathrm{ESR}}} \tag{152}$$

is [for $\Delta\omega_0 = 0$ and the no-saturation condition of equation (58)]

$$R^{-1} = \Omega_p/\Omega_{o,\,p} + [(1 + \Delta\omega_p^2 T_p^2)/T_p\Omega_{o,\,p}] \, dp^{-2} \tag{153}$$

and

$$R_\infty^{-1} = \Omega_p/\Omega_{o,\,p}$$
$$= \frac{1 + D(p)b''}{D(p)b''} = [D(p)b'']^{-1} + 1 \tag{154}$$

Here, R_∞ is defined in the same manner as the asymptotic R of equation (93). Equation (154) shows how Heisenberg spin exchange is effective in enabling significant ELDOR reduction factors. Equation (154) has been confirmed experimentally, as shown in Figure 11.

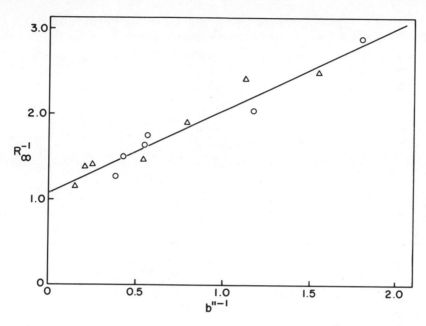

Figure 11. Linear least-squares fit for R_∞^{-1} versus b''^{-1} for aqueous PADS solutions at 24°C; $\triangle$ = data for two hyperfine line separation ($\tilde{M} = 1$ line observed and $\tilde{M} = -1$ line pumped); $\bigcirc$ = data for one hyperfine line separation ($M = 0$ line observed and $M = -1$ line pumped). The slope and intercept are 1.03 ± 0.08 and 1.07 ± 0.07, respectively, as compared to values of unity theoretically predicted. (From Eastman *et al.*[25])

Now for ENDOR and a single nucleus of spin $I = \frac{1}{2}$ with $W_x = 0$, one has

$$\Omega_e = \frac{1}{W_e} \frac{[2W_e + (W_n + \omega_{HE}/2)]}{W_e + (W_n + \omega_{HE}/2)} \tag{155a}$$

$$\Omega_n = \frac{1}{W_n} \frac{[2W_n + (W_e + \omega_{HE}/2)]}{[W_e + \omega_{HE}/2 + W_n]} \tag{155b}$$

$$\Omega_{e,n} = (W_e + W_n + \omega_{HE}/2)^{-1} \tag{155c}$$

and

$$\xi_e^r(d_n \to \infty) = \frac{W_n}{(2W_n + W_e + \omega_{HE}/2)(W_e + W_n + \omega_{HE}/2)} \tag{156}$$

In general, if $W_n = 0$ and $W_x = 0$, then $\xi_e^r(d_n \to \infty) = 0$ even for more than one magnetic nucleus. Thus Heisenberg exchange is *not* an effective ENDOR mechanism; i.e., it is ENDOR "inactive," although it is ELDOR "active." The manner in which Heisenberg exchange suppresses ENDOR enhancements is illustrated in Figure 12 representing a concentration-dependent study, while the T/η dependence of the exchange contribution to the ENDOR linewidths as predicted by equations (141), (136), and (138a) for strong exchange are illustrated in Figure 13.

In the case of triple resonance, one has

$$\Omega_{n,\,n'} = \frac{1}{W_n} \frac{W_e + \omega_{\text{HE}}/2}{W_e + W_n + \omega_{\text{HE}}/2} \tag{157}$$

$$\xi_e^r(d_n, d_{n'} \to \infty) = \frac{1}{W_e + W_n + \omega_{\text{HE}}/2} \tag{158}$$

and

$$E = \frac{W_e}{W_e + W_n + \omega_{\text{HE}}/2} \tag{159}$$

so that maximum enhancements accrue from $W_e \gg W_n, \omega_{\text{HE}}$.

More detailed models of Heisenberg spin exchange are summarized elsewhere.[27,28] Also, another concentration-dependent relaxation mechanism, which can become important at reduced temperatures, is that of intermolecular dipolar interactions between electron spins. Its ENDOR effects are discussed in some detail by Leniart *et al.*[29]

In concluding the last two sections, we note that the characteristic behavior of the various relaxation mechanisms as differently manifested in linewidth, saturation, ENDOR, and ELDOR is a useful approach for separating out the many possible components of relaxation in a particular paramagnetic system. We present in Table 2 a simplified summary of these characteristics.

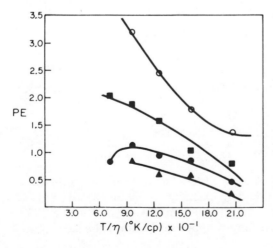

Figure 12. The percent enhancement at infinite rf power as a function of T/η for solutions of DSQ in DME; ▲ = concentration of $9.9 \times 10^{-4} M$; ● = concentration of $6.3 \times 10^{-4} M$; ■ = concentration of $3.0 \times 10^{-4} M$; ○ = concentration of $1.5 \times 10^{-4} M$.

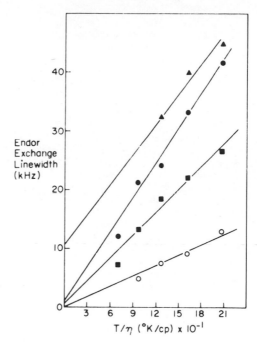

Figure 13. The exchange contribution to the ENDOR linewidth as a function of T/η for solutions of DSQ in DME. $\blacktriangle = 9.9 \times 10^{-4}\ M$; $\bullet = 6.3 \times 10^{-4}\ M$; $\blacksquare - 3.0 \times 10^{-4}\ M$; $\bigcirc = 1.5 \times 10^{-4}\ M$.

6. *General Approach*

One finds that, in general, multiple resonance ESR experiments in liquids may be expressed in the matrix form equation (46) with formal solution given by equations (48)–(51).[6] In this formal solution, **Z** is a vector in the space of all induced transitions and **X** is a vector in the space of all spin eigenstates. The only requirement is that a *raising* convention apply. This is the requirement that all induced transitions in the space of **Z** are those in which there is (are) increase(s) in spin quantum number but *no* decrease(s) in spin quantum number (see Ref. 9). This requirement is often met for ENDOR and ELDOR experiments, but it sometimes requires neglect of some multiple quantum transitions. If it is not met, then a somewhat more complex form of equations (40) and (48)–(51) could become necessary. Also, in summary, the validity of the general relaxation equation (15) for well-separated hyperfine lines requires that

$$|\gamma_e B_0|,\ |\gamma_i B_0|,\ |\gamma_e \bar{a}_i|,\ \tau_c^{-1} \gg \varepsilon(t),\ |R| \tag{160}$$

where τ_c refers to the relevant correlation time(s). We outline below a "diagram method" for constructing the matrices needed for equations (46) and (48)–(51).

Table 2. ESR Linewidth

	Linewidths		
Mechanism	Nuclear spin dependence	Field frequency dependence	Temperature and viscosity
Intramolecular			
G-Isotropic—secular only	None	Quadratic	As τ_c
Dipole-isotropic			
Secular	M^2	None	As τ_c
Nonsecular	M^2	$[1 + \omega_0^2 \tau_c^2]^{-1}$	As $\tau_c[1 + \omega_0^2 \tau_c^2]^{-1}$
Isotropic G-dipole			
X-Term—secular only	M	Linear	As τ_c
G-Anisotropic			
Secular	None	Quadratic	$\tau_R \propto \eta/T$
Nonsecular	None	$\omega_0^2[1 + \omega_0^2 \tau_R^2]^{-1}$	$\tau_R[1 + \omega_0^2 \tau_R^2]^{-1}$
Dipole-anisotropic			
Secular $(S_z I_z)$	M^2	None	τ_R
Pseudosecular $(S_z I_\pm)$	M^2	None	τ_R
Nonsecular			
$S_\pm I_z$	M^2	$[1 + \omega_0^2 \tau_R^2]^{-1}$	$\tau_R[1 + \omega_0^2 \tau_R^2]^{-1}$
$S_\pm I_\pm, S_\pm I_\mp$	M^2	$[1 + \omega_0^2 \tau_R^2]^{-1}$	$\tau_R[1 + \omega_0^2 \tau_R^2]^{-1}$
Anisotropic G-dipole X-term			
Secular	M	Linear	τ_R
Nonsecular	M	$\omega_0[1 + \omega_0^2 \tau_R^2]^{-1}$	$\tau_R[1 + \omega_0^2 \tau_R^2]^{-1}$
Quadrupolar			
Secular	No width contribution	—	τ_R
Pseudosecular	M^2 and M^4	None	τ_R
Spin rotation			
Secular	None	None	$\tau_J \propto T/\eta$
Nonsecular	None	None	$\tau_J \propto T/\eta$
Intramolecular spin–orbit processes			
Secular	None	None	Independent
Nonsecular	None	None	Independent
Zero-field splitting $S > \frac{1}{2}$			
Secular	None	None	τ_R
Nonsecular	None	$[1 + b^2 \omega_0^2 \tau_R^2]^{-1}$	$\tau_R[1 + b^2 \omega_0^2 \tau_R^2]^{-1}$
Intermolecular			
Heisenberg spin exchange (+electron transfer)	Symmetric dependence on D_M	None	$T/\eta[1 + (J\tau_1)^{-2}]^{-1}$ $\tau_1 \propto \eta/T$ for HE
Dipole-Dipole			
Secular $S_{1z} S_{2z} + S_{1\pm} S_{2\mp}$ like	Symmetric dependence on D_M	None	τ_t
Pseudosecular $S_{1\pm} S_{2\mp}$ unlike	Symmetric dependence on D_M	None	τ_t
Nonsecular $S_{1\pm} S_{2z} + S_{1\pm} S_{2\pm}$	Symmetric dependence on D_M	$[1 - \frac{3}{8}(2b\omega_0 \tau_t)^{1/2}]$ (lowest order in ω_0) $b = 1$ or 2	$\tau_t[1 - \frac{3}{8}(2b\omega_0 \tau_t)^{1/2}]$ (lowest order in ω_0)

and Relaxation Mechanisms

Saturation (Ω)		ELDOR	ENDOR		
Activity	Nuclear spin dependence	Activity	Activity	Linewidth contribution	Enhancement (%) (maximum)
None	None	None	None	None	None
None	—	None	None	Yes	—
W_{x_1}	Yes	None unless $W_{x_2} \neq 0$	W_{x_1}	Yes	Goes as W_{x_1}/W_e
None	—	None	None	None	—
None	—	None	None	None	—
W_e	None	W_e	W_e	Yes	W_e
None	—	None	None	Yes	—
W_n	Yes	W_n reduction	W_n	Yes	$W_n/W_e \sim 1$
W_e	M^2	W_e	W_e	Yes	W_e
W_{x_1}, W_{x_2}	Yes	W_x enhancement	W_{x_1}, W_{x_2}	Yes	Goes as W_x/W_e
None	—	None	None	None	None
W_e	M	W_e	W_e	Yes	W_e
None	—	None	None	Yes	—
W_x	Yes	W_n reduction	W_n	Yes	$W_n/W_e \sim 1$
None	—	None	None	None	—
W_e	None	W_e	W_e	Yes	W_e
None	—	None	None	None	—
W_e	None	W_e	W_e	W_e	W_e
None	—	None	None	None	—
W_e type	None	W_e type	W_e type	Yes	W_e type
ω_{EX}	Some dependence on D_M	ω_{EX} reduction	No	Yes	Decreases enhancements
None	—	None	None	None	—
Analagous to ω_{EX}	Some dependence on D_M	Reduction analogous to ω_{EX}	No	Yes	Decreases enhancements
W_e	Some dependence on D_M	W_e	W_e	Yes	W_e

We exclude in the present discussion any effects from high-frequency and/or large-amplitude modulation of the dc field. They are discussed for ELDOR in Chapter 5. Some discussion of such effects for ENDOR is given in Ref. 29 and Chapter 2.

6.1. *Diagram Method*

An energy level diagram such as the one in Figure 6 for $S = \frac{1}{2}$ and $I = 1$ is constructed.[9] (In this example, ω_e is close to ω_{ab}, and ω_n is close to $\omega_{bd} = \omega_{df}$).

6.1.1. *Forming the Transition Moment Matrix* **d**

a. Draw an arrow between each pair of states for which the energy difference $\hbar\omega_\lambda$ is nearly resonant with an applied radiation field and for which a "transition moment" d_λ or d_η (see below) exists (i.e., the "allowed" transitions). Each arrow should point to the state of increasing quantum number (i.e., the raising convention).

b. An element of the **d** matrix is labeled $d_{\lambda, i}$, where λ represents an induced transition and i is a particular eigenstate. This matrix element is minus (plus) d_λ if λ includes that eigenstate and if the arrow for that transition points toward (away from) the eigenstate. Otherwise it is zero.

c. For an allowed ESR transition near resonance, we have

$$d_\lambda \equiv \tfrac{1}{2}\gamma_e \beta_e \langle -\tfrac{1}{2}, \{M_I\} | S_- | \tfrac{1}{2}, \{M_I\} \rangle \tag{161}$$

where d_λ is the "transition moment" for the λth ESR transition corresponding to a particular nuclear configuration abbreviated as $\{M_I\}$. In a similar manner, we define NMR transition moments

$$d_r \equiv \tfrac{1}{2}\gamma_n B_r [J_r(J_r + 1) - M_r(M_r + 1)]^{1/2} \tag{162}$$

where the subscript r refers to a particular set of equivalent nuclear spins each of spin I that are "excited" by the rf field, and

$$J_r = \sum_{k \text{ in } r} I_k \quad \text{with} \quad J_n = J_r, J_r - 1, \text{ etc.} \tag{163}$$

as allowed by the vector addition of the nuclear spins.

d. Now we introduce the notation that the higher letter is the one of increasing quantum number (i.e., $j > i$ so j is of greater quantum number). Thus if we let λ refer to the $i \leftrightarrow j$ transition, then

$$d_{\lambda, i} = d_{(ij), i} = d_{ij} \tag{164a}$$

$$d_{\lambda, j} = d_{(ij), j} = -d_{ij} \tag{164b}$$

$$d_{\lambda, k} = 0 \quad \text{for } k \neq i, j \tag{164c}$$

[Note that following the discussion below equation (23'), we need only consider those d_λ, d_η to be nonzero if the λth, ηth transitions are "excited" by the rf field. In the example of Figure 6, $d_{(ab), a} = -d_{(ab), b} = \frac{1}{2}\gamma_e \beta_e$,

$$d_{(bd), b} = -d_{(bd), d} = \tfrac{1}{2}\gamma_n \beta_n \langle 1, 0 | J_- | 1, 1 \rangle = \frac{\sqrt{2}}{2}\gamma_n \beta_n \qquad (165a)$$

$$d_{(df), d} = -d_{(df), f} = \tfrac{1}{2}\gamma_n \beta_n \langle 1, -1 | J_- | 1, 0 \rangle = \frac{\sqrt{2}}{2}\gamma_n \beta_n \qquad (165b)$$

while all other $d_{\lambda, k}$ are zero.] [Actually equations (162) and (165) represent only the | zero-order NMR transition moments, which must, in general, be corrected according to equation (61).]

6.1.2. *Saturation Matrix* S

We now use equation (49b) as the definition of the S matrix. It is a matrix in the space of the induced transitions. Consider the $S_{\lambda, \eta}$ element. And let us suppose that λ is the $i \leftrightarrow j$th transition while η is the $k \leftrightarrow l$th transition. Then

$$S_{\lambda, \eta} = S_{(ij), (kl)} = 2 \sum_{m, n} d_{(ij), m}(\mathbf{W}^p)^{-1}_{mn} d^{p \, \mathrm{tr}}_{n, (kl)}$$

$$= 2d_{ij}d_{kl}[(\mathbf{W}^p)^{-1}_{ik} + (\mathbf{W}^p)^{-1}_{jl} - (\mathbf{W}^p)^{-1}_{il} - (\mathbf{W}^p)^{-1}_{jk}] \qquad \text{for } p \neq i, j, k, l$$

$$\tag{166}$$

where we have used the above rules for the $\mathbf{d}$ matrix. We now introduce the cofactors of $\mathbf{W}$ (see the Appendix) to get

$$S_{\lambda, \eta} = \frac{2d_{ij}d_{kl}}{AC}[(C^p_{ki} - C^p_{li}) + (C^p_{lj} - C^p_{kj})], \qquad \text{where } i \neq j \text{ and } p \neq i, j \quad (167)$$

Then, by use of identity (2) of the Appendix [equation (A.22)], we have

$$S_{\lambda, \eta} = \frac{2d_{ij}d_{kl}}{AC}[C^l_{ki} + C^k_{lj}] = \frac{2d_{ij}d_{kl}}{AC}[C^l_{ki} - C^l_{kj}] \qquad (168)$$

which by identities (3) and (5) of the Appendix [equation (A.22)], respectively, give

$$S_{\lambda, \eta} = \frac{2d_{ij}d_{kl}}{C}C_{kl, \, ij} = \frac{2d_{ij}d_{kl}}{C}C_{ij, \, kl}$$

$$\equiv d_{ij}d_{kl}\Omega_{ij, \, kl} \qquad (169)$$

Now suppose $p = l$. Then $d^{l\,\text{tr}}_{l(rs)} = 0$ for all r, s. Then, instead of equation (166), we have

$$S_{\lambda,\eta} = 2d_{ij}d_{kl}[(W^l)^{-1}_{ik} - (W^l)^{-1}_{jk}]$$

$$= \frac{2d_{ij}d_{kl}}{AC}[C^l_{ki} - C^l_{kj}] = d_{ij}d_{kl}\Omega_{ij,kl} \tag{170}$$

which is the same result as equation (169). These results then prove equation (51).

Note that from identity (5) of the Appendix [equation (A.22)], it follows that $S_{\lambda,\eta} = S_{\eta,\lambda}$ so that the **S** matrix is symmetric. It also follows from the properties of **d** (viz., there are no nonvanishing transition moments between pairs of states coupled by multiple quantum transition) that there are no elements of **S** involving multiple quantum transitions; λ and η must each refer to an allowed single-quantum transition.

6.1.3. *Coherence Matrix* **K**

The diagonal elements of **K** are obtained from

$$-[\dot{\sigma} + i(\mathscr{H}_0, \sigma)] \to \mathbf{K}^{(d)} \tag{171}$$

The only nonvanishing terms are for $\sigma_{\alpha\beta}$, where $E_\alpha \neq E_\beta$ and $\alpha \leftrightarrow \beta$, is an induced transition (not necessarily an allowed transition). The nondiagonal elements of **K** are obtained from:

$$[\varepsilon(t), \sigma] \to \mathbf{K}^{(n)} \tag{172}$$

where only those terms involving $\sigma_{\alpha\beta}(E_\alpha \neq E_\beta)$ and where $\alpha \leftrightarrow \beta$ is an induced transition need be retained. [The terms diagonal in σ in this equation lead to $\mathbf{d}\chi + \mathbf{Q}$ in equation (46a).]

a. Consider those pairs of states connected by an arrow [step (a) for forming the **d** matrix]. Such an arrow represents a single quantum transition corresponding to a steady-state, nonzero value for Z_λ, where Z_λ fulfills the raising convention (see rule 1 for the **d** matrix). For example, in Figure 6, they are $Z_{ba} \equiv Z_e$, $Z_{db} \equiv Z_1$; $Z_{fd} = Z_2$.

b. Now consider all pairs of levels connected by two contiguous arrows (e.g., a and d, b and f in Figure 6). These lead to two quantum transitions, with which are associated nonvanishing Z_λ (e.g., Z_{da} and $Z_{fb} \equiv Z_c$ in Figure 6). The pairs of levels connected by three contiguous arrows are the triple quantum transitions (e.g., Z_{fa}), etc.

c. The diagonal elements of **K** associated with the n-tuple quantum transitions are obtained by adding the $\Delta\omega_\lambda$ for the **n** consecutive single quantum transitions; e.g., $(\omega_n - \omega_{bd}) + (\omega_n - \omega_{df})$ for Z_{fb}.

d. The only off-diagonal elements of $\mathbf{K}$ for the n-tuple transitions are with $(n - 1)$-tuple transitions and with $(n + 1)$-tuple transitions, such that the $(n \pm 1)$-tuple transition plus or minus a single quantum transition η equals the n-tuple transition. The matrix element is just $\pm d_\eta$, where the plus (minus) sign is used when the arrow for the η_kth transition points toward (away from) the state in common with the other constituent transition (either an $n - 1$ or an n-tuple transition) (e.g., $K_{ba,\,da} = +d_n$; $K_{db,\,da} = -d_e$; $K_{da,\,fa} = +d_n$).

6.1.4. *The* $\mathbf{Q}$ *Vector*

$\mathbf{Q}$ is a vector in the space of induced transitions. Its elements are given simply by

$$Q_\lambda = q\omega_\lambda d_\lambda \tag{173}$$

where, again, $q = \hbar/kTA$, and ω_λ and d_λ are the resonance frequencies and transition moments for the λth transition. Thus, Q_λ is nonzero only for allowed transitions.

6.1.5. *Linewidth Matrix* $\mathbf{R}$

The $\mathbf{R}$ matrix is a matrix in the space of the induced transitions. It is determined by the general expressions for the linewidths (see Section 4.4.2 and Freed and Fraenkel[12]). It is only necessary to include linewidths for those transitions that couple into the problem of interest. This will usually imply that (1) it is some allowed or forbidden transition induced by $\mathbf{d}$, and/or (2) it is coupled to such a transition by off-diagonal elements of $\mathbf{R}$.

6.2. *Effects of Degenerate States and Transitions*

Very often in ESR spectroscopy, one encounters multiple hyperfine lines resulting from degenerate states. This is often the case if there are several (completely) equivalent nuclei in the radical. (The distinction between equivalent and completely equivalent nuclei is discussed by Freed and Fraenkel.[12] We shall, for simplicity here, only consider completely equivalent nuclei.) It is then possible to sum over such degeneracies so as to reduce the size of the matrices defined by equations (46) and (48)–(51).[30] Let us refer to a set of completely equivalent nuclei as the rth set and label the degenerate states and transitions by κ corresponding to a particular set of values of the $\{J_r\}$ and $\{M_r\}$. That is, a particular nuclear state is described by the configuration $|\{J_r\kappa\}, \{M_r\kappa\}\rangle$, where for the rth set of equivalent nuclei, the operators $J_r = \sum_{i\text{ in }r} I_i$ and $J_{rz} = \sum_{i\text{ in }r} I_{iz}$, where the sum is over all individual equivalent spins. The curly brackets refer to the collection of J_r

and M_r eigenvalues, respectively, for all the different sets of completely equivalent nuclei. (In the case of a particular NMR transition, we of course mean an $M_r \leftrightarrow M_r \pm 1$ transition). Thus, we define

$$Z_\lambda^{av} \equiv \sum_\kappa Z_{\lambda\kappa} \tag{174a}$$

$$X_\lambda^{av} \equiv \sum_\kappa X_{\lambda\kappa\pm} \tag{174b}$$

which are, respectively, sums over the z components for the λth degenerate transition, and the diagonal density matrix elements (actually their deviations from thermal equilibrium value) for the states between which these transition occur. (Note that we have *not* summed over all transitions, etc., corresponding to a particular hyperfine line regardless of differences in $\{J_r\}$ value. This was discussed in Ref. 30.) It was shown in Ref. 30 that the solution may be rewritten as

$$\mathbf{Z}^{s''} = (\mathbf{M}^s)^{-1}(-\mathbf{R}^s)^{-1}\mathbf{Q}^s \tag{175}$$

with

$$\mathbf{M}^s = \mathbf{1} + (\mathbf{R}^s\mathbf{K}^s)^2 + (-\mathbf{R}^s)^{-1}\mathbf{S}^s \tag{176}$$

and

$$\mathbf{S}^s = 2\mathbf{d}(\mathbf{W}^{j,s})^{-1}\mathbf{d}^{\text{tr }j} \tag{177}$$

Here,

$$\mathbf{Z}^s \equiv \mathbf{D}^{-1/2}\mathbf{Z}^{av} \tag{178}$$

where $\mathbf{D}^{1/2}$ is a diagonal matrix whose elements are made up from the degeneracies $D(\lambda)$ of the λth transition (corresponding to a particular configuration $\{J_r, M_r\}$).

$$D_{\lambda,\eta}^{1/2} \equiv [D(\lambda)]^{1/2}\,\delta_{\lambda,\eta} \tag{179a}$$

so

$$(\mathbf{D}^{1/2})_{\lambda,\eta}^{-1} \equiv (D^{-1/2})_{\lambda,\eta} = [D(\lambda)]^{-1/2}\,\delta_{\lambda,\eta} \tag{179b}$$

One has $\mathbf{Q}^s = \mathbf{D}^{1/2}\mathbf{Q}$. One also has

$$R_{\lambda,\eta}^s = (R_{\lambda,\lambda}^d - D(\lambda)R_{\lambda,\lambda}^{non})\,\delta_{\lambda,\eta} + R_{\lambda,\eta}^{non}[D(\lambda)D(\eta)]^{1/2} \tag{180}$$

where the diagonal elements $R_{\lambda\kappa,\lambda\kappa}^d$ obey

$$R_{\lambda\kappa,\lambda\kappa}^d = R_{\lambda,\lambda}^d \tag{181a}$$

for all the κ in λ (since the nuclei are completely equivalent), and the nondiagonal elements $R_{\lambda\kappa,\eta\rho}^{non}$, which only come from exchange effects (for the same reason), obey

$$R_{\lambda\kappa,\eta\rho}^{non} = R_{\lambda,\eta}^{non} = \omega_{EX}(2/A) \qquad \lambda_\kappa \neq \eta_\rho \tag{181b}$$

Note that for consistency we have

$$[R_{\lambda, \lambda}^{d\, \text{EX}} - D(\lambda)R_{\lambda, \lambda}^{\text{non, EX}}] = -\omega_{\text{EX}}[1 - D(\lambda)/\tfrac{1}{2}A] \qquad (182)$$

[This was not clearly given in Ref. 30, where the $D(\lambda)$ was left out on the left-hand side.] One also has

$$K_{\lambda\kappa, \eta\rho}^{s} = K_{\lambda, \eta}\, \delta_{\kappa, \rho} \qquad (183)$$

In the absence of exchange, one has

$$W_{\alpha, \beta}^{s} = W_{\alpha, \beta} = W_{\alpha\kappa, \beta\rho} \qquad \text{for all } \kappa \text{ and } \rho \qquad (184)$$

In the presence of exchange, it is more convenient to go over to the approach based upon equation (147). Then if we assume the sufficiency conditions given above equation (147), we have

$$\hat{W}_{\lambda, \eta}^{s} = \{[\hat{W}_{\lambda, \lambda}^{d} - D(\lambda)W_{\lambda, \lambda}^{\text{non, EX}}]\, \delta_{\lambda, \eta} + \hat{W}_{\lambda, \eta}^{\text{non, EX}}[D(\lambda)D(\eta)]^{1/2} + \hat{W}_{\lambda, \eta}^{\text{non, END}}\} \qquad (185)$$

where

$$[\hat{W}_{\lambda, \lambda}^{d,\, \text{EX}} - D(\lambda)\hat{W}_{\lambda, \lambda}^{\text{non, EX}}] = 2\omega_{\text{HE}}(\tfrac{1}{2} - D(\lambda)/A) \qquad (186a)$$

and

$$\hat{W}_{\lambda, \eta}^{\text{non, EX}} = -2\omega_{\text{HE}}/A \qquad (186b)$$

(which is also more clearly given than in Ref. 30).

When we use this second approach, then equation (177) becomes

$$\mathbf{S}^{s} = 4(\hat{\mathbf{d}}, \tilde{\mathbf{d}})\begin{bmatrix} (\hat{\mathbf{W}}^{s})^{-1} & \mathbf{0} \\ \mathbf{0} & (\mathbf{W}^{j,\, s})^{-1} \end{bmatrix}\begin{pmatrix} \hat{\mathbf{d}}^{\text{tr}} \\ \hat{\mathbf{d}}^{\text{tr}} \end{pmatrix} = 4[\hat{\mathbf{d}}(\hat{\mathbf{W}}^{s})^{-1}\hat{\mathbf{d}}^{\text{tr}} + \tilde{\mathbf{d}}(\tilde{\mathbf{W}}^{j,\, s})^{-1}\tilde{\mathbf{d}}^{\text{tr}\, j}] \qquad (187)$$

where the submatrices $\hat{\mathbf{d}}$ and $\tilde{\mathbf{d}}$ (as well as $\hat{\mathbf{W}}^{s}$ and $\tilde{\mathbf{W}}^{j,\, s}$) are defined in the basis of linear combinations: $\hat{\chi}_{\lambda} = \chi_{\lambda+} - \chi_{\lambda-}$; $\tilde{\chi}_{\lambda} = \chi_{\lambda+} + \chi_{\lambda-}$. [Note that the singularities in $\hat{\mathbf{W}}$ due to equation (29) now appear in $\tilde{\mathbf{W}}$ and it may be treated by the methods of the Appendix.] The construction of $\hat{\mathbf{d}}$ and $\tilde{\mathbf{d}}$ are similar to that of $\mathbf{d}$, and was given explicitly in Ref. 30.

In either approach, we may define the symmetrized saturation parameter by

$$S_{\lambda, \eta}^{s} \equiv d_{\lambda}d_{\eta}\Omega_{\lambda, \eta}^{s} \qquad (188)$$

It then follows from the above, that

$$Z_{\lambda\kappa} = Z_{\lambda} \qquad \text{independent of } \kappa \text{ in } \lambda \qquad (189a)$$

$$\chi_{\lambda\kappa\pm} = \chi_{\lambda\pm} \qquad \text{independent of } \kappa \text{ in } \lambda \qquad (189b)$$

so

$$Z_{\lambda}^{(s)} = D(\lambda)^{1/2}Z_{\lambda} \qquad (189c)$$

and

$$\chi_{\lambda}^{(s)} = D(\lambda)^{1/2}\chi_{\lambda} \qquad (189d)$$

and we observe, for the λth transition, $Z_\lambda^{av} = D(\lambda)^{1/2} Z_\lambda^{(s)}$. One may now obtain solutions as in the nondegenerate case.

In the case when equivalent nuclei are not completely equivalent, there will, in general, be off-diagonal couplings due to the END terms in both $\mathbf{W}$ and $\mathbf{R}$ between degenerate states and ESR transitions belonging to different values of J_r. This complicates the analysis,[6] but by the use of general symmetry arguments[31a] these difficulties may be minimized.*

7. Average ENDOR and ELDOR

The general multiple-level problem necessarily involves computer simulations. However, under certain limiting conditions, it is possible to obtain simple analytic expressions. This matter was discussed in detail in Ref. 30. We only wish to summarize the results in their simplest form here. We must make the following assumptions:

1. $\mathbf{K}^s$ is diagonal, i.e., the rf and microwave powers are weak enough that (ENDOR) coherence effects are negligible.

2. $\mathbf{R}^s$ is approximately diagonal (except for exchange coupling of degenerate transitions); i.e., (a) the END terms are not a very large component of the ESR or ENDOR linewidths and (b) ω_{EX} is small enough that the different ESR lines remain well separated.

7.1. Average ENDOR

Under the above conditions, it is possible to obtain a simple "average ENDOR" or "average ELDOR" result as though one had the simple four-level systems discussed earlier but with modified parameters. In particular, one has [see equation (78)] for average ENDOR of spins $I = \frac{1}{2}$:

$$E_v = \frac{\xi_{e,v}}{\Omega_{e,v} - \xi_{e,v}} \approx \frac{\xi_{e,v}}{\Omega_{e,v}} \approx \frac{\frac{1}{2} n_v d_{n_v}^2 h(b'') W_e\{0\}^{-2}}{1 + \Delta\omega_{n_v}^2 T_{2,n}^2 + \alpha d_{n_v}^2 b_v h(b'') W_e\{0\}^2} \quad (190)$$

where the vth set of n_v equivalent nuclei are near resonance, and from equation (61), $d_{n_v} = \frac{1}{2}\gamma_n(1 \pm r_n a_{n_v})B_\eta$, where $r_n \equiv (\gamma_e/\gamma_n)(1/2B_0) \approx 1/10$ at X band for protons; $h(b'') = 1 + \frac{1}{2}\omega_{EX}/W_e\{0\}$ $W_e\{0\}$ is the electron spin-flip rate that is independent of quantum number M_v, $b_v = \frac{1}{2}j_{vv}^D(0)/W_e\{0\}$ [see equation (95)]; $T_{2,n_v} \approx [W_e\{0\}h(b'')]^{-1}$. This is the very lowest-order result in the limit that

$$3n_v b_v / h(b'') \ll 1 \quad (191)$$

---

* Reference 30, footnote 25; and Ref. 31b. One, for example, avoids the problem of pseudotransition probabilities.

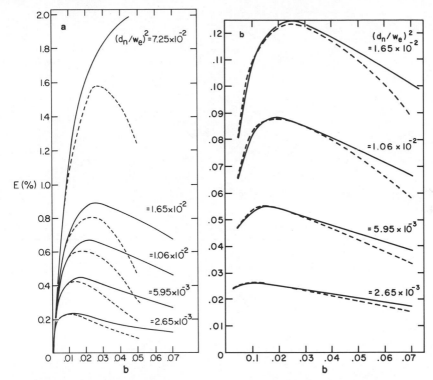

Figure 14. Comparison of the percent enhancements predicted by the average ENDOR approximation (dashed lines) and by the exact solutions (solid lines) for the case of four equivalent spins of $I = \frac{1}{2}$ given as a function of b for different values of $[d_n/W_e(0)]^2$. (a) The $M = 0$ ESR line is saturated, $b'' = 0$, $\Delta\omega_e = \Delta\omega_n = 0$, and $d_e^2 T_{2e}(M)\Omega_e(M) = 1$. (From Freed *et* al,[30]) (b) The $M = 0$ ESR line is saturated, $b'' = 0.08$, $\Delta\omega_e = \Delta\omega_n = 0$, and $d_e^2 T_{2e}(M)\Omega_e(M) = 1$.

The factor $\alpha = 1$ for a radical containing a single set of equivalent nuclei, and in the presence of other equivalent nuclei with small W_{n_r} ($\ll W_{n_v}$), but in the opposite limit of $W_{n_r} \gg W_{n_v}$, it could be reduced to about $\frac{1}{2}$ or $\frac{1}{3}$.[29] Thus, if the NMR transitions are not saturated, i.e., $d_{n_v}^2 \ll b_v h(b'')W_e$, the ratios of the ENDOR peak heights vary as $n_v d_{n_v}^2$ for this case of nuclear spins of $\frac{1}{2}$. In this limit, then, one could determine n_v, which is helpful in assigning the ENDOR transitions. The range of validity of the average ENDOR approach is shown in Figure 14, where the more complete formulas for average ENDOR are compared with the computer-calculated exact results for a case of four equivalent protons. The form of the average ENDOR expressions (e.g., that of a simple four-level system) has been found to be useful also in cases where the detailed average ENDOR theory is not applicable.[29] This is illustrated in

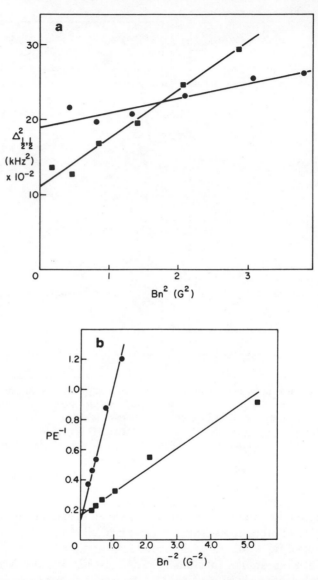

Figure 15. (a) Plots of $\Delta^2_{1/2, 1/2}$ as a function of B^2_n (as determined by the rf coherence effect) for PBSQ (●) and DSQ (■) in ethanol at $T = -30°C$. (b) Plot of PE^{-1} as a function of B^{-2}_n (as determined by the rf coherence effect) for PBSQ (●) and DSQ (■) in ethanol at $T = -30°C$.

Figure 15, where, for complex spin systems, the linear dependence of the square of the ENDOR width and the inverse dependence of the ENDOR enhancement on B_n^2 as predicted for the simple case of equations (76)–(78) is shown.

7.2. *Average ELDOR*

In the case of average ELDOR, we consider the results in the limit as $d_p^2 \to \infty$, and $d_o^2 \to 0$, [see equations (59) and (93)]. Then

$$R(d_p^2 \to \infty) \approx \frac{(D_p\{M_r'\}/D_o\{M_r\})^{1/2}\langle \Omega_{o,\,p}^s\{M_r, M_r'\}\rangle}{\langle \Omega_p^s\{M_r'\}\rangle} \qquad (192)$$

where the terms in angular brackets are the respective "average" saturation parameters associated with the pumped hyperfine line (nuclear configuration $\{M_r'\}$) and the observing line (nuclear configuration $\{M_r\}$). Then, to lowest order in the b_r and for $I = \frac{1}{2}$, one has

$$R(d_p^2 \to \infty) \approx (D_p\{M_r'\}\frac{\omega_{\text{EX}}}{AW_e} + \frac{1}{2}\sum_v [b_v/h(b'')][(\frac{1}{2}n_v \pm M_v)\delta_{M_v, M_v \mp 1}] \qquad (193)$$

The $\delta_{M_v, M_v \mp 1}$ terms are taken as nonzero only for combinations of $\{M_r\}$ and $\{M_r'\}$ such that only the vth set of equivalent nuclei differ in the observing and pump spin configurations by $\Delta M_v = \pm 1$, and all other sets of equivalent nuclei have the same M_r values. This configuration corresponds to an ELDOR "fundamental line." Overtone ELDOR lines, wherein the $\{M_r'\}$ of the pump line differs more significantly from the $\{M_r\}$ of the observing line, would yield reductions, which come in higher powers of the b_r, and are thus insignificant for $b_r/h(b'') \ll 1$. Thus, in the absence of exchange, in the limits for which average ELDOR applies, only fundamentals will be observed. However, if exchange is dominant, then the ELDOR reduction will depend simply on $D_p\{M_r'\}(\omega_{\text{EX}}/AW_e)$ and there are no selection rule restrictions upon which lines are observed or pumped.

8. *Saturation and Double Resonance in the Slow-Tumbling Region*

8.1. *General Considerations*

We would now like to briefly show how the general theory for saturation and multiple resonance in the motional narrowing region may be reformulated to cover the case of slow tumbling. We shall only consider a simple line for this purpose.[5,32a] Details for multiple lines are given elsewhere.[32-34] The basis for the slow tumbling theory is the stochastic Liouville expression for $\dot{\sigma}(\Omega, t)$[5,33,35-37]:

$$\dot{\sigma}(\Omega, t) = -i[\mathscr{H}_0^x + \varepsilon(t)^x + \mathscr{H}_1(\Omega)^x + iR' - i\Gamma_\Omega][\sigma(\Omega, t) - \sigma_{eq}(\Omega)] \qquad (194)$$

Here, Γ_Ω is the Markov operator for the motional process; we will specifically assume it is rotational tumbling, with Ω representing the Euler angles between the fixed laboratory coordinate frame and a molecule-based coordinate frame. Thus, the Euler angles Ω are fluctuating in time. Note that the expression is written for a $\dot{\sigma}(\Omega, t)$, which is both a spin density operator as well as a classical probability function in the values of the random variable Ω. We can recover the ordinary spin density matrix we have used until now by averaging over orientations:

$$\sigma(t) = \int d\Omega \; \sigma(\Omega, t)P_{eq}(\Omega) \equiv \langle P_{eq}(\Omega) | \sigma(\Omega, t) | P_{eq}(\Omega) \rangle \qquad (195)$$

where $P_{eq}(\Omega)$ is the equilibrium distribution of orientations and a convenient bra–ket notation is introduced. We have also included in equation (194) a term R', which is that part of the relaxation matrix that is orientation independent.

We note that the Markov operator $\Gamma\Omega$ has associated with it the expression

$$\frac{\partial}{\partial t} P(\Omega, t) = -\Gamma_\Omega P(\Omega, t) \qquad (196)$$

where $P(\Omega, t)$ is the probability of finding Ω at a particular state at time t. The process is assumed to be stationary, so that Γ is time independent, and also we have

$$\Gamma_\Omega P_{eq}(\Omega) = 0 \qquad (197)$$

We again can make use of equations (14), (15), and (18), for the power absorption, except that now,

$$Z'' = \int d\Omega \; Z(\Omega)'' P_{eq}(\Omega) \qquad (198)$$

where

$$Z(\Omega)'' = \sigma_{ba}(\Omega, t)e^{-i\omega t} \qquad (199)$$

We now introduce the normalized eigenfunctions of Γ_Ω, the $G_{KM}^L(\Omega)$:

$$\Gamma_\Omega G_{KM}^L(\Omega) = \tau_L^{-1} G_{KM}^L(\Omega) \qquad (200)$$

where $\tau_L^{-1} = RL(L+1)$, with R the Brownian rotational diffusion coefficient, which we have assumed is isotropic. Note that $\tau_2 = \tau_R$, which plays the principal role in motional narrowing theory. Also,

$$G_{KM}^L(\Omega) = \left[\frac{(2L+1)}{8\pi^2}\right]^{1/2} \mathscr{D}_{KM}^L(\Omega) \tag{201}$$

where $\mathscr{D}_{KM}^L(\Omega)$ are the Wigner rotation matrices or the generalized spherical harmonics. We then expand in this orthonormal basis set:

$$Z(\Omega) = \sum_{L, K, M} C_{KM}^L G_{KM}^L(\Omega) \tag{202a}$$

$$\chi(\Omega) = \sum_{L, K, M} b_{KM}^L G_{KM}^L(\Omega) \tag{202b}$$

where the C_{KM}^L and b_{KM}^L are expansion coefficients to be solved for. They are, in general, functions of $\Delta\omega$. For an isotropic liquid, $P_{eq}(\Omega) = 1/8\pi^2$, so the ESR signal is proportional to

$$\frac{1}{8\pi^2} \int \text{Im } Z(\Omega, \Delta\omega) \, d\Omega = \left(\frac{1}{8\pi^2}\right)^{1/2} \text{Im } C_{0, 0}^0(\Delta\omega) \tag{203}$$

8.2. Saturation: A Simple Line

We now take as our orientation-dependent perturbation[5]

$$\mathscr{H}_1(\Omega) = \mathscr{F}\mathscr{D}_{00}^2(\Omega)S_z \tag{204a}$$

with

$$\mathscr{F} = \tfrac{2}{3}(\beta_e \beta_0/\hbar)(g_{\parallel} - g_{\perp}) \tag{204b}$$

That is, we are assuming an axially symmetric g tensor, and we only consider the secular contribution. (The inclusion of the nonsecular contributions is discussed by Freed et al.[5]) We now take spin matrix elements of equation (194), as was done in equations (12) and (19). Then we use the eigenfunction expansions given by equation (202), premultiply through these equations by $G_{K'M'}^{L'}(\Omega)^*$, and integrate over Ω.[17] This yields a set of coupled algebraic equations for the C_{KM}^L and the b_{KM}^L:

$$[(\Delta\omega) - i(T_2^{-1} + \tau_L^{-1})]C_{00}^L(\Delta\omega) - \sum_{L'} \kappa_{L, L'} C_{0, 0}^{L'}(\Delta\omega) + 2^{1/2}db_{0, 0}^L$$

$$= q\omega_0 d\delta_{L, 0} \tag{205}$$

and

$$-i(T_1^{-1} + \tau_L^{-1})b_{0, 0}^L(\Delta\omega) + 2^{1/2}d \, \text{Im } C_{0, 0}^L(\Delta\omega) = 0 \tag{206}$$

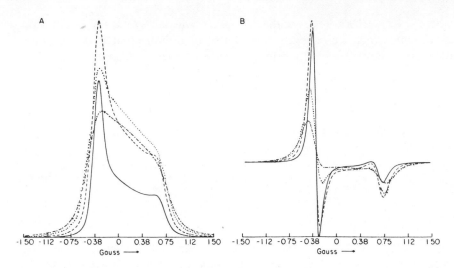

Figure 16. Saturation of single line with rotationally invariant T_1, as a function of B_1 for $|\mathscr{F}|/R = 100$. The different values of $B_1/2$ are 0.01 G (——), 0.025 G (– – –), 0.050 G (···), and 0.75 G (·). These correspond to intensity factors of 1414, 1052, 580, and 352, respectively. $g_{\parallel} = 2.00235$, $g_{\perp} = 2.00310$, $T_2 = T_1 = (2W_e)^{-1}$, and $(2/3^{1/2})T_2^{-1}/|\gamma_e| = 0.02$ G. (A) absorption; (B) derivative. (Intensity factor is integrated area under absorption curve in relative units.) (From Freed et al.[5])

Here, the orientation-independent T_2^{-1} and T_1^{-1} arise from R', also

$$\kappa_{L, L'} = [(2L + 1)(2L' + 1)]^{1/2} \begin{pmatrix} L & 2 & L' \\ 0 & 0 & 0 \end{pmatrix} \mathscr{F} \tag{207}$$

where $\begin{pmatrix} L & 2 & L' \\ 0 & 0 & 0 \end{pmatrix}$ is a $3j$ symbol[38] that obeys the triangle rule so that $L' = L$ or $L \pm 2$. All other symbols are as previously defined. We must now solve this coupled set of equations to obtain $\mathrm{Im}\, C_{0,\,0}^0(\Delta\omega)$ to obtain the spectrum, and this may readily be accomplished on a computer. One merely truncates the coupled equations at a high enough value of L to guarantee convergence. A typical computer simulation is shown in Figure 16.

We can, however, arrange these coupled equations in a convenient matrix array we have seen before by letting

$$(-T_2^{-1} + E_L)\delta_{L,\,L'} = R_{L,\,L'} \tag{208a}$$

$$\Delta\omega + \kappa_{L,\,L'} = K_{L,\,L'} \tag{208b}$$

also

$$(T_1^{-1} + \tau_L^{-1})\delta_{L,\,L'} = \hat{W}_{L,\,L'} \tag{208c}$$

and

$$q\omega_0 d\delta_{L,0} = Q_L \tag{208d}$$

Also, we can introduce a $\hat{\mathbf{d}}$ matrix and a $\hat{\mathbf{d}}^{\text{tr}}$ matrix, which by equations (205) and (206) only couple a C_{00}^L with a $b_{0,0}^L$.

In fact, once we make these substitutions we see that we again get the same formal matrix structure as equations (48) and (49), but with equation (49b) replaced by equation (187) and equation (48c) modified accordingly. Also, the $C_{0,0}^L$ form the $\mathbf{Z}$ vector, while the $b_{0,0}^L$ form the $\hat{\chi}$ vector. Note that in the present application, the $\hat{\mathbf{W}}$ matrix is already diagonal, as in the $\mathbf{R}$ matrix. The off-diagonal couplings arise from the $\mathbf{K}$ matrix.

8.3. ELDOR: A Simple Line

We can exploit this approach to consider ELDOR on this simple line. Here, we must replace equations (205) and (206) to yield[32]

$$[\Delta\omega_\alpha - i(T_2^{-1} + \tau_L^{-1})]C_{0,0}^L(\alpha) - \sum_{L'} \kappa_{L,L'} C_{00}^{L'}(\alpha) + 2^{1/2} dp b_{0,0}^L$$

$$= q\omega_e d_\alpha \delta_{L,0} \tag{209}$$

and

$$-i[T_1^{-1} + \tau_L^{-1}]b_{0,0}^L + \sqrt{2} \sum_{\alpha=o,p} d_\alpha \text{Im} C_{0,0}^L(\alpha) = 0 \tag{210}$$

where $\alpha = o$ or p, referring to observing or pumping signals. The observing absorption line shapes are given by $C_{0,0}^0(0) = C_{0,0}^0(0, \Delta\omega_o, \Delta\omega_p)$. This again yields our standard form for the coupled matrix equations. We show in Figure 17 a computer simulation of such a slow-tumbling ELDOR experiment for this case of a simple line.

The more complex equations one obtains when there are several spin eigenlevels are discussed elsewhere.[32-34] Also, ENDOR in the slow-tumbling region may be treated by similar methods.

The physically new feature here may be appreciated by first realizing that $b_{0,0}^0$, which is the average saturation, is just relaxed by T_1^{-1}, but the $b_{0,0}^L$ for $L > 0$, which represent nonspherically symmetric components of the saturation, are relaxed by the combination $T_1^{-1} + \tau_L^{-1}$; i.e., the rotational motion actually transfers the saturation, which is introduced at one point in the line by the pumping field to other points in the line, which may be observed with a weak observing field. This effect becomes more important as the ratio τ_L^{-1}/T_1^{-1} increases (i.e., the rotational motion slows down).

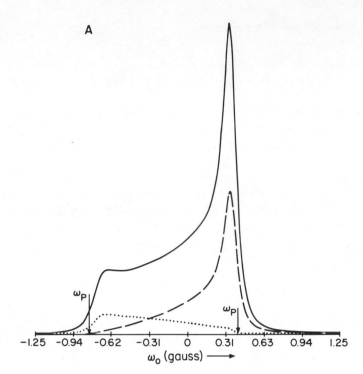

A

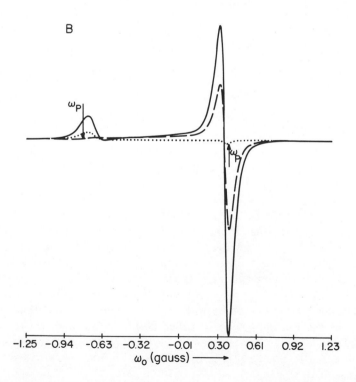

B

Appendix. General Properties of the Transition Probability Matrix $\mathbf{W}$

We first write the cofactor C_{ii} of the $\mathbf{W}$ matrix:

$$
C_{ii} \equiv
\begin{array}{c}
\quad\quad\quad 1 \quad\quad\quad\quad 2 \quad\quad\quad\; i \\
\begin{array}{c} 1 \\ \\ 2 \\ \\ \\ i \\ \\ \end{array}
\left|
\begin{array}{ccccc}
\sum_k W_{1k} & \cdots & -W_{12} & \cdots & 0 \quad \cdots \\
\\
-W_{21} & \cdots & +\sum_k W_{2k} & \cdots & 0 \quad \cdots \\
\\
\vdots & & \vdots & & \vdots \\
\\
0 & \cdots & 0 & \cdots & 1 \quad \cdots \\
\\
\vdots & & \vdots & & \vdots
\end{array}
\right|
\end{array}
\tag{A.1}
$$

Add all columns to the jth except the ith and use the property of $\mathbf{W}$ that the sums of all the columns are zero. Then add all rows to the jth except the ith and use the property of $\mathbf{W}$ that the sums of all the rows are zero (note $\mathbf{W}$ is symmetric). Then one has

$$
C_{ii} =
\begin{array}{c}
\quad\quad\quad\quad\quad\quad\quad\quad\quad\quad j \quad\quad\quad\; i \\
\begin{array}{c} \\ \\ \\ \\ j \\ \\ \\ i \\ \\ \end{array}
\left|
\begin{array}{ccccc}
\sum_k W_{1k} & -W_{12} & \cdots & W_{1i} & \cdots \; 0 \; \cdots \\
\\
-W_{21} & \sum_k W_{2k} & \cdots & W_{21} & \cdots \; 0 \; \cdots \\
\\
\vdots & \vdots & & \vdots & \vdots \\
\\
W_{1i} & W_{2i} & \cdots & \sum_k W_{ki} & \cdots \; 0 \; \cdots \\
\\
\vdots & \vdots & & \vdots & \vdots \\
\\
0 & 0 & \cdots & 0 & \cdots \; 1 \; \cdots \\
\\
\vdots & \vdots & & \vdots & \vdots
\end{array}
\right|
\end{array}
\tag{A.2}
$$

Figure 17. Observing frequency-sweep ELDOR line shapes for an axially symmetric g tensor undergoing isotropic brownian rotational diffusion with $\tau_R = 2.3 \times 10^{-6}$ sec. (A) absorption, (B) first derivative; (——) pure ESR, ($\cdots$) pump on and $\omega_p|\gamma_e| = 0.4$ G; (– – –) pump on and $\omega_p|\gamma_e| = -0.8$ G. All have $g_\parallel = 2.0235$, $g_\perp = 2.00310$, $B_0 = 3300$ G, $T_1^{-1} = 1.76 \times 10^{+5}$ sec^{-1}, $(2/3)^{1/2}T_1^{-1}|\gamma_e| = 0.02$ G. (From Bruno and Freed.[32])

Now permute the ith and jth rows, then the ith and jth columns. It immediately follows that $C_{ii} = C_{jj}$.

Now consider C_{ij}:

$$C_{ij} \equiv \begin{array}{c} \\ \\ \\ \\ j \\ \\ \\ i \\ \\ \end{array} \begin{vmatrix} \sum_k W_{1k} & -W_{12} & 0 & -W_{1i} & \cdots \\ -W_{21} & \sum_k W_{2k} & 0 & -W_{2i} & \cdots \\ \vdots & \vdots & \vdots & \vdots & \\ -W_{j1} & -W_{j2} & \cdots\ 0\ \cdots & -W_{ji} & \cdots \\ \vdots & \vdots & \vdots & \vdots & \\ 0 & 0 & \cdots\ 1\ \cdots & 0 & \cdots \\ \vdots & \vdots & \vdots & \vdots & \end{vmatrix} \qquad (A.3)$$

Now add all columns to the ith except the jth. Then

$$C_{ij} = \begin{array}{c} \\ \\ \\ \\ j \\ \\ \\ i \\ \\ \end{array} \begin{vmatrix} \sum_k W_{1k} & -W_{12} & \cdots\ 0\ \cdots & W_{1j} & \cdots \\ -W_{21} & \sum_k W_{2k} & \cdots\ 0\ \cdots & W_{2j} & \cdots \\ \vdots & \vdots & \vdots & & \\ -W_{j1} & -W_{j2} & \cdots\ 0\ \cdots & -\sum_k W_{jk} & \cdots \\ \vdots & \vdots & \vdots & & \\ 0 & 0 & \cdots\ 1\ \cdots & 0 & \cdots \\ \vdots & \vdots & \vdots & \vdots & \end{vmatrix} \qquad (A.4)$$

Then exchange the ith and jth columns (an odd permutation) to show that $C_{ij} = C_{ii}$.

It thus follows that

$$C_{ij} = C_{ii} = C_{jj} = C \qquad \text{for all } i \text{ and } j \qquad (A.5)$$

$$\therefore \quad |W^k| = \sum_j C_{kj} = AC \qquad (A.6)$$

where A is the dimension of **W**.

We now consider

$$
C_{jl}^i \equiv
\begin{array}{c}
\\
\\
\\
\\
i \\
j \\
\\
\end{array}
\left|
\begin{array}{cccc}
& & & l \\
\sum_k W_{1k} & -W_{12} & \cdots & 0 & \cdots \\
-W_{21} & \sum_k W_{2k} & \cdots & 0 & \cdots \\
\vdots & \vdots & & \vdots \\
1 & 1 & \cdots & 0 & \cdots \\
0 & 0 & \cdots & 1 & \cdots \\
\vdots & \vdots & & \vdots
\end{array}
\right|
\qquad (A.7)
$$

and

$$
C_{il}^j \equiv
\begin{array}{c}
\\
\\
\\
\\
i \\
\\
j \\
\\
\end{array}
\left|
\begin{array}{cccc}
& & & l \\
\sum_k W_{1k} & -W_{12} & \cdots & 0 & \cdots \\
-W_{21} & \sum_k W_{2k} & \cdots & 0 & \cdots \\
\vdots & \vdots & & \vdots \\
0 & 0 & \cdots & 1 & \cdots \\
\vdots & \vdots & & \vdots \\
1 & 1 & \cdots & 0 & \cdots \\
\vdots & \vdots & & \vdots
\end{array}
\right|
\qquad (A.8)
$$

Now exchange the ith and jth rows (an odd permutation) of either one to show

$$C_{jl}^i = -C_{il}^j \qquad (A.9)$$

We now wish to prove that

$$C_{il}^k = C_{jl}^k - C_{ji}^k \tag{A.10}$$

We start with

$$
C_{il}^k \equiv
\begin{array}{c}
\\
\\
\\
k \\
\\
j \\
\\
i \\
\\
\end{array}
\begin{vmatrix}
\sum_m W_{1m} & -W_{12} & \cdots & -W_{1k} & \cdots & 0 & \cdots & -W_{1i} & \cdots \\
-W_{21} & \sum_m W_{2m} & \cdots & -W_{2k} & \cdots & 0 & \cdots & -W_{2i} & \cdots \\
\vdots & \vdots & & \vdots & & \vdots & & \vdots & \\
1 & 1 & & 1 & & 0 & & 1 & \cdots \\
\vdots & \vdots & & \vdots & & \vdots & & \vdots & \\
-W_{j1} & -W_{j2} & \cdots & -W_{jk} & \cdots & 0 & \cdots & -W_{ji} & \cdots \\
\vdots & \vdots & & \vdots & & \vdots & & \vdots & \\
0 & 0 & & 0 & \cdots & 1 & \cdots & 0 & \cdots \\
\vdots & \vdots & & \vdots & & \vdots & & \vdots & \\
\end{vmatrix}
\tag{A.11}
$$

(with columns labeled k, l, i)

One now adds all rows to the jth except the kth and ith to obtain
$C_{il}^k =$

$$
\begin{array}{c}
\\
\\
\\
k \\
\\
l \\
\\
i \\
\\
\end{array}
\begin{vmatrix}
\sum_m W_{1m} & -W_{12} & \cdots & -W_{1k} & \cdots & 0 & \cdots & -W_{1i} & \cdots \\
-W_{21} & \sum_m W_{2m} & \cdots & -W_{2k} & \cdots & 0 & \cdots & -W_{2i} & \cdots \\
\vdots & \vdots & & \vdots & & \vdots & & \vdots & \\
1 & 1 & \cdots & 1 & & \cdots 0 \cdots & & 1 & \cdots \\
\vdots & \vdots & & \vdots & & \vdots & & \vdots & \\
W_{k1}+W_{i1} & W_{k2}+W_{i2} & \cdots & -\sum_m W_{km}+W_{ki} & \cdots & 0 & \cdots & -\sum_m W_{im}+W_{ki} & \cdots \\
\vdots & \vdots & & \vdots & & \vdots & & \vdots & \\
0 & 0 & \cdots & 0 & & \cdots 1 \cdots & & 0 & \cdots \\
\vdots & \vdots & & \vdots & & \vdots & & \vdots & \\
\end{vmatrix}
$$

(with columns labeled k, l, i)

$$\tag{A.12}$$

Thus, C_{il}^k may be written as the sum of two determinants, in which the first determinant has as it lth row:

$$\left[W_{ki} \quad W_{k2} \quad \cdots \quad -\sum_m W_{km} \quad \cdots \quad 0 \quad \cdots \quad W_{ki} \quad \cdots \right]$$

while the second has as its lth row:

$$\left[W_{i1} \quad W_{i2} \quad \cdots \quad W_{ik} \quad \cdots \quad 0 \quad \cdots \quad -\sum_m W_{im} \quad \cdots \right]$$

while all other elements are as in equation (A.12). Now, by permuting the jth and kth rows of the first determinant, we find that it equals C_{il}^j, which, from equation (A.9), is equal to $-C_{jl}^i$. Then, by permuting the ith and kth rows of the second determinant, we find that it equals C_{jl}^k. It thus follows that $C_{il}^k = C_{jl}^k - C_{jl}^i$.   QED

We now wish to prove the relation

$$C_{ki}^l - C_{kj}^l = AC_{kl,\,ij} \tag{A.13}$$

Let us first consider

$$C_{kj}^l \equiv \begin{array}{c} \\ \\ \\ \\ k \\ \\ \\ l \\ \\ \end{array} \begin{vmatrix} \sum_m W_{1m} & -W_{12} & \cdots & \overset{i}{-W_{1i}} & \cdots & \overset{j}{0} & \cdots \\ -W_{21} & \sum_m W_{2m} & \cdots & -W_{2i} & \cdots & 0 & \cdots \\ \vdots & \vdots & & \vdots & & \vdots & \\ 0 & 0 & \cdots & 0 & \cdots & 1 & \cdots \\ \vdots & \vdots & & \vdots & & \vdots & \\ 1 & 1 & \cdots & 1 & \cdots & 0 & \cdots \\ \vdots & \vdots & & \vdots & & \vdots & \end{vmatrix} \tag{A.14}$$

Now add all the columns of C^l_{kj} to the ith except for the jth and then permute the ith and jth columns:

$$
C^l_{kj} = \quad
\begin{array}{c}
\\
\\
\\
\\
k \\
\\
l \\
\\
\end{array}
\left|
\begin{array}{cccc}
\overset{i}{} & & \overset{i}{} & \overset{j}{} \\
\sum_m W_{1m} & -W_{12} & \cdots \; 0 \; \cdots & -W_{1j} \; \cdots \\[2mm]
-W_{21} & \sum_m W_{2m} & \cdots \; 0 \; \cdots & -W_{2j} \; \cdots \\[2mm]
\vdots & \vdots & \vdots & \vdots \\[2mm]
0 & 0 & \cdots \; 1 \; \cdots & 0 \; \cdots \\[2mm]
\vdots & \vdots & \vdots & \vdots \\[2mm]
1 & 1 & \cdots \; 0 \; \cdots & 1-A \; \cdots \\[2mm]
\vdots & \vdots & \vdots & \vdots \\
\end{array}
\right|
\qquad (A.15)
$$

Now consider

$$
C^l_{kj} \equiv \quad
\begin{array}{c}
\\
\\
\\
\\
k \\
\\
l \\
\\
\end{array}
\left|
\begin{array}{cccc}
& & \overset{i}{} & \overset{j}{} \\
\sum_m W_{1m} & -W_{12} & \cdots \; 0 \; \cdots & -W_{1j} \; \cdots \\[2mm]
-W_{21} & \sum_m W_{2m} & \cdots \; 0 \; \cdots & -W_{2j} \; \cdots \\[2mm]
\vdots & \vdots & \vdots & \vdots \\[2mm]
0 & 0 & \cdots \; 1 \; \cdots & 0 \; \cdots \\[2mm]
\vdots & \vdots & \vdots & \vdots \\[2mm]
1 & 1 & \cdots \; 0 \; \cdots & 1 \; \cdots \\[2mm]
\vdots & \vdots & \vdots & \vdots \\
\end{array}
\right|
\qquad (A.16)
$$

We now perform the operation $C_{ki}^l - C_{kj}^l$ by simply subtracting the elements of the jth columns of equation (A.15) from (A.16) to obtain

$$
C_{ki}^l - C_{kj}^l = \quad
\begin{array}{c}
\\ \\ \\ k \\ \\ l \\ \\
\end{array}
\begin{array}{c}
 \\
\end{array}
\left|
\begin{array}{ccccccccc}
 & & & & & i & & j & \\
\sum_m W_{1m} & \cdots & -W_{12} & \cdots & 0 & \cdots & 0 & \cdots \\
-W_2 & \cdots & \sum_m W_{2m} & \cdots & 0 & \cdots & 0 & \cdots \\
\vdots & & \vdots & & \vdots & & \vdots & \\
0 & \cdots & 0 & \cdots & 1 & \cdots & 0 & \cdots \\
\vdots & & \vdots & & \vdots & & \vdots & \\
1 & \cdots & 1 & \cdots & 0 & \cdots & A & \cdots \\
\vdots & & \vdots & & \vdots & & \vdots & \\
\end{array}
\right|
$$

$$
= A \times \quad
\begin{array}{c}
\\ \\ \\ k \\ \\ l \\ \\
\end{array}
\left|
\begin{array}{ccccccccc}
 & & & & & i & & j & \\
\sum_m W_{1m} & & -W_{12} & \cdots & 0 & \cdots & 0 & \cdots \\
-W_{21} & & +\sum_m W_{2m} & \cdots & 0 & \cdots & 0 & \cdots \\
\vdots & & \vdots & & \vdots & & \vdots & \\
0 & & 0 & \cdots & 1 & \cdots & 0 & \cdots \\
\vdots & & \vdots & & \vdots & & \vdots & \\
0 & & 0 & \cdots & 0 & \cdots & 1 & \cdots \\
\vdots & & \vdots & & \vdots & & \vdots & \\
\end{array}
\right|
\qquad \text{(A.17)}
$$

$$
\equiv \quad AC_{kl,\,ij} = AC_{lk,\,ji} = -AC_{kl,\,ji} = -A_{lc,\,ij} \qquad \text{(A.17')}
$$

The last equalities follow from simple permutations of equations (A.17). QED

It also follows from equation (A.17) that

$$
C_{kl,\,ij} = C_{ij,\,kl} \qquad \text{(A.18)}
$$

because we can interchange rows and columns since $\mathbf{W}$ is symmetric.

Another important identity is

$$C_{ik,\,ij} - C_{il,\,ij} = C_{lk,\,ij} \qquad (A.19)$$

The proof is as follows. We have from equation (A.13) that

$$AC_{ik,\,ij} = C_{ii}^k - C_{ij}^k \qquad (A.20a)$$

$$AC_{il,\,ij} = C_{ii}^l - C_{ij}^l \qquad (A.20b)$$

Therefore,

$$
\begin{aligned}
A(C_{ik,\,ij} - C_{il,\,ij}) &= -(C_{ii}^l - C_{ii}^k) + (C_{ij}^l - C_{ij}^k) \\
&= -C_{ki}^l + C_{kj}^l = -AC_{kl,\,ij} = AC_{lk,\,ij} \qquad (A.21)
\end{aligned}
$$

Furthermore, one may show, utilizing a theorem of Ledermann,[39] that C, $C_{ij,\,ij}$, and $C_{ij,\,ik}$ are always positive quantities, but no *a priori* statement can be made about $C_{ij,\,kl}$ where $i \neq k$, $j \neq l$.

We now summarize these important identities:

(1) $C_{ij} = C_{ii} = C_{jj} = C$ for all i, j

(2) $C_{jl}^i = -C_{il}^j = C_{jl}^k - C_{il}^k,$ $i \neq j,\, k$ and $j \neq k$

(3) $C_{ki}^l - C_{kj}^l = AC_{kl,\,ij},$ $l \neq k$ and $i \neq j$ (A.22)

(4) $C_{ik,\,ij} - C_{il,\,ij} = C_{lk,\,ij},$ $l \neq k$ and $i \neq j,\, k,\, l$

(5) $C_{lk,\,ij} = -C_{lk,\,ij} = C_{lk,\,ji} = C_{ij,\,kl},$ $l \neq k$ and $i \neq j$

Notation

The numbers in parentheses next to each definition represent equation numbers.

a, b, a', b'	Specific eigenstates of spin Hamiltonian (8)
$a, \bar{a}, \bar{a}_n, a(t)$	Isotropic hyperfine splittings; the overbar explicitly indicates a time average, the subscript n indicates the splitting for the nth set of nuclei (42, 61, 120)
a	Hydrodynamic radius of radical of interest (114, 138)
A	Total number of spin eigenstates of spin Hamiltonian $\mathscr{H}_0$ (6)
$\mathbf{A}$	Inverse moment of inertia tensor (114)
b	Dimensionless ratio: W_n/W_e (94')
b''	Dimensionless ratio: ω_{HE}/AW_e (149')
$b_{km}^L(\Delta\omega)$	Coefficient in the expansion of $\chi(\Omega)$ in eigenfunctions of the diffusion operator; it is still a function of $\Delta\omega$ (202)
B_1	Strength of the rotating rf or microwave field (1, 11)
B_o, B_p, B_e, B_n	Value of B_1 for observing or pumping mode in ELDOR (41) and the value for the microwave and rf modes in ENDOR (41)
B_0	Strength of the dc magnetic field (27)

$C_{lk} = C$	lkth cofactor of $\mathbf{W}$ and is equal to C independent of values of l and k (34, 37)
$C_{ij,kl}$	Double cofactor of $\mathbf{W}$ (35)
C_{ki}^l	kith cofactor of $\mathbf{W}^l$ (36)
C	Spin-rotational constant of the radical (113)
$C_{KM}^L(\Delta\omega)$	Coefficient in the expansion of $Z(\Omega)$ in eigenfunctions of the diffusion operator; it is still a function of $\Delta\omega$ (202).
$d = \frac{1}{2}\gamma_e B_1$	Induced transition moment due to applied rf or microwave field (13)
$d_o, d_p, d_e,$ d_n, d_λ	Value of d for observing or pumping mode in ELDOR (44) and the value for the microwave and rf modes in ENDOR (61, 64); λ represents the value of d for the λth induced transition (173)
d_{n_0}	Nuclear spin transition moment neglecting the high-field correction (61)
d_n^{crit}	Critical value of d_n for coherence splittings in ENDOR (83)
$\mathbf{d}, \mathbf{d}^{\text{tr}}$	Matrix of transition moments and its transpose (46)
$\hat{\mathbf{d}}, \tilde{\mathbf{d}}$	Partitioned components of $\mathbf{d}$ (187)
d	Interaction distance for exchange (138)
$D^{(m)}$	mth irreducible-tensor component of the electron–nuclear dipolar coefficients (105)
D	Coefficient for translational diffusion of a radical (138)
$D(\lambda)$	Degeneracy of the λth transition (140)
$\mathbf{D}^{1/2}$	Diagonal matrix whose elements are the $[D(\lambda)]^{1/2}$ (178)
$\mathscr{D}_{KM}^L(\Omega)$	Generalized spherical harmonics (Wigner rotation matrices) (201)
E	Enhancement factor in ENDOR (78)
f	Debye–Hückel correction to rate of bimolecular collisions (138)
$\mathscr{F} = \frac{2}{3}(\beta_e B_o/\hbar)(g_{\parallel} - g_{\perp})$	(204)
$\mathbf{g}, \Delta\mathbf{g}$	g tensor and deviation of g tensor from free electron value g_e [below (114)]
$g_s, \bar{g}_s$	g factor equal to Tr g. The overbar explicitly indicates a time average (111, 125)
$g_k, g_{\parallel}, g_{\perp}, g^{(m)}$	g tensor components $k = x$, y, or z in its principal axis system, $g_{\parallel} = g_z$ and $g_{\perp} = g_x = g_y$ for axial symmetry; $g^{(m)}$ are irreducible tensor components of g (111, 112, 125)
$G_{km}^L(\Omega)$	Normalized eigenfunctions of Γ_Ω, the diffusion operator (200)
$h(b'')$	$= 1 + \frac{1}{2}\omega_{EX}/W\{0\}$ (190)
$\mathscr{H}_0, \mathscr{H}_0^x$	Zero-order spin Hamiltonian, and the superscript x indicates the superoperator form, e.g., $\mathscr{H}_0^x\sigma = [\mathscr{H}_0, \sigma]$ (5, 131)
$\mathscr{H}_i(t), \langle\mathscr{H}_1(t)\rangle, \mathscr{H}_1^x$	Randomly modulated perturbation term in the spin Hamiltonian, its ensemble averaged value, and its superoperator form [below (27, 123, 194)]
$\mathscr{H}_J, \mathscr{H}_J^x$	Heisenberg spin-exchange term in the spin Hamiltonian and its superoperator form (132, 133)
I	Nuclear spin quantum number for single nucleus [above (85)]
I	moment of inertia of radical (113)
$j^D(\omega), j^{G_2}(\omega), j^{DG_2}(\omega),$ $j^Q(\omega), j^I(\omega), J^{G_0}(\omega),$ $J^{IG_0}(\omega)$	Spectral density as a function of ω from random modulation of: D, electron–nuclear dipolar interaction (END) (103); G_2, g tensor (111); DG_2, cross-term between END and g tensor (112); Q, quadrupolar interaction (106); I, isotropic dipolar (118); G_0, g shift (125); IG_0, cross-term between isotropic dipolar and g-shift (126)
J	Total nuclear spin quantum number for group of equivalent nuclear spins [above (65)]

J	Twice the exchange integral (133)
k	Boltzmann's constant (6)
$\mathbf{K}, \mathbf{K}^s$	Coherence matrix and its symmetrized form in the presence of degenerate transitions (46, 176)
L	Principal quantum number for generalized spherical harmonics (200)
M_0	Equilibrium magnetization (1)
$M_x, M_y,$ $M_\pm = M_x \pm iM_y$	Components of magnetization in the laboratory frame (2)
$\tilde{M}_x, \tilde{M}_y$	Components of magnetization in the rotating frame (1)
M_s	Quantum number for projection of electron spin on z axis (8)
M, M_a, M_b	Quantum number for projection of nuclear spin on z axis; the subscripts a or b refer to the ath or bth spin state [above (103, 124)]
$\{M_I\}$	A particular configuration of the nuclear spin states in a multi-nuclear spin problem (161)
$\mathbf{M}, \mathbf{M}^s$	Basic matrix in the solution of multiple resonance; the superscript s indicates the symmetrized form in the presence of degeneracies (49, 175)
$\mathscr{M}_\pm$	Quantum mechanical operator for the magnetization (3)
n_v	Number of equivalent nuclei in the vth set of equivalent nuclei (190)
$\mathscr{N}$	Concentration of radicals with electron spin $S = \frac{1}{2}$ (3)
P	Power absorbed from the resonant field (14)
P_i	Population of the ith state $= \sigma_{ii}$ [below (59)]
$P(\Omega, t), P_{eq}(\Omega)$	Classical time-dependent distribution function for orientation of radical and its time-independent equilibrium value (195, 196)
$\mathscr{P}, \mathscr{P}^x$	Operator that permutes electron spins and its superoperator form (135–137)
$q = \hbar/kTA$	(24)
$\mathbf{Q}$	Vector, in transition space, of the driving terms (47)
eQ	Quadrupole coupling constant (107)
r'	Radial distance of the unpaired electron with respect to a nucleus (105)
r_n	Dimensionless parameter measuring importance of high-field correction to d_n [below (190)]
$R, R_{\alpha\alpha'\beta\beta'}$	Relaxation matrix and its general four-indexed matrix element (5)
$R_{ba, ba}, R_{\lambda\lambda} = R_\lambda$	Diagonal element of relaxation matrix for the linewidth of the transition between states a and b (10); also its general form for the λth transition (180)
$R_{bb, aa}, R_{aa, aa}$	Relaxation matrix elements that yield the transition probabilities (19)
$\mathbf{R}, \mathbf{R}^s, R_{\lambda, \eta}$	Matrix yielding the (coupled) linewidths, its symmetrized form, and the matrix element coupling the λth and ηth transitions (46, 175, 180)
R	Reduction factor in ELDOR (93)
R'	That part of the relaxation matrix R that does not arise from rotational tumbling (194)
R	Rotational diffusion coefficient (200)
$S_\pm$	Raising and lowering operators for the electron spin (4)
$\mathbf{S}, \mathbf{S}^s, S_{\lambda, \eta}$	Saturation matrix, its symmetrized form, and the matrix element coupling the λth and ηth transitions (49, 51, 176)
t	Time (2)

T	Kelvin temperature (6)
T_1	Longitudinal spin–relaxation time for a simple two-level spin system (1)
$T_2^{-1}, T_{2o}^{-1}, T_{2p}^{-1},$ $T_e^{-1}, T_n^{-1}, T_x^{-1}$	Simple linewidths contained in **R**, where o and p are for observing and pumping transitions and e, n, and x are for electron spin, nuclear spin, and cross-transitions, and the subscript 2 is often dropped (1, 44, 64)
$T'_{2ab} - 1$	Secular contribution to the linewidth for the transition between states a and b (122)
Tr, Tr$_s$	Trace over spin states (and symmetrized trace for interacting dimers) (4, 13)
U, U_a	A vector and its element as used in solving for the saturation of a simple line (30)
$\mathbf{U}^l$	Obtained from **U** by replacing the lth element by zero (30')
V_{ba}	Line-shape term for simple transition between a and b (31)
V'_{ij}	V is the electrostatic potential at the nucleus and $V'_{ij}, i, j = x', y'$, or z', indicates differentiation of V with respect to i and j (108)
W_{ab}, $Wa_{a\gamma}$	Transition probability from state b to state a (and from arbitrary state γ to state a) leading to spin relaxation (20)
W_e, W_n, W_x	Transition probabilities for: e, pure electron spin flips; n, pure nuclear spin flips; and x, combined electron spin and nuclear spin flips; cross-transitions (26, 93)
W_e^{SR}, $W_e\{0\}$	Spin-rotational contribution to W_e and the value of W_e for electron spin flips in the nuclear configuration specified by $\{0\}$ (109, 190)
W, $\mathbf{W}^l$, $\mathbf{W}^{l,s}$	Transition probability matrix; the modified transition probability matrix obtained by replacing the lth row of **W** with ones; the symmetrized form of $\mathbf{W}^l$ when one sums over degenerate states (30, 30', 177)
$\hat{\mathbf{W}}$, $\tilde{\mathbf{W}}$, $\hat{\mathbf{W}}^s$, $\tilde{\mathbf{W}}^s$	Partitioned submatrices of a rearranged **W** and their symmetrized forms when one sums over degenerate states (147, 187)
$Y_{2,m}(\theta', \phi')$	Second-rank spherical harmonics (105)
$Z = Z' + iZ''$, Z^*	Function describing the induced transition, where Z' is the dispersive component while Z'' is the absorptive component and Z^* is the complex conjugate (15, 17, 18, 26)
Z''_o, Z''_p	Absorptive components for observing and pumping modes in ELDOR (44, 45)
Z''_{ENDOR}, Z''_{ESR}	Absorptive components in presence and absence of resonant NMR field in ENDOR (78)
$\mathbf{Z} = \mathbf{Z}' + i\mathbf{Z}''$	Vector of induced transitions (46)
$\mathbf{Z}^{\text{av}}$, $\mathbf{Z}^s$	Vector of elements Z_λ^{av} involving a sum over all κ elements of $Z_{\lambda\kappa}$ belonging to the λth degenerate transition; $\mathbf{Z}^s$ is the symmetrized form of $\mathbf{Z}^{\text{av}}$ (174, 178)
α, α', β, β'	Arbitrary spin states (5b)
α	Correction factor in average ENDOR (190)
β_e	Bohr magneton (111)
γ_e, γ_n	Gyromagnetic ratio: e, electronic; n, nuclear (1, 60)
Γ_Ω	Markovian (diffusion) operator for the motional process (194)
$\Delta_e \equiv \Delta\omega_e$, $\Delta_n \equiv \Delta\omega_n$	(62)
$\varepsilon(t)$	Term in the spin Hamiltonian that includes the interaction of the spins with the oscillating fields (5)
$[\nabla\varepsilon]^{(m)}$	Irreducible tensor components of the electric-field gradient (108)
η	Solvent viscosity (114)

θ' — Polar angle for defining the orientation in the molecular frame of the radial vector from the nucleus to the unpaired electron distribution (105)

κ — Label of degenerate nuclear spin states corresponding to the same value of J and M [above (103)]

$\kappa_{L,L'}$ — L, L'th matrix element for "coherence matrix" in slow tumbling (205)

ξ_o — Term representing effect of ELDOR on observing ESR signal for simple case (54)

ξ_e, ξ_e^r — Term representing effect of ENDOR on observing ESR signal for simple case; superscript r implies the NMR transition is exactly on resonance (69, 75)

ρ — Spin-density matrix for an interacting pair of dimers (131)

$\sigma(t)$, $\sigma_{\alpha\alpha'}$, $\sigma_{\alpha\alpha'}^*$ — Spin-density matrix for the radicals, the α–α'th matrix element, and its Hermitian conjugate (3, 5, 17)

σ_{eq}, $\sigma_{eq,\,\alpha\alpha'}$ — Equilibrium spin-density matrix and its $\alpha\alpha'$th matrix element (5)

$\sigma(\Omega, t)$ — Combined molecular classical orientation distribution function and spin-density matrix (194)

$\sigma_{eq}(\Omega)$ — Equilibrium value of $\sigma(\Omega, t)$ (194)

τ_c — Correlation function for random process (27)

τ_R, τ_L — Correlation functions for rotational reorientation (104, 200)

τ_J — Correlation function for angular momentum relaxation (113)

τ_1 — Mean lifetime of interacting radical pairs (132)

τ_2 — Mean time between bimolecular encounters (131)

ϕ' — Azimuthal angle for defining the orientation in the molecular frame of the radial vector from the nucleus to the unpaired electron distribution (105)

χ, χ_a — Deviation of σ from σ_{eq} and its diagonal element for state a (222)

$\boldsymbol{\chi}$ — Column vector of elements χ_i (30)

$\chi_{\alpha\pm}$, $\chi_\pm$ — Value of χ for state $M_s = \pm$ and $\{M_I\} = \alpha$; also $\chi_\pm$ is a normalized sum of $\chi_{\alpha\pm}$ (142, 143)

$\hat{\boldsymbol{\chi}}$ — A partitioned subvector of $\boldsymbol{\chi}$ whose λth element equals $\chi_{\lambda+} - \chi_{\lambda-}$ [(142), below (187)]

$\chi_{\lambda\pm}^{av}$ — Sum over all κ elements of $\chi_{\lambda\kappa\pm}$ belonging to the same degenerate state. $\lambda_\pm$ (174)

ψ_e — Electronic wave function of the unpaired electron (105)

$\omega_{\alpha\alpha'}$; ω_λ — Larmor frequency for $\alpha \leftrightarrow \alpha'$ transition and for λth transition (5, 173)

ω_0 — Larmor frequency for ESR transition (12)

ω_o, ω_p — Frequency of applied observing and pumping microwave fields in ELDOR (44)

ω_e, ω_n — Frequency of applied ESR and NMR fields in ENDOR (62)

$\omega^{a,\,b}(t)$ — Fluctuating component of frequency difference between states a and b (122)

ω_{HE}, ω_{EX} — Heisenberg exchange frequency and general exchange frequency including both Heisenberg exchange and chemical exchange [(135), below (144)]

$\Delta\omega \equiv \omega - \omega_e$ — $= \omega - \omega_0$, with ω the applied microwave frequency (1, 16, 44)

$\Delta\omega_\lambda = \omega_i - \omega_\lambda$ — Difference between frequency of ith rotating field and Larmor frequency for λth transition that is nearly resonant with ω_i [below (173)]

$\Delta\omega_o, \Delta\omega_p$ Near-resonant frequency difference for observing and pumping fields in ELDOR (44)

$\Delta\omega_e, \Delta\omega_n$ Near-resonant frequency difference for ESR and NMR fields in ENDOR (see Δ_e, Δ_n)

$\Omega_{ij, kl}\Omega_{\lambda, \eta}$ General saturation parameter coupling the $(i \leftrightarrow j)$th transition to the $(k \leftrightarrow l)$th transition and coupling the λth and ηth transitions (35, 41)

$\Omega_{ba, ba} = \Omega_{ba}$ Diagonal saturation parameter, which plays a role related to T_1 for steady-state saturation of the $a \leftrightarrow b$ transition (31, 39)

Ω Euler angles defining the transformation between laboratory and molecular coordinate frames (194)

ACKNOWLEDGMENT

We wish to thank NSF for partial support of this work through grants no. CHE-75-00938 and CHE-77-26996.

References

1. A. Abragam, *The Principles of Nuclear Magnetism*. Oxford University Press, London (1961).
2. A. G. Redfield, *Adv. Mag. Res.* **1**, 1 (1965).
3. J. H. Freed, *J. Chem. Phys.* **49**, 376 (1968).
4. L. T. Muus and P. W. Atkins (eds.), *Electron-Spin Relaxation in Liquids*, Plenum Press, New York (1972).
5. J. H. Freed, G. V. Bruno, and C. F. Polnaszek, *J. Phys. Chem.* **75**, 3385 (1971).
6. J. H. Freed, *J. Chem. Phys.* **43**, 2312 (1965).
7. J. H. Freed, *J. Phys. Chem.* **71**, 38 (1967).
8. J. S. Hyde, J. C. W. Chien, and J. H. Freed, *J. Chem. Phys.* **48**, 4211 (1968).
9. J. H. Freed, D. S. Leniart, and J. H. Freed, *J. Chem. Phys.* **47**, 2762 (1968).
10. J. H. Freed, *J. Chem. Phys.* **50**, 2271 (1969).
11. G. Rist and J. H. Freed, unpublished results.
12. J. H. Freed and G. K. Fraenkel, *J. Chem. Phys.* **39**, 326 (1963).
13. P. W. Atkins, in *Electron-Spin Relaxation in Liquids* (L. T. Muus and P. W. Atkins, eds.), Chapter XI, Plenum Press, New York (1972).
14. G. K. Fraenkel, *J. Phys. Chem.* **71**, 139 (1967).
15. J. H. Freed, *J. Chem. Phys.* **41**, 2077 (1964).
16. J. H. Freed and G. K. Fraenkel, *J. Chem. Phys.* **41**, 3623 (1964).
17. J. H. Freed, in *Electron-Spin Relaxation in Liquids* (L. T. Muus and P. W. Atkins, eds.), Chapters VIII and XVIII, Plenum Press, New York (1972).
18. D. Kivelson, in *Electron-Spin Relaxation in Liquids* (L. T. Muus and P. W. Atkins, eds.), Chapter X, Plenum Press, New York (1972).
19. G. R. Luckhurst, in *Electron-Spin Relaxation in Liquids* (L. T. Muus and P. W. Atkins, eds.), Chapter XV, Plenum Press, New York (1972).
20. S. A. Goldman, G. V. Bruno, C. F. Polnaszek, and J. H. Freed, *J. Chem. Phys.* **56**, 716 (1972).
21. J. S. Hwang, R. P. Mason, L. P. Hwang, and J. H. Freed, *J. Phys. Chem.* **79**, 489 (1975).
22. C. F. Polnaszek and J. H. Freed, *J. Phys. Chem.* **79**, 2283 (1975).

23. J. H. Freed, *J. Chem. Phys.* **66**, 4183 (1977).
24. M. P. Eastman, R. G. Kooser, M. R. Das, and J. H. Freed, *J. Chem. Phys.* **51**, 2690 (1969).
25. M. P. Eastman, G. V. Bruno, and J. H. Freed, *J. Chem. Phys.* **52**, 321 (1970).
26. J. I. Kaplan, *J. Chem. Phys.* **28**, 278, 462 (1958); S. Alexander, *J. Chem. Phys.* **37**, 966, 974 (1962).
27. J. H. Freed, in *Chemically-Induced Magnetic Polarization* (L. T. Muus, T. W. Atkins, K. A. McLauchlin, and J. B. Pedersen, eds.), Chapter 19, D. Reidel, Dordrecht, Holland (1977).
28. S. A. Goldman, J. B. Pedersen, and J. H. Freed, to be published.
29. D. S. Leniart, H. D. Connor, and J. H. Freed, *J. Chem. Phys.* **63**, 165 (1975).
30. J. H. Freed, D. S. Leniart, and H. D. Connor, *J. Chem. Phys.* **58**, 3089 (1973).
31a. N. C. Pyper, *J. Mol. Phys.* **21**, 1 (1971); **22**, 433 (1971); B. C. Sanctuary, *J. Chem. Phys.* **64**, 4352 (1976).
31b. J. H. Freed, unpublished results.
32a. G. V. Bruno and J. H. Freed, *Chem. Phys. Lett.* **25**, 328 (1974).
32b. G. V. Bruno, Ph.D. Thesis, Cornell University, Ithaca, New York (1973).
33. J. H. Freed, *J. Phys. Chem.* **78**, 1155 (1974).
34. J. S. Hyde, M. D. Smigel, L. R. Dalton, and L. A. Dalton, *J. Chem. Phys.* **62**, 1655 (1975).
35. J. H. Freed, *Ann. Rev. Phys. Chem.* **23**, 265 (1972).
36. A. J. Vega and D. Fiat, *J. Chem. Phys.* **60**, 579 (1974).
37. L. P. Hwang and J. H. Freed, *J. Chem. Phys.* **63**, 118 (1975).
38. A. R. Edmonds, *Angular Momentum in Quantum Mechanics*. Princeton University Press, Princeton, New Jersey (1957).
39. W. Ledermann, *Proc. Cambridge Phil. Soc.* **46**, 581 (1950).

4

Solution ENDOR

Neil M. Atherton

1. Introduction

ENDOR is one of the oldest double resonance techniques, the first experiments on solids having been reported by Feher in 1956.[1] In fluid solutions, the nuclear spin relaxation times are shorter than in solids, and higher radiofrequency powers are required to observe ENDOR, for the nuclear transitions must be driven at a rate comparable to that due to relaxation. This requirement makes the instrumentation for solution ENDOR more difficult than that for solids and explains, at least to some extent, why the application of the technique to solutions has lagged somewhat behind that to solids.

The first solution ENDOR signals were obtained by Cederquist from metal–ammonia solutions.[2] Unfortunately, this work has never been published in the conventional sense and so is probably not as widely known and discussed as it deserves. It was not until 1964 that Hyde and Maki[3] reported signals from free radicals in solution, with a fuller account describing the spectrometer and results for a range of systems appearing a year later.[4] Since that time, the field has expanded steadily but quite slowly. A fair assessment of the present state of affairs is probably that solution ENDOR is well established and reasonably well understood, but there are many applications yet to be perceived and made. It would be surprising if experiments still to be done did not yield some unlooked-for results.

In ENDOR, one monitors the level of an ESR absorption and records the change in it when a nuclear resonance transition is driven by a second

Neil M. Atherton • Department of Chemistry, The University of Sheffield, Sheffield, England

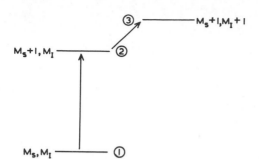

Figure 1. Schematic characterization of energy levels and transitions involved in ENDOR.

radiofrequency field. The bare bones of the experiment are indicated in Figure 1, where the transition 1–2 is the ESR ($\Delta M_S = 1$) and 2–3 the nuclear resonance ($|\Delta M_I| = 1$). We will be thinking about the reasons for doing solution ENDOR experiments, and the results, in two ways. The first of these is that one achieves resolution enhancement by doing the double resonance experiment, so that the hyperfine coupling constant can often be measured more precisely than from the conventional ESR spectrum. Section 2 of this chapter will deal with this aspect of solution ENDOR: we will consider the theory of the ENDOR frequencies and discuss examples of cases where their measurement has yielded information of interest for various reasons. This aspect deals with the time-averaged properties of the system under study. The second way is concerned with its dynamical behavior. Referring again to Figure 1, it is not difficult to see that the ENDOR response must depend critically on the relaxation behavior of the system. The levels shown in Figure 1 are, of course, only part of a multilevel system, and relaxation between states 1 and 2 occurs by many routes other than the direct one. The ENDOR response arises because driving the 2–3 transition changes the overall relaxation rate between states 1 and 2. The theory of saturation and solution ENDOR has been developed extensively by Freed and his collaborators, and is discussed in Chapter 3. In Section 3, we will discuss relaxation effects in solution ENDOR using a physical approach. We will illustrate how some observations show up the importance of particular relaxation mechanisms and examine how ENDOR may assist in relaxation studies.

The dual view of solution ENDOR outlined above may be somewhat artificial, but it is convenient. In building up our account of the subject we will make use of examples from the literature to illustrate and amplify the points to be made but will not attempt extensive literature coverage. The field has been fully covered by literature reviews in continuing series,[5–10] and the information may rapidly be retrieved. We will also omit discussion of experimental considerations, as this topic is covered in Chapter 2.

2. Analysis of ENDOR Frequencies

2.1. Transition Frequencies

The Hamiltonian appropriate for discussing the ENDOR frequencies in a radical in which there is hyperfine coupling to just one nucleus comprises the electron and nuclear Zeeman terms and the magnetic hyperfine coupling:

$$H = g\beta \mathbf{B} \cdot \mathbf{S} - \gamma \mathbf{B} \cdot \mathbf{I} + a' \mathbf{I} \cdot \mathbf{S} \tag{1}$$

To first-order in the hyperfine interaction, the eigenvalues are

$$E = g\beta B_0 M_S - \gamma B_0 M_I + a' M_S M_I \tag{2}$$

and it is convenient to write this in (circular) frequency units as

$$E/h = v_e M_S - v_n M_I + a M_S M_I \tag{3}$$

Here, v_e $(= g\beta B_0)$ is the electron resonance frequency for no hyperfine coupling, v_n $(= \gamma B_0)$ is the resonance frequency for the "free" nucleus, and the prime has been dropped from the coupling constant to signify the change of units. The selection rule on ENDOR transitions is $\Delta M_S = 0$, $\Delta M_I = 1$, and so the ENDOR frequencies are

$$v_{\text{ENDOR}} = |v_N \pm a/2| \tag{4}$$

where the upper sign refers to the $M_S = -\frac{1}{2}$ manifold if the coupling constant is positive and to the $M_S = +\frac{1}{2}$ manifold if the coupling is negative. The sign cannot, of course, be determined from the simple spectrum.

For protons at X band, the free nuclear frequency is about 14 MHz, so if the hyperfine coupling is less than 28 MHz, the ENDOR spectrum comprises two lines separated by the hyperfine coupling and symmetrically placed about the free nuclear frequency. To date, this situation has obtained for all the proton ENDOR spectra that have been observed in solution; the radicals studied have been relatively stable species with some aromatic group or groups and couplings greater than 28 MHz (1 mT, 10 G) are not common in such species. No doubt, as more solution ENDOR work is done and experimentalists get more confident and venturesome, radicals with larger proton couplings (e.g., aliphatic radicals prepared in flow systems) will be studied. When the hyperfine coupling is larger than $2v_n$, the spectrum comprises two lines separated by $2v_n$, and the sum of their frequencies is the hyperfine coupling. However, the first-order energies equation (3) will not generally give the ENDOR frequencies very satisfactorily in this situation. Analysis of the hyperfine interaction to higher order is discussed later in this section, and for the time being, we will assume the coupling to be small and the first-order solution of the wave equation to be valid.

For hyperfine coupling to one proton, the ESR and the ENDOR spectra are not very different; they both consist of two lines separated by the hyperfine coupling constant. The only immediately obvious advantage of ENDOR is that one might hope to be able to resolve a smaller hyperfine splitting than one can in conventional ESR. For example, the smallest hyperfine splitting one could hope to resolve using an ESR spectrometer with 100-kHz field modulation would be 196 kHz (7 μT, 70 mG), while in ENDOR, one would hope to be able to obtain some resolution of a coupling of somewhat less than 100 kHz. However the advantage of ENDOR becomes more dramatically apparent when one considers coupling to a group of equivalent nuclei.

Provided the hyperfine coupling is small, then the ESR and ENDOR frequencies for a radical in which there is hyperfine coupling to a group of equivalent nuclei can be simply calculated using the maximum resultant nuclear spin (we are not interested in the degeneracies of the ESR transitions). Equation (3) becomes

$$E/h = v_e M_S - v_n M_F + a M_S M_F \tag{5}$$

where M_F takes the $(2F + 1)$ values $+F$, $(F - 1)$, ..., $-F$ and is the z-component of the maximum resultant nuclear spin

$$\mathbf{F} = \sum_k \mathbf{I}_k \tag{6}$$

The important point is that the frequencies of the $\Delta M_F = 1$ ENDOR transitions are still given by equation (4); there are still only two lines in the ENDOR spectrum, while the ESR spectrum contains $(2kI + 1)$. This point was very eloquently made by Hyde, in his first detailed discussion of solution ENDOR observations, using triphenylmethyl as an example.[4] This radical contains two sets of six equivalent protons at the *ortho* and *meta* positions of the three rings and a set of three equivalent protons at the *para* positions of the three rings. The ESR spectrum thus contains $(7 \times 7 \times 4) = 196$ lines, which fall over a frequency range of about 70 MHz. In contrast, the ENDOR spectrum contains six lines, occurring over about 20 MHz. It is clear that the ENDOR technique has the greater inherent resolution, even if the linewidths are assumed to be the same as those in ESR.

Quite generally, we expect ENDOR to be useful for obtaining precise measurements of hyperfine coupling constants when these are not readily obtained from the ESR spectrum. We consider some examples to illustrate the point:

a. If one hyperfine coupling is a near-integral multiple of another, there are accidental near-degeneracies in the ESR transition frequencies, but the transitions are well separated in the ENDOR spectrum. The radical anion of 9,10-anthraquinone provides an example, illustrated in Figure 2. There are

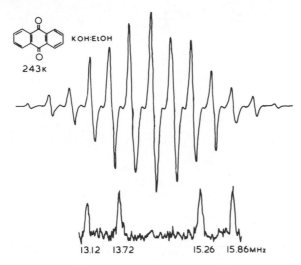

Figure 2. Esr and ENDOR spectra of the anthraquinone radical anion. The whole ENDOR spectrum is shown, the free proton frequency being 14.49 MHz.

two sets of four equivalent protons in this radical, and the moderate-quality ESR spectrum shown contains only nine lines instead of the twenty-five expected. However, the two couplings are very clear in the ENDOR spectrum. In this case, since there are only two couplings, no real problem arises with the ESR spectrum, but the example does indicate rather well the difference between the two sorts of spectrum.

b. Often small differences in hyperfine coupling are not clearly resolved in ESR spectra. Figure 3 illustrates this sort of case with the radical anion of 9,10-phenanthraquinone. Symmetry indicates four pairs of protons, but the ESR spectrum has the gross structure of a quintet of quintets, indicating that the pairs of couplings are nearly equivalent in pairs. The differences are quite well resolved in the ENDOR spectrum.

c. Often a small hyperfine coupling may be very difficult to resolve in ESR, at best appearing only very incompletely resolved. This is so for the *tert*-butyl protons in the radical anion of 2,5-di-*tert*-butyl-*p*-benzoquinone. Figure 4 shows an ENDOR spectrum for these protons obtained from an ESR spectrum in which the coupling was not even partially resolved.

d. Large radicals of low symmetry may have so many couplings that useful resolution can never be achieved in the ESR spectrum, and one cannot even obtain tentative values to begin a computer simulation. A recent literature example is provided by the radical anion of 6-methyl-2-phenyl-5-azacycl[3.2.2]azine studied by Gerson *et al.*[11] The ESR spectrum is expected to have 10,368 lines from coupling to twelve protons and two nitrogen nuclei, and the spectrum obtained is of little immediate use.

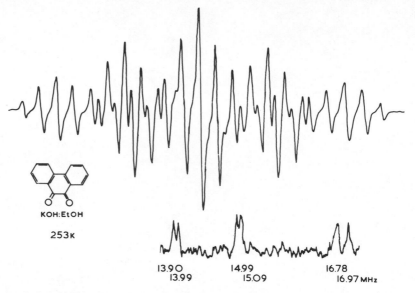

KOH:EtOH

253к

13.9O   14.99   16.78
   13.99   15.O9   16.97 MHz

Figure 3. ESR and ENDOR spectra of the phenanthraquinone radical anion. The free proton frequency in 14.49 MHz and the low-frequency lines corresponding to the two larger couplings are not shown in the ENDOR spectrum.

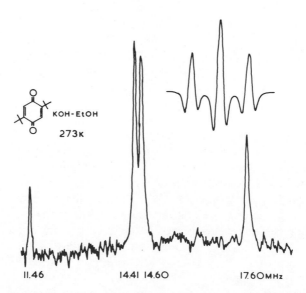

KOH-EtOH

273к

11.46   14.41 14.6O   17.6Omhz

Figure 4. ESR (second-derivative) and ENDOR spectra of the 2,5-di-*tert*-butylquinone radical anion.

However, the proton ENDOR spectrum gave very clearly all eight different proton couplings, and with this information, the nitrogen couplings were obtained from the ESR spectrum.

Equation (3) only gives a good description of the ESR and ENDOR frequencies when the hyperfine coupling is small. In general, the exact eigenvalues for a single electron coupled to a single nucleus are given by the Breit–Rabi formula,[12] but for many purposes, a perturbation expansion to second order is adequate. The second-order correction to be added to the right-hand side of equation (3) is

$$E/h = +(M_S a^2/2v_e)\{I(I + 1) - M_I(M_I + 2M_S)\} \tag{7}$$

Here we have neglected v_n compared to v_e in the denominator. As mentioned above, radicals with large proton hyperfine couplings have not yet been studied with solution ENDOR, and the second-order term has been negligible in the analysis of such spectra.[13] However, ^{14}N ($I = 1$) ENDOR has been observed for nitroxides, and it was found that the transition frequencies could not be satisfactorily accounted for using first-order theory.[14] It is instructive to consider what effect the second-order correction should have on the ENDOR spectrum. Higher-order hyperfine effects are of course well known in the ENDOR of solids,[15] but there is an extra aspect to the problem that has become apparent for solutions, and we will review briefly the effect on the transition frequencies.

Figure 5a shows how the second-order correction modifies the hyperfine levels when $I = 1$: the separations within a particular electron spin state are no longer equal. Shifts are defined in terms of a correction s,

$$s = a^2/4v_e \tag{8}$$

In an ENDOR experiment in which one monitors an extreme hyperfine line ($M_I = \pm 1$), one expects only the ENDOR transitions having a level in common with that being saturated to contribute to the spectrum, and the spectra should be simple but with one line shifted from its first-order frequency, as shown in Figure 5b. However, if the $M_I = 0$ ESR transition is saturated, one may expect all ENDOR frequencies to appear, and in principle the spectrum should show a "second-order splitting." In the original nitroxide work, the second-order splitting was not resolved, as the lines were relatively broad, but the transition frequencies could be measured with sufficient precision for second-order effects to be discerned.[14] However, it was found that the observed frequencies could not be accounted for by invoking a departure from a first-order hyperfine effect. The discrepancy can be accounted for by invoking a contribution to the resonance frequency from the dynamic frequency shift.[16,17] This is very difficult to measure from a conventional ESR spectrum and in fact has only been observed under rather

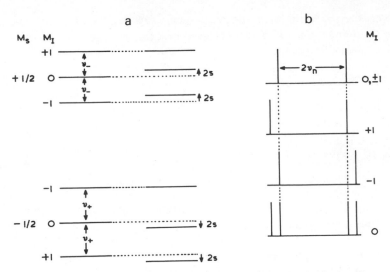

Figure 5. (a) Modification of the hyperfine levels in an $S = \frac{1}{2}$, $I = 1$ system when the coupling is treated to second order. The first-order levels are shown on the left, and there are just two ENDOR frequencies, $\nu_{\pm} = \frac{1}{2}a \pm \nu_n$. The corrected levels are shown on the right, and now there are four ENDOR frequencies. The unit of second-order shift s is defined in equation (8). (b) Comparison of ENDOR spectra when the hyperfine interaction is treated to first and second order. The first-order spectrum is shown at the top and is independent of which hyperfine line is saturated. The three spectra below are for saturation of each separate hyperfine line when the coupling is treated to second order.

special circumstances.[18] The indication that a dynamic frequency shift may be obtained from ENDOR in unremarkable circumstances raises some interesting possibilities.

The fact that there should be a dynamic contribution to resonance frequencies is well known from relaxation theory, and the ESR case has been discussed in detail by Fraenkel.[16,17] In terms of Redfield theory,[19] the shift comes from the imaginary part of the relaxation matrix; it is the sine Fourier transform of the correlation function,

$$\Delta\omega \sim \int_0^{\infty} \overline{\langle \alpha | \mathscr{H}(t) | \beta \rangle \langle \beta | \mathscr{H}(t + \tau) | \alpha \rangle} \sin \omega\tau \, d\tau \qquad (9)$$

where α and β label the two states involved in the transition, $\mathscr{H}(t)$ is the Hamiltonian of the fluctuating interactions causing the relaxation, and the bar denotes the ensemble average. If the correlation time is relatively long, then the dynamic shift is simply related to the mean square fluctuation in resonance frequency and is independent of the correlation time. This is in contrast to linewidth and transition probability terms in which the mean square frequency fluctuation always appears multiplied by the correlation time or a function of it. It is important to be able to separate the two factors if one wishes to test models of the dynamic processes causing the interactions to fluctuate, and so dynamic frequency shift data are a valuable complement to relaxation data.

To take account of the dynamic shift in the second-order terms in the hyperfine interaction one should make the substitution

$$a^2 = \bar{a}^2 + \langle(\delta a)^2\rangle \tag{10}$$

in equations (7) and (8). Here a is the isotropic hyperfine coupling and $\langle(\delta a)^2\rangle$ the mean square fluctuation in the hyperfine splitting. A value of $\langle(\delta a)^2\rangle$ was deduced from the ENDOR data on the nitroxides,[14] and values have also been obtained from alkali metal ENDOR of complexes with di-o-mesitoylbenzene.[20,21] However, no comparison with theoretical values has yet been made. It does seem that this is an aspect of solution ENDOR that might well be worth pursuing further, and one would like to see this done.

2.2. Assignment of Transitions

The assignment of hyperfine coupling constants to positions in a molecule does not follow immediately upon their measurement, and this is the case no matter how the data are obtained. However, in ESR, one does not normally have any difficulty in deciding how many nuclei, say protons, belong to a particular value of the coupling constant; the information is contained in the relative intensities of the ESR lines. This is not always true with a solution ENDOR spectrum; the relative intensities of the lines are not necessarily in proportion to the number of nuclei contributing. Under certain limiting or idealized conditions they may be, but because the ENDOR intensities are so critically relaxation dependent, and because nuclei with different coupling constants may have different relaxation rates or even mechanisms, then in general they will not be. Knowledge of the number of nuclei having a particular coupling is a useful start to assigning the coupling to a position in the molecule, and it is a nuisance that this information may be obscured in the ENDOR spectrum. Detailed analysis of the way in which the ENDOR intensities respond to varying physical conditions should, in principle, provide the information, but in practice it has not been used unsupported. In this section, we review the various other methods that have been used.

Perhaps the most obvious method is to make a trial assignment based on experience and run a computer simulation of the ESR spectrum. This method was appreciated very early on and used, for example, in work on triarylmethyl radicals[22] and on some semiquinone ions.[23] It is perfectly sound and should be unambiguous, though tedious under some circumstances (e.g., if the lines in the ESR spectrum do not all have the same width).

A method requiring some chemical labor is to examine partially deuterated species. The effect on an ESR spectrum of replacing ^1H by ^2H is not necessarily helpful so far as resolution is concerned; one reduces the hyperfine splitting and increases the number of lines. However, the effect on the ENDOR spectrum is much more direct, because the ENDOR frequencies depend on the free nuclear frequency as well as the hyperfine coupling. For example, in an X-band experiment where the free proton frequency is 14 MHz, a ^1H coupling of 3 MHz should give ENDOR at 12.5 and 15.5 MHz, while the ENDOR frequencies for ^2H having the same spin density at the nucleus would be 0.65 and 3.65 MHz. The study of partially deuterated samples enabled Hyde to measure and assign all seven proton hyperfine couplings in the 2,4,6-triphenylphenoxy radical.[24] The ENDOR spectrum of the fully protonated radical is not fully resolved, and so, in addition to giving the assignments, deuteration meant that the couplings could be measured accurately—to ± 10 kHz, in fact. Another case where partial deuteration was very strikingly used is a study by Möbius' group on the radical anion and cation of rubrene (tetraphenyltetracene).[25] Here again, all the proton couplings were precisely measured and unambiguously assigned.

The third distinct method of assignment that has been used depends on the so-called nuclear–nuclear coherence effect.[26] A detailed account of coherence effects is outside the scope of this chapter, so we will simply indicate the phenomenon and its use. The effect may manifest itself if there is a hyperfine manifold of equally spaced levels so that, at the relatively high rf power levels used for solution ENDOR, double quantum transitions occur to a significant extent. One has such a ladder if there is hyperfine coupling to a single nucleus having $I \geq 1$ or, and this is the important case that concerns us, there is coupling to a group of two or more equivalent protons (or other nuclei with $I = \frac{1}{2}$). In a double quantum transition, the component of the nuclear spin angular momentum changes by two units through the simultaneous absorption of two coherent quanta. One does not observe double quantum transitions directly in ENDOR, but they can affect the appearance of the spectrum. The results depend on the ESR line being monitored; if a degenerate transition is saturated, then a splitting of the ENDOR transition may occur at high rf power but not at low. In contrast, for a nondegenerate ESR transition, such as an extreme hyperfine line from a group of equivalent nuclei, no splitting occurs at high rf power. Details of the splitting in the ENDOR depend in a complicated way on the relaxation, rf power, and so forth, and the full theoretical analysis is complicated but undoubtedly correct.[26,27] Fortunately, if one simply needs to decide if a particular ENDOR line belongs to a set of equivalent nuclei or a single nucleus, a qualitative observation is all that is required.

In practice, the method has not been used in quite the manner implied by the above description. Clearly, if one can unambiguously saturate a particular hyperfine component, then one probably can tell already which ENDOR lines belong to groups of nuclei and which not. The technique that has been used is to monitor an overmodulated ESR spectrum so that many hyperfine lines contribute to each ENDOR line. The nuclear–nuclear coherence effect then manifests itself as a broadening with increasing rf power of those ENDOR that correspond to hyperfine coupling to groups of two or more equivalent nuclei. Work on the radical anions of phenylnaphthalenes[28] and phenylcyclazines[11] illustrates the power and usefulness of this method of assignment. As yet it has been relatively little used but should certainly be counted as a standard technique, and no doubt it will find increasing application.

To conclude discussion of this topic, we must note that the signals observed in electron–nuclear–nuclear triple resonance experiments may give a clearer indication of the numbers of contributing protons than do ordinary ENDOR signals.[29,30] Chapter 14 by Möbius and Biehl discusses these triple resonance experiments.

2.3. *Applications of Proton ENDOR*

By far the greatest part of solution ENDOR work has been done on protons. In this section, we pick out a few examples that seem of particular interest in that they have wider implications than just the measurement of hyperfine couplings.

The theory of molecular electronic structure has been very successful in accounting for proton hyperfine coupling constants. For nearly twenty years, there has been a very fruitful interaction between theory and experiment in this area, from the earliest work using Hückel theory, through the development of semiempirical calculations taking account of all the valence electrons and *ab initio* methods. Since the availability of solution ENDOR made it possible to measure unambiguously hyperfine couplings in complicated radicals, σ–π delocalization has been given renewed attention. The question arises when one has coupled aromatic systems that are not constrained to be coplanar, such as a phenyl-substituted aromatic radical, with the radical anion of biphenyl as one of the simplest examples. If the systems are not coplanar, then, in the framework of π-electron theory, the conjugation is decreased, and one takes account of this by reducing the resonance integral for the bond between the two systems. The amount of reduction needed to account for the observed couplings may be interpreted to give the deviation from coplanarity by assuming the angular dependence of the integral. There are cases where this procedure seems to work satisfactorily, a recent example being 9-arylxanthyl radicals,[31] but there are others where it does not. In particular, π-electron theory predicts that the magnitudes of the couplings in a phenyl substituent should be in the order *para* > *ortho* > *meta*, whereas in the anion and cation of rubrene the order is *meta* > *para* = *ortho.*[25] This sort of observation can be accounted for by assuming that in the noncoplanar system there is conjugation into the σ-system of the substituent, and of course one needs to take account of all the valence electrons to describe this. There are now several cases where ENDOR measurements quite definitely indicate the necessity of invoking this σ–π delocalization. In addition to rubrene,[25] the 9-phenylanthracene[32] and methyl-substituted biphenyl anions[33] are particularly clear examples. One has the feeling that the literature on this topic does not form a very coherent picture; perhaps a thorough check of all available data with a reassessment of the quality of agreement between (π-electron) theory and experiment would prove interesting.

The geometry dependence of couplings to β protons in substituted aromatic radicals has been known for years.[34] One observes a thermally averaged coupling, and if internal motions are hindered, the coupling may be temperature dependent.[35] If the ESR spectrum is complicated, it may be quite a task to sort out the temperature dependence, and ENDOR can be very useful. There are several instances of cases where temperature-dependent couplings have been monitored using ENDOR and interpreted qualitatively in terms of hindered rotation or torsional oscillation—e.g., the 4-formyl-2,6-di-*tert*-butyl-[36] and 4-ethoxy-2,6-di-*tert*-butylphenoxy[37] radicals. Quantitative analysis of the temperature dependence enables energy barrier heights or separations between thermally accessible states to be estimated. Thus, Bauld *et al.* have estimated barrier heights to rotation in several cyclopropyl-substituted radicals,[38] while Nemoto *et al.* have estimated the energy separation between two conformations of the 4,4'-diisopropylbiphenyl anion.[39] Again, von Borczyskowski *et al.* have characterized quantitatively rates of conformational interconversion for several silacyclopentadienyl anions.[40]

If one has two β protons on the same carbon atom, then they may have different time-averaged locations with respect to the remainder of the radical and different hyperfine couplings. Such differences have been resolved by ENDOR—e.g., in a substituted triphenylmethyl radical,[41] where the single ENDOR line observed for the two nuclei at relatively high temperature was observed to split at lower temperature. Observations such as this give an indication of equilibrium geometries. Similar behavior was observed for the γ protons in the cyclohexyl ring of the 4-cyclohexyl-2,6-di-*tert*-butylphenoxy radical[42] and taken to indicate the importance of a direct hyperconjugative mechanism for transferring spin density to γ protons. Superficially equivalent α protons may become distinct if the radical as a whole does not have the same symmetry as that which applies locally and if internal motion is not completely free and rapid; this effect shows up in the ENDOR spectra in some of the large, stable radicals—*bis*-diphenylene-β-phenylallyl,[43,44] diphenylpicrylhydrazyl,[44] and picryl-*N*-aminocarbazyl.[44]

Variations of hyperfine couplings in radicals and radical ions in solution can also reflect intermolecular interactions; many cases of solvent dependence of couplings and of ion-pairing effects have been reported. Solution ENDOR should certainly help in the collection of data in this area, for, if the ESR spectrum is not well resolved, it is not always easy to follow changes in all the individual couplings. Measurements of the couplings in *tert*-butylphenylnitroxide in ethanol and in toluene[45] illustrate this point, but as yet, ENDOR has hardly been applied in this field. One would like to see an increase in activity here, for the study of solute–solvent and solute–solute interactions seems interesting and important.

The development of ESR parallels that of many other branches of spectroscopy in that interest has spread from the physical basis to the chemical applications. One may expect solution ENDOR to follow a similar pattern, though more practitioners are needed to accelerate the process. An indication of one sort of application where the advantages of ENDOR over conventional ESR are apparent is provided by studies of anion radicals of semiquinones.[46–48] In alkaline alkanols, the anions of quinones are readily formed, but the ESR spectra can rapidly become complicated by the appearance of secondary radicals formed by alkoxide attack on the substrate. The ESR spectra of two or more overlapping radicals can be difficult to interpret, but in ENDOR experiments, one can saturate different features and obtain the ENDOR spectra of individual radicals. For *p*-benzoquinone in propan-2-ol, no fewer than four secondary radicals were characterized and identified in this way.[48]

2.4. *ENDOR of Other Nuclei*

In the earliest solution ENDOR experiments on sodium in liquid ammonia, Cederquist[2] observed signals from ^{23}Na, ^{14}N, and ^{15}N, as well as from ^{1}H. There is rapid exchange in these solutions, the ESR spectrum is a sharp line, and the ENDOR spectra consisted of single lines at the free nuclear frequencies. More recent work has been on systems where the hyperfine coupling could be measured, and the ENDOR frequencies have been in accord with the analysis of Section 2.1. In addition to conventional spectra from ^{23}Na and ^{14}N, signals have also been reported for ^{2}H, ^{13}C, ^{19}F, and other alkali metals. As yet, the number of papers is small; the subject is just at the early stage, where it is known that the experiments are possible. Before mentioning the observations, we will have to consider briefly a new aspect of ENDOR, the transition moment for the nuclear transitions.

The strength of the interaction between a nuclear spin and an rf field depends on the magnitude of the nuclear moment, and a relative measure for different nuclei can be taken from the various resonance frequencies for the

Table 1

Typical Free Nuclear Frequencies for X-Band ENDOR[a]

Nucleus	Frequency (MHz^{-1})	Nucleus	Frequency (MHz^{-1})
^1H	14.00	^{19}F	13.17
^2H	2.15	^{23}Na	3.70
^7Li	5.44	^{31}P	5.67
^{13}C	3.52	^{85}Rb	1.35
^{14}N	1.01	^{87}Rb	4.58
^{15}N	1.42	^{133}Cs	1.84

[a] Prepared from data in *Handbook of Chemistry and Physics*, 55th edition (R. C. Weast, ed.), Chemical Rubber Company Press, Cleveland (1974).

free nuclei in a constant magnetic field. Table 1 gives these frequencies for the nuclei of interest here at a field strength appropriate for X-band experiments. The transition probabilities depend on the squares of the interaction with the rf field, and so one expects it to be much easier to observe ENDOR of ^1H than of, say, ^{14}N, with a given rf field. Up to a point, this argument is sound, but for a nucleus whose hyperfine coupling is large compared to its Zeeman energy, the apparent difficulty is ameliorated through the operation of the so-called rf enhancement effect of the hyperfine interaction.

This effect, which has been described independently in a variety of ways,[15,49–52] is omitted in the naive consideration of the transition probabilities implied in the previous paragraph. A simple physical picture, due to Geschwind,[50] is that in an ENDOR experiment, the hyperfine field of the electron at the nucleus has a component at the radiofrequency. This arises through the interaction of the electron moment with the rf field.[52] Analysis shows that, to a first approximation, the value of the rf field strength B_2 to be used in calculating transition probabilities is not the nominal value but an effective value

$$B_2^{eff} = B_2(1 \pm a/2v_n) \qquad (11)$$

where a is the hyperfine coupling, v_n the free nuclear frequency, and the $\pm$ sign refers to the sign of M_s. The rf enhancement factor, the quantity in brackets in equation (11), can rapidly become surprisingly large; for example, for ^{14}N at X band, it is 20 for a coupling of 38 MHz.

Double resonance work on ^{14}N has so far been limited to nitroxide radicals,[14,53] and for isotropic solutions, the interest has centered on relaxation mechanisms. However, spectra have also been obtained for anisotropic systems of partially oriented radicals dissolved in a nematic liquid crystal.[54] There have, of course, been many magnetic resonance studies using nematic liquid crystals in the past decade, but this was a novel application for ENDOR.

It enabled an estimate of the quadrupole coupling to ^{14}N to be made. This interaction does not affect the ESR or the ENDOR frequencies in isotropic solution, and for a crystal, the ESR spectrum is only affected in second order. It affects the ENDOR spectrum from a crystal more directly, but few instances have been analyzed. The demonstration that one can gain information about this coupling relatively simply from solution studies, even if they are extraordinary solutions, is most valuable and one would welcome further investigations of this technique.

The ESR spectra obtained from di-*o*-mesitoylbenzene reduced with alkali metals show much larger metal hyperfine splittings than one usually obtains from ion pairs.[55] ENDOR spectra of the ^{7}Li, ^{23}Na,[20] and $^{85,87}Rb$[21] nuclei have been observed for these systems, the high-frequency lines occurring in the frequency range where signals corresponding to small proton couplings occur. There is no difficulty in measuring the magnitudes of the metal coupling constants from the ESR spectra, and the main interest in the ENDOR spectra has been in the relaxation aspect. This is also true of the first ^{13}C ENDOR to be reported,[56] and to some extent, of more recent work[75]; we will return to these matters in Section 3.2.

In all the systems mentioned so far in this section, the nuclei have had small magnetogyric ratios but relatively large hyperfine couplings, and it seems clear that the rf enhancement effect has been significant. Recently Möbius and his colleagues have studied nuclei with small moments and small couplings.[72] This is an important development, as it demonstrates the technical feasibility of detecting solution ENDOR signals at 1 MHz or less in X-band experiments. ENDOR of ^{2}H in partially deuterated perinaphthenyl in nematic phase IV was observed, and the out-of-plane component of the quadrupole coupling measured.[73] The value is the same (within experimental error) as that observed in the diamagnetic ground states of even alternant hydrocarbons. This is consistent with the expected uniformity of π electron charge distribution in alternant systems. Alkali ENDOR for the ion pairs of the biphenyl and fluorenone radical anions with lithium, sodium, rubidium, and cesium has also been observed.[72,74] The results indicate different hyperfine couplings from those inferred from the contact shifts of the alkali NMR. The latter are, of course, obtained from concentrated solutions, and presumably intermolecular effects in these account for the difference from the dilute solutions used for the ENDOR measurements.

The last nucleus for mention here is ^{19}F. The magnetic moment is almost as large as that of ^{1}H, and so one would hope to be able to study small couplings using a spectrometer built for proton ENDOR without benefit of the rf enhancement. This turns out to be so, though there are only two reports so far. Couplings to α-fluorines in several partially fluorinated benzophenone radical anions have been measured by Möbius and co-workers.[57] The ESR spectra of these radicals are quite complex, but with precise ^{1}H and ^{19}F couplings available from ENDOR, one can be confident about experimental data to be compared with the predictions of electronic structure theory. As with so many other systems studied by solution ENDOR, the relaxation aspect again proved of interest here. ENDOR of the β-fluorines in the perfluorohexamethylbenzene radical anion has also been reported but not discussed in detail.[22]

3. Analysis of ENDOR Intensities and Linewidths

3.1. Relaxation and ENDOR

Anyone who attempts a solution ENDOR experiment must very quickly become aware of how relaxation dependent the response is. One often finds that good spectra can only be obtained over a rather narrow range of conditions and that these optimum conditions vary for different radicals in a particular solvent and for a particular radical in different solvents. We have already indicated in Section 1 that the ENDOR response arises from the change in relaxation rate that occurs when the nuclear transition is driven, and we now examine this theme in more detail.

One is used to thinking about spin–lattice relaxation and saturation in terms of the spin–lattice relaxation time T_1. All the textbooks on magnetic resonance show that, for a two-level system,

$$T_1^{-1} = W_\uparrow + W_\downarrow \tag{12}$$

where $W_\uparrow$ and $W_\downarrow$ are the probabilities of upward and downward transitions resulting from the spin–lattice interaction. If the lattice is treated classically, then the two probabilities are equal and one has

$$T_1^{-1} = 2W \tag{13}$$

Under these conditions, one keeps the books straight by imposing phenomenologically that the populations relax to their Boltzmann values at thermal equilibrium, so that the magnetization along the direction of the applied field relaxes to a nonzero equilibrium value. This procedure is justified by proper consideration of the lattice.[19]

In the multilevel systems of interest in ENDOR, there are, in general, nonzero probabilities for all conceivable transitions, and the rate of approach to thermal equilibrium depends on all of them. If one is concerned with a transition between a pair of levels, say, i and j, then T_1 is replaced by a relaxation parameter, which is a function of all the transition probabilities.[52,58–60] Physically, one can think of the relaxation parameter as the resistance to transitions between the two states, and it is helpful to consider the problem in terms of an analog resistance network.[61] We consider a simple example to illustrate this point and show how the relaxation parameter can be expressed in terms of the various transition probabilities.

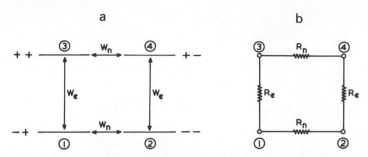

Figure 6. (a) Levels and transition probabilities for a model $S = \frac{1}{2}$, $I = \frac{1}{2}$ system. The levels are labeled with the signs of $(M_S M_I)$.   (b) Resistance network analog.

For simplicity, we consider an $S = \frac{1}{2}$, $I = \frac{1}{2}$ system in which only the pure $\Delta M_S = 1$ and pure $\Delta M_I = 1$ transitions have nonzero probabilities. The levels and probabilities are shown in Figure 6a, while Figure 6b shows the resistance network analog. If we are considering the 1–3 ESR transition, then we want the resistance between points 1 and 3 of the equivalent circuit, R_{13}. From inspection of Figure 6b, this is

$$R_{13} = \frac{R_e(R_e + 2R_n)}{2(R_e + R_n)} \tag{14}$$

If we now write this in terms of the conductances W_e and W_n and use equation (13), we obtain as the effective T_1 for the 1–3 transition

$$(T_1)_{13} = \frac{W_n + 2W_e}{4W_e(W_e + W_n)} \tag{15}$$

This analysis gives a rather clear physical insight into relaxation in a multilevel system and can, in principle, be extended to a more complex level scheme. However, the analysis rapidly becomes cumbersome, even for the four-level system, when cross-relaxation (W_{14}, $W_{23} \neq 0$) is admitted.

The required result can more generally be expressed in terms of cofactors and double cofactors of the determinant of the matrix of transition probabilities. The matrix is written in the basis of eigenstates of the system, and for the case considered above (Figure 6a) is given as

	1	2	3	4
1	$-(W_n + W_e)$	W_n	W_e	0
2	W_n	$-(W_n + W_e)$	0	W_e
3	W_e	0	$-(W_n + W_e)$	W_n
4	0	W_e	W_n	$-(W_n + W_e)$

$$\tag{16}$$

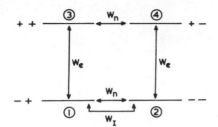

Figure 7. Levels and transition probabilities for the model $S = \frac{1}{2}, I = \frac{1}{2}$ system when the 1–2 transition is driven with rf in an ENDOR experiment. In the resistance network analog (Figure 6b), W_I^{-1} is like an extra resistance in parallel to R_n between points 1 and 2.

This matrix is singular, and care is needed with its inversion, but that will not concern us here. The saturation parameter for transition i–j in a multilevel system is defined as[52]

$$\Omega_{ij,\,ij} = 2C_{ij,\,ij}/C \tag{17}$$

where $C_{ij,\,ij}$ is the double cofactor obtained by crossing out rows and columns i and j from the determinant of the **W** matrix, and C is any cofactor. For the example we have been using one has

$$\Omega_{13,\,13} = \frac{W_n + 2W_e}{W_e(W_e + W_n)} \tag{18}$$

and comparing this result with equation (15), we find for the effective spin–lattice relaxation time for the 1–3 transition*

$$(T_1)_{13} = \tfrac{1}{4}\Omega_{13,\,13} \tag{19}$$

Now we must see how the relaxation rate is affected when we use a radiofrequency field to drive a nuclear resonance transition. We continue to use our specific simple example and suppose that we excite the 1–2 transition. The picture in Figure 6 must now be modified to that shown in Figure 7, where the radiation-induced transition probability is denoted by W_I. The saturation parameter is now

$$\Omega'_{13,\,13} = \frac{2[(W_n + W_e)(W_I + W_n) + W_e W_n]}{W_e[(W_I + W_n)(W_e + 2W_n) + W_e W_n]} \tag{20}$$

This expression reduces to equation (18) for $W_I = 0$. Clearly, the modification to the saturation parameter depends on how the rate of the radiation-induced nuclear transitions compares with that of the lattice-induced ones.

If we suppose that the ESR line under examination has a Bloch line shape, then it is easy to see that the intensity of absorption at a particular

---

* Stephen and Fraenkel[59] define $\Omega_{ij,\,ij} = C_{ij,\,ij}/2C$, and in this case, one has the immediate identity $\Omega_{ij,\,ij} \equiv (T_1)_{ij}$. We have used Freed's definition,[52] equation (17).

microwave frequency (magnetic field) will change if T_1 changes. For exact resonance, the absorption is, in standard notation,

$$v = \frac{\gamma B_1 T_2 M_0}{1 + \gamma^2 B_1^2 T_1 T_2} \tag{21}$$

and the change Δv resulting from a change ΔT_1 in T_1 is

$$\Delta v = \gamma B_1 T_2 M_0 \left[\frac{\gamma^2 B_1^2 T_1 T_2}{(1 + \gamma^2 B_1^2 T_1 T_2)^2} \right] \frac{\Delta T_1}{T_1} \tag{22}$$

This is the ENDOR signal.

In a relatively early paper, Allendoefer and Maki[62] used this type of phenomenological analysis to interpret their observations on the 2,4,6-tri-*tert*-butylphenoxy radical. They were able to give a consistent interpretation of the dependence of the signals on microwave power, rf power, temperature, and solvent viscosity. They also considered the effect of Zeeman modulation and accounted for the reduced ENDOR intensity observed for protons with small hyperfine coupling. More recently, the phenomenological approach has been used by Atherton and Day in a discussion of α and β protons in π radicals.[63] These authors formulated relaxation mechanisms explicitly.

In equation (22), the dependence of the ENDOR signals on the radio-frequency is contained in ΔT_1, which depends on W_I [cf. equations (18) and (20)]. In their work, Atherton and Day[63] obtained W_I by assuming that the rf absorption had a Bloch line shape and calculated the appropriate T_1 and T_2 from the assumed relaxation mechanisms and then in fact assumed exact resonance and only calculated the peak ENDOR signal intensities. The phenomenological analysis is perhaps most useful for doing this, for seeing how ENDOR works, and gaining insight into the effects of particular relaxation mechanisms. For calculating ENDOR line shapes, it seems better to use the general matrix equations given by Freed,[52] which can be programmed quite generally for application to a variety of systems.

An analytical expression for the ENDOR line shape does in fact come from Freed's analysis,[52] and since it displays very clearly the dependence of the signal on relaxation, we conclude this section with a brief discussion of it. The result is that under certain limiting conditions, for exact resonance of the ESR transition ij, the ENDOR signal for nuclear transition jk is

$$\Delta v = q\omega_e d_e \left(\frac{\Omega_{ij,\, jk}}{\Omega_{ij,\, ij}} \right)^2 \frac{d_n^2 T_n}{1 + (\Delta_n T_n)^2 + \Lambda T_n \Omega_{jk,\, jk} d_n^2} \tag{23}$$

where, in standard notation, d_e and d_n are $\frac{1}{2}\gamma B$ for the electron and nuclear moments interacting, respectively, with the microwave and rf fields, Δ_n measures the distance of the radiofrequency from exact resonance, T_n is the line-width of the nuclear transition, and $q\omega_e$ relates to the thermal equilibrium magnetization. The factor Λ in the saturation term is

$$\Lambda = 1 - \frac{\Omega_{ij,\,jk}^2}{\Omega_{ij,\,ij}\Omega_{jk,\,jk}} \tag{24}$$

where $\Omega_{ij,\,ij}$ and $\Omega_{jk,\,jk}$ are the saturation parameters for the electron and nuclear spin transitions, and $\Omega_{ij,\,jk}$ is a cross-relaxation parameter obtained from the $\mathbf{W}$ matrix by deleting rows i and j and columns j and k and using the resulting double cofactor in equation (17). Equation (23) shows that the ENDOR should have a Bloch line shape with a relaxation-dependent intensity factor. It looks helpful if one wants to think about ENDOR effects in a physical kind of way using the equivalent electric circuit analog type of approach. The saturation parameters for the electron and nuclear transitions can be measured directly on an analog circuit [cf. equation (14)], and it has recently been shown that one can also directly obtain an analog value for the cross-saturation parameter from measurements on an equivalent circuit.[64]

In this section, we have tried to use a physically oriented approach to illuminate the relaxation dependence of solution ENDOR in a general kind of way and to complement more formal discussion.[52] We turn now to illustrate how particular relaxation mechanisms may manifest themselves in ENDOR spectra.

3.2. Relaxation Mechanisms

The theory of spin relaxation in solution is well understood. In general, one would expect all the magnetic interactions in a radical in solution to fluctuate to some extent so that all relaxation mechanisms might make some contribution, and indeed, this does make the understanding of the absolute values of ESR linewidths difficult.[65] However, particular mechanisms can manifest themselves in distinctive ways; for example, modulation of the electron Zeeman and dipolar hyperfine interactions through tumbling of the radical gives rise to the familiar M_I dependence of the ESR linewidths,[66] and this feature may enable such mechanisms to be studied individually. Similarly, it turns out that different relaxation mechanisms have distinctive effects on the ENDOR response, and in this section we will consider qualitatively how this comes about. We will concern ourselves with the strengths of the ENDOR signals for saturation of various hyperfine lines.

The simplest case of a relaxation mechanism that should have a distinctive effect on the ENDOR is modulation of the isotropic hyperfine interaction. If the relaxation Hamiltonian for a single interacting nucleus is written in the form

$$H(t) = a(t)S_z I_z + \tfrac{1}{2}a(t)[S_+ I_- + S_- I_+] \tag{25}$$

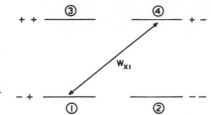

Figure 8. Flip-flop relaxation in an $S = \frac{1}{2}$, $I = \frac{1}{2}$ system. The levels are labeled with the signs of $(M_S M_I)$.

one can see straight away that the only type of transition induced is the so-called flip-flop transition $(M_S M_I) \leftrightarrow (M_S \pm 1, M_I \mp 1)$ in which both electron and nuclear spin states are changed, but in opposite directions. The level diagram in Figure 8, drawn for $I = \frac{1}{2}$, helps to make clear the expected effect on the ENDOR spectrum. If the 1–3 ESR transition is saturated, then when an rf field drives the 3–4 transition, spins can get back into level 1 via W_x, and one expects an ENDOR response. However, irradiating the 1–2 transition does not help desaturate level 3, since no new relaxation route is opened up—there should be no ENDOR response. For saturation of the 2–4 ESR transition, irradiation of 1–2 should enable W_x to desaturate level 4, giving rise to an ENDOR signal, while irradiation of the 3–4 transition should not. The prediction then is very clear: If the high-field ESR line is monitored, the high-frequency ENDOR signal should be strong; if the low-field ESR line is monitored, the low-frequency ENDOR signal should be strong. This conclusion is independent of the sign of the hyperfine coupling constant.

Hyperfine couplings to β protons in π radicals may fluctuate as a result of internal motions, giving a time dependence to the dihedral angle defining the disposition of the proton with reference to the π system.[34] The case of the 2,6-di-*tert*-butyl-4-cyclohexylphenoxy radical seems to be fairly clear cut. There is a single β proton in the cyclohexyl group in this radical, and the coupling to it shows a strong temperature dependence ascribed to hindered rotation.[67] The ENDOR spectrum[68] shows just the behavior predicted above. There is an implied suggestion[63] that modulation of the isotropic coupling may be more generally important in determining the ENDOR behavior of β protons, for example, in the anion radicals of methyl-substituted quinones. However, detailed analysis of the ENDOR response of the duroquinone anion indicated that it was not important in that radical.[69] Again, the alkali metal ENDOR of the complexes of di-*o*-mesitoyl benzene shows the behavior expected for strong flip-flop relaxation,[20,21] but recent work by Van der Drift and his colleagues shows that modulation of the isotropic metal couplings is not important in these systems.[76]

A familiar relaxation mechanism to all who have given any consideration to ESR linewidths is, of course, the electron–nuclear dipolar (END)

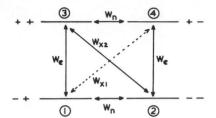

Figure 9. END-induced relaxation in a $S = \frac{1}{2}$, $I = \frac{1}{2}$ system. The levels are labeled with the signs of $(M_S M_I)$. W_{x2} is the flip-flop process, W_{x1} the flip-flop, and the latter is drawn dashed to emphasize that it is slower than the former.

interaction; this anisotropic coupling fluctuates as the radical tumbles in solution. The relaxation Hamiltonian is

$$H(t) = S_\alpha A'_{\alpha\beta}(t) I_\beta \qquad \alpha, \beta = x, y, z$$

and this gives rise to pure electron spin $(M_S, M_I) \leftrightarrow (M_S \pm 1, M_I)$, pure nuclear spin $(M_S, M_I) \leftrightarrow (M_S, M_I \pm 1)$, flip-flop $(M_S, M_I) \leftrightarrow (M_S \pm 1, M_I \mp 1)$, and flop-flop $(M_S, M_I) \leftrightarrow (M_S \pm 1, M_I \pm 1)$ transitions. Expressions for these various transition probabilities are given by Freed[52] and have been summarized for special cases by others.[63,70] The pure nuclear spin relaxation rate is independent of M_S, and so there is no effect from it that would make the high- and low-frequency ENDOR lines for a particular ESR transition have different intensities. However, the two cross-relaxation rates are different, the flip-flop process being slower than the flop-flop by a factor of 6. The situation is summarized for an $S = \frac{1}{2}, I = \frac{1}{2}$ system in Figure 9, and qualitative analysis of the effect on ENDOR shows that the relative intensities vary in the opposite sense to that expected for modulation of the isotropic hyperfine coupling. For saturation of the high-field ESR transition, the ENDOR line at low frequency should be more intense than that at high frequency. Under limiting conditions, the maximum intensity ratio should be $9 : 1$, though in practice, relaxation by other routes and by other mechanisms will operate to reduce it.[21]

The spectra of nitroxide radicals have the qualitative behavior described above,[14,53] and the END mechanism is clearly of major importance in these systems. This is hardly surprising, since they are essentially nitrogen-centered radicals and one expects a healthy dipolar coupling. Similarly, α-fluorine nuclei should have considerable dipolar coupling, and the ENDOR spectra of the fluoranil radical anion do indeed show the type of effects expected on the basis of the above discussion.[57] The comparative temperature behavior of the ^1H and ^{19}F signals in fluorine-substituted benzophenone anions[57] can also be understood in terms of stronger dipolar relaxation for the fluorine.

These two examples show how the operation of two familiar basic relaxation mechanisms can be illustrated rather strikingly and satisfyingly by ENDOR. Detailed quantitative comparisons of observed effects with theory are very difficult[69] because of the complexities of theory and experiment, but so far as can be seen at present, spin relaxation theory and solution ENDOR observations have a happy mutual relationship. The former explains the latter and the latter gives added credence to the former.

In ESR, of course, there has always been a great deal of interest in chemical exchange as a relaxation mechanism. Roughly speaking, exchange is not helpful for ENDOR, as it tends to wipe out the signals. But there is one process that has a distinctive effect and that may have been observed. This is so-called unpolarized exchange.[71] It could occur if there were exchange between radicals and diamagnetic precursors in a system such that the nuclear spin relaxation in the diamagnetic species were independent of their occasional excursions to paramagnetism. Such a process effectively causes nuclear relaxation within a given electron spin state, and so it is rather like the END mechanism. It has been pointed out that it could explain the temperature dependence of the metal ENDOR from the complexes of di-*o*-mesitoylbenzene with rubidium.[21] The difficulty is in seeing how the conditions required for unpolarized exchange to occur could obtain. This is an interesting problem that should receive further attention.

3.3. Linewidths

Studies of linewidths in ESR spectra have been enormously fruitful and indeed the principal means of studying electron spin relaxation. ENDOR linewidths should be similarly informative but as yet have been exploited but little. The factors determining the linewidth are related to or are the same as those determining the ENDOR mechanism and those we have been discussing. In particular, the END interaction is probably the dominant contributor. This is expected from Freed's analysis[52] and has been clearly established for some quinone radical anions by the very detailed studies of Leniart, Connor, and Freed.[69] Just as in ESR, so in ENDOR may exchange processes make a contribution to the linewidth, and measurement of the additional linebroadening enables the rate of the exchange process to be measured. This application of ENDOR has been made, but only in a few cases. Das, Connor, Leniart, and Freed[23] have studied the rate of the hindered motion of the long side chain in the radical anion of ubiquinone and showed that the kinetic data obtained from ENDOR agreed well with those obtained from ESR. Von Borczyskowski, Möbius, and Plato[40] have studied hindered rotation of the phenyl rings in 2,5-diphenylsilacyclopentadiene radical anion. They obtained data at both the fast and slow exchange limits, and the kinetic conclusions were in excellent agreement.

These papers[23,40] certainly show the applicability of ENDOR to the study of kinetic processes of chemical interest, but perhaps somewhat more care is needed than in the more familiar ESR studies. In the first place, one has to be careful to measure the unsaturated linewidth; this is obtained by using widths extrapolated to zero rf power in the kinetic analysis. The second point is that, as temperature is the usual variable, one has to be sure that one is truly measuring the exchange contribution to the width as a function of it. This has been done by assuming that the temperature dependence of the nonexchange part of the width was reflected by the variation in width of some other line for a coupling not being modulated by the exchange process.

4. Conclusion

In the author's view, the solution ENDOR field is wide open for exploitation. So few of the very many possible systems have been studied, that one can put almost anything into an ENDOR spectrometer and come up with something interesting. The ability to measure precisely hyperfine coupling constants and small changes in them offers tremendous scope for studying radical geometries, solvation, ion-pairing, and so forth, while the interpretation of ENDOR effects using spin relaxation theory cannot but enhance our knowledge and understanding of molecular dynamics. It is to be hoped that in the next few years the number of practitioners will increase and thus the potential of the technique will be more fully realized.

ACKNOWLEDGMENT

I am grateful to Professor Klaus Möbius for a preprint of reference 73 and for other information about the work cited in references 72–76.

References

1. G. Feher, *Phys. Rev.* **103**, 834–835 (1956).
2. A. Cederquist, Ph.D. thesis, Washington University, St. Louis, Missouri (1963).
3. J. S. Hyde and A. H. Maki, *J. Chem. Phys.* **40**, 3117–3118 (1964).
4. J. S. Hyde, *J. Chem. Phys.* **43**, 1806–1818 (1965).
5. N. M. Atherton, in *Electron Spin Resonance* (R. O. C. Norman, ed.), Vol. 1, pp. 32–46 (1973); Vol. 2, pp. 36–51 (1974); Vol. 3, pp. 23–34 (1976), Specialist Periodical Reports, The Chemical Society, London. K. Möbius, also in *Electron Spin Resonance*, Vol. 4, pp. 16–29 (1977).
6. A. L. Kwiram, *Ann. Rev. Phys. Chem.* **22**, 133–170 (1971).
7. J. H. Freed, *Ann. Rev. Phys. Chem.* **23**, 265–310 (1972).
8. J. S. Hyde, *Ann. Rev. Phys. Chem.* **25**, 407–435 (1974).
9. A. L. Kwiram, in *Magnetic Resonance* (C. A. McDowell, ed.), *M.T.P Int. Rev. Sci. Phys. Chem.*, Ser. 1, Vol. 4, pp. 271–316, Butterworths' Publications, London (1972).
10. R. D. Allendoerfer, in *Magnetic Resonance* (C. A. McDowell, ed.), *M.T.P. Int. Rev. Sci. Phys. Chem.*, Ser. 2, Vol. 4, pp. 29–53, Butterworths' Publications, London (1975).
11. F. Gerson, J. Jachimowicz, K. Möbius, R. Biehl, J. S. Hyde, and D. S. Leniart, *J. Mag. Res.* **18**, 471–484 (1975).

12. G. Breit and I. I. Rabi, *Phys. Rev.* **38**, 2082–2083 (1931).
13. A. H. Maki, R. D. Allendoerfer, J. C. Danner, and R. T. Keys, *J. Amer. Chem. Soc.* **90**, 4225–4231 (1968).
14. D. S. Leniart, J. C. Vedrine, and J. S. Hyde, *Chem. Phys. Lett.* **6**, 637–640 (1970).
15. A. Abragam and B. Bleaney, *Electron Paramagnetic Resonance of Transition Ions*, Oxford University Press, Oxford England (1970).
16. G. K. Fraenkel, *J. Chem. Phys.* **42**, 4275–4298 (1965).
17. G. K. Fraenkel, *J. Phys. Chem.* **71**, 139–171 (1967).
18. R. D. Allendoerfer and P. H. Rieger, *J. Chem. Phys.* **46**, 3410–3418 (1967); R. J. Faber and G. K. Fraenkel, *J. Chem. Phys.* **47**, 2462–2476 (1967).
19. A. G. Redfield, *Adv. Mag. Res.* **1**, 1–32 (1965).
20. N. M. Atherton and B. Day, *J. Chem. Soc., Faraday II* **69**, 1801–1807 (1973).
21. H. van Willigen, M. Plato, R. Biehl, K. P. Dinse, and K. Möbius, *Molec. Phys.* **26**, 793–809 (1973).
22. R. D. Allendoerfer and A. H. Maki, *J. Amer. Chem. Soc.* **91**, 1088–1094 (1969).
23. M. R. Das, H. D. Connor, D. S. Leniart, and J. H. Freed, *J. Amer. Chem. Soc.* **92**, 2258–2268 (1970).
24. J. S. Hyde, *J. Phys. Chem.* **71**, 68–73 (1967).
25. R. Biehl, K. P. Dinse, K. Möbius, M. Plato, H. Kurreck, and U. Mennenga, *Tetrahedron* **29**, 363–368 (1973).
26. J. H. Freed, D. S. Leniart, and J. S. Hyde, *J. Chem. Phys.* **47**, 2762–2773 (1967).
27. K. P. Dinse, K. Möbius, and R. Biehl, *Z. Naturforsch.* **28a**, 1069–1080 (1973).
28. K. P. Dinse, R. Biehl, K. Möbius, and M. Plato, *J. Mag. Res.* **6**, 444–452 (1972).
29. J. H. Freed, *J. Chem. Phys.* **50**, 2271–2272 (1968).
30. K. P. Dinse, R. Bichl, and K. Möbius, *J. Chem. Phys.* **61**, 4335–4341 (1974).
31. Y. Yamada, S. Toyoda, and K. Ouchi, *J. Phys. Chem.* **78**, 2512–2515 (1974).
32. R. Biehl, M. Plato, K. Möbius, and K. P. Dinse, in *Magnetic Resonance and Related Phenomena: Proceedings of the 17th AMPERE Congress, 1972* (V. Hovi, ed.), pp. 423–426, North-Holland, Amsterdam (1973).
33. T. C. Christidis and F. W. Heineken, *Chem. Phys.* **2**, 239–244 (1973).
34. C. Heller and H. M. McConnell, *J. Chem. Phys.* **32**, 1535–1539 (1960).
35. E. W. Stone and A. H. Maki, *J. Chem. Phys.* **37**, 1326–1333 (1962).
36. R. D. Allendoerfer and D. J. Eustace, *J. Phys. Chem.* **75**, 2765–2769 (1971).
37. N. M. Atherton, A. J. Blackhurst, and I. P. Cook, *Trans. Faraday Soc.* **67**, 2510–2515 (1971).
38. N. L. Bauld, J. D. McDermed, C. E. Hudson, Y. S. Rim, J. Zoeller, R. D. Gordon, and J. S. Hyde, *J. Amer. Chem. Soc.* **91**, 6666–6676 (1969).
39. F. Nemoto, F. Shimoda, and K. Ishizu, *Chem. Lett.*, 693–697 (1974).
40. C. von Borczyskowski, K. Möbius, and M. Plato, *J. Mag. Res.* **17**, 202–211 (1975).
41. J. S. Hyde, R. Breslow, and C. DeBoer, *J. Amer. Chem. Soc.* **88**, 4763–4764 (1966).
42. R. F. Adams and N. M. Atherton, *Molec. Phys.* **17**, 673–676 (1969).
43. K. Watanabe, J. Yamauchi, H. Ohya-Nishiguchi, Y. Deguchi, and K. Ishizu, *Chem. Lett.* **1974**, 489–492 (1974).
44. N. S. Dalal, D. E. Kennedy, and C. A. McDowell, *Chem. Phys. Lett.* **30**, 186–189 (1975).
45. K. Ishizu, H. Nagai, K. Mukai, M. Kohno, and T. Yamamoto, *Chem. Lett.*, 1261–1264 (1973).
46. N. M. Atherton and A. J. Blackhurst, *J. Chem. Soc. Faraday II* **68**, 470–475 (1972).
47. Y. Kotake and K. Kuwata, *Bull. Chem. Soc. Japan*, **45**, 2663 (1972).
48. N. M. Atherton and P. A. Henshaw, *J. Chem. Soc. Perkin II*, 258–260 (1975).
49. D. H. Whiffen, *Molec. Phys.* **10**, 595–596 (1966).
50. S. Geschwind, in *Hyperfine Interactions* (A. J. Freeman and R. B. Frankel, eds.), pp. 225–286, Academic Press, New York (1967).
51. E. R. Davies and T. R. Reddy, *Phys. Lett.* **31A**, 398–399 (1970).

52. J. H. Freed, *J. Chem. Phys.* **43**, 2312–2332 (1965).
53. N. M. Atherton and M. Brustolon, *Molec. Phys.* **32**, 23–31 (1976).
54. K. P. Dinse, K. Möbius, M. Plato, R. Biehl, and H. Haustein, *Chem. Phys. Lett.* **14**, 196–200 (1972).
55. B. J. Herold, A. F. Neiva Correia, and H. dos Santos Veiga, *J. Amer. Chem. Soc.* **87**, 2661–2665 (1965).
56. K. P. Dinse, K. Möbius, R. Biehl, and M. Plato, in *Magnetic Resonance and Related Phenomena: Proceedings of the 17th AMPERE Congress, 1972* (V. Hovi, ed.), pp. 419–422, North-Holland, Amsterdam, 1973.
57. W. Lubitz, K. P. Dinse, K. Möbius, and R. Biehl, *Chem. Phys.* **8**, 371–383 (1975).
58. J. P. Lloyd and G. E. Pake, *Phys. Rev.* **94**, 579–591 (1954).
59. M. J. Stephen and G. K. Fraenkel, *J. Chem. Phys.* **32**, 1435–1444 (1960).
60. M. J. Stephen, *J. Chem. Phys.* **34**, 484–489 (1961).
61. F. Bloch, *Phys. Rev.* **102**, 104–135 (1956).
62. R. D. Allendoerfer and A. H. Maki, *J. Mag. Res.* **3**, 398–410 (1970).
63. N. M. Atherton and B. Day, *Molec. Phys.* **27**, 145–158 (1974).
64. N. M. Atherton and P. A. Kennedy, *Chem. Phys. Lett.* **43**, 186–188 (1976).
65. B. G. Segal, A. Reymond, and G. K. Fraenkel, *J. Chem. Phys.* **51**, 1336–1352 (1969).
66. J. H. Freed and G. K. Fraenkel, *J. Chem. Phys.* **39**, 326–348 (1963).
67. N. M. Atherton and R. S. F. Herding, *Nature* **198**, 987–988 (1963).
68. N. M. Atherton and B. Day, *Chem. Phys. Lett.* **15**, 428–430 (1972).
69. D. S. Leniart, H. D. Connor, and J. H. Freed, *J. Chem. Phys.* **63**, 165–199 (1975).
70. N. M. Atherton, *Electron Spin Resonance*, Ellis Horwood, Chichester, England (1973).
71. J. H. Freed, *J. Phys. Chem.* **71**, 38–51 (1967).
72. K. Möbius, *ENDOR and TRIPLE resonance on radicals in solution*, Paper presented at the Electron Spin Resonance Symposium, University of Nijmegen, Holland, August 1976.
73. R. Biehl, W. Lubitz, K. Möbius, and M. Plato, *J. Chem. Phys.* **66**, 2074–2078 (1977).
74. W. Lubitz, R. Biehl, and K. Möbius, *J. Mag. Res.* **27**, 411 (1977).
75. B. Kirste, H. Kurreck, W. Lubitz, and K. Schubert, *J. Amer. Chem. Soc.* **100**, 2292–2299 (1978).
76. E. Van der Drift and K. Möbius, personal communications.

5

Modulation Effects in Multiple Electron Resonance Spectroscopy

Lauraine A. Dalton and Larry R. Dalton

1. Introduction

In the broadest sense, modulation effects can be considered to represent those perturbations or influences on spectroscopic line shapes that arise either (1) from applied coherent (frequency, amplitude, or phase) modulation of the electromagnetic radiation incident upon the sample or modulation of the energy levels (resonance condition) as found, for example, with Stark modulation in microwave spectroscopy[1-6] or Zeeman modulation in magnetic resonance[7-19] or (2) from modulation of the populations of energy levels or of the phase coherence of precessing electric or magnetic dipoles arising from stochastic lattice fluctuations.[19-28] Thus, modulation effects can be classified as either applied and coherent or as molecular and stochastic. Earlier in this volume, Freed has discussed the effects of stochastic lattice fluctuations upon electron and nuclear spin relaxation rates and hence upon magnetic resonance and multiple magnetic resonance line shapes. While we shall explicitly consider both types of modulation effects, we shall do so from the standpoint of examining the importance of applied coherent modulation in determining magnetic resonance and multiple res-

Lauraine A. Dalton • Department of Molecular Biology, Vanderbilt University, Nashville, Tennessee. *Present Address:* Department of Chemistry, State University of New York at Stony Brook, Stony Brook, New York   *Larry R. Dalton* • Departments of Chemistry and Biochemistry, Vanderbilt University, Nashville, Tennessee; Department of Chemistry, State University of New York at Stony Brook, Stony Brook, New York

onance responses. There are three aspects to the utilization of coherent modulation and accompanying phase-sensitive detection:

1. The initial motivation for the employment of modulation was, almost certainly, to convert dc signals to higher frequencies and thus to discriminate against the $1/f$ noise from detectors employed in microwave and magnetic resonance experiments and at high microwave field intensities to discriminate against low-frequency source noise. The main advantage of conversion of a dc signal to a high ac frequency followed by phase-sensitive detection at the fundamental or higher harmonic of that frequency is that, generally speaking, ac amplification is less troublesome than dc amplification, and significant noise reduction is accomplished employing narrow bandwidths and variable time constant filter circuits. We note in turn that narrow bandwidths are possible because of the extremely high-Q filter circuits that are an integral part of phase-sensitive amplifiers.

2. The employment of modulation to effect more precise definition of resonance frequencies represents a second type of application. When low-amplitude, low-frequency Zeeman modulation, Stark modulation, or frequency modulation of the electromagnetic radiation incident upon the sample is employed, detection of each higher harmonic of the applied modulation corresponds to the display of a higher derivative of the spectrum obtained in the absence of modulation; e.g., first harmonic (fundamental) detection yields a spectrum that approximates the first derivative of the spectrum in the absence of modulation, detection at the second harmonic of the applied modulation yields a spectrum which approximates the second derivative, and so forth. An example is presented in Figure 1, which demonstrates how the measurement of hyperfine interactions is aided by detection of higher harmonics. This use of modulation can perhaps be most significant in the analysis of the resonance spectra of disordered solids (powder patterns).

3. More recently modulation has been employed to probe molecular dynamics (stochastic molecular modulation effects). Stated briefly, resonance and double resonance spectra obtained employing applied modulation and phase-sensitive detection depend upon the ratios of applied and molecular modulation frequencies, and employment of applied frequencies of the order of molecular frequencies facilitates the measurement of the molecular rates. The measurement of molecular modulation or relaxation rates by pulsing the applied electromagnetic radiation and monitoring the transient response is well known.[29,30] Periodic modulation can in an analogous manner be considered to evoke a transient response and thus provide a time base for the measurement of molecular rates. It is in this last sense that modulation is of the most interest in understanding the detailed responses observed in single and multiple resonance experiments and to which we shall devote our primary attention in this chapter.

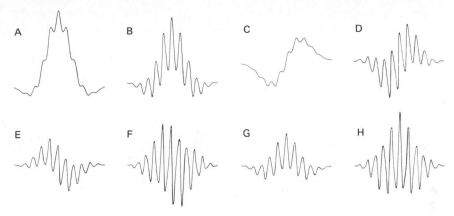

Figure 1. Computer simulations of EPR signals at the first and second harmonics of the Zeeman modulation are shown to demonstrate the employment of Zeeman modulation to improve spectral resolution. The signals and relative gains are as follows: (A) $\alpha(1)$, the in-phase dispersion signal at the first harmonic of the modulation, relative gain = 2; (B) $\beta(1)$ the out-of-phase dispersion signal, relative gain = 25; (C) $\gamma(1)$, the in-phase absorption signal, relative gain = 1; (D) $\delta(1)$, the out-of-phase absorption signal, relative gain = 25; (E) $\alpha(2)$, the in-phase dispersion signal at the second harmonic of the modulation, relative gain = 10; (F) $\beta(2)$, the out-of-phase dispersion signal, relative gain = 40; (G) $\gamma(2)$, the in-phase absorption signal, relative gain = 10; (H) $\delta(2)$, the out-of-phase absorption signal, relative gain = 40. The computer simulations are of the low-field $[\nu(^{14}N) = +1]$ hyperfine component of a nitroxide (2,2,6,6-tetramethyl-4-piperidinol-1-oxyl) EPR spectrum. Relevant parameters include: $\bar{a}(\alpha\text{-CH}_3, \text{equatorial}) = -0.456$ G; $\bar{a}(\alpha\text{-CH}_3, \text{axial}) = 0.00$ G; $\bar{a}(\beta\text{-CH}_2, \text{equatorial}) = -0.559$ G; $\bar{a}(\beta\text{-CH}_2, \text{axial}) = -0.299$ G; $T_{2e}(\text{effective}) = 2.7 \times 10^{-7}$ sec; $T_{1e}(\text{effective}) = 1.3 \times 10^{-6}$ sec; an observing microwave field intensity of 0.017 G, a peak-to-peak Zeeman modulation amplitude of 0.24 G, and a modulation frequency ω_s of $2\pi \times 10^5$ rad/sec. (From Percival et al.[52])

Our discussion will focus upon commonly observed Zeeman modulation effects. Obviously, for many types of experiments, amplitude, frequency, or phase modulation of the applied electromagnetic fields is desirable. A detailed discussion of these subjects is beyond the scope of the present chapter and the reader is referred to Dalton[31] and Günthard.[32] Also, although many of the remarks of this section are applicable to other spectroscopic experiments, our presentation will be limited to a consideration of electron paramagnetic resonance and double resonance.

Figure 2 is a block diagram of a simple EPR spectrometer. In the conventional mode of operation, the signal response arises from phase-sensitive detection with respect to the microwave electromagnetic radiation field and with respect to an applied Zeeman modulation field. There are four unique components of the signal response detected at a given harmonic of the microwave and modulation frequencies. These are (1) the signal component in phase with the microwaves (called the dispersion signal) and in phase with

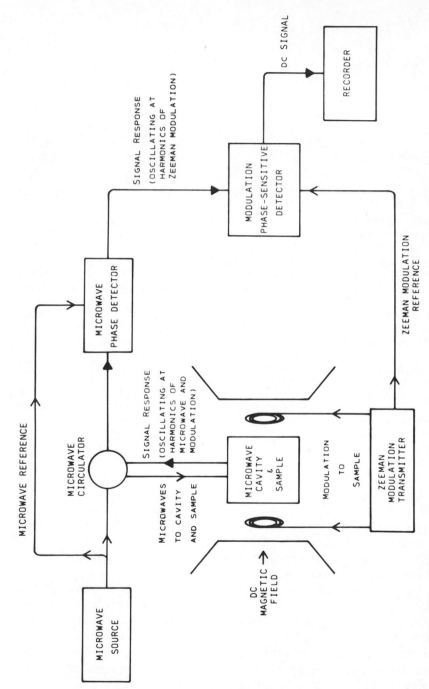

Figure 2. Block diagram of an electron paramagnetic resonance spectrometer showing the coding and decoding of the signal response.

the Zeeman modulation, (2) the signal component in phase with the microwaves but 90° out of phase (phase quadrature) with the Zeeman modulation, (3) the component out of phase with the microwaves (called the absorption signal) and in phase with the Zeeman modulation, and (4) the signal component out of phase with respect to both the microwaves and the Zeeman modulation. For the sake of brevity we shall denote these $\alpha(k)$, $\beta(k)$, $\gamma(k)$, and $\delta(k)$, respectively, where k denotes the harmonic detected. With conventional EPR spectrometers, the microwave frequencies employed are sufficiently high that an appreciable signal exists only at the first harmonic (fundamental). On the other hand, Zeeman modulation frequencies are usually in the range 10 to 10^6 Hz and thus for finite modulation amplitudes appreciable signal will exist at the higher harmonics. Hence, the index k in $\alpha(k), \ldots, \delta(k)$ can be understood to refer to the signal component at the kth Zeeman modulation harmonic with detection at the first microwave harmonic.

At low Zeeman modulation frequencies and amplitudes, the signal components 90° out of phase with respect to the driving modulation field, $\beta(k)$ and $\delta(k)$, are small and approach zero as the modulation frequency goes to

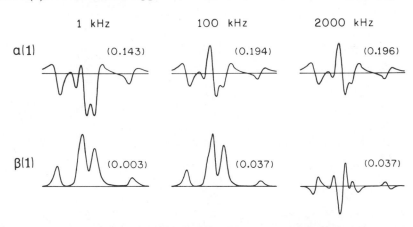

Figure 3. Components of the total electron spin response as a function of Zeeman modulation frequencies for frequencies of 1, 100, and 2000 kHz. Signal amplitudes in arbitrary units are given in parentheses; amplitudes are measured as the distance between maximum and minimum displacements of the signal. The in- and out-of-phase dispersion signals detected at the first harmonic of the Zeeman modulation [$\alpha(1)$ and $\beta(1)$, respectively] shown in this figure were computed employing the following parameters: $g_{zz} = 2.00241$, $g_{xx} = g_{yy} = 2.00741$; $A_{zz}(^{14}\text{N}) = 35.0$ G, $A_{xx}(^{14}\text{N}) = A_{yy}(^{14}\text{N}) = 7.0$ G; $T_{1e} = 6.6 \times 10^{-6}$ sec; $T_{2e} = 2.4 \times 10^{-8}$ sec; a microwave field intensity h_o of 0.25 G and a microwave frequency of $\omega_0 = 2\pi(9.45 \times 10^9)$ Hz; a Zeeman modulation amplitude H_s of 2.5 G corresponding to a peak-to-peak modulation amplitude of 5.0 G. Calculations were carried out for an isotropic Brownian diffusion model and a rotational correlation time τ_2 of 10^{-7} sec. Parameters were chosen to give a crude simulation of the spectra of maleimide spin-labeled human oxyhemoglobin A in aqueous 20% glycerol solutions at 5°C.

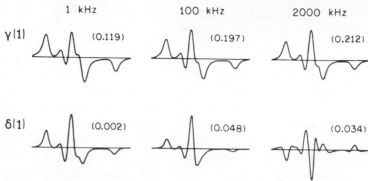

Figure 4. The variation of the in- and out-of-phase absorption signals detected at the first harmonic of the modulation [γ(1) and δ(1), respectively] with Zeeman modulation frequency. Parameters are the same as in Figure 3.

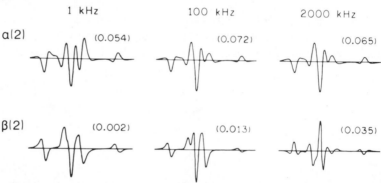

Figure 5. The variation of the in- and out-of-phase dispersion signals detected at the second harmonic of the modulation with Zeeman modulation frequency. Parameters are the same as in Figure 3.

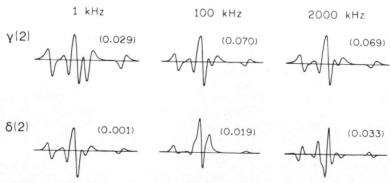

Figure 6. The variation of the in- and out-of-phase absorption signals at the second harmonic of the modulation as a function of Zeeman modulation frequency. Parameters are the same as in Figure 3. The same units are employed in Figures 3–6 so that quantitative comparisons can be made among these figures.

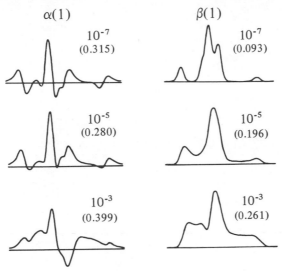

$\alpha(1)$ $\beta(1)$

10^{-7} (0.315) 10^{-7} (0.093)

10^{-5} (0.280) 10^{-5} (0.196)

10^{-3} (0.399) 10^{-3} (0.261)

Figure 7. The computer-simulated dispersion signals at the first harmonic of the Zeeman modulation and in phase with the modulation $\alpha(1)$ and in phase quadrature $\beta(1)$. Spectra are calculated for a Zeeman modulation frequency of 100 kHz and for isotropic Brownian rotational correlation times ranging from 10^{-7} to 10^{-3} sec as indicated in the figure. Also indicated in the figure (in parentheses) are the signal amplitudes calculated between points of maximum positive and negative excursions. The same parameters were employed in calculating spectra as given in Figure 3, with the exception that we attempted to account for the effects of unresolved hyperfine interactions by setting $T_{2e} = 2.0 \times 10^{-7}$ sec and employing the Gaussian convolution function $g(H_o) = A \exp[\frac{1}{2}(H_o - H^{res})^2/4.52]$ to broaden spectra. This procedure improves agreement between experimental and theoretical spectra for long correlation times (e.g., $\tau_2 = 10^{-3}$ sec). (From Perkins et al.[111])

zero. However, for modulation frequencies comparable to molecular relaxation rates, the in-phase quadrature signals $\beta(k)$ and $\delta(k)$ can be comparable to or even greater in amplitude than the in-phase signals $\alpha(k)$ and $\gamma(k)$. Figures 3–11 show computer simulations of a ^{14}N nitroxide spin label showing the dependence of all eight signal components at the first and second harmonics of the Zeeman modulation upon modulation frequency and amplitude and upon rotational correlation time (reflecting the modulation of magnetic interactions by molecular dynamics, i.e., by the isotropic rotational diffusion of the spin-labeled molecule). Two points are dramatically illustrated by these figures. The first is that when the ratio of applied modulation to molecular modulation frequencies is near unity, phase quadrature signal intensities are comparable to in-phase signal intensities. Second, it is to be noted that although all eight signal components $\alpha(1), \ldots, \delta(2)$ are determined in each instant by one set of applied and molecular modulation conditions, not all signal components are equally sensitive to a given variable (e.g., energy transfer or relaxation rate). This latter observation has

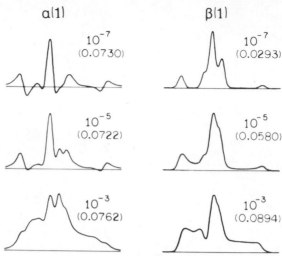

Figure 8. The computer-simulated dispersion signal at the first harmonic of the Zeeman modulation and in phase with the modulation $\alpha(1)$ and the dispersion signal at the first harmonic of the Zeeman modulation and out-of-phase with the modulation $\beta(1)$. Spectra are calculated for a Zeeman modulation frequency of 100 kHz ($\omega_s = 2\pi \times 10^5$ Hz) and a peak-to-peak modulation amplitude of 1 G ($H_s = 0.5$ G). Spectra were computed for isotropic Brownian rotational correlation times ranging from 10^{-7} to 10^{-3} sec as indicated in the figure. Also indicated in the figure (in parentheses) are the signal amplitudes calculated between points of maximum positive and negative excursions. Other computational parameters have the same values as in Figure 7. Comparison of spectra in Figure 7 with the corresponding spectra in Figure 8 demonstrates the dependence upon Zeeman modulation amplitude. (From Perkins et al.[111])

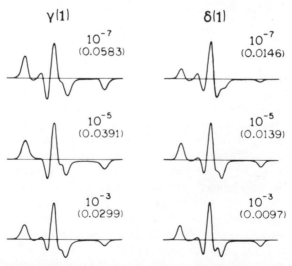

Figure 9. Computer-simulated $\gamma(1)$ and $\delta(1)$ spectra calculated for a Zeeman modulation frequency of 100 kHz and a peak-to-peak modulation amplitude of 1 G. Other parameters are as given in Figure 8. (From Perkins et al.[111])

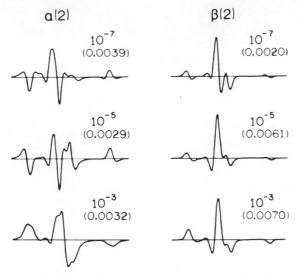

Figure 10. Computer-simulated $\alpha(2)$ and $\beta(2)$ spectra. Computational parameters have the same values as in Figure 8. (From Perkins *et al.*[111])

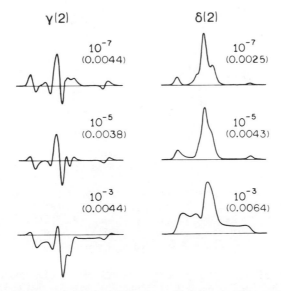

Figure 11. Computer-simulated $\gamma(2)$ and $\delta(2)$ spectra. Computational parameters have the same values as in Figure 8. A comparison of the computed spectra shown in Figures 3–11 demonstrates the dependence of the various signal components at the first and second harmonics of the Zeeman modulation upon Zeeman modulation frequency, modulation amplitude, and the frequency of the stochastic modulation (i.e., upon rotational correlation time). These spectra indicate the distribution of spin response with respect to frequency and phase relative to the driving microwave and modulation fields. (From Perkins *et al.*[111])

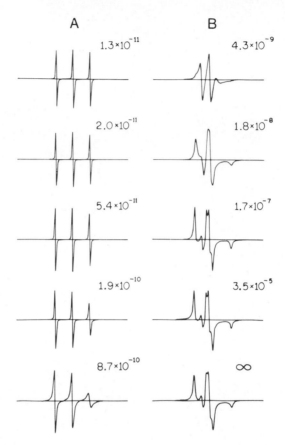

Figure 12. Variation of conventional EPR spectra $\gamma(1)$ with rotational correlation time (indicated in the figure). These spectra were recorded employing the experimental model system of perdeutero-2,2,6,6-tetramethy×-4-piperidinol-1-oxyl (D-TANOL) in perdeutero-*sec*-butyl-benzene (D-SBB), and spectra were recorded at the following temperatures and rotational correlation times: $+24°C$ $(1.3 \times 10^{-11}$ sec), $0.6°C$ $(2.0 \times 10^{-11}$ sec), $-33.0°C$ $(5.4 \times 10^{-11}$ sec), $-58.5°C$ $(1.9 \times 10^{-10}$ sec), $-77.2°C$ $(8.7 \times 10^{-10}$ sec), $-88.5°C$ $(4.3 \times 10^{-9}$ sec), $-95.7°C$ $(1.8 \times 10^{-8}$ sec), $-104.5°C$ $(1.7 \times 10^{-7}$ sec), $-119.9°C$ $(3.5 \times 10^{-5}$ sec), $-160°C$ $(\tau_2 \sim \infty)$. The spectra presented in column A were recorded employing a 100-G scan range, while those in column B were recorded employing a 200-G scan range. All spectra were recorded employing a 100-kHz Zeeman modulation field of 0.2-G peak-to-peak amplitude and an incident microwave power of 0.5 mW. Computer analysis shows that D-TANOL in D-SBB is best described by jump or free diffusion rather than by a Brownian diffusion model.

been employed to facilitate an important extension[33,34] of the commonly employed spin-labeling technique[27,35-41] of defining biological structures, interactions, and functions. Until 1972, the molecular dynamics of spin-labeled biomolecules were defined by examining the $\gamma(1)$ signal under conditions of small Zeeman modulation and microwave field amplitudes such that the observed signal approximated the first derivative of the $\gamma(0)$ signal (the absorption signal obtained without Zeeman modulation). As is demonstrated for the experimental model system of perdeutero-2,2,6,6-tetramethyl-4-piperidinol-1-oxyl (D-TANOL) in perdeutero-*sec*-butylbenzene (D-SBB) shown in Figure 12, such $\gamma(1)$ spectra are from a practical standpoint insensitive to molecular motion characterized by rotational correlation times longer than 3×10^{-7} sec. On the other hand, other signal components, particularly $\beta(1)$ and $\delta(2)$ and to a lesser extent $\alpha(1)$, exhibit good sensitivity to molecular dynamics for correlation times as long as 10^{-3} sec, as is shown in Figures 13 and 14. By increasing the microwave field intensity until resonance transitions are partially saturated, the sensitivity of $\gamma(1)$ signals to slow motion can be increased somewhat, as is shown in Figure 15; however, the sensitivity never approaches that of the $\beta(1)$ and $\delta(2)$ signals. As many proteins and supramolecular complexes are characterized by correlation times in the micro- to millisecond range, examination and analysis of $\beta(1)$ and $\delta(2)$ signals represents an important biophysical measurement capability.[42] Obviously, these signals can be understood only by an explicit consideration of both applied and molecular modulation events.

For a system as complex as a slowly diffusing spin label, the amplitude and spectral shape of a component of the total signal such as $\beta(1)$ will depend upon many molecular and instrumental parameters, and the derivation of analytical expressions to describe the signal components is impractical. Analytical expressions can, however, be derived[33] for a system giving rise to a resonance line unbroadened by magnetic anisotropy and meeting the following criteria: (1) electron spin–lattice relaxation is the only saturation transfer mechanism, (2) electron spin–lattice and electron spin–spin relaxation times are equal, (3) the Zeeman modulation amplitude is less than the microwave field intensity, and (4) the condition of saturation exists. For such a case, the phase quadrature dispersion signal $\beta(1)$ can be shown to be proportional to $\omega_s T_1/(1 + \omega_s^2 T_1^2)$, where ω_s is the Zeeman modulation frequency and T_1 is the electron spin–lattice relaxation time.[33] Thus, a plot of signal amplitude of the $\beta(1)$ component versus ω_s yields a symmetric curve with a maximum at $\omega_s = (T_1)^{-1}$ and affords a convenient means of measuring the molecular relaxation rate. The behavior is necessarily more complex for systems with several molecular processes modulating the induced saturation; however, it can be noted that the faster events will in general dominate the response.

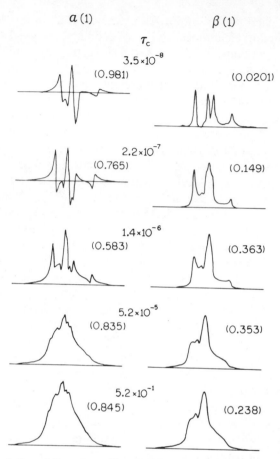

Figure 13. Variation of dispersion, $\alpha(1)$ and $\beta(1)$, EPR spectra with rotational correlation time is shown for a 10^{-4} *M* solution of D-TANOL in D-SBB. Signal amplitudes (in arbitrary units) calculated between points of maximum positive and negative excursions are given in parentheses. Spectra were recorded employing a 100-kHz Zeeman modulation field of 1.6-G peak-to-peak amplitude and an incident microwave power of 15 mW. Signal amplitudes can be compared within a figure such as 13 but not between figures such as 13 and 14.

In *e*lectron–*n*uclear *d*ouble *r*esonance (ENDOR) and *e*lectron–*e*lectron *d*ouble *r*esonance (ELDOR) experiments, an attempt to minimize applied modulation effects is often made. For example, Varian ENDOR (E-700 and E-1700) spectrometers are designed to operate employing 35-Hz Zeeman modulation, a frequency significantly less than commonly encountered molecular relaxation rates. The newly developed Bruker ENDOR (B-ER 200EN) spectrometer avoids Zeeman modulation (and thus some signal reduction and line broadening) by employing frequency modulation of and

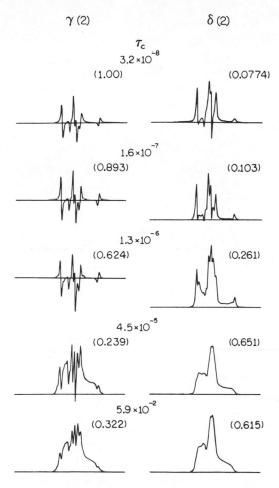

Figure 14. Variation of second harmonic absorption, $\gamma(2)$ and $\delta(2)$, EPR spectra with rotational correlation time is shown for a 10^{-4} M solution of D-TANOL in D-SBB. Signal amplitudes (in arbitrary units) calculated between points of maximum positive and negative excursions are given in parentheses. Spectra were recorded employing a 50-kHz Zeeman modulation field of 1.6-G peak-to-peak amplitude and an incident microwave power of 15 mW.

phase-sensitive detection with respect to the modulation of the applied radio-frequency field. Moreover, Hyde *et al.*[43,44] and Dorio and Chien[45–47] have observed that applied modulation effects in certain ELDOR experiments can be minimized by employing 270-Hz Zeeman modulation; it was found that this modulation frequency afforded adequate signal-to-noise (whereas substantially lower frequencies did not) and greatly simplified the analysis of ELDOR spectra. However, when modulation frequencies approaching

$$\gamma(1) \qquad\qquad\qquad \delta(1)$$

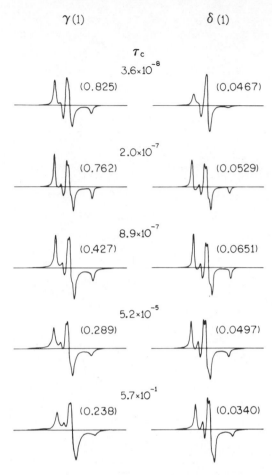

Figure 15. Variation of first harmonic absorption, $\gamma(1)$ and $\delta(1)$, EPR spectra with rotational correlation time is shown for a 10^{-4} M solution of D-TANOL in D-SBB. Signal amplitudes are given in parentheses. Spectra were recorded employing a 100-kHz Zeeman modulation field of 1.6-G peak-to-peak amplitude and an incident microwave power of 15 mW.

molecular relaxation rates (e.g., 100 kHz) are employed, ELDOR line shapes and signal amplitudes are observed to depend strongly upon modulation conditions.[48–52] As is shown in Figure 16, Zeeman modulation can also be employed to enhance the sensitivity of ELDOR to molecular motion. Again an important technological advance appears possible by employing applied modulation in a meaningful manner, namely, to enhance the sensitivity of spectra to molecular modulation effects. In a similar manner when Zeeman modulation frequencies of the order of 100 kHz are employed, ENDOR re-

sponses are observed to depend upon modulation conditions.[53] For example, Zarinov, Meiklyar, and Falin[53] have observed ENDOR signal
amplitudes for $^{47}Ti^{2+}$ in CaF_2 to go from a small positive signal to a large
negative signal as the amplitude of the applied 100-kHz Zeeman modulation
field is varied from 1 to 10 G. For this latter condition, the intensity of the
ENDOR signal amplitude amounted to approximately 10% of the EPR signal.

In Section 3 we discuss a variety of modulation effects for two classes of
systems, namely, spin labels and spin-labeled biomolecules[31,42,54,55] in
solution and solid-state organic radicals.[51,56]

First, in Section 2, we present a mathematical formalism capable of
explaining Zeeman modulation effects in EPR, ELDOR, and ENDOR. Modulation of electromagnetic radiation fields, including the 70-kHz modulation
employed in commercial spectrometers for automatic frequency control

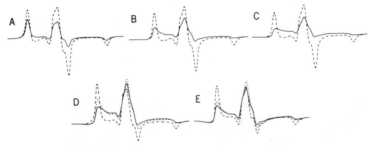

Figure 16. Passage ELDOR spectra for a 10^{-4} M solution of D-TANOL in D-SBB are shown
as a function of temperature and hence of rotational correlation time. (A) The dashed-line
spectrum is the $[\gamma(1), h_p = 0]$ EPR spectrum recorded at a temperature of $-110°C$ employing
a 100-kHz Zeeman modulation field of 4.0-G peak-to-peak amplitude and an incident microwave observing power of 1 mW. The solid-line spectrum is the $[\gamma(1), h_p \neq 0]$ EPR spectrum
recorded in the presence of an incident pump power of 400 mW and a microwave pump–
microwave observer frequency separation of $+4.5$ MHz. (B) The dashed line spectrum is the
$[\gamma(1), h_p = 0]$ EPR recorded at a temperature of $-120°C$ employing a 100-kHz Zeeman
modulation field at 4.0-G peak-to-peak amplitude and an incident microwave observing
power of 1 mW. The solid-line spectrum is the $[\gamma(1), h_p \neq 0]$ EPR recorded in the presence
of an incident pump power of 400 mW and a microwave pump–microwave observer frequency
separation of $+4.5$ MHz. (C) Same conditions as in A and B apply, except that the
sample temperature is $-130°C$. (D) The $[\gamma(1), h_p \neq 0]$ spectra recorded as described in A
through C are superimposed. Spectra are given for the following temperatures: $-110°C$
(dashed line), $-120°C$ (dotted line), $-130°C$ (solid line). (E) The $[\gamma(1), h_p \neq 0]$ spectra
recorded employing a 10-kHz Zeeman modulation field of 4-G peak-to-peak amplitude are
shown for temperatures of $-110°C$ (dashed line), $-120°C$ (dotted line), and $-130°C$ (solid
line). Spectra were recorded for an incident observing microwave power of 1 mW and an
incident pump power of 400 mW. The pump–observer frequency separation of $+4.5$ MHz
was employed. In order to avoid thermal differences due to microwave heating, all spectra
were recorded with a microwave pump power of 400 mW incident upon the cavity. For
$[\gamma(1), h_p = 0]$ spectra, the pump frequency was adjusted so as to be well off resonance.

(AFC) of the microwave source, will be ignored. Although we have analyzed such effects as well as the problems of variation of electromagnetic radiation and modulation field amplitudes and phases over samples of finite dimensions, such considerations would carry us too far afield in the present discussion. Because of space limitations, we have restricted the number of theoretical and experimental spectra shown. More comprehensive treatment of specific topics is available elsewhere.[34,54,57-67]

2. Theory

Formally, a complete description of EPR, ELDOR, and ENDOR responses (spectra) for simple nitroxide (spin label) and hydrocarbon radicals in high magnetic fields and recorded employing Zeeman modulation and phase-sensitive detection can be calculated by solving the spin density matrix equation,

$$\dot{\sigma}(\Omega, t)$$
$$= -i[\mathscr{H}(\Omega, t), \sigma(\Omega, t)] - \Gamma_R\{\sigma(\Omega, t) - \sigma^0(\Omega, t)\} - \Gamma_\Omega\{\sigma(\Omega, t) - \sigma^0(\Omega, t)\} \tag{1}$$

where $\sigma(\Omega, t)$ is the spin density matrix, $\mathscr{H}(\Omega, t)$ is the intramolecular time- and orientation-dependent spin Hamiltonian, Γ_Ω is a Markovian motional operator affecting only the orientational variable Ω, Γ_R describes spin relaxation arising from the modulation of spatial coordinates other than Ω, and $\sigma^0(\Omega, t)$ is the equilibrium spin density matrix. Neglecting electron–electron dipolar and nuclear quadrupolar interactions, $\mathscr{H}(\Omega, t)$ may be expressed as

$$\mathscr{H}(\Omega, t) = \mathscr{H}_0 + \mathscr{H}_1(\Omega) + \varepsilon(t) \tag{2}$$

where the time- and orientation-independent Hamiltonian is

$$\mathscr{H}_0 = (\bar{g}\beta_e/\hbar)H_0 S_z - \sum_i \{\gamma_{n_i} H_0 I_{z_i} + \gamma_e \bar{a}_i[S_z I_{z_i} + \tfrac{1}{2}(S_+ I_{-_i} + S_- I_{+_i})]\} \tag{3}$$

including isotropic electron Zeeman, nuclear Zeeman, and electron–nuclear hyperfine interactions; γ_e and γ_{n_i} are the electron and nuclear (for the ith nuclear species) gyromagnetic ratios, H_0 is the dc magnetic field, and $\bar{a}_i$ is the isotropic hyperfine interaction for the ith species.

$\mathscr{H}_1(\Omega)$ describes the anisotropic electron Zeeman and electron–nuclear hyperfine interactions. The coordinates Ω relate the laboratory (defined by H_0) and molecular (principal axes systems of the magnetic interactions) frames. Assuming the principal axes of the electron Zeeman and principal axes of all electron–nuclear hyperfine interactions are coincident, we write for $\mathscr{H}_1(\Omega)$,

$$\mathscr{H}_1(\Omega)$$

$$= \mathscr{D}^2_{0,0}(\Omega)\left\{F'_0 S_z + \sum_i A'_i[S_z I_{z_i} - \tfrac{1}{4}(S_+ I_{-_i} + S_- I_{+_i})]\right\}$$

$$+ [\mathscr{D}^2_{0,2}(\Omega) + \mathscr{D}^2_{0,-2}(\Omega)]\left\{F'_2 S_z + \sum_i A'_{2_i}[S_z I_{z_i} - \tfrac{1}{4}(S_+ I_{-_i} + S_- I_{+_i})]\right\}$$

$$+ [\mathscr{D}^2_{1,2}(\Omega) + \mathscr{D}^2_{1,-2}(\Omega)]\left[F_2 S_+ + \sum_i A_{2_i}(S_z I_{+_i} + S_+ I_{z_i})\right]$$

$$- [\mathscr{D}^2_{-1,2}(\Omega) + \mathscr{D}^2_{-1,-2}(\Omega)]\left[F_2 S_- + \sum_i A_{2_i}(S_z I_{-_i} + S_- I_{z_i})\right]$$

$$- \sum_i [\mathscr{D}^2_{2,0}(\Omega)S_+ I_{+_i} + \mathscr{D}^2_{-2,0}(\Omega)S_- I_{-_i}]A_i$$

$$+ \sum_i \{[\mathscr{D}^2_{1,0}(\Omega)I_{+_i} - \mathscr{D}^2_{-1,0}(\Omega)I_{-_i}]S_z + [\mathscr{D}^2_{1,0}(\Omega)S_+ - \mathscr{D}^2_{-1,0}(\Omega)S_-]I_{z_i}\}A_i$$

$$+ [\mathscr{D}^2_{1,0}(\Omega)S_+ - \mathscr{D}^2_{-1,0}(\Omega)S_-]F_0$$

$$- \sum_i ([\mathscr{D}^2_{2,2}(\Omega) + \mathscr{D}^2_{2,-2}(\Omega)]S_+ I_{+_i} + [\mathscr{D}^2_{-2,2}(\Omega) + \mathscr{D}^2_{-2,-2}(\Omega)]S_- I_{-_i})A_{2_i}$$

$$\tag{4}$$

where

$$F_0 = -\tfrac{1}{2}/6^{1/2}[2g_{zz} - g_{xx} - g_{yy}]\hbar^{-1}\beta_e H_0$$

$$F'_0 = \tfrac{1}{3}[2g_{zz} - g_{xx} - g_{yy}]\hbar^{-1}\beta_e H_0$$

$$F_2 = -\tfrac{1}{4}[g_{xx} - g_{yy}]\hbar^{-1}\beta_e H_0$$

$$F'_2 = (1/6^{1/2})[g_{xx} - g_{yy}]\hbar^{-1}\beta_e H_0$$

$$A_i = \frac{-|\gamma_e|}{2/6^{1/2}}(2A_{zz_i} - A_{xx_i} - A_{yy_i})$$

$$A'_i = \tfrac{1}{3}|\gamma_e|(2A_{zz_i} - A_{xx_i} - A_{yy_i})$$

$$A_{2_i} = -\tfrac{1}{4}|\gamma_e|(A_{xx_i} - A_{yy_i})$$

$$A'_{2_i} = \frac{|\gamma_e|}{6^{1/2}}(A_{xx_i} - A_{yy_i})$$

Note also that $\bar{g} = \tfrac{1}{3}[g_{xx} + g_{yy} + g_{zz}]$ and $\bar{a}_i = \tfrac{1}{3}[A_{xx_i} + A_{yy_i} + A_{zz_i}]$.

The time-dependent Hamiltonian $\varepsilon(t)$ describes the interaction of the spins with all electromagnetic radiation and Zeeman modulation fields,

$$\varepsilon(t) = d_o(S_+ e^{-i\omega_o t} + S_- e^{i\omega_o t}) + d_p(S_+ e^{-i\omega_p t} + S_- e^{i\omega_p t})$$

$$+ \sum_i d_{n_i}(I_{+_i} e^{-i\omega_n t} + I_{-_i} e^{i\omega_n t}) + d_n'(S_+ e^{-i\omega_n t} + S_- e^{i\omega_n t})$$

$$+ (d_s S_z + \sum_i d_{s_i}' I_{z_i})(e^{i\omega_s t} + e^{-i\omega_s t}) \tag{5}$$

where $d_o = \frac{1}{2}\gamma_e h_o$, $d_p = \frac{1}{2}\gamma_e h_p$, $d_{n_i} = \frac{1}{2}\gamma_{n_i} h_n$, $d_n' = \frac{1}{2}\gamma_e h_n$, $d_s = \frac{1}{2}\gamma_e H_s$, and $d_{s_i}' = \frac{1}{2}\gamma_{n_i} H_s$; ω_o, ω_p, ω_n, and ω_s are, respectively, the frequencies of the microwave observing field, the microwave pump field, the radiofrequency field, and the Zeeman modulation field with amplitudes h_o, h_p, h_n, and H_s. Equation (5) is applicable to EPR and passage EPR ($h_n = 0$, $h_p = 0$), ENDOR ($h_p = 0$), and ELDOR ($h_n = 0$) experiments. The term "microwave observing field" refers to the microwave field that stimulates the spin response that is phase-detected and displayed; the intensity of the microwave observing field is often sufficient to introduce saturation or partial saturation of the observed resonance. "Microwave pump field" refers to the second microwave field employed in ELDOR experiments; in such experiments, the pump field intensity is usually substantially greater than that of the microwave observer.

A convenient basis set for the solution of equation (1) are the eigenfunctions of $\mathcal{H}_0$, $|\alpha, v\rangle$, where α and v denote the high field electron and nuclear spin states, respectively. However, in general form, the solution of equation (1) requires solution of an infinite set of coupled equations, which is, of course, impractical. We are thus required to find approximate solutions that provide adequate characterization of particular experimental situations. The more general solutions of equation (1) are simply too mathematically tedious to warrant detailed discussion in this chapter; however, it is legitimate to ask if the density matrix formalism of equation (1) is capable of quantitatively describing reasonably complex experimental situations. We believe this question has been answered satisfactorily by the work of B. H. Robinson and Dalton.[61]

Let us briefly review the details of a representative calculation. Freed and co-workers[28,62,65,68-80] have discussed the treatment of the terms $\mathcal{H}_0$, $\mathcal{H}_1(\Omega)$, Γ_R, Γ_Ω for a number of experimental systems. We have also reviewed this matter elsewhere[33]; thus, in this chapter we shall focus our attention on those manipulations of the density matrix equation necessary when $\varepsilon(t)$ is explicitly considered. A simple but appropriate example of characteristic manipulations is the analysis of ELDOR spectra for rapidly tumbling (the fast-motion limit where motional frequencies are much greater than the frequencies of anisotropic magnetic interactions) nitroxide radicals. For this case, equation (2) can be written as

$$\mathcal{H}(\Omega, t) = \mathcal{H}(t) = \mathcal{H}_0 + \varepsilon(t) \tag{6}$$

Neglecting proton hyperfine interactions, we write

$$\mathcal{H}_0 = (\bar{g}\beta_e/\hbar)H_0 S_z - \gamma_n H_0 I_z - \gamma_e \bar{a}S_z I_z \tag{7}$$

We shall evaluate equation (1) in terms of the reduced spin density matrix $\chi = \sigma - \sigma^\circ$. Invoking the high-field and high-temperature approximations, we write the equilibrium spin density matrix as

$$\sigma^0 = N^{-1} - q\mathcal{H}_0 \tag{8}$$

where $q = \hbar/NkT$, N being the total number of spin states (e.g., $N = 6$ for ^{14}N, $N = 4$ for ^{15}N). Using these approximations, equation (1) can be written as

$$
\begin{aligned}
i\left(\frac{d}{dt} + \Gamma_R\right)\chi_{\alpha\alpha'}^{vv'} &= \{[\omega(\alpha, v) - \omega(\alpha', v')] \\
&\quad + [d_s(m_\alpha - m_{\alpha'}) + d_s'(m_v - m_{v'})](e^{i\omega_s t} + e^{-i\omega_s t})\}\chi_{\alpha\alpha'}^{vv'} \\
&\quad + \{(d_o e^{-i\omega_o t} + d_p e^{-i\omega_p t})[\chi_{\alpha-1, \alpha'}^{vv'} - \chi_{\alpha, \alpha'+1}^{vv'}] \\
&\quad + (d_o e^{i\omega_o t} + d_p e^{i\omega_p t})[\chi_{\alpha+1, \alpha'}^{vv'} - \chi_{\alpha, \alpha'-1}^{vv'}]\} \\
&\quad \pm q[\omega(\alpha, v) - \omega(\alpha', v')](d_o e^{\mp i\omega_o t} + d_p e^{\mp i\omega_p t}) \\
&\quad \times (\pm\delta_{\alpha, \alpha'\pm 1})(\delta_{v, v'})
\end{aligned}
\tag{9}
$$

where $\omega(\alpha, v) = (\bar{g}\beta_e/\hbar)H_0 m_\alpha - \gamma_n H_0 m_v - \gamma_e \bar{a}m_\alpha m_v$. Let us consider the various allowed matrix elements starting with the off-diagonal elements where we set $\alpha = \beta$, $\alpha' = \alpha$, and $v = v' = v$ [α and β denote the two electron spin states and v denotes nuclear spin states ($+1, 0, -1$ for ^{14}N and $+\frac{1}{2}, -\frac{1}{2}$ for ^{15}N)]. Equation (9) thus becomes

$$
\begin{aligned}
i\left(\frac{d}{dt} + \Gamma_R\right)\chi_{\beta\alpha}^{vv} &= [-\omega_v - d_s(e^{i\omega_s t} + e^{-i\omega_s t})]\chi_{\beta\alpha}^{vv} \\
&\quad + (d_o e^{i\omega_o t} + d_p e^{i\omega_p t})[\chi_{\alpha\alpha}^{vv} - \chi_{\beta\beta}^{vv} - q\omega_v]
\end{aligned}
\tag{10}
$$

where we have defined $-\omega_v = \omega(\beta, v) - \omega(\alpha, v) = -[(\bar{g}\beta_e/\hbar)H_0 - \gamma_e \bar{a}m_v]$. Since in equation (10) $\chi_{\beta\alpha}^{vv}$ is coupled to the difference between the diagonal elements $(\chi_{\alpha\alpha}^{vv} - \chi_{\beta\beta}^{vv})$ rather than to a particular diagonal element, it is useful to construct the corresponding difference equation. First setting $\alpha = \alpha' = \alpha$ to produce

$$i\left(\frac{d}{dt} + \Gamma_R\right)\chi_{\alpha\alpha}^{vv} = \chi_{\beta\alpha}^{vv}(d_o e^{-i\omega_o t} + d_p e^{-i\omega_p t}) - \chi_{\alpha\beta}^{vv}(d_o e^{i\omega_o t} + d_p e^{i\omega_p t}) \tag{11}$$

then $\alpha = \alpha' = \beta$, and finally, subtracting the resulting two equations produces the desired diagonal difference equation

$$i\left(\frac{d}{dt} + \Gamma_R\right)(\chi_{\alpha\alpha}^{vv} - \chi_{\beta\beta}^{vv}) = 2\chi_{\beta\alpha}^{vv}(d_o e^{-i\omega_o t} + d_p e^{-i\omega_p t}) - 2\chi_{\alpha\beta}^{vv}(d_o e^{i\omega_o t} + d_p e^{i\omega_p t})$$

(12)

In solving the differential equations (10) and (12), we assume a solution of the form

$$\chi_{\alpha\alpha'}^{vv'} = \sum_{n, l, r} Z_{\alpha\alpha'}^{vv'}(n, l, \pm r) \exp[i(n\omega_o + l\omega_p \pm r\omega_s)t]$$

(13)

or

$$\chi_{\alpha\alpha'}^{vv'} = \sum_k Z_{\alpha\alpha'}^{vv'(k)} e^{i(k\omega)t}$$

(14)

We further define

$$Y_v^{(k)} = Z_{\alpha\alpha}^{vv(k)} - Z_{\beta\beta}^{vv(k)}$$

(15)

Introducing these expressions into equations (10) and (12), we obtain

$$i[i(k\omega) + \Gamma_R]Z_{\beta\alpha}^{v(k)}$$
$$= \{- \omega_v Z_{\beta\alpha}^{v(k)} - d_s(Z_{\beta\alpha}^{v(n, l, \pm r - 1)} + Z_{\beta\alpha}^{v(n, l, \pm r + 1)})$$
$$+ [d_o Y_v^{(n - 1, l, \pm r)} + d_p Y_v^{(n, l - 1, \pm r)}]$$
$$- q\omega_v[d_o \delta_{n, 1} \delta_{l, 0} \delta_{r, 0} + d_p \delta_{n, 0} \delta_{l, 1} \delta_{r, 0}]\} e^{i(k\omega)t}$$

(16a)

and

$$i[i(k\omega) + \Gamma_R]Y_v^{(k)}$$
$$= \{2d_o[Z_{\beta\alpha}^{v(n + 1, l, \pm r)} - Z_{\alpha\beta}^{v(n - 1, l, \pm r)}] + 2d_p[Z_{\beta\alpha}^{v(n, l + 1, \pm r)} - Z_{\alpha\beta}^{v(n, l - 1, \pm r)}]\} e^{i(k\omega)t}$$

(16b)

The specific system of equations we need to solve for the time-independent response can be obtained by multiplying equations (16a) and (16b) by $e^{-i(k'\omega)t}$ and integrating over time. The index k' picks out the particular microwave and modulation harmonics of interest.

Before proceeding we need to consider two additional facets of the calculation. Specifically we need to examine the Hermiticity of χ to see how $Z_{\beta\alpha}$ and $Z_{\alpha\beta}$ are related, and we need to define the relaxation matrix Γ_R. The density matrix equation is most simply

$$\dot{\sigma} = -i[\mathcal{H}, \sigma]$$

(17)

then

$$\dot{\sigma}^{T*} = -i[\mathcal{H}^{T*}, \sigma^{T*}]$$

(18)

But since $\mathscr{H}$ is Hermitian, by definition,

$$\dot{\sigma}^{T*} = -i[\mathscr{H}, \sigma^{T*}] \tag{19}$$

Since σ^{T*} is a solution to the same equation as $\sigma, \sigma^{T*} = \sigma$ or σ is Hermitian. It follows that $\chi = \sigma - \sigma^0$ is Hermitian as well. Using the Hermiticity of χ we can see that

$$\chi_{\alpha\alpha'}^{vv'} = \sum_k Z_{\alpha\alpha'}^{vv'(k)} e^{i(k\omega)t} = \chi_{\alpha'\alpha}^{v'v*} = \sum_k Z_{\alpha'\alpha}^{v'v(-k)*} e^{-i(k\omega)t} \tag{20}$$

Comparing the summations term for term, we find

$$Z_{\alpha\alpha'}^{vv'(k)} = Z_{\alpha'\alpha}^{v'v(-k)*} \tag{21}$$

Examples relevant to our calculations are

$$Z_{\alpha\beta}^{vv(k)} = Z_{\beta\alpha}^{vv(-k)*} \tag{22}$$

and

$$Z_{\alpha\alpha}^{vv(k)} = Z_{\alpha\alpha}^{vv(-k)*} \tag{23}$$

The relaxation matrix Γ_R is defined in terms of spin–spin (T_{2e}^{-1}) and spin–lattice (W_e) relaxation rates. Differing nuclear states are coupled by nuclear relaxation (W_n) processes; in the present example, we neglect the coupling of nuclear states by Heisenberg spin exchange as is appropriate for the consideration of dilute solutions of spin labels. We also neglect cross-relaxation (simultaneous electron and nuclear spin flip) processes.

Off-diagonal elements of Γ_R reflect the loss of phase coherence and are defined as

$$\Gamma_R \chi_{\alpha\alpha'}^{vv'} = T_{2e}^{-1} \chi_{\alpha\alpha'}^{vv'} \tag{24}$$

We shall neglect the effect of nuclear relaxation processes upon the phase coherence of the electron spins.

Diagonal elements can be associated with populations of the spin levels. Population of the levels can be influenced either by dissipating excitation energy directly to the lattice (W_e) or following a nuclear spin flip (W_n). The coupling of the diagonal elements effected by Γ_R can be written as (for the $v = 1$ ^{14}N nuclear spin state)

$$\Gamma_R \chi_{\alpha\alpha}^{v=1} = W_e \chi_{\alpha\alpha}^{v=1} + W_n \chi_{\alpha\alpha}^{v=1} - W_e \chi_{\beta\beta}^{v=1} - W_n \chi_{\alpha\alpha}^{v=0} \tag{25a}$$

and

$$\Gamma_R \chi_{\beta\beta}^{v=1} = W_e \chi_{\beta\beta}^{v=1} + W_n \chi_{\beta\beta}^{v=1} - W_e \chi_{\alpha\alpha}^{v=1} - W_n \chi_{\beta\beta}^{v=0} \tag{25b}$$

From equations (25a) and (25b), we write for the diagonal difference equation

$$\Gamma_R(\chi_{\alpha\alpha}^{v=1} - \chi_{\beta\beta}^{v=1}) = (2W_e + W_n)(\chi_{\alpha\alpha}^{v=1} - \chi_{\beta\beta}^{v=1}) - W_n(\chi_{\alpha\alpha}^{v=0} - \chi_{\beta\beta}^{v=0}) \tag{26}$$

Similarly,

$$\Gamma_R \chi_{\alpha\alpha}^{v=0} = W_e \chi_{\alpha\alpha}^{v=0} + 2W_n \chi_{\alpha\alpha}^{v=0} - W_e \chi_{\beta\beta}^{v=0} - W_n \chi_{\alpha\alpha}^{v=1} - W_n \chi_{\alpha\alpha}^{v=-1} \quad (27a)$$

$$\Gamma_R \chi_{\beta\beta}^{v=0} = W_e \chi_{\beta\beta}^{v=0} + 2W_n \chi_{\beta\beta}^{v=0} - W_e \chi_{\alpha\alpha}^{v=0} - W_n \chi_{\beta\beta}^{v=1} - W_n \chi_{\beta\beta}^{v=-1} \quad (27b)$$

Then, subtracting equation (27b) from (27a), we obtain

$$\Gamma_R(\chi_{\alpha\alpha}^{v=0} - \chi_{\beta\beta}^{v=0})$$
$$= [2W_e + 2W_n](\chi_{\alpha\alpha}^{v=0} - \chi_{\beta\beta}^{v=0}) - W_n(\chi_{\alpha\alpha}^{v=1} - \chi_{\beta\beta}^{v=1}) - W_n(\chi_{\alpha\alpha}^{v=-1} - \chi_{\beta\beta}^{v=-1})$$
$$(28)$$

Finally, we obtain

$$\Gamma_R(\chi_{\alpha\alpha}^{v=-1} - \chi_{\beta\beta}^{v=-1}) = [2W_e + W_n](\chi_{\alpha\alpha}^{v=-1} - \chi_{\beta\beta}^{v=-1}) - W_n(\chi_{\alpha\alpha}^{v=0} - \chi_{\beta\beta}^{v=0})$$
$$(29)$$

As we noted earlier, observed signals are detected at the first harmonic of the observing microwave frequency; for the high (9 GHz) microwave frequencies involved in common experimental situations, we can neglect all higher harmonics. We are interested, then, in the equations of (16a) that represent the first harmonics of the observer and pump microwave frequencies, i.e., $k = k_o = (1, 0, \pm r)$ and $k = k_p = (0, 1, \pm r)$. The Zeeman modulation harmonic identifier r has been left general, because for the modulation amplitudes and frequencies employed in many experiments, there exists extensive coupling between harmonics. The generalized format will enable us to work more readily with this coupling.

Introducing equations (26), (28), and (29) into equation (16a), we obtain for the microwave observer

$$i[i(\omega_o \pm r\omega_s) + T_{2e}^{-1}]Z_{\beta\alpha}^{v(1, 0, \pm r)}$$
$$= -\omega_v Z_{\beta\alpha}^{v(1, 0, \pm r)} - d_s[Z_{\beta\alpha}^{v(1, 0, \pm r-1)} + Z_{\beta\alpha}^{v(1, 0, \pm r+1)}] + d_o Y_v^{(0, 0, \pm r)}$$
$$+ d_p Y_v^{(1, -1, \pm r)} - q\omega_v d_o \delta_{r, 0} \quad (30a)$$

and for the microwave pump

$$i[i(\omega_p \pm r\omega_s) + T_{2e}^{-1}]Z_{\beta\alpha}^{v(0, 1, \pm r)}$$
$$= -\omega_v Z_{\beta\alpha}^{v(0, 1, \pm r)} - d_s[Z_{\beta\alpha}^{v(0, 1, \pm r-1)} + Z_{\beta\alpha}^{v(0, 1, \pm r+1)}] + d_o Y_v^{(-1, 1, \pm r)}$$
$$+ d_p Y_v^{(0, 0, \pm r)} - q\omega_v d_p \delta_{r, 0} \quad (30b)$$

In order to make the system of equations more manageable, we again invoke the high-field (thus high microwave frequency) approximation and argue that the terms $Y_v^{(1, -1, \pm r)}$ and $Y_v^{(-1, 1, \pm r)}$ will be negligible for commonly encountered experimental conditions. This simplification is significant in that each equation now couples to the same diagonal difference element $Y_v^{(0, 0, \pm r)}$. Choosing $k = (0, 0, \pm r)$ for the diagonal difference equation, we write

$$i[i(\pm r\omega_s) + \Gamma_R]Y_v^{(0, 0, \pm r)}$$

$$= 2d_o[Z_{\beta\alpha}^{v(1, 0, \pm r)} - Z_{\alpha\beta}^{v(-1, 0, \pm r)}] + 2d_p[Z_{\beta\alpha}^{v(0, 1, \pm r)} - Z_{\alpha\beta}^{v(0, -1, \pm r)}] \quad (31)$$

Recalling that $Z_{\alpha\beta}^{v(k)} = Z_{\beta\alpha}^{v(-k)*}$, we write

$$i[i(\pm r\omega_s) + \Gamma_R]Y_v^{(0, 0, \pm r)}$$

$$= 2d_o[Z_{\beta\alpha}^{v(1, 0, \pm r)} - Z_{\beta\alpha}^{v(1, 0, \mp r)*}]2d_p[Z_{\beta\alpha}^{v(0, 1, \pm r)} - Z_{\beta\alpha}^{v(0, 1, \mp r)*}] \quad (32)$$

By defining $\Delta_{ov} = \omega_o - \omega_v$ and $\Delta_{pv} = \omega_p - \omega_v$ and letting $v = 1$ (i.e., the low-field ^{14}N EPR line), equations (30a,b) and (32) can be written as

$$(\Delta_{o1} \pm r\omega_s - iT_{2e}^{-1})Z_{\beta\alpha}^{1(1, 0, \pm r)} + d_o\,Y_1^{(0, 0, \pm r)}$$

$$= d_s[Z_{\beta\alpha}^{1(1, 0, \pm r-1)} + Z_{\beta\alpha}^{1(1, 0, \pm r+1)}] + q\omega_1 d_o\delta_{r, 0} \quad (33a)$$

$$(\Delta_{p1} \pm r\omega_s - iT_{2e}^{-1})Z_{\beta\alpha}^{1(0, 1, \pm r)} + d_p\,Y_1^{(0, 0, \pm r)}$$

$$= d_s[Z_{\beta\alpha}^{1(0, 1, \pm r-1)} + Z_{\beta\alpha}^{1(0, 1, \pm r+1)}] + q\omega_1 d_p\delta_{r, 0} \quad (33b)$$

$$[\mp r\omega_s + i(2W_e + W_n)]Y_1^{(0, 0, \pm r)} - iW_n\,Y_0^{(0, 0, \pm r)}$$

$$= 2d_o[Z_{\beta\alpha}^{1(1, 0, \pm r)} - Z_{\beta\alpha}^{1(1, 0, \mp r)*}] + 2d_p[Z_{\beta\alpha}^{1(0, 1, \pm r)} - Z_{\beta\alpha}^{1(0, 1, \mp r)*}] \quad (33c)$$

Four unique signals exist at a given microwave and modulation harmonic:

1. *The in-phase dispersion:*

$$\alpha_v(1, 0, r) = Z_{\beta\alpha}^{\prime v(1, 0, r)} + Z_{\beta\alpha}^{\prime v(1, 0, -r)} \quad (34a)$$

2. *The phase-quadrature dispersion:*

$$\beta_v(1, 0, r) = Z_{\beta\alpha}^{\prime\prime v(1, 0, r)} - Z_{\beta\alpha}^{\prime\prime v(1, 0, -r)} \quad (34b)$$

3. *The in-phase absorption:*

$$\gamma_v(1, 0, r) = Z_{\beta\alpha}^{\prime\prime v(1, 0, r)} + Z_{\beta\alpha}^{\prime\prime v(1, 0, -r)} \quad (34c)$$

4. *The phase-quadrature absorption:*

$$\delta_v(1, 0, r) = Z_{\beta\alpha}^{\prime v(1, 0, r)} - Z_{\beta\alpha}^{\prime v(1, 0, -r)} \quad (34d)$$

where we have indicated the decomposition of Z into real and imaginary parts as $Z = Z' + iZ''$. The EPR spectrum or total electron spin response is the sum over all nuclear spin states, namely,

$$\alpha(1, 0, r) = \sum_v \alpha_v(1, 0, r) \quad (35a)$$

$$\beta(1, 0, r) = \sum_v \beta_v(1, 0, r) \quad (35b)$$

$$\gamma(1, 0, r) = \sum_v \gamma_v(1, 0, r) \quad (35c)$$

$$\delta(1, 0, r) = \sum_v \delta_v(1, 0, r) \quad (35d)$$

We are now faced with the task of expressing equation (33) in terms of the detected signals. If, after explicitly writing out equation (33a) for $+r$ and $-r$, we alternately add and subtract these equations, we obtain for the sum

$$(\Delta_{o1} - iT_{2e}^{-1})[Z_{\beta\alpha}^{1(1,0,r)} + Z_{\beta\alpha}^{1(1,0,-r)}] + r\omega_s[Z_{\beta\alpha}^{1(1,0,r)} - Z_{\beta\alpha}^{1(1,0,-r)}]$$
$$+ d_o[Y_1^{(0,0,r)} + Y_1^{(0,0,-r)}]$$
$$= d_s[Z_{\beta\alpha}^{1(1,0,r-1)} + Z_{\beta\alpha}^{1(1,0,-r-1)} + Z_{\beta\alpha}^{1(1,0,r+1)} + Z_{\beta\alpha}^{1(1,0,-r+1)}]$$
$$+ 2q\omega_1 d_o \delta_{r,0} \tag{36a}$$

and for the difference

$$(\Delta_{o1} - iT_{2e}^{-1})[Z_{\beta\alpha}^{1(1,0,r)} - Z_{\beta\alpha}^{1(1,0,-r)}] + r\omega_s[Z_{\beta\alpha}^{1(1,0,r)} + Z_{\beta\alpha}^{1(1,0,-r)}]$$
$$+ d_o[Y_1^{(0,0,r)} - Y_1^{(0,0,-r)}]$$
$$= d_s[Z_{\beta\alpha}^{1(1,0,r-1)} - Z_{\beta\alpha}^{1(1,0,-r-1)} + Z_{\beta\alpha}^{1(1,0,r+1)} - Z_{\beta\alpha}^{1(1,0,-r+1)}] \tag{36b}$$

By separating $Z_{\beta\alpha}^{1(k)}$ into real and imaginary parts and blocking these parts as in equation (34), we write equation (36) as

$$(\Delta_{o1} - iT_{2e}^{-1})[\alpha_1(1,0,r) + i\gamma_1(1,0,r)] + r\omega_s[\delta_1(1,0,r) + i\beta_1(1,0,r)]$$
$$+ d_o[Y_1^{(0,0,r)} + Y_1^{(0,0,-r)}]$$
$$= d_s[\alpha_1(1,0,r-1) + i\gamma_1(1,0,r-1) + \alpha_1(1,0,r+1)$$
$$+ i\gamma_1(1,0,r+1)] + 2q\omega_1 d_o \delta_{r,0} \tag{37a}$$

and

$$(\Delta_{o1} - iT_{2e}^{-1})[\delta_1(1,0,r) + i\beta_1(1,0,r)] + r\omega_s[\alpha_1(1,0,r) + i\gamma_1(1,0,r)]$$
$$+ d_o[Y_1^{(0,0,r)} - Y_1^{(0,0,-r)}]$$
$$= d_s[\delta_1(1,0,r-1) + i\beta_1(1,0,r-1) + \delta_1(1,0,r+1)$$
$$+ i\beta_1(1,0,r+1)] \tag{37b}$$

The next step is to separate equations (37a) and (37b) into real and imaginary parts, but first we need to consider the terms $[Y_1^{(0,0,r)} + Y_1^{(0,0,-r)}]$ and $[Y_1^{(0,0,r)} - Y_1^{(0,0,-r)}]$. From our discussion of the Hermiticity of the matrix elements, we note that $Z_{\alpha\alpha}^{\nu\nu(k)} = Z_{\alpha\alpha}^{\nu\nu(-k)*}$; it follows that

$$Z_{\alpha\alpha}^{1(k)} + Z_{\alpha\alpha}^{1(-k)} = 2\,\text{Re}\,Z_{\alpha\alpha}^{1(k)} = 2Z_{\alpha\alpha}'^{1(k)}. \tag{38}$$

Similarly,

$$Z_{\alpha\alpha}^{1(k)} - Z_{\alpha\alpha}^{1(-k)} = 2\,\text{Im}\,Z_{\alpha\alpha}^{1(k)} = 2iZ_{\alpha\alpha}''^{1(k)} \tag{39}$$

By combining these with the corresponding expressions for $Z_{\beta\beta}^{1(k)}$, we find

$$[Y_1^{(0,0,r)} + Y_1^{(0,0,-r)}] = 2[Z_{\alpha\alpha}'^{1(0,0,r)} - Z_{\beta\beta}'^{1(0,0,r)}] \tag{40a}$$

$$[Y_1^{(0,0,r)} - Y_1^{(0,0,-r)}] = 2i[Z_{\alpha\alpha}''^{1(0,0,r)} - Z_{\beta\beta}''^{1(0,0,r)}] \tag{40b}$$

that is, the bracketed terms are either completely real or completely imaginary. We redefine these for notational simplicity as $Y_{1+}^{(r)}$ and $iY_{1-}^{(r)}$, respectively. Equation (37) can now be separated into the equations

$$\Delta_{o1}\alpha_1(1, 0, r) + T_{2e}^{-1}\gamma_1(1, 0, r) + r\omega_s\delta_1(1, 0, r) + d_o\,Y_{1+}^{(r)}$$
$$- d_s[\alpha_1(1, 0, r - 1) + \alpha_1(1, 0, r + 1)] = 2q\omega_1 d_o\delta_{r, 0} \qquad (41a)$$

$$- T_{2e}^{-1}\alpha_1(1, 0, r) + r\omega_s\beta_1(1, 0, r) + \Delta_{o1}\gamma_1(1, 0, r)$$
$$- d_s[\gamma_1(1, 0, r - 1) + \gamma_1(1, 0, r + 1)] = 0 \qquad (41b)$$

$$r\omega_s\alpha_1(1, 0, r) + T_{2e}^{-1}\beta_1(1, 0, r) + \Delta_{o1}\delta_1(1, 0, r)$$
$$- d_s[\delta_1(1, 0, r - 1) + \delta_1(1, 0, r + 1)] = 0 \qquad (41c)$$

$$\Delta_{o1}\beta_1(1, 0, r) + r\omega_s\gamma_1(1, 0, r) - T_{2e}^{-1}\delta_1(1, 0, r) + d_o\,Y_{1-}^{(r)}$$
$$- d_s[\beta_1(1, 0, r - 1) + \beta_1(1, 0, r + 1)] = 0 \qquad (41d)$$

It should be apparent that by applying the same operational procedure to equation (33b), we can generate the analogous set of pump equations:

$$\Delta_{p1}\alpha_1(0, 1, r) + T_{2e}^{-1}\gamma_1(0, 1, r) + r\omega_s\delta_1(0, 1, r) + d_p\,Y_{1+}^{(r)}$$
$$- d_s[\alpha_1(0, 1, r - 1) + \alpha_1(0, 1, r + 1)] = 2q\omega_1 d_p\delta_{r, 0} \qquad (42a)$$

$$- T_{2e}^{-1}\alpha_1(0, 1, r) + r\omega_s\beta_1(0, 1, r) + \Delta_{p1}\gamma_1(0, 1, r)$$
$$- d_s[\gamma_1(0, 1, r - 1) + \gamma_1(0, 1, r + 1)] = 0 \qquad (42b)$$

$$r\omega_s\alpha_1(0, 1, r) + T_{2e}^{-1}\beta_1(0, 1, r) + \Delta_{p1}\delta_1(0, 1, r)$$
$$- d_s[\delta_1(0, 1, r - 1) + \delta_1(0, 1, r + 1)] = 0 \qquad (42c)$$

$$\Delta_{p1}\beta_1(0, 1, r) + r\omega_s\gamma_1(0, 1, r) - T_{2e}^{-1}\delta_1(0, 1, r) + d_p\,Y_{1-}^{(r)}$$
$$- d_s[\beta_1(0, 1, r - 1) + \beta_1(0, 1, r + 1)] = 0 \qquad (42d)$$

Returning to the diagonal difference equation (33c), we write the equations for $r = +r$ and $r = -r$. Summing, we find that

$$- r\omega_s[Y_1^{(r)} - Y_1^{(-r)}] + i(2W_e + W_n)[Y_1^{(r)} + Y_1^{(-r)}] - iW_n[Y_0^{(r)} + Y_0^{(-r)}]$$
$$- i4d_o[Z_{\beta\alpha}''^{1(1, 0, r)} + Z_{\beta\alpha}''^{1(1, 0, -r)}] - i4d_p[Z_{\beta\alpha}''^{1(0, 1, r)} + Z_{\beta\alpha}''^{1(0, 1, -r)}] = 0 \qquad (43)$$

or on introduction of the newly defined parameters

$$r\omega_s\,Y_{1-}^{(r)} - (2W_e + W_n)Y_{1+}^{(r)} + W_n\,Y_{0+}^{(r)} + 4d_o\gamma_1(1, 0, r) + 4d_p\gamma_1(0, 1, r) = 0 \qquad (44)$$

In the same way, the difference equation gives

$$(2W_e + W_n)Y_{1-}^{(r)} + r\omega_s\,Y_{1+}^{(r)} - W_n\,Y_{0-}^{(r)} + 4d_o\delta_1(1, 0, r) + 4d_p\delta_1(0, 1, r) = 0 \qquad (45)$$

These 10 equations—(41a–d), (42a–d), (44), and (45)—are the equations we must solve for a particular component (harmonic) of the signal response for the low-field ($v = 1$) electron spin resonance transition. There are, of course, comparable sets for the other two ^{14}N hyperfine lines (electron spin transitions corresponding to $v = 0$ and $v = -1$). Since the three transitions are coupled by the nuclear spin flip (W_n) transitions of the relaxation matrix, we cannot solve each system of 10 equations in closed form; rather, we must solve the complete set of 30 equations simultaneously. In addition, we must consider the coupling to the other modulation harmonics. More precisely, we must be concerned with all those harmonics that feed in a significant contribution to the desired signal response (harmonic).

The computational difficulty resulting from harmonic coupling is much the same here as in the single resonance calculations, which have been discussed in detail elsewhere.[33,81–83] It is sufficient to note that for a particular harmonic, each group of 30 equations may be blocked into the general matrix form

$$\mathbf{D}_s \mathbf{S}_{r-1} + \mathbf{M} \mathbf{S}_r + \mathbf{D}_s \mathbf{S}_{r+1} = \mathbf{S}_{\text{OLN}} \qquad (46)$$

and further, these equations may be assembled into the single tridiagonal supermatrix equation,

$$
\begin{pmatrix}
\mathbf{M}_0 & \mathbf{D}_s & 0 & \cdot & \cdot & \cdot \\
\mathbf{D}_s & \mathbf{M}_1 & \mathbf{D}_s & \cdot & \cdot & \cdot \\
0 & \mathbf{D}_s & \mathbf{M}_2 & \cdot & \cdot & \cdot \\
\cdot & \cdot & \cdot & & & \\
\cdot & \cdot & \cdot & & & \\
\cdot & & & & &
\end{pmatrix}
\begin{pmatrix}
\mathbf{S}_0 \\
\mathbf{S}_1 \\
\mathbf{S}_2 \\
\cdot \\
\cdot \\
\cdot
\end{pmatrix}
=
\begin{pmatrix}
\mathbf{S}_{\text{OLN}_0} \\
\mathbf{S}_{\text{OLN}_1} \\
\mathbf{S}_{\text{OLN}_2} \\
\cdot \\
\cdot \\
\cdot
\end{pmatrix}
\qquad (47)
$$

$\mathbf{D}_s$ is the elemental matrix describing the coupling of signals of interest (r) to the higher and lower harmonics ($r \pm 1$). It is diagonal in blocks as shown in Table 1. $\mathbf{M}_r$ is the matrix block for any harmonic r. It can be blocked as

$$
\mathbf{M}_r =
\begin{pmatrix}
\mathbf{M}_1 & \mathbf{W}_N & 0 \\
\mathbf{W}_N & \mathbf{M}_0 & \mathbf{W}_N \\
0 & \mathbf{W}_N & \mathbf{M}_{-1}
\end{pmatrix}
\qquad (48)
$$

$\mathbf{M}_v$ (i.e., $\mathbf{M}_1$, $\mathbf{M}_0$, $\mathbf{M}_{-1}$) describes the coupling between the pump and observer microwave fields for a particular nuclear state. For example, $\mathbf{M}_1$ is given as

	(1, 0, r)				(0, 0, r)		(0, 1, r)			
	α_1	β_1	γ_1	δ_1	Y_{1+}	Y_{1-}	δ_1	γ_1	β_1	α_1
	T_{2e}^{-1}	$r\omega_s$	Δ_{o1}	0	0	0	0	0	0	0
	$r\omega_s$	T_{2e}^{-1}	0	Δ_{o1}	0	0	0	0	0	0
	Δ_{o1}	0	T_{2e}^{-1}	$r\omega_s$	d_o	0	0	0	0	0
	0	Δ_{o1}	$r\omega_s$	$-T_{2e}^{-1}$	0	d_o	0	0	0	0
$\mathbf{M}_1 =$	0	0	$4d_o$	0	$-(2W_e + W_n)$	$r\omega_s$	0	$4d_p$	0	0
	0	0	0	$4d_o$	$r\omega_s$	$(2W_e + W_n)$	$4d_p$	0	0	0
	0	0	0	0	0	d_p	$-T_{2e}^{-1}$	$r\omega_s$	Δ_{p1}	0
	0	0	0	0	d_p	0	$r\omega_s$	T_{2e}^{-1}	0	Δ_{p1}
	0	0	0	0	0	0	Δ_{p1}	0	T_{2e}^{-1}	$r\omega_s$
	0	0	0	0	0	0	0	Δ_{p1}	$r\omega_s$	$-T_{2e}^{-1}$

$$(49)$$

Table 1

(1, 0, r)				(0, 0, r)		(0, 1, r)			
α_v	β_v	γ_v	δ_v	Y_{v+}	Y_{v-}	δ_v	γ_v	β_v	α_v
0	0	$-d_s$	0	0	0	0	0	0	0
0	0	0	$-d_s$	0	0	0	0	0	0
$-d_s$	0	0	0	0	0	0	0	0	0
0	$-d_s$	0	0	0	0	0	0	0	0
0	0	0	0	0	0	0	0	0	0
0	0	0	0	0	0	0	0	0	0
0	0	0	0	0	0	0	0	$-d_s$	0
0	0	0	0	0	0	0	0	0	$-d_s$
0	0	0	0	0	0	$-d_s$	0	0	0
0	0	0	0	0	0	0	$-d_s$	0	0

$\mathbf{W}_N$ contains the coupling between the diagonal differences. It is

$$
\mathbf{W}_N =
\begin{array}{c}
\begin{array}{c}
\overbrace{\quad(1,\,0,\,r)\quad}^{\alpha_v\ \ \beta_v\ \ \gamma_v\ \ \delta_v} \quad
\overbrace{\quad(0,\,0,\,r)\quad}^{Y_{v+}\ \ \ \ Y_{v-}} \quad
\overbrace{\quad(0,\,1,\,r)\quad}^{\delta_v\ \ \gamma_v\ \ \beta_v\ \ \alpha_v}
\end{array}\\[4pt]
\left(
\begin{array}{cccccccccc}
0 & 0 & 0 & 0 & 0 & 0 & 0 & 0 & 0 & 0\\
0 & 0 & 0 & 0 & 0 & 0 & 0 & 0 & 0 & 0\\
0 & 0 & 0 & 0 & 0 & 0 & 0 & 0 & 0 & 0\\
0 & 0 & 0 & 0 & 0 & 0 & 0 & 0 & 0 & 0\\
0 & 0 & 0 & 0 & W_n & 0 & 0 & 0 & 0 & 0\\
0 & 0 & 0 & 0 & 0 & -W_n & 0 & 0 & 0 & 0\\
0 & 0 & 0 & 0 & 0 & 0 & 0 & 0 & 0 & 0\\
0 & 0 & 0 & 0 & 0 & 0 & 0 & 0 & 0 & 0\\
0 & 0 & 0 & 0 & 0 & 0 & 0 & 0 & 0 & 0\\
0 & 0 & 0 & 0 & 0 & 0 & 0 & 0 & 0 & 0
\end{array}
\right)
\end{array}
\tag{50}
$$

S_{OLN} is a solution vector and is made up of the solutions to equations (41a–d), (42a–d), (44), and (45) and their analogs for the other nuclear indices. Equation (47) can be solved by adapting common matrix inversion techniques, e.g., the Gauss and Gauss–Jordan elimination techniques, to the supermatrix problem.[33,34,81–83]

An equivalent supermatrix master equation can be derived starting with the phenomenological Bloch equations, as we have demonstrated elsewhere.[49] At the present level of sophistication, equivalence of the density matrix and Bloch equations is readily demonstrated. Thomas and McConnell[84] have also employed a Bloch equation formulation with explicit consideration of applied Zeeman and stochastic molecular modulation to compute resonance spectra for slowly tumbling spin labels.

Returning to our discussion of the ELDOR experiment performed on rapidly tumbling ^{14}N nitroxides, we note that ELDOR experiments are commonly performed as field-swept or as frequency-swept experiments. In field-swept ELDOR experiments the frequencies of the microwave pump and observer remain fixed throughout the experiment while the dc magnetic field is slowly swept through the resonance spectrum. The frequency difference between the pump and observer is chosen to correspond to some meaningful spectral parameter, such as an electron–nuclear hyperfine interaction, or a multiple thereof. Thus, as the magnetic field is swept, the transitions of interest become resonant simultaneously; more explicitly, the field-swept experiment matches corresponding points in the pumped and observed lines at each instant in time.

In frequency-swept ELDOR experiments, it is common practice to fix the dc magnetic field and the microwave observer frequency so that some particular point on the observed resonance line (e.g., the line center or point for which $\Delta_{ov} = 0$) is monitored as the microwave pump frequency is varied through other resonance lines. In such an experiment, one observes the influence of all parts of the pumped line on a single observer point.

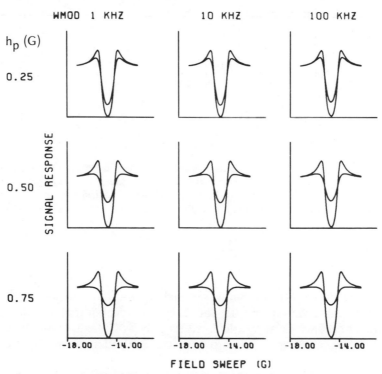

Figure 17. Computer-simulated field-swept dispersion, $\alpha(1)$, ELDOR spectra for a hypothetical nitroxide radical system as a function of Zeeman modulation frequency and microwave pump field intensity. Each of the nine graphs shows the superposition of EPR spectra recorded in the presence and absence of the microwave pump field. Other parameters employed in the calculations include $T_{2e} = 2.75 \times 10^{-7}$ sec, $W_e = 1.0 \times 10^6$ Hz, $W_n = 1.0 \times 10^7$ Hz, $h_o = 0.075$ G, $H_s = 0.6$ G. The coupling to eleven r states was considered. The pump–observer frequency separation remained fixed at $\omega_p - \omega_o = \gamma_e \bar{a}$. Note the shape of the $[\alpha(1), h_p \neq 0]$ spectra for $h_p = 0.75$ G.

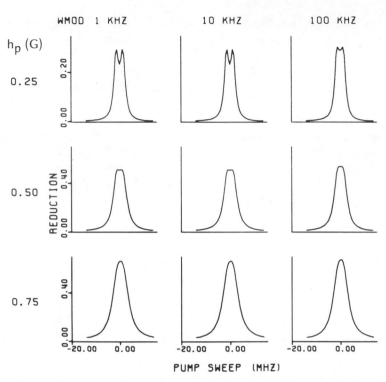

Figure 18. Computer-simulated frequency-swept dispersion, $\alpha(1)$, ELDOR spectra for a hypothetical nitroxide radical system as a function of Zeeman modulation frequency and microwave pump field intensity. ELDOR spectra are shown as a function of the sweep of the pump frequency $(\omega_p - \omega_o)$ maintaining the microwave observer frequency (ω_o) and dc magnetic field fixed so that $\Delta_o = 0$. ELDOR spectra are plotted as reductions that give the ratio of the ELDOR response, $[\alpha(1), h_p = 0] - [\alpha(1), h_p \neq 0]$, to the conventional EPR, $[\alpha(1), h_p = 0]$. Other parameters employed in the calculations are the same as in Figure 17.

In Figures 17–21, we show selected computer-simulated field- and frequency-swept ELDOR spectra for a hypothetical nitroxide spin label system. These spectra were calculated employing the equations derived in the preceding paragraphs. We emphasize that in the derivation discussed the fast-tumbling limit was assumed, i.e., the effects of $\mathscr{H}_1(\Omega)$ and Γ_Ω are represented only by the inclusion of W_n terms in Γ_R. Moreover proton hyperfine interactions were neglected, so that the calculations are valid only for proton deficient nitroxides such as peroxylamine disulfonate (PADS) or for proton-containing nitroxides for which the proton hyperfine lines are strongly coupled by either proton nuclear relaxation or Heisenberg spin exchange.

In Figures 17–21, the variation of selected ELDOR signals with Zeeman modulation frequency ω_s and microwave pump field intensity h_p is

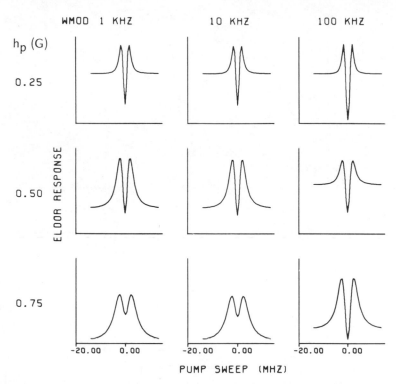

Figure 19. Computer-simulated frequency-swept, out-of-phase dispersion, $\beta(1)$, ELDOR spectra. The values of parameters employed in the computation of spectra are the same as given in Figure 17.

demonstrated. For the molecular relaxation rates employed in these calculations, modulation effects are shown to be important only for frequencies greater than 10 kHz. It is noted that dispersion ELDOR line shapes appear to be more sensitive to Zeeman modulation frequency and to the microwave pump field intensity than are the absorption ELDOR line shapes. However, as can be seen from Figure 22, absorption ELDOR amplitudes (reduction factors) do vary with Zeeman modulation frequency for frequencies greater than 10 kHz. These figures also draw attention to the fact that plots of the reciprocals of the ELDOR reduction factors R^{-1} versus h_p^{-2} for various Zeeman modulation frequencies all approach the same reduction factor at infinite pump power, providing that a nonsaturating microwave observer is employed. Thus, while individual reduction factors depend in a complex manner upon ω_s, H_s, h_p, the various molecular relaxation rates, and upon how the ELDOR measurement was performed, the ELDOR reduction factors at infinite pump power (R_∞) are dependent only upon molecular relaxation rates. This fact was first noted by Freed and co-workers,[86] who also noted

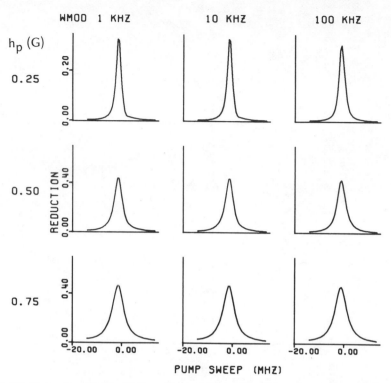

Figure 20. Computer-simulated frequency-swept absorption, $\gamma(1)$, ELDOR spectra as a function of Zeeman modulation frequency and microwave pump field intensity. ELDOR spectra are shown as a function of the sweep of the pump frequency maintaining the microwave observer frequency and the dc magnetic field fixed so that a maximum or a minimum ($\Delta_o \sim \pm 1/2^{1/2}T_{2e}$) of an absorption line is monitored. The relationship $\Delta_o \sim \pm 1/2^{1/2}T_{2e}$ is only approximate, as Zeeman modulation and the intense microwave pump affect the observed line; these effects are taken into account in calculations, and reduction factors are computed at the maximum or minimum of the observed absorption line. Most experimental ELDOR spectra reported in the literature for this configuration appear symmetrical, as shown in the bottom line of this figure. We have measured ELDOR spectra for aqueous solutions of peroxylamine disulfonate at low modulation amplitudes and microwave powers and have observed the unsymmetrical lineshapes shown in the top line of this figure. Parameters employed in computing spectra include, in addition to those given in the figure, the following: $T_{2e} = 2.75 \times 10^{-7}$ sec, $W_e = 1.0 \times 10^6$ Hz, $W_n = 1.0 \times 10^6$ Hz, $h_o = 0.075$ G, $H_s = 0.6$ G.

that errors in measurement of R_∞ could result from the nonlinearity of the plots of R^{-1} versus h_p^{-2} obtained from certain ELDOR experiments.

The problem of measuring R_∞ (and thus the ratio of molecular relaxation rates) appeared to be that of finding a technique for measuring ELDOR reduction factors that would yield a linear plot of R^{-1} versus h_p^{-2}. Such a quest is the subject of a recent article by Van der Drift, Mehlkopf, and

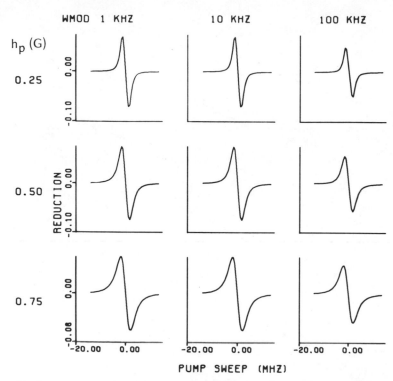

Figure 21. Computer-simulated frequency-swept absorption, $\gamma(1)$, ELDOR spectra as a function of the sweep of the pump frequency while maintaining the microwave observer frequency and the dc magnetic field fixed so that $\Delta_o = 0$ [the crossing point or exact resonance of the $\gamma(1)$ line is monitored]. Parameters employed in the calculations include: $T_{2e} = 2.75 \times 10^{-7}$ sec, $W_e = 1.0 \times 10^6$ Hz, $W_n = 1.0 \times 10^6$ Hz, $h_o = 0.075$ G, $H_s = 0.6$ G.

Smidt.[87] These workers examined the R^{-1} versus h_p^{-2} behavior of absorption, $\gamma(1)$, ELDOR reduction factors measured in field-swept experiments, frequency-swept experiments, and in an experiment involving synchronous sweeping of the dc magnetic field and the pump microwave frequency such that the pumped transition was always at exact resonance. They observed that, unlike conventional field-swept and frequency-swept absorption ELDOR, this latter experiment resulted in reduction factors yielding a straight-line plot for R^{-1} versus h_p^{-2}. To arrive at this conclusion, Van der Drift *et al.*[87] considered experiments involving Zeeman modulation frequencies less than molecular relaxation rates, modulation amplitudes less than resonance linewidths, and systems with well-separated homogeneous electron spin resonance lines. For these conditions, simple analytical expressions for R can be obtained from equation (47) for the several ELDOR sweep modes and commonly employed methods of defining reduction factors;

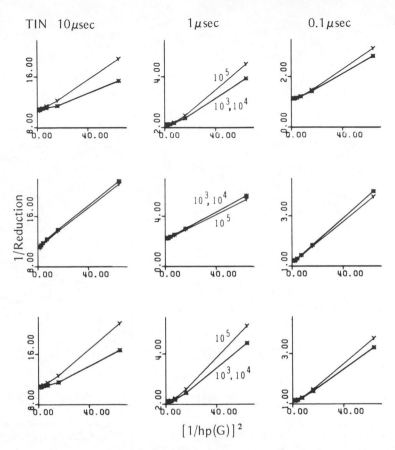

$$[1/\mathrm{hp(G)}]^2$$

Figure 22. Plots of reciprocal ELDOR reduction factors versus the reciprocal of the square of the microwave pump field intensity for three characteristic experiments (display modes). (*Top row*) Results are shown for a field-swept, first harmonic, in-phase absorption ELDOR experiment with the microwave observer–microwave pump frequency separation maintained equal to the nitrogen hyperfine interaction frequency. Also, $\Delta_o = \Delta_p \sim \pm 1/3^{1/2} T_{2e}$. (*Middle row*) Results are shown for the synchronous magnetic field–pump microwave frequency-swept experiment proposed by Van der Drift, Mehlkopf, and Smidt[87] or the equivalent experiments discussed in the text. The crucial conditions are that the in-phase absorption signal be detected and the reduction factor be determined for the conditions $\Delta_o \sim \pm 1/3^{1/2} T_{2e}$ and $\Delta_p = 0$. (*Bottom row*) Results are shown for the conventional pump frequency-swept experiment where the in-phase absorption signal is detected. Reduction factors are calculated for $\Delta_o = 0$, $\Delta_p \sim \pm 1/3^{1/2} T_{2e}$. Results are shown for three characteristic nuclear relaxation rates, namely, $W_n = 10^5$ Hz (*left column*), $W_n = 10^6$ Hz (*middle column*), $W_n = 10^7$ Hz (*right column*). In each element of this matrix of graphs, ELDOR reduction factors are presented for three Zeeman modulation frequencies ω_s, namely, $2\pi \times (10^3, 10^4, 10^5$ Hz). The data for modulation frequencies of $2\pi \times 10^3$ and $2\pi \times 10^4$ are essentially indistinguishable. The frequency for a particular set of data is explicitly denoted in the middle column. Other parameters employed in the computation of data shown include $T_{2e} = 2.75 \times 10^{-7}$ sec, $W_e = 10^6$ Hz, $h_o = 0.075$ G, and $H_s = 0.01$ G.

Table 2

Limiting Form Analytical Equations for ELDOR Reduction Factors at Low Zeeman Modulation Frequencies and Amplitudes

Type of ELDOR experiment	$(R_\infty/R)^a$
Field-swept, first harmonic, in-phase absorption[b] $\Delta_o = \Delta_p \sim \pm 1/3^{1/2} T_{2e}$	$\dfrac{1}{3S} \dfrac{(4 + 3S)^2}{(8 + 3S)}$
Field-swept, first harmonic, in-phase absorption $\Delta_o \sim \pm 1/3^{1/2} T_{2e}, \quad \Delta_p = 0$	$\dfrac{1 + S}{S}$
Field-swept, first harmonic, in-phase dispersion $\Delta_o = \Delta_p = 0$	$\dfrac{1 + S}{S}$
Pump frequency-swept, first harmonic, in-phase absorption $\Delta_o = 0, \quad \Delta_p \sim \pm 1/3^{1/2} T_{2e}$	$\dfrac{(1 + S)^{3/2}}{S}$
Pump frequency-swept, first harmonic, in-phase absorption $\Delta_o = \Delta_p \sim \pm 1/3^{1/2} T_{2e}$	$\dfrac{1}{3S} \dfrac{(4 + 3S)^2}{(8 + 3S)}$
Observer frequency-swept, first harmonic, in-phase absorption $\Delta_o \sim \pm 1/3^{1/2} T_{2e}, \quad \Delta_p = 0$	$\dfrac{1 + S}{S}$
Synchronous magnetic field–pump frequency-swept, first harmonic, in-phase absorption $\Delta_o \sim \pm 1/3^{1/2} T_{2e}, \quad \Delta_p = 0$	$\dfrac{1 + S}{S}$
Pump frequency-swept, first harmonic, in-phase dispersion $\Delta_o = \Delta_p = 0$	$\dfrac{1 + S}{S}$

[a] These expressions apply for the conditions of well-separated, homogeneous resonance lines, Zeeman modulation frequencies approaching zero (much less than molecular relaxation rates), and Zeeman modulation amplitudes much less than resonance linewidths. Moreover, the linewidths of the pumped and observed resonances are assumed to be equal and the microwave observing field is assumed to be nonsaturating. In the above, we have defined the saturation factor S as $S = 4d_p^2 T_{1_{\text{eff}}} T_{2_{\text{eff}}}$, where $d_p = \frac{1}{2}\gamma_e h_p$.

[b] The maximum of a first harmonic absorption, $\gamma(1)$, resonance line depends upon the microwave and Zeeman modulation field amplitudes as well as upon the spin–spin relaxation rate T_{2e}^{-1}. Thus, for commonly employed instrument settings, the absorption maximum is only approximately equal to $3^{-1/2} T_{2e}^{-1}$.

these analytical expressions are summarized in Table 2. It is to be noted that for the conditions imposed to derive the expressions shown in Table 2, signals at a given Zeeman modulation harmonic can be calculated as derivatives of the signal at the zeroth harmonic (the signal in the absence of Zeeman modulation). This relationship is demonstrated for the condition of $d_p = 0$ as follows: For this condition, equation (47) can be written as

$$
\begin{pmatrix}
-T_{2e}^{-1} & r\omega_s & \Delta_{ov} & 0 & 0 & 0 \\
r\omega_s & T_{2e}^{-1} & 0 & \Delta_{ov} & 0 & 0 \\
\Delta_{ov} & 0 & T_{2e}^{-1} & r\omega_s & d_o & 0 \\
0 & \Delta_{ov} & r\omega_s & -T_{2e}^{-1} & 0 & d_o \\
0 & 0 & 4d_o & 0 & -W_{\text{eff}}^v & r\omega_s \\
0 & 0 & 0 & 4d_o & r\omega_o & W_{\text{eff}}^v
\end{pmatrix}
\begin{pmatrix}
\alpha_v(1,0,r) \\
\beta_v(1,0,r) \\
\gamma_v(1,0,r) \\
\delta_v(1,0,r) \\
Y_{v+}^{(r)} \\
Y_{v-}^{(r)}
\end{pmatrix}
$$

$$
=
\begin{pmatrix}
d_s[\gamma_v(1,0,r+1) + \gamma_v(1,0,r-1)] \\
d_s[\delta_v(1,0,r+1) + \delta_v(1,0,r-1)] \\
d_s[\alpha_v(1,0,r+1) + \alpha_v(1,0,r-1)] + 2q\omega_v d_o \delta_{r,0} \\
d_s[\beta_v(1,0,r+1) + \beta_v(1,0,r-1)] \\
0 \\
0
\end{pmatrix}
\tag{51}
$$

where

$$
W_{\text{eff}}^v = \frac{2W_e(4W_e^2 + 8W_e W_n + 3W_n^2)}{(4W_e^2 + 6W_e W_n + W_n^2)}
$$

for $v = +1, -1$, and

$$
W_{\text{eff}}^v = \frac{2W_e(2W_e + 3W_n)}{(2W_e + W_n)}
$$

for $v = 0$. If modulation amplitudes are small enough that coupling from higher harmonics can be neglected, and if modulation frequencies are low enough to warrant neglect of terms in $r\omega_s$, equation (51) can be written for the signals at the first harmonic of the Zeeman modulation as

$$
\begin{pmatrix}
-T_{2e}^{-1} & \Delta_{ov} & 0 & \vdots & 0 & 0 & 0 \\
\Delta_{ov} & T_{2e}^{-1} & d_o & \vdots & 0 & 0 & 0 \\
0 & 4d_o & -W_{\text{eff}}^v & \vdots & 0 & 0 & 0 \\
\hdashline
0 & 0 & 0 & \vdots & T_{2e}^{-1} & \Delta_{ov} & 0 \\
0 & 0 & 0 & \vdots & \Delta_{ov} & -T_{2e}^{-1} & d_o \\
0 & 0 & 0 & \vdots & 0 & 4d_o & W_{\text{eff}}^v
\end{pmatrix}
\begin{pmatrix}
\alpha_v(1, 0, 1) \\
\gamma_v(1, 0, 1) \\
Y_{v+}^{(1)} \\
\beta_v(1, 0, 1) \\
\delta_v(1, 0, 1) \\
Y_{v-}^{(1)}
\end{pmatrix}
$$

$$
=
\begin{pmatrix}
d_s \gamma_v(1, 0, 0) \\
d_s \alpha_v(1, 0, 0) \\
0 \\
0 \\
0 \\
0
\end{pmatrix}
\tag{52}
$$

The upper left-hand block of equation (52) can be solved to yield

$$
\alpha_v(1, 0, 1) = d_s \left[-T_{2e} \gamma_v(1, 0, 0) + \frac{\Delta_{ov} T_{2e}^2 [\alpha_v(1, 0, 0) + T_{2e} \Delta_{ov} \gamma_v(1, 0, 0)]}{1 + T_{2e}^2 \Delta_{ov}^2 + 4d_o^2 (W_{\text{eff}}^v)^{-1} T_{2e}} \right]
\tag{53a}
$$

and

$$
\gamma_v(1, 0, 1) = d_s \left[\frac{T_{2e}[\alpha_v(1, 0, 0) + T_{2e} \Delta_{ov} \gamma_v(1, 0, 0)]}{1 + T_{2e}^2 \Delta_{ov}^2 + 4d_o^2 (W_{\text{eff}}^v)^{-1} T_{2e}} \right]
\tag{53b}
$$

The matrix master equation for the signals in the absence of modulation is

$$
\begin{pmatrix}
-T_{2e}^{-1} & \Delta_{ov} & 0 \\
\Delta_{ov} & T_{2e}^{-1} & d_o \\
0 & 4d_o & -W_{\text{eff}}^v
\end{pmatrix}
\begin{pmatrix}
\alpha_v(1, 0, 0) \\
\gamma_v(1, 0, 0) \\
Y_{v+}^{(0)}
\end{pmatrix}
=
\begin{pmatrix}
0 \\
2q\omega_v d_o \\
0
\end{pmatrix}
\tag{54}
$$

which yields upon solution

$$\alpha_v(1, 0, 0) = \frac{2q\omega_v d_o \Delta_{ov} T_{2e}^2}{1 + T_{2e}^2 \Delta_{ov}^2 + 4d_o^2 (W_{eff}^v)^{-1} T_{2e}} \qquad (55a)$$

and

$$\gamma_v(1, 0, 0) = \frac{2q\omega_v d_o T_{2e}}{1 + T_{2e}^2 \Delta_{ov}^2 + 4d_o^2 (W_{eff}^v)^{-1} T_{2e}} \qquad (55b)$$

Differentiation of these signals with respect to Δ_{ov} produces

$$\frac{d}{d\Delta_{ov}} (\alpha_v(1, 0, 0)) = \alpha_v(1, 0, 0)' = \frac{2q\omega_v d_o T_{2e}^2}{1 + T_{2e}^2 \Delta_{ov}^2 + 4d_o^2 (W_{eff}^v)^{-1} T_{2e}}$$
$$- \frac{4q\omega_v d_o \Delta_{ov}^2 T_{2e}^4}{[1 + T_{2e}^2 \Delta_{ov}^2 + 4d_o^2 (W_{eff}^v)^{-1} T_{2e}]^2} \qquad (56a)$$

and

$$\frac{d}{d\Delta_{ov}} (\gamma_v(1, 0, 0)) = \gamma_v(1, 0, 0)' = \frac{-4q\omega_v d_o \Delta_{ov} T_{2e}^3}{[1 + T_{2e}^2 \Delta_{ov}^2 + 4d_o^2 (W_{eff}^v)^{-1} T_{2e}]^2} \qquad (56b)$$

Equation (55) also provides the signals at the zeroth harmonic of the Zeeman modulation that we need to complete the solution of equation (53). Upon substitution of equations (55a,b) into equations (53a,b), we obtain

$$\alpha_v(1, 0, 1) = d_s \left[\frac{4q\omega_v d_o \Delta_{ov}^2 T_{2e}^4}{[1 + T_{2e}^2 \Delta_{ov}^2 + 4d_o^2 (W_{eff}^v)^{-1} T_{2e}]^2} \right.$$
$$\left. - \frac{2q\omega_v d_o T_{2e}^2}{1 + T_{2e}^2 \Delta_{ov}^2 + 4d_o^2 (W_{eff}^v)^{-1} T_{2e}} \right] \qquad (57a)$$

which is identically $-d_s \alpha_v(1, 0, 0)'$. Similarly,

$$\gamma_v(1, 0, 1) = d_s \left[\frac{4q\omega_v d_o \Delta_{ov} T_{2e}^3}{[1 + T_{2e}^2 \Delta_{ov}^2 + 4d_o^2 (W_{eff}^v)^{-1} T_{2e}]^2} \right] \qquad (57b)$$

which is $-d_s \gamma_v(1, 0, 0)'$. In like manner, the derivative approximation can be shown to hold for EPR signals for the condition that $d_p \neq 0$ and hence for ELDOR reduction factors. Freed and co-workers,[86] Van der Drift et al.,[87] and other workers[50] have obtained several of the expressions given in Table 2 by computing derivatives of the zeroth harmonic absorption signals.

From Table 2, it is clear that linear plots of R^{-1} versus h_p^{-2} can be obtained from each of the sweep modes when the appropriate choice of spectral points for the measurement of ELDOR reduction factors is made. However, of the several sweep modes considered, the synchronous magnetic field–microwave pump frequency-swept experiment and ELDOR experiments where dispersion detection is employed are the most convenient. For example, although keeping the microwave pump and dc magnetic field fixed such that $\Delta_p = 0$ and sweeping the microwave observer frequency is theoretically equivalent to the synchronous magnetic field–pump frequency-swept experiment, this former mode is unsatisfactory in practice as it is desirable to utilize a narrow-band observing channel to optimize signal-to-noise ratio.

When comparing absorption and dispersion experiments, it is well to note that dispersion experiments are more sensitive to klystron frequency modulation noise than are absorption experiments and are thus unsuited for experiments at microwave observing power levels where source noise dominates crystal detector noise.

The question must be asked as to how well the limiting form expressions given in Table 2 describe real systems, or equivalently, how rapidly do the limiting approximations become invalid as (1) the rotational correlation time is increased, (2) the Zeeman modulation amplitude and frequency are increased, and (3) inhomogeneous rather than homogeneous EPR lines are considered.

Peroxylamine disulfonate (PADS) in glycerol–water solutions is convenient for testing the dependence upon rotational correlation time as relaxation parameters have been determined by Goldman, Bruno, and Freed.[72] Obviously, the limiting form equations cannot be simply applied to the "slow" $(\tau_2 > 3 \times 10^{-7}$ sec) or to the "intermediate" $(3 \times 10^{-7} > \tau_2 > 3 \times 10^{-9}$ sec) motion regions, as the lineshapes are significantly different from those predicted by the limiting form equations. The question to be answered is what is the slowest correlation time in the "fast" $(3 \times 10^{-9} > \tau_2 > 10^{-11}$ sec) motion region for which the limiting forms are valid. To answer this, consideration must also be given to the sensitivity of various ELDOR parameters (i.e., the slopes and intercepts of graphs of R^{-1} versus d_p^{-2}) to the various relaxation rates. From Table 2 we have, for the experiments of greatest interest,

$$R = R_\infty \left(\frac{S}{1 + S} \right) \tag{58a}$$

or

$$\frac{1}{R} = \frac{1}{R_\infty} \left(1 + \frac{1}{S} \right) \tag{58b}$$

In the notation of Hyde, Chien, and Freed,[85] equations (58) can be written as

$$\frac{1}{R} = \frac{\Omega_p}{\Omega_{op}} + \frac{1}{d_p^2 T_{2e} \Omega_{op}} \tag{59}$$

For an $S = \frac{1}{2}$, $I = 1$ nitroxide system where the central line is pumped and the low-field line is observed, we have (in our notation)

$$\frac{1}{R(0, +1)} = \frac{2W_e + W_n}{W_n} + \left\{ d_p^2 T_{2e} \left[\frac{2W_n}{W_e(2W_e + 3W_n)} \right] \right\}^{-1} \qquad (60)$$

Measurement of the slope and intercept from a graph of $1/R(0, +1)$ versus d_p^{-2} appears to permit determination of W_e and W_n, provided T_{2e} is known from linewidth measurements. However, as can be seen from Table 3, the slope (i.e., Ω_{op}) and intercept (i.e., Ω_p/Ω_{op}) are rather insensitive to changes in W_n for values of W_n greater than W_e. For this condition ($W_n > W_e$), ELDOR experiments such as we are presently considering are unsatisfactory either for the measurement of molecular relaxation rates or for rigorously testing the agreement between theory and experiment. Unfortunately, W_n is greater than W_e for the slower motions of the "fast" motion region for PADS solutions. We have carried out analysis of graphs of R^{-1} versus h_p^{-2} for dilute solutions of PADS characterized by correlation times in the range 3×10^{-11} to 3×10^{-9} sec. The limiting form equations of Table 2, when modified to take into account the effects of motion producing unequal linewidths for the three EPR transitions [i.e., $T_{2e}^{-1} \rightarrow T_{2e}^{-1}(v)$], yield results that agree reasonably well with experimental results over this range, providing that peak-to-peak Zeeman modulation amplitudes are kept to 0.5 or less of the smallest EPR linewidths (measured as the full width between points of maximum slope of the EPR absorption). This observation must, of course, be viewed in light of the preceding remarks about the sensitivity of Ω_{op} and Ω_p/Ω_{op} to changing relaxation rates for $W_n > W_e$.

Table 3

The Sensitivity of Slopes (Ω_{op}) and Intercepts (Ω_p/Ω_{op}) of Graphs of $1/R(0, +1)$ versus d_p^{-2} to Changes in Molecular Relaxation Rates[a]

W_n (sec^{-1})	Ω_{op} (sec)	Ω_p/Ω_{op}
1×10^4	3.88×10^{-8}	101
5×10^4	1.74×10^{-7}	21.0
1×10^5	3.08×10^{-7}	11.0
5×10^5	8.00×10^{-7}	3.00
1×10^6	1.00×10^{-6}	2.00
5×10^6	1.25×10^{-6}	1.20
1×10^7	1.29×10^{-6}	1.10
5×10^7	1.32×10^{-6}	1.02

[a] Calculated from equation (60) with $W_e = 5 \times 10^5$ Hz.

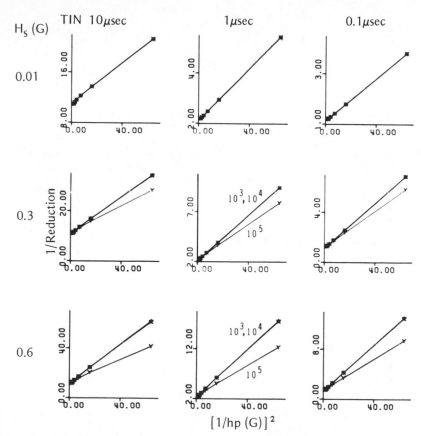

Figure 23. Plots of reciprocal in-phase dispersion (field- or frequency-swept with $\Delta_o = \Delta_p = 0$) ELDOR reduction factors versus the reciprocal of the square of the microwave pump field intensity for three values of the Zeeman modulation amplitude: $H_s = 0.01$ G, (*top row*) $H_s = 0.3$ G, (*middle row*) $H_s = 0.6$ G, (*bottom row*). Other parameters have the same values as in Figure 22.

The behavior of graphs of R^{-1} versus h_p^{-2} as a function of Zeeman modulation amplitude is shown in Figures 22–23. The theory presented in this chapter, when modified to take into account v-dependent T_{2e} values, permits excellent simulation of Zeeman modulation amplitude and frequency effects for PADS solutions. This fact, together with the comparison presented in Table 4 of the results of calculations that explicitly consider modulation conditions and results calculated from the limiting equations of Table 2, permits estimation of the errors that arise when employing the limiting form equations to analyze ELDOR data for conditions of finite Zeeman modulation amplitudes and frequencies. From Table 4, it is clear that greater errors result for the faster W_n rates. Rather interestingly, for

Table 4

Comparison of the Results of Exact Calculation with Results from the Limiting Form Equations for Analysis of Graphs of $1/R(0, +1)$ versus d_p^{-2} for Systems Consisting of Homogeneous EPR Lines[a]

	Synchronous magnetic field–pump frequency-swept, in-phase, first harmonic absorption with $\Delta_o \sim \pm 1/3^{1/2} T_{2e}, \Delta_p = 0$				Field- or frequency-swept, in-phase, first harmonic dispersion with $\Delta_o = 0, \Delta_p = 0$			
	$\omega_s = 2\pi \times 10^3$ Hz		$\omega_s = 2\pi \times 10^5$ Hz		$\omega_s = 2\pi \times 10^3$ Hz		$\omega_s = 2\pi \times 10^5$ Hz	
$2H_s$ (G)	$\dfrac{T_{1e}(A)[b]}{T_{1e}(E)}$	$\dfrac{W_n(A)}{W_n(E)}$	$\dfrac{T_{1e}(A)}{T_{1e}(E)}$	$\dfrac{W_n(A)}{W_n(E)}$	$\dfrac{T_{1e}(A)}{T_{1e}(E)}$	$\dfrac{W_n(A)}{W_n(E)}$	$\dfrac{T_{1e}(A)}{T_{1e}(E)}$	$\dfrac{W_n(A)}{W_n(E)}$
	$W_n = 10^5$ sec^{-1}, $T_{1e} = (2W_e)^{-1} = 10^{-6}$ sec for exact calculation							
0.02	0.975	0.996	0.985	1.01	0.980	1.00	0.980	1.00
0.60	0.416	2.31	0.526	1.78	0.423	2.34	0.606	1.56
1.20	0.168	5.42	0.415	1.93	0.085	7.67	0.390	1.39
	$W_n = 10^6$ sec^{-1}, $T_{1e} = (2W_e)^{-1} = 10^{-6}$ sec for exact calculation							
0.02	0.896	0.970	0.954	0.944	0.927	0.972	0.927	0.972
0.60	0.393	2.14	0.581	1.36	0.425	2.08	0.565	1.42
1.20	0.161	4.47	0.109	3.57	0.218	3.20	0.363	1.64
	$W_n = 10^7$ sec^{-1}, $T_{1e} = (2W_e)^{-1} = 10^{-6}$ sec for exact calculation							
0.02	0.820	1.02	0.900	0.794	0.869	0.959	0.869	0.959
0.60	0.369	1.50	0.545	0.680	0.413	1.62	0.529	0.787
1.20	0.150	2.15	0.325	0.530	0.217	1.22	0.328	0.648

[a] Graphs of $1/R(0, +1)$ versus d_p^{-2} are calculated employing equation (47) where modulation amplitude and frequency effects are explicitly considered. Parameters employed in the calculation include $T_{2e}(+1) = T_{2e}(0) = T_{2e}(-1) = 2.75 \times 10^{-7}$ sec, $T_{1e} = 10^{-6}$ sec, $h_o = 0.075$ G. The values of other parameters employed are indicated in the table. These plots were then analyzed to determine T_{1e} and W_n employing the limiting form equations discussed in the text and assuming T_{2e} to be known.
[b] The notation $T_{1e}(A)/T_{1e}(E)$ is a shorthand notation for $T_{1e}(\text{Approximate})/T_{1e}(\text{Exact})$.

small modulation amplitudes, errors at Zeeman modulation frequencies of 10^5 Hz are often less than those at 10^3 Hz. For modulation amplitudes less than resonance linewidths, the discrepancy between exact and approximate calculations is not very serious; conversely, for modulation amplitudes exceeding resonance linewidths, the errors can be quite large. It is, of course, a simple matter to determine when modulation amplitudes are significantly less than resonance linewidths for homogeneous lines, as modulation amplitudes are readily calibrated from sideband splittings or from voltages introduced into a coil placed in the microwave cavity. However, it is not sufficient to assume that if the EPR lineshapes are not significantly distorted that the modulation amplitude is insignificant. For inhomogeneous lines, overmodulation effects (relative to intrinsic linewidths) are not so easily determined as for homogeneous lines, since T_{2e} can no longer be calculated from the apparent (or envelope) resonance linewidth. T_{2e} values can be obtained from analysis of the EPR spectra only if all hyperfine couplings leading to the inhomogeneous broadening are known.

We have also examined the typical spin label system 2,2,6,6-tetramethyl-4-piperidinol-1-oxyl (TANOL), where inhomogeneous broadening due to proton hyperfine interactions is present. Appreciable errors were found to result from employment of the equations of Table 2 for this case, even in the limit of very small modulation amplitudes. It was also found that the synchronous magnetic field–pump frequency-swept absorption and dispersion experiments, which were most attractive for homogeneous lines, lose their advantages for the inhomogeneous lines of TANOL. Of course, each example of inhomogeneous broadening is unique, depending upon the precise hyperfine couplings, intrinsic T_{2e}, and any relaxation processes such as Heisenberg spin exchange that couple the superhyperfine lines. Thus, the system of perdeuterated 2,2,6,6-tetramethyl-4-piperidone-1-oxyl (D-TANONE) can appear to behave similar to a homogeneously broadened system, even though this system is obviously inhomogeneously broadened by the small deuteron hyperfine interactions.

To this point, our discussion has focused upon in-phase signals. As is shown in Figure 24, certain frequency-swept ELDOR signal amplitudes can maximize for phase angles other than zero. Thus, when applied and molecular modulation frequencies are comparable, consideration of modulation conditions can be important for the optimization of signal-to-noise. More importantly, the analysis of the modulation frequency and phase dependence of ELDOR signals provide important additional criteria for the determination of molecular relaxation rates from ELDOR measurements. For example, W_e and W_n (and not just the ratio of these rates) can be determined even for those cases where T_{2e} has not been determined from EPR linewidth measurements. As is evident from Figure 24 and our preceding comments regarding the sensitivity of determining molecular rates for experiments where modulation effects are avoided, the induction and analysis of Zeeman modulation effects can be particularly helpful in determining rates for the cases where $W_n > W_e$. It is to be noted that the modulation conditions of frequency, amplitude, and phase can be determined with high accuracy, in contrast to certain other parameters, such as the microwave pump field intensity, which is determined only with some difficulty. Thus, the precise simulation of modulation effects can provide an excellent criterion for the precise determination of molecular relaxation rates.

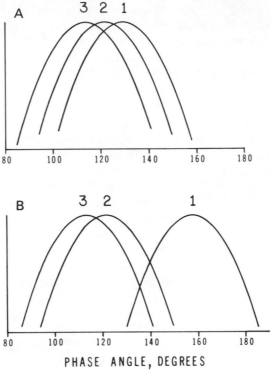

Figure 24. The variation of signal intensity of the pump frequency-swept absorption ELDOR signal recorded with the microwave observer frequency and dc magnetic field fixed so that $\Delta_o = 0$ is shown as a function of phase angle between the driving modulation field and the reference voltage. The line shapes of such ELDOR signals do not change with phase angle; however, consideration of the present figure indicates that signal intensities do vary with phase angle, and moreover, maximum intensities are often observed for phase angles of 80° to 160°. Parameters employed in the calculations for homogeneous resonance lines include $T_{2e} = 2.7 \times 10^{-7}$ sec, $h_p = 0.5$ G, $h_o = 0.075$ G, $\omega_s = 2\pi \times 10^5$ rad/sec, and $H_s = 0.01$ G. (A) all spectra were computed for $W_e = 10^6$ Hz and W_n values of: (1) 10^5 Hz, (2) 10^6 Hz, (3) 10^7 Hz. (B) spectra were computed for the following values: (1) $W_e = W_n = 10^5$ Hz, (2) $W_e = W_n = 10^6$ Hz, (3) $W_e = W_n = 10^7$ Hz.

Although in this chapter we have explicitly considered only ELDOR experiments on rapidly tumbling spin labels, the mathematical formulation is easily extended to treat other double (ENDOR) and triple[88-93] resonance experiments.[94] Indeed, for systems for which the high-field approximation applies and for which the total number of spins are conserved during the experiment [optical detection of magnetic resonance (ODMR) experiments, for example, violate both of these conditions], the density matrix treatment outlined represents a convenient means of quantitatively analyzing the spin response.

3. Examples of Zeeman Modulation Effects

3.1. Modulation Effects in the Resonance and Double Resonance Spectra of Spin Labels

3.1.1. Examples of Direct Saturation Transfer by Rotational Diffusion

Modulation of anisotropic magnetic interactions [represented by $\mathcal{H}_1(\Omega)$ in equation (2)] by molecular rotation [represented by Γ_Ω in equation (1)] has a dramatic effect upon magnetic resonance and double resonance spectra. A typical example of the effect upon conventional EPR, $\gamma(1)$, spectra is shown in Figure 12, which illustrates that conventional EPR spectra are sensitive to rotational diffusion characterized by correlation times in the range 10^{-11} to 3×10^{-7} sec. Appropriate choice of Zeeman modulation and phase-sensitive detection conditions can be utilized to enhance the sensitivity to molecular motion, e.g., detection of $\beta(1)$ and $\delta(2)$ spectra permit measurement of substantially slower motion, as is demonstrated in Figures 13 and 14 and in a number of research publications.[31,33,34,42,54,55,81–84,95–100] Indeed, rotational correlation times as long as 10^{-3} sec appear accessible to measurement. Field-swept ELDOR measurements performed employing Zeeman modulation amplitudes greater than the frequency separation of the pump and observing microwave fields and Zeeman modulation frequencies of the order of molecular relaxation rates permit determination of rates in the slow tumbling region, as is evident from Figure 16. The preceding experiments are obviously dependent upon the combined action of applied and molecular modulation conditions in determining the passage of spin transitions or packets through resonance; they are thus commonly referred to as "passage" experiments.[101–105]

It is appropriate to ask if there exist double resonance techniques that are sensitive to slow molecular diffusion but that are independent of Zeeman modulation or "passage" effects. The answer is, at least theoretically, yes. Bruno and Freed[73] have shown theoretically that conventional, $\gamma(0)$, ELDOR experiments without Zeeman modulation are sensitive to slow rotational diffusion. Hyde *et al.*,[43] Thomas,[106] McConnell,[107] and Dalton *et al.*[33] have also advanced theoretical arguments predicting that such stationary ELDOR (or saturation transfer dependent) experiments should be sensitive to rotational correlation times as long as 10^{-4} sec. As analysis of stationary ELDOR is mathematically much simpler than is the analysis of passage experiments, it is obviously important to ascertain if modulation-independent resonance signals are experimentally realizable. We have attempted to carry out such an evaluation, choosing as model systems 10^{-3} M solutions of 2,2,6,6-tetramethyl-4-piperidone-1-oxyl (TANONE), 2,2,6,6-tetramethyl-4-

piperidinol-1-oxyl (TANOL) and 17β-hydroxy-4',4'-dimethylspiro[5α-androstane-3,2'-oxazolidin]-3'-oxyl (HDA) in sec-butylbenzene (SBB). As Heisenberg spin exchange (which depends upon radical concentrations) is also an ELDOR mechanism, it was felt necessary to maintain spin label concentrations below 3×10^{-3} M. We would also note that as radical concentrations employed in spin labeling of biomolecules seldom result in final spin label concentrations greater than 10^{-3} M, it is believed that for any ELDOR experiment to be of general practical utility, detection of millimolar spin label concentrations is prerequisite.

For our 10^{-3} M solutions of nitroxides, we found frequency-swept ELDOR experiments to fail because of poor signal-to-noise ratios when attempted without Zeeman modulation, when attempted employing low-frequency amplitude modulation of the pumping microwaves and phase detecting at the modulation frequency, or when attempted employing low Zeeman modulation frequencies (35, 270 Hz) and low Zeeman modulation amplitudes (less than 1 G). We would note that although the last two experiments employ modulation, the results can be adequately analyzed starting with a mathematical formulation that does not explicitly consider modulation. In these experiments, modulation serves two important purposes: (1) to convert the signal to a sufficiently high frequency before microwave detection so that low-frequency ($1/f$) noise of the detector does not interfere, and (2) to provide baseline stability. Obviously, all three experiments fail because of inadequate discrimination against $1/f$ noise; microwave pump modulation also fails to provide adequate baseline stability. We note that baseline stability depends on the " transfer of modulation " principle that modulation appear on the detecting microwaves only as a consequence of interactions within the spin system and not by any direct process. In ELDOR, the transfer of modulation principle is satisfied if the isolation between pump and observing modes is very high. The 40-dB isolation of our ELDOR cavity is inadequate. We were, on the other hand, successful in obtaining meaningful ELDOR spectra employing 270-Hz Zeeman modulation fields of large amplitudes (e.g., 14 G).[43] We would also note that similar results have been obtained by Dorio and Chien.[45-47] Such spectra, while independent of Zeeman modulation frequency, are somewhat dependent upon Zeeman modulation amplitude. As shown by Hyde et al.,[43] analysis of such experiments requires a detailed treatment of modulation terms in $\varepsilon(t)$. Stationary ELDOR spectra are predicted to depend strongly upon microwave pump power. The obvious conclusion from our studies is that experiments that omit Zeeman modulation are impractical with current ELDOR instrumentation. A detailed mathematical consideration of Zeeman modulation is necessary for ELDOR experiments suitable for measurement of rotational correlation times for slowly diffusing spin labels. In this regard, it is just as convenient to perform experiments at high Zeeman modulation frequency

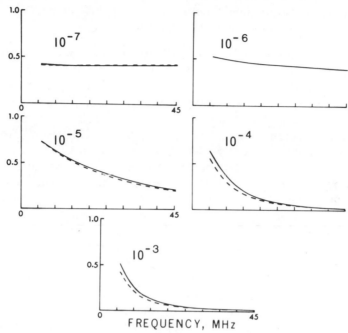

Figure 25. The variation of the computer-simulated frequency-swept absorption, $\gamma(1)$, ELDOR spectra calculated for Zeeman modulation frequencies of 10^5 (solid lines) and 270 (dashed lines) Hz as a function of isotropic Brownian diffusion rotation correlation time (shown in the figure). Other parameters employed in the calculations include $T_{1e} = 6.6 \times 10^{-6}$ sec, $T_{2e} = 2 \times 10^{-7}$ sec, $h_o = 0.01$ G, $h_p = 0.3$ G, and $H_s = 0.1$ G.

and small Zeeman modulation amplitude as the converse experiment just described. Calculated spectra for Zeeman modulation frequencies of 270 and 10^5 Hz are shown in Figure 25. These spectra indicate the magnitude of the variation of ELDOR reductions with Zeeman modulation frequency in the slow motion region and further indicate that spectra for high and low Zeeman modulation frequencies are computed with equal ease when molecular and applied modulation effects are both treated explicitly. The data show that the dependence of ELDOR reduction factors on Zeeman modulation frequency causes greater relative errors in the measurement of correlation time for long correlation times (i.e., of the order of 10^{-4} to 10^{-3} sec). Indeed, for measurement of rotational correlation times in the range 10^{-7} to 10^{-5} sec, neglect of Zeeman modulation frequency effects can result in acceptably small errors.

In summary, we note that rotational correlation times for slowly tumbling nitroxides can be determined by measurement of $\beta(1)$ and $\delta(2)$ single resonance signals and by ELDOR experiments that utilize Zeeman modulation.

3.1.2. Examples of Saturation Transfer by Nuclear Relaxation and the Effects of Unresolved Hyperfine Interactions

From quantum mechanical considerations, one would expect the unpaired electron of nitroxide radicals to experience finite electron–nuclear hyperfine interactions with nearby methyl and methylene protons as well as with the nitrosyl nitrogen, and indeed, such interactions have been observed by NMR,[108,109] EPR,[109–112] and ENDOR[49,52,111] spectroscopy.

Although proton hyperfine interactions are always present, the observability of such interactions depends upon a number of conditions. At high radical concentrations and temperatures such that Heisenberg spin exchange rates are greater than proton hyperfine frequencies but less than the nitrogen hyperfine interaction frequency, the proton lines will be effectively coupled by exchange, and the EPR lines corresponding to the three nitrogen spin states will behave as homogeneously broadened EPR lines. For molecular rotational frequencies of the order of anisotropic proton interaction frequencies, proton nuclear relaxation may effectively couple the proton superhyperfine lines, and the nitrogen EPR lines (envelopes) will appear homogeneously or pseudohomogeneously broadened. However, for dilute solutions of nitroxides characterized by rotational correlation times from 5×10^{-11} to 10^{-9} sec, significant nitrogen nuclear relaxation can exist without the coupling of proton superhyperfine lines by either Heisenberg spin exchange or by proton relaxation. For this situation, the nitrogen EPR lines behave as inhomogeneously broadened lines. Typical dispersion and absorption EPR spectra showing the low-field $(v = +1)$ nitrogen hyperfine component are presented in Figures 26 and 27.

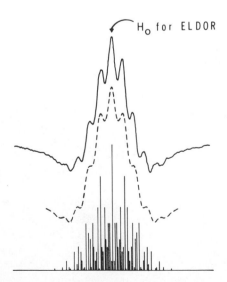

H_0 for ELDOR

Figure 26. Experimental and theoretical dispersion, $\alpha(1)$, EPR spectra for the low-field $[v(^{14}N) = +1]$ hyperfine component of the electron resonance of a 1.5×10^{-3} M solution of TANOL in SBB at $-63°C$. The experimental spectrum is indicated by a solid line, while the theoretical spectrum is indicated by a dashed line. The stick diagram indicates the transition frequencies and relative intensities (degeneracies) of the proton superhyperfine lines. The arrow indicates spin packets monitored by the microwave observer when recording the frequency-swept ELDOR spectra shown in Figure 28. The parameters employed in the computation of spectra are the same as given in Figure 1. (From Percival et al.[52])

Figure 27. Experimental (solid line) and theoretical (dashed line) absorption, $\gamma(1)$, EPR
spectra for the low-field $[\nu(^{14}N) = +1]$ hyperfine component of the electron resonance of a
1.5×10^{-3} M solution of TANOL in SBB
at $-63°C$. The arrow at point A indicates the
position of the microwave observer when recording the frequency-swept ELDOR spectra
shown in Figure 29, while the arrow at point B
indicates the position of the microwave observer
for the ELDOR data presented in Figure 24 and
that discussed in the text (p. 218). The parameters employed in the computation of spectra
are the same as given in Figure 1. (From Percival
et al.[52])

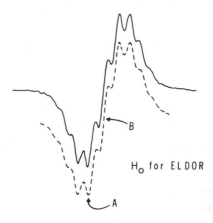

As is demonstrated in Figures 28 and 29, the width of frequency-swept
ELDOR lines corresponds to the width of individual proton superhyperfine
lines rather than to the envelope width of nitrogen hyperfine lines. The
widths of the individual spin packets determined by spin–spin relaxation
effects are of the order of 0.2 to 0.4 G; thus, even the 0.24 G peak-to-peak
Zeeman modulation field employed to obtain the ELDOR spectra shown in

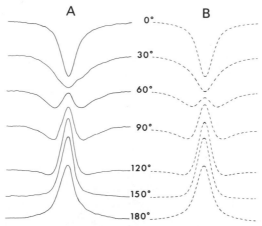

Figure 28. The variation of the pump frequency-swept dispersion ELDOR detected at the first
harmonic of the modulation with reference phase angle for 1.5×10^{-3} M TANOL in SBB
at $-63°C$ (A) and the corresponding computer simulated spectra (B). The experimental
spectra were obtained with a modulation frequency of 10^5 Hz and amplitude of 0.24 G. The
microwave pump and observer power settings were 860 mW and 5 mW, respectively. The
position of the microwave observer is indicated in Figure 26. The theoretical spectra were
computed employing a nitrogen nuclear relaxation rate of 2.6×10^5 Hz and assuming proton
relaxation rates are insignificant. Other parameters include an effective electron spin–lattice
relaxation time of 1.3×10^{-6} sec, an effective electron spin–spin relaxation time of 2.7×10^{-7}
sec, and a rotational correlation time of 2×10^{-10} sec. Other parameters are the same as
given in Figure 1. (From Percival *et al.*[52])

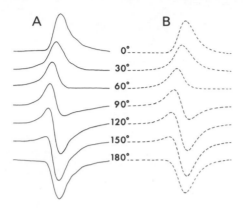

Figure 29. Variation of the pump frequency-swept absorption ELDOR detected at the first harmonic of the modulation with reference phase angle for 1.5×10^{-3} M TANOL in SBB at $-63°C$ (A) and the corresponding computer-simulated spectra (B). Other conditions are the same as in Figure 28, except that the position of the microwave observer is as shown in Figure 27, point A. (From Percival *et al.*[52])

Figures 28 and 29 is sufficient to cause some broadening of ELDOR lines. Broadening is also observed due to the microwave pump field intensity for fields comparable to spin packet widths. The precise effect of Zeeman modulation amplitude upon the line shapes of frequency-swept ELDOR is complex, depending upon spectral overlap determined by coupling constants and intrinsic linewidths and upon the portion of the spectrum monitored; however, providing that molecular and instrumental parameters are known or can be reasonably estimated, spectral simulation can be carried out and employed to refine the determination of spectroscopic parameters and molecular relaxation rates, as is shown in Figures 28 and 29.

From Figures 28 and 29, it is clear that out-of-phase signal intensities are comparable to in-phase signal intensities. Indeed, frequency-swept absorption ELDOR signals recorded[52] observing the center of a nitrogen hyperfine envelope (point B in Figure 27) show a maximum for a reference phase to driving modulation angle of 120°. Such considerations are, of course, of importance in optimizing signal-to-noise and in the measurement of spectroscopic parameters (e.g., hyperfine couplings) in frequency-swept ELDOR experiments.

The dependence of ELDOR signals upon phase angle can be employed to determine both electron and nitrogen nuclear spin–lattice relaxation rates and not just the ratio of rates as is the case when low-frequency Zeeman modulation is employed.

It is also of interest to consider the effects of superhyperfine interactions upon field-swept ELDOR experiments. Plots of experimental reciprocal reduction factors versus reciprocal pump power have been compared to corresponding computer-simulated theoretical plots.[52] The effect upon absorption ELDOR reduction factors is minimal; however, the dependence of dispersion reduction factors upon Zeeman modulation frequency is greatly increased for inhomogeneous resonance lines. A similar observation holds for the synchronous magnetic field–pump frequency-swept experiment. The strong dependence of such experiments upon modulation frequency for inhomogeneous resonance lines suggests the need for caution in applying limiting forms (such as shown in Table 2) of general theoretical equations.

As a final comment, we would note that we have carried out simulation of ELDOR responses for solutions of TANOL and TANONE in SBB for correlation times in the range 5×10^{-11} to 10^{-9} sec where we have explicitly treated the effect of jump diffusion modulating the anisotropic magnetic interactions. For such calculations, quantitative simulations of ELDOR spectra were obtained without the introduction of phenomenological nuclear relaxation rates into Γ_R. This result establishes that motional modulation of pseudosecular interactions is the dominant nuclear relaxation mechanism for this correlation time region.

We have already mentioned that ENDOR measurements (for example, those performed with a Varian E-1700 spectrometer) commonly involve low-frequency (34.7 Hz) Zeeman modulation and high-frequency (1 kHz) amplitude modulation of the radiofrequency field with double decoding of the spin response. We have recorded typical nitroxide spectra employing such modulation; we have also recorded spectra employing, in addition to Zeeman modulation, a 50-kHz frequency (ΔF) modulation of the radiofrequency field. This additional modulation results in a derivativelike display.[111] Increasing the frequency of ΔF modulation results in broadening of the ENDOR linewidths, as expected. We have found that maximum ENDOR enhancements are observed when the reference phase of the 34.7-Hz phase-sensitive detector is in phase with the driving Zeeman modulation field. This is not surprising, considering that this modulation frequency is well removed from the molecular relaxation rates. For systems such as nitroxides, modulation effects are expected to be less dramatic for ENDOR experiments as compared to ELDOR experiments, since the magnitude of interaction of the Zeeman modulation fields with electron and nuclear spins is proportional to the magnetic moment of the spin system, so less effect results from interaction of nuclear spins with the Zeeman modulation field than for electron spin systems (pumped and observed for the ELDOR experiment). The preceding statement should not be construed to mean that modulation effects are unimportant in ENDOR for all systems. That is clearly not the case for systems where strong passage conditions apply.

3.2. Modulation Effects in the Resonance and Double Resonance Spectra of Solid-State Organic Radicals

3.2.1. The ·CH_2COO^- Radical

Although the mechanisms of saturation transfer differ for organic radicals in the solid state and in solution, single and double resonance spectra for both systems will be sensitive to modulation conditions when applied and molecular modulation rates are competitive. The mathematical formalism described in this chapter is suited equally well for the description of either system, particularly as we treat relaxation processes phenomenologically in terms of a Redfield-type relaxation matrix so that the intrinsic differences between solid- and liquid-state mechanisms do not force modification of our theoretical formalism other than to dictate appropriate adjustment of the values for the phenomenological rates.

In this section, we apply, with appropriate modification, the mathematical formalism of the theory section (Section 2) to the analysis of the Zeeman modulation frequency-dependent ELDOR spectra of the $\cdot CH_2COO^-$ radical in radiation-damaged single crystals of zinc acetate. Kispert and co-workers[51] report frequency-swept absorption ELDOR detected with the phase-sensitive detector reference in phase with the Zeeman modulation and recorded observing the crossing point ($\Delta_{o4} = 0$) of the high-field EPR line. Specifically, they observed the ELDOR reduction factors for $\cdot CH_2COO^-$ to go from positive to negative as the Zeeman modulation frequency was varied from 100 kHz to 1 kHz. Moreover, reduction factors measured pumping different EPR lines were observed to invert at different modulation frequencies, i.e., were observed to exhibit different dependences upon modulation frequency.

Before we proceed to a discussion of the simulation of ELDOR spectra, it is appropriate to consider the EPR spectra of the $\cdot CH_2COO^-$ radical. The EPR spectra are characterized by hyperfine splittings arising from interaction of the unpaired electron with two magnetically inequivalent protons that can undergo exchange by rotation about the C—C bond. If we assign coupling constants a_1 and a_2 where $a_1 > a_2$ to the protons of the $\cdot CH_2COO^-$ radical, the allowed electron spin transitions occur at (1) $(a_1 + a_2)/2$, (2) $(a_1 - a_2)/2$, (3) $(a_2 - a_1)/2$, and (4) $(-a_1 - a_2)/2$. The action of proton exchange can be realized by considering a classic two-site jump or equilibrium process such as shown in Figure 30. Since the exchange accomplishes a switching of coupling constants for the two protons, transitions 1 and 4 are not affected, as they represent states in which the proton spins have like signs. It is the inner two transitions, 2 and 3, that are affected by the proton exchange. In the slow exchange limit, $\Delta\omega = (a_1 - a_2) - (a_2 - a_1) \gg 1/\tau$, where $1/\tau$ is the exchange frequency, the EPR spectrum appears as a quartet with equal line intensities (1 : 1 : 1 : 1). At the fast exchange limit, $\Delta\omega \ll 1/\tau$,

Figure 30. The $\cdot CH_2COO^-$ radical undergoes an exchange process, presumably rotation about the C—C bond, at a rate $1/\tau$. As such, a classic two-site jump model in which protons H_a and H_b interchange coupling constants, a_1 and a_2, may be utilized. (From Perkins et al.[113])

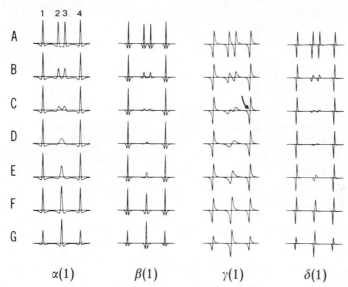

Figure 31. The four computer-simulated EPR spectra of the $\cdot CH_2COO^-$ radical detected at the first harmonic of the Zeeman modulation field. The signals presented are, from left to right, the in-phase dispersion $\alpha(1)$, the out-of-phase dispersion $\beta(1)$, the in-phase absorption $\gamma(1)$, and the out-of-phase absorption $\delta(1)$. The parameters employed in calculations include: $T_{1e} = T_{1n} = 6.6 \times 10^{-6}$ sec, $T_{2e} = 5.0 \times 10^{-8}$ sec, $h_o = 0.01$ G, $\omega_s = 2\pi \times 10^5$ Hz, $a_1 = 25.34$ G, and $a_2 = 17.48$ G. The exchange times are, for spectra A through G, respectively: 5.99×10^{-6}, 7.5×10^{-8}, 3.5×10^{-8}, 1.64×10^{-8}, 9.0×10^{-9}, 3.0×10^{-9}, and 1.0×10^{-9} sec. The positioning of the microwave observer for the computation of ELDOR spectra discussed in the text and shown in Figure 32 is indicated by an arrow. (From Perkins *et al.*[113])

the interior transitions are averaged and the EPR spectrum appears as a 1 : 2 : 1 triplet. Theoretical simulation of the effects of exchange from the fast to slow motion limits is shown in Figure 31.

In addition to the exchange process just discussed, we must also consider electron spin–spin (T_{2e}^{-1}), electron spin–lattice (W_e), and proton nuclear spin–lattice (W_n) relaxation processes. Of these additional processes, only electron spin–spin relaxation significantly affects conventional EPR line shapes (EPR recorded in the presence of a nonsaturating microwave field). On the other hand, all four processes will influence saturation and spectral diffusion of saturation, thus determining ELDOR responses. Of these, we note that proton spin–lattice relaxation connects the z-component magnetizations (diagonal elements) of states separated by a single proton spin quantum number (transition). As such, this process couples electron spin transitions 1 and 4 to transitions 2 and 3. In the calculations shown here, we assume that transitions 1 through 4 are characterized by the same electron spin–lattice and electron spin–spin relaxation rates.

In previous work,[113] we presented computer simulations of the four frequency-swept ELDOR signals detected at the first harmonic of the Zeeman modulation. The dependence of these signals upon Zeeman modulation frequency was shown. In all of these computer simulations, the microwave observer was taken as positioned at exact resonance of the high-field electron spin transition (i.e., $\Delta_{o4} = 0$).

Let us first consider the in-phase absorption, $\gamma(1, 0, 1)$, frequency-swept ELDOR signal as it was this response experimentally observed by Kispert and co-workers.[51] The computer-simulated spectra[113] shown in Figure 32, like the experimental spectra,[51] invert (change from positive to negative reduction factors) as the Zeeman modulation frequency is varied from 100 to 1 kHz. It is further to be noted that the ELDOR response arising from pumping the low-field electron spin transition (Δ_{p1}) inverts at a different Zeeman modulation frequency than that at which the inner ELDOR signals (Δ_{p2} and Δ_{p3}) invert.

The two out-of-phase (or phase-quadrature) ELDOR signals, $\beta(1, 0, 1)$ and $\delta(1, 0, 1)$, are also predicted to respond quite strongly to changes in Zeeman modulation frequency.[113] Of interest is the manner in which the out-of-phase dispersion signal, $\beta(1, 0, 1)$, evolves from a predominantly "dispersionlike" signal at the higher modulation frequencies to an "absorptionlike" signal at lower frequencies. Of greater importance, however, is the fact that both $\beta(1, 0, 1)$ and $\delta(1, 0, 1)$ signals should be as easily detected as the $\alpha(1, 0, 1)$ and $\gamma(1, 0, 1)$ signals, since their computed

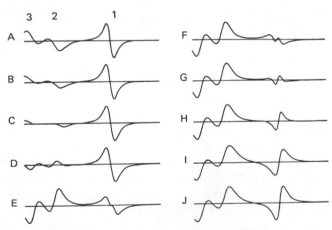

Figure 32. Computer-simulated frequency-swept ELDOR $\gamma(1, 0, 1)$ spectra as a function of ω_s are shown. Calculation parameters are as given in Figure 31. The values of ω_s (in rad/sec) for spectra A through J, respectively, are 6.28×10^5, 5.06×10^5, 4.45×10^5, 3.8×10^5, 2.86×10^5, 2.62×10^5, 2.37×10^5, 1.88×10^5, 6.28×10^4, and 6.28×10^3. The observer was positioned at exact resonance of transition 4, as shown by the arrow in Figure 31, and the pump was swept through transitions 3, 2, and 1.

intensities are competitive even though their EPR counterparts certainly are not comparable. Finally, we observe that the $\alpha(1, 0, 1)$ signal is quite insensitive to variation in Zeeman modulation frequency.

3.2.2. Modulation Effects in the ENDOR of Solid-State Materials

Wolbarst[114] has reported the experimental investigation of the dependence of ENDOR enhancements for F-center resonances upon Zeeman and radiofrequency amplitude modulation conditions. He employed 1900-Hz amplitude modulation of the radiofrequency field and Zeeman modulation for modulation frequencies in the range 2 to 100 kHz; however, phase-sensitive detection was carried out only with respect to the modulated radiofrequency field. Nevertheless, enhancement factors, defined as the ratio of the ENDOR signal with modulation to that without, were observed to increase both with Zeeman modulation frequency and with Zeeman modulation amplitude for amplitudes up to 0.7 G. The enhancement factors appeared to increase logarithmically with Zeeman modulation frequency. Following discussions with Hyde and Allendoerfer, Wolbarst mentions that a possible explanation of the Zeeman modulation dependence is the bringing of more spin packets of the inhomogeneous line into contact with the microwave field as the frequency (hence sideband separation) and amplitude is increased. Wolbarst also investigated the dependence of enhancement factors upon the rate at which the radiofrequency is chopped and observed the enhancement factors to increase exponentially with chopping rate.

Zarinov, Meiklyar, and Falin[53] report the observation of Zeeman modulation effects for the dispersion ENDOR spectra of $^{47}Ti^{2+}$ in CaF_2 at 4.2°K. Explicitly, they observed that the ENDOR signals recorded employing 100-kHz Zeeman modulation and phase-sensitive detection at the fundamental varied from small positive to large negative signals as the Zeeman modulation amplitude was increased from 1 to 10 G. The large negative ENDOR enhancement observed for the 10 G field amounted to 10% of the EPR signal.

3.2.3. Examples of Orientation and Species Selection

In the preceding section, we mentioned the possibility of Zeeman modulation affecting double resonance spectra by bringing adjacent lines into interaction (resonance) with the microwave observing field. This type of effect is particularly critical for the ENDOR and ELDOR spectra of randomly oriented solid-state radicals. For such systems, different portions of the EPR "powder" spectrum can be associated with different molecular orientations, and if molecular orientation coupling mechanisms such as molecular motion and spin exchange are absent, ENDOR and ELDOR spectra will be "single-crystal-like," i.e., arise from nuclear spin transitions that can be associated with the small orientational ensemble of radicals interacting directly with the microwave observing field. This feature of ENDOR and ELDOR spectra is referred to as "orientation selection" and has been demonstrated experimentally by Hyde and co-workers,[115-122] Allendoerfer,[123] Fritz *et al.*,[124] and by others. Even when low-frequency Zeeman modulation is employed, increasing modulation amplitude can have a dramatic effect upon double resonance spectra by bringing more spin packets (hence more molecular orientations) into resonance with the observing microwave field. As resonance fields change most slowly with changing molecular orientation for spectral turning points corresponding to principal elements of magnetic tensors, Zeeman modulation amplitude variation will have the least effect for monitoring these spectral positions, and analyses neglecting modulation[116,117,123,124] often give satisfactory simulation of double resonance spectra. On the other hand, Zeeman modulation amplitude will be most effective in bringing into resonance a large range of molecular orientations for spectral positions between turning points. To observe ENDOR spectra for such EPR spectral regions, Hyde *et al.*[125] have suggested replacing the low-frequency Zeeman modulation with 180° modulation of the phase of the resonance microwave signal with respect to the reference microwave signal. As such modulation does not move the observer off resonance, it is predicted to yield ENDOR intensities four times greater than with Zeeman modulation, and indeed Hyde *et al.*[125] have observed signals that are 2.5 to 4 times stronger.

ACKNOWLEDGEMENTS

It is a pleasure to acknowledge the substantial contributions that Professors J. S. Hyde and A. L. Kwiram have made to our studies of modulation effects. Dr. Norval Galloway is to be thanked for calculating the data presented in Table 4. This work has benefited from support provided by the National Science Foundation (GP-42998X and CHE-7701018).

References

1. W. E. Good, *Phys. Rev.* **70**, 213–218 (1946).
2. C. H. Townes, *Phys. Rev.* **70**, 665–671 (1946).
3. R. H. Hughes and E. B. Wilson, Jr., *Phys. Rev.* **71**, 562–563 (1947).
4. B. P. Dailey, *Phys. Rev.* **72**, 84–85 (1947).
5. R. J. Watts and D. Williams, *Phys. Rev.* **72**, 1122–1123 (1947).
6. R. Karplus, *Phys. Rev.* **73**, 1027–1034 (1948).
7. N. Bloembergen, E. M. Purcell, and R. V. Pound, *Phys. Rev.* **73**, 679–712 (1948).
8. F. Bloch, W. W. Hansen, and M. Packard, *Phys. Rev.* **70**, 474–485 (1946).
9. B. Smaller, *Phys. Rev.* **83**, 812–820 (1951).

10. J. H. Arnold and M. E. Packard, *J. Chem. Phys.* **19**, 1608–1609 (1951).
11. M. M. Perlman and M. Bloom, *Phys. Rev.* **88**, 1290–1291 (1952).
12. E. R. Andrew, *Phys. Rev.* **91**, 425 (1953).
13. B. Smaller and E. L. Yasaitis, *Rev. Sci. Instr.* **24**, 991–992 (1953).
14. K. Halbach, *Helv. Phys. Acta* **27**, 259–282 (1954); *Phys. Rev.* **119**, 1230–1233 (1960).
15. P. S. Hubbard, Jr., and T. J. Rowland, *J. Appl. Phys.* **28**, 1275–1281 (1957).
16. G. B. Benedek and T. Kushida, *Phys. Rev.* **118**, 46–57 (1959).
17. J. D. Macomber and J. S. Waugh, *Phys. Rev.* **A140**, 1494–1497 (1965).
18. C. P. Poole, Jr., *Electron Spin Resonance*, Wiley-Interscience, New York (1967).
19. A. Abragam, *The Principles of Nuclear Magnetism*, Oxford University Press, London (1961).
20. G. E. Pake, *Paramagnetic Resonance*, W. A. Benjamin, New York (1962).
21. C. P. Slichter, *Principles of Magnetic Resonance*, Harper and Row, New York (1963).
22. I. V. Aleksandrov, *The Theory of Nuclear Magnetic Resonance*, Academic Press, New York (1966).
23. A. Carrington and A. D. McLachlan, *Introduction to Magnetic Resonance*, Harper and Row, New York (1967).
24. L. T. Muus and P. W. Atkins (eds.), *Electron Spin Relaxation in Liquids*, Plenum Press, New York (1972).
25. K. E. Shuler (ed.), *Stochastic Processes in Chemical Physics*, Wiley Interscience, New York (1969).
26. N. M. Atherton, *Electron Spin Resonance*, Halsted Press, London (1973).
27. L. J. Berliner (ed.), *Spin Labeling: Theory and Applications*, Academic Press, New York (1976).
28. J. H. Freed, *Ann. Rev. Phys. Chem.* **23**, 265–310 (1972).
29. K. J. Standley and R. A. Vaughan, *Electron Spin Relaxation Phenomena in Solids*, Plenum Press, New York (1969).
30. A. A. Manenkov and R. Orbach (eds.), *Spin–Lattice Relaxation in Ionic Solids*, Harper and Row, New York (1966).
31. L. R. Dalton, *Saturation Transfer Spectroscopy*, Wiley-Interscience, New York (to be published).
32. H. H. Günthard, *Ber. Bunsen-Ges.* **78**, 1110–1115 (1974).
33. L. R. Dalton, B. H. Robinson, L. A. Dalton, and P. Coffey, in *Advances in Magnetic Resonance* (J. S. Waugh, ed.), Vol. 8, pp. 149–259, Academic Press, New York (1976).
34. D. D. Thomas, L. R. Dalton, and J. S. Hyde, *J. Chem. Phys.* **65**, 3006–3024 (1976).
35. C. L. Hamilton and H. M. McConnell, in *Structural Chemistry and Molecular Biology* (A. Rich and N. Davidson, eds.), pp. 115–149, W. H. Freeman, San Francisco (1968).
36. H. M. McConnell and B. G. McFarland, *Quart. Rev. Biophys.* **3**, 91–136 (1970).
37. P. C. Jost and O. H. Griffith, in *Methods in Pharmacology: Vol. II. Physical Methods* (C. Chignell, ed.), pp. 223–276, Appleton-Century-Crofts, New York (1972).
38. P. C. Jost, A. S. Waggoner, and O. H. Griffith, in *Structure and Function of Biological Membranes* (L. I. Rothfield, ed.), pp. 83–144, Academic Press, New York (1971).
39. I. C. P. Smith, in *Biological Applications of Electron Spin Resonance Spectroscopy* (H. M. Swartz, J. R. Bolton, and D. C. Borg, eds.), pp. 483–539, Wiley-Interscience, New York (1972).
40. F. S. Axel, *Biophys. Struct. Mechanism* **2**, 181–218 (1976).
41. G. I. Likhtenstein, *Spin Labeling Methods in Molecular Biology*, Wiley-Interscience, New York (1976).
42. J. S. Hyde, in *Methods in Enzymology. Enzyme Structure. Part F* (C. H. W. Hirs and S. N. Timasheff, eds.), Vol. 49G, No. 19, pp. 480–511, Academic Press, New York (1978).
43. J. S. Hyde, M. D. Smigel, L. R. Dalton, and L. A. Dalton, *J. Chem. Phys.* **62**, 1655–1667 (1975).

44. M. D. Smigel, L. R. Dalton, J. S. Hyde, and L. A. Dalton, *Proc. Nat. Acad. Sci. USA* **71**, 1925–1929 (1974).
45. M. M. Dorio and J. C. W. Chien, *Macromolecules* **8**, 734–739 (1975).
46. M. M. Dorio and J. C. W. Chien, *J. Mag. Res.* **20**, 114–123 (1975).
47. M. M. Dorio, Ph.D. Thesis, University of Massachusetts, Amherst (1975).
48. B. H. Robinson, J.-L. Monge, L. A. Dalton, and L. R. Dalton, *Chem. Phys. Lett.* **28**, 169–175 (1974).
49. L. A. Dalton, J.-L. Monge, L. R. Dalton, and A. L. Kwiram, *Chem. Phys.* **6**, 166–182 (1974).
50. M. M. Dorio and J. C. W. Chien, *J. Chem. Phys.* **62**, 3963–3967 (1975).
51. C. Mottley, K. Chang and L. D. Kispert, *J. Mag. Res.* **19**, 130–143 (1975).
52. P. W. Percival, J. S. Hyde, L. A. Dalton, and L. R. Dalton, *J. Chem. Phys.* **62**, 4332–4342 (1975).
53. M. M. Zarinov, V. P. Meiklyar, and M. L. Falin, *Fiz. Tverd. Tela* **17**, 3438–3440 (1975) [English transl.: *Sov. Phys. Solid State* **17**, 2251–2252 (1976)].
54. J. S. Hyde and L. R. Dalton, in *Spin Labeling II: Theory and Applications* (L. J. Berliner, ed.), pp. 1–70, Academic Press, New York (1979).
55. D. D. Thomas, in *Trends in Biochemical Science*, Vol. 2, Elsevier, Amsterdam (1977).
56. J. S. Hyde, *J. Chem. Phys.* **43**, 1806–1818 (1965).
57. B. H. Robinson, Ph.D. Thesis, Vanderbilt University, Nashville, Tennessee (1975).
58. N. B. Galloway, Ph.D. Thesis, Vanderbilt University, Nashville, Tennessee (1977).
59. S. M. Moskow, unpublished results.
60. R. C. Perkins, Jr., Ph.D. Thesis, Vanderbilt University, Nashville, Tennessee (1977).
61. B. H. Robinson and L. R. Dalton, *Chem. Phys.* **36**, 207–237 (1979).
62. S. A. Goldman, G. V. Bruno, C. F. Polnaszek, and J. H. Freed, *J. Chem. Phys.* **56**, 716–735 (1972).
63. P. A. Egelstaff, *J. Chem. Phys.* **53**, 2590–2598 (1970).
64. E. N. Ivanov, *Zhur. Eksp. Teor. Fiz.* **45**, 1509–1517 (1963) [English translation: *Sov. Phys. JETP* **18**, 1041–1045 (1964)].
65. J. H. Freed, in *Spin Labeling: Theory and Applications* (L. J. Berliner, ed.), pp. 53–132, Academic Press, New York (1976).
66. P. Coffey, B. H. Robinson, and L. R. Dalton, *Chem. Phys. Lett.* **35**, 360–366 (1975).
67. P. W. Percival and J. S. Hyde, *J. Mag. Res.* **23**, 249–257 (1976).
68. J. H. Freed, in *Electron Spin Relaxation in Liquids* (L. T. Muus and P. W. Atkins, eds.), pp. 387–409, Plenum Press, New York (1972).
69. J. H. Freed, G. V. Bruno, and C. F. Polnaszek, *J. Phys. Chem.* **75**, 3385–3399 (1971).
70. S. A. Goldman, G. V. Bruno, and J. H. Freed, *J. Phys. Chem.* **76**, 1858–1860 (1972).
71. C. F. Polnaszek, G. V. Bruno, and J. H. Freed, *J. Chem. Phys.* **58**, 3185–3199 (1973).
72. S. A. Goldman, G. V. Bruno, and J. H. Freed, *J. Chem. Phys.* **59**, 3071–3091 (1973).
73. G. V. Bruno and J. H. Freed, *Chem. Phys. Lett.* **25**, 328–332 (1974).
74. G. V. Bruno and J. H. Freed, *J. Phys. Chem.* **78**, 935–940 (1974).
75. R. P. Mason and J. H. Freed, *J. Phys. Chem.* **78**, 1321–1323 (1974).
76. R. P. Mason, C. F. Polnaszek, and J. H. Freed, *J. Phys. Chem.* **78**, 1324–1329 (1974).
77. J. S. Hwang, R. P. Mason, L.-P. Hwang, and J. H. Freed, *J. Phys. Chem.* **79**, 489–511 (1975).
78. C. F. Polnaszek and J. H. Freed, *J. Phys. Chem.* **79**, 2283–2306 (1975).
79. J. H. Freed, in *Electron Spin Relaxation in Liquids* (L. T. Muus and P. W. Atkins, eds.), pp. 503–530, and references contained therein, Plenum Press, New York (1972).
80. J. H. Freed, D. S. Leniart, and H. D. Connor, *J. Chem. Phys.* **58**, 3089–3105 (1973).
81. B. H. Robinson, L. R. Dalton, L. A. Dalton, and A. L. Kwiram, *Chem. Phys. Lett.* **29**, 56–64 (1974).
82. L. R. Dalton, P. Coffey, L. A. Dalton, B. H. Robinson, and A. D. Keith, *Phys. Rev.* **A11**, 488–498 (1975).

83. P. Coffey, B. H. Robinson, and L. R. Dalton, *Molec. Phys.* **31**, 1703–1715 (1976).
84. D. D. Thomas and H. M. McConnell, *Chem. Phys. Lett.* **25**, 470–475 (1974).
85. J. S. Hyde, J. C. W. Chien, and J. H. Freed, *J. Chem. Phys.* **48**, 4211–4226 (1968).
86. M. P. Eastman, G. V. Bruno, and J. H. Freed, *J. Chem. Phys.* **52**, 321–327 (1970).
87. E. Van der Drift, A. F. Mehlkopf, and J. Smidt, *Chem. Phys. Lett.* **36**, 385–389 (1975).
88. R. J. Cook and D. H. Whiffen, *Proc. Phys. Soc. London* **84**, 845–848 (1964).
89. J. H. Freed, *J. Chem. Phys.* **50**, 2271–2271 (1969).
90. N. S. Dalal and C. A. McDowell, *Chem. Phys. Lett.* **6**, 617–619 (1970).
91. J. A. R. Coope, N. S. Dalal, C. A. McDowell, and R. Srinivasan, *Molec. Phys.* **24**, 403–415 (1972).
92. K. P. Dinse, R. Biehl, and K. Möbius, *J. Chem. Phys.* **61**, 4335–4341 (1974).
93. R. Biehl, M. Plato, and K. Möbius, *J. Chem. Phys.* **63**, 3515–3522 (1975).
94. L. R. Dalton and A. L. Kwiram, unpublished results.
95. J. S. Hyde and L. R. Dalton, *Chem. Phys. Lett.* **16**, 568 572 (1972).
96. J. S. Hyde and L. R. Dalton, in *4th International Biophysics Congress—Symposial Papers*, pp. 687–702. Pushchino, Moscow (1973).
97. J. S. Hyde and D. D. Thomas, *Ann. N.Y. Acad. Sci.* **222**, 680–692 (1973).
98. D. D. Thomas, J. C. Seidel, J. S. Hyde, and J. Gergely, *Proc. Nat. Acad. Sci. USA* **72**, 1729–1733 (1975).
99. D. D. Thomas, J. C. Seidel, J. Gergely, and J. S. Hyde, *J. Supramolec. Struct.* **3**, 376–390 (1975).
100. C. Mailer and B. M. Hoffman, *J. Phys. Chem.* **80**, 842–846 (1976).
101. A. M. Portis, *Phys. Rev.* **100**, 1219–1221 (1955).
102. A. M. Portis, *Technical Note No. 1*, Sarah Mellon Scaife Radiation Laboratory, University of Pittsburgh, Pittsburgh, Pennsylvania, unpublished.
103. J. S. Hyde, *Phys. Rev.* **119**, 1483–1492 (1960).
104. M. Weger, *Bell System Tech. J.* **39**, 1013–1112 (1960).
105. G. Feher, *Phys. Rev.* **114**, 1219–1244 (1959).
106. D. D. Thomas, Ph.D. Thesis, Stanford University, Palo Alto, California (1975); see particularly Appendix A, pp. 92–94.
107. H. M. McConnell, in *Spin Labeling: Theory and Applications* (L. J. Berliner, ed.), pp. 525–560. Academic Press, New York (1976); see particularly Appendix A.
108. R. Briere, H. Lemaire, A. Rassat, P. Rey, and A. Rousseau, *Bull. Soc. Chim. France* **12**, 4479–4484 (1967).
109. R. W. Kreilick, *J. Chem. Phys.* **46**, 4260–4264 (1967).
110. C. C. Whisnant, S. Ferguson, and D. B. Chesnut, *J. Phys. Chem.* **78**, 1410–1415 (1974).
111. R. C. Perkins, Jr., T. Lionel, B. H. Robinson, L. A. Dalton, and L. R. Dalton, *Chem. Phys.* **16**, 393–403 (1976).
112. T. B. Marriott, S. P. Van, and O. H. Griffith, *J. Mag. Res.* **24**, 41–52 (1976).
113. R. C. Perkins, Jr., L. R. Dalton, and L. D. Kispert, *J. Mag. Res.* **26**, 25–33 (1977).
114. A. B. Wolbarst, *Solid State Commun.* **18**, 1193–1195 (1976).
115. G. H. Rist and J. S. Hyde, *J. Chem. Phys.* **49**, 2449–2451 (1968).
116. G. H. Rist and J. S. Hyde, *J. Chem. Phys.* **52**, 4633–4643 (1970).
117. J. S. Hyde, G. H. Rist, and L. E. Göran Eriksson, *J. Phys. Chem.* **72**, 4269–4276 (1968).
118. W. H. Walker, J. Salach, M. Gutman, T. P. Singer, J. S. Hyde, and A. Ehrenberg, *FEBS Lett.* **5**, 237–240 (1969).
119. L. E. Göran Eriksson, J. S. Hyde, and A. Ehrenberg, *Biochem. Biophys. Acta* **192**, 211–230 (1969).
120. L. E. Göran Eriksson, A. Ehrenberg, and J. S. Hyde, *Eur. J. Biochem.* **17**, 539–543 (1970).
121. J. S. Hyde, R. C. Sneed, Jr., and G. H. Rist, *J. Chem. Phys.* **51**, 1404–1416 (1969).
122. D. J. Lowe and J. S. Hyde, *Biochim. Biophys. Acta* **377**, 205–210 (1975).
123. R. D. Allendoerfer, *Chem. Phys. Lett.* **17**, 172–174 (1972).

124. J. Fritz, R. Anderson, J. Fee, G. Palmer, R. H. Sands, J. C. M. Tsibris, I. C. Gunsalus, W. H. Orme-Johnson, and H. Beinert, *Biochim. Biophys. Acta* **253**, 110–113 (1971).
125. J. S. Hyde, T. Astlind, L. E. Göran Eriksson, and A. Ehrenberg, *Rev. Sci. Instr.* **41**, 1598–1600 (1970).

6

Disordered Matrices

Larry Kevan and P. A. Narayana

1. ENDOR in Disordered Matrices

ENDOR in disordered matrices—polycrystalline, glassy, and amorphous media—is a rapidly developing area because it has been recognized that considerable information can be obtained. For many systems of chemical, technological, and biochemical interest, single crystals do not exist or are not readily obtainable. Thus, it is particularly important to develop methods of investigating paramagnetic species in disordered systems. Extraction of geometric and electronic structural information from ESR spectra in disordered solids is generally difficult, and double resonance methods like ENDOR and ELDOR can often be of considerable aid. Indeed, in some cases, the interpretation of the double resonance spectra is much easier. In this chapter, we wish to demonstrate the types of information obtainable by electron magnetic double resonance methods applied to radicals in disordered solids.

The type of ENDOR spectra to be expected in disordered solids is illustrated by Figure 1. In the liquid phase, the triphenylmethyl radical in toluene shows three narrow ENDOR lines above the free proton frequency corresponding to the meta, ortho, and para ring protons. In the frozen glassy matrix, two changes occur. One, the liquid-phase ENDOR lines broaden and become very weak, often undetectable. This is due to an average of absorptions from different orientations of radicals with respect to the magnetic field, which depends on the magnitude of the hyperfine anisotropy. For α-protons, this anisotropy is about half of the hyperfine coupling A, so that the principal values of the hyperfine tensor for α-protons occur near $A/2$, A,

---

Larry Kevan and P. A. Narayana • Department of Chemistry, Wayne State University, Detroit, Michigan

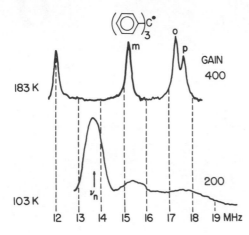

Figure 1. Proton ENDOR spectra of triphenylmethyl radical in the liquid (183°K) and glassy (103°K) phases of toluene. The three high-frequency peaks are assigned as indicated to the meta (m), ortho (o), and para (p) protons, respectively. A matrix ENDOR line at the free proton frequency v_n appears when the solution is frozen. Adapted from J. S. Hyde, G. H. Rist, and L. E. G. Eriksson, *J. Phys. Chem.* **72**, 4269 (1968).

and $3A/2$. The ENDOR spectra are thus expected to extend over the range of $A/2$ to $3A/2$, but with some buildup of intensity at the three principal values of the hyperfine tensor corresponding to those molecules with their respective principal axes oriented along the magnetic field. β protons have considerably less anisotropy than α protons and generally exhibit much stronger ENDOR lines in disordered solids. In particular, β protons in rotating methyl groups show quite strong ENDOR lines.

The second feature of Figure 1 is the appearance in the solid of a strong ENDOR line at the free nuclear frequency. This line has been called both a "distant" ENDOR line[1] and a "matrix" ENDOR line.[2] The distinction has been discussed in a recent book on magnetic double resonance.[3] The best criterion for distinction between "distant" and "matrix" ENDOR signals appears to be the response time of the ENDOR signal. Distant ENDOR signals decay relatively slowly on a nuclear spin–lattice relaxation time scale, while matrix ENDOR signals decay more rapidly on an electron spin–lattice relaxation time scale. We believe that observations of ENDOR signals at high radiofrequency power with high-frequency modulation in the ENDOR detection system, as typically found in commercial spectrometers, are probably best identified as matrix ENDOR lines. These lines are due to almost purely dipolar coupling of the unpaired electron with surrounding (i.e., matrix) magnetic nuclei.

We find it convenient to separately consider the information available from the anisotropic ENDOR lines (i.e., those occurring away from the free nuclear frequency) and from the matrix ENDOR lines in disordered matrices.

1.1. Anisotropic ENDOR

We will consider two different cases of anisotropic ENDOR. In the first case, it is possible to select only certain orientations of radicals with respect to the magnetic field. It is then possible to obtain spectra from disordered solids that are single-crystal-like. In the second case, all orientations contribute to the ENDOR spectra and one has a powder-type ENDOR spectrum.

Single-crystal-like ENDOR spectra can be obtained by setting the magnetic field at so-called turning points in the ESR spectra where g or hyperfine anisotropy is evident. A turning point refers to a magnetic field at which one orientation of a g or hyperfine tensor dominates sufficiently to cause a significant change in shape of an idealized Gaussian or Lorentzian ESR line. In general, the larger the anisotropy, the better the selection of one orientation of radicals.

There are several cases where strong magnetic anisotropy can lead to single-crystal-like ENDOR spectra: (a) The anisotropy of the hyperfine tensor of one nucleus is larger than for all other nuclei and larger than the g anisotropy. This occurs for nitroxide radicals, where the nitrogen anisotropy dominates the ESR spectrum. It may then be possible to determine the hyperfine tensors of protons in the radical, providing their tensors are parallel to the nitrogen tensor. (b) The g anisotropy dominates all hyperfine interactions. One can then set the magnetic field at $g_{\parallel}$ or $g_{\perp}$ to obtain $A_{\parallel}$ and $A_{\perp}$ for various magnetic nuclei if the g and A tensors have parallel axes. Even if the axes are not quite parallel, it is sometimes possible to deduce the angle of rotation between the two tensors in one plane by doing simulations of the ESR spectra. In many metal ion complexes, particularly iron and cobalt complexes, g anisotropy dominates. For example, this is the case in ferrimyoglobin and hemoglobin. (c) Even when the hyperfine anisotropy of one nucleus and the g anisotropy are of the same order of magnitude, if their axes of largest anisotropy coincide, it is possible to obtain hyperfine tensors for other nuclei. This case is illustrated by copper complexes, where the g tensor and copper hyperfine tensor axes coincide and hyperfine tensors of ligand nitrogen and proton nuclei can often be determined. (d) The anisotropy of the electron-spin–electron-spin interaction is larger than the g or hyperfine anisotropies. This results in a triplet state spectrum dominated by the electron dipolar coupling tensor D. By setting the magnetic field at $D_{\parallel}$ or $D_{\perp}$ positions, one may be able to determine hyperfine tensors by ENDOR.

An example of case (a) where the anisotropy of one nucleus is very large is shown by the 1,3,6,8-tetra-*tert*-butylcarbazyl radical in perdeuterated toluene at 77°K.[4] The ESR spectrum is a triplet dominated by the parallel component of the nitrogen hyperfine anisotropy ($A_{\parallel} = 20.52$ G and $A_{\perp} = 0.08$ G). By setting the magnetic field at the high-field line, molecules with the magnetic field parallel to the nitrogen p orbital and perpendicular to the molecular plane are selected. ENDOR then shows a single-crystal-like spectrum shown in Figure 2a corresponding to $A_{\parallel} = A_z$ proton components. The two outer lines are assigned to the 2 and 7 protons and the two inner lines to the rotating *tert*-butyl protons by analogy to isotropic couplings determined in solution. The $A_{\perp}$ proton hyperfine components cannot be clearly determined like $A_{\parallel}$, because there is no magnetic field where only

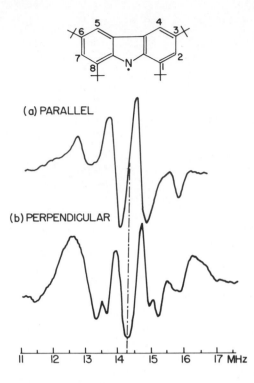

(a) PARALLEL

(b) PERPENDICULAR

Figure 2. The ENDOR spectrum of the 1,3,6,8-tetra-*tert*-butylcarbazyl radical. (a) Magnetic field set at the high-field extremum of the ESR spectrum to select those molecules oriented with their nitrogen p orbital parallel to the magnetic field. (b) Magnetic field set near the center of the ESR spectrum. The dashed line indicates the free proton frequency. Adapted from R. D. Allendoerfer, *Chem. Phys. Lett.* **17**, 172 (1972).

the perpendicular components contribute. When the magnetic field is set on the center ESR line, there are contributions from all orientations of the radical. By varying the magnetic field on the central ESR line, the ENDOR spectrum in Figure 2b is obtained, which shows additional weak lines near 13.6 and 15.0 MHz. These lines are assigned to one $A_\perp(A_y)$ for the 2 and 7 protons. From the isotropic coupling measured for the 2 and 7 protons in solution, the remaining tensor component A_x can be obtained, so that the complete proton hyperfine tensor is determined.

An example of case (c) is shown by copper picolinate doped into zinc picolinate tetrahydrate.[5] The ESR is a typical copper spectrum showing g anisotropy with the $g_\perp$ component clearly split into a quartet by the $A_\perp$ component of the copper $(I = \frac{3}{2})$ hyperfine interaction. The $A_\perp$ coupling as well as the quadrupole interaction of the ligand nitrogen is revealed in a single-crystal-like ENDOR spectrum obtained at the magnetic field corresponding to one of the $g_\perp$ components. The $A_\parallel$ coupling is more difficult to uncover. When the observing magnetic field is moved to the typically poorly resolved $g_\parallel$ component of the ESR spectrum, the ENDOR spectra broaden and some new lines appear that can be assigned to $A_\parallel$ components of the nitrogen hyperfine coupling.

The radicals formed in irradiated polycrystalline carboxylic acids do not show any strong, characteristic hyperfine anisotropy, so their ENDOR spectra are expected to be of the powder type. In addition, at 4°K, the cross relaxation times T_x are shorter than the electron spin–lattice relaxation times, so one expects the ENDOR spectrum to be the same regardless of which part of the ESR spectrum is saturated.[6] To analyze such ENDOR spectra, it is necessary to simulate the powder ENDOR spectra and to obtain the coupling constants from the best fit.

A simulation method has been developed by Dalton and Kwiram that seems quite successful.[6] First, the ENDOR transition frequencies are calculated to second order, assuming certain hyperfine parameters. Second, the relative ENDOR amplitudes from given hyperfine interactions are calculated, partly phenomenologically, by considering various relaxation pathways and calculating transition moments with the second-order wave functions. Third, powder ENDOR spectra are simulated by summing spectra from different orientations at 0.3° to 1° intervals. The relative signal amplitude for a given proton hyperfine interaction is given by

$$\chi''(\theta, \phi) = \frac{S[1 \mp B'(\theta, \phi)/2v_p]^2 R}{1 + S[1 \mp B'(\theta, \phi)/2v_p]^2} \tag{1}$$

where $S = 0.25\gamma_p^2 H_2^2 T_{1n} T_{2n}$ is taken as an isotropic constant, v_p is the frequency at which a proton ENDOR line is observed, $R = 0.5(1 - \alpha_+)$ is also taken as an isotropic constant, with α_+ a relaxation parameter that depends on electron, nuclear, and cross-relaxation transition probabilities, and B' is the $S_x I_x$ coefficient in the hyperfine Hamiltonian, which is the isotropic coupling for purely isotropic hyperfine interactions. S and R are parameters, while B' and v_p depend upon the ENDOR frequencies. The simulations are quite good for β protons. Figure 3 shows simulated and experimental

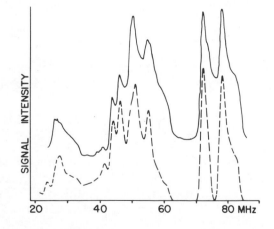

Figure 3. The proton ENDOR spectrum at 4.2°K of the $CH_3\dot{C}HCOOH$ radical in polycrystalline L-alanine γ-irradiated at room temperature. The solid line is the observed spectrum and the dashed line is a computer-simulated spectrum based on equation (1). Adapted from D. Becker, Ph.D. Thesis, University of Washington, Seattle, Washington (1976).

spectra for the radical in irradiated L-alanine. Both the positions and the relative intensities are reproduced rather well by the simulation. Analysis gives both parallel and perpendicular hyperfine components for radicals of two different conformations, so rather detailed information can be obtained. In the general case, where both α and β protons are present, the ENDOR features due to the α protons are suppressed because of their larger anisotropy.

If one performs ENDOR at higher magnetic field than the typical X-band field (3.3 kG), it is possible to improve the resolution of strongly coupled protons like α protons.[7] This occurs because the ENDOR frequencies are a function of the free nuclear frequency, which depends on magnetic field. For weakly coupled protons where the separation between pairs of ENDOR lines at X band is given by the hyperfine splitting A, there is no increase in resolution on going to higher field, because A is independent of field (see Chapter 4). But for strongly coupled protons, the separation of a pair of ENDOR lines is given by $2v_p$ (v_p being the free nuclear frequency) at low field and by A at high field. Since under these conditions $A > 2v_p$, the spectral resolution can be increased by two to three times by going to higher field.

This principle has been demonstrated for α-proton couplings of radicals in irradiated glutaric acid by comparing ENDOR spectra at 9 GHz and 35 GHz.[7] For this practical case, $A > 27$ MHz defines a strongly coupled proton at 9 GHz, and a twofold increase in spectral resolution is typical at 35 GHz.

Analysis of powder ENDOR spectra by simulation seems easier than for powder ESR spectra. The ENDOR transition frequencies associated with a given hyperfine interaction are independent of the other magnetic interactions, while this is not true for ESR spectra. The number of transitions per unit frequency interval or spectral density is less for powder ENDOR than for powder ESR. The ENDOR linewidths are less than ESR linewidths. Forbidden transitions complicate the ESR but not the ENDOR spectra.

1.2. Matrix ENDOR

The most prominent ENDOR line in most solid disordered systems is the matrix ENDOR line occurring at the free nuclear frequency. This line has been interpreted as being due to a purely dipolar interaction between an unpaired electron and surrounding magnetic nuclei in the matrix.[2] In liquids, this line is averaged to zero by rapid tumbling of the radical. Thus, matrix ENDOR can be used to probe molecular motion as a function of temperature and to compare the degree of molecular motion in different phases and different matrices at the same temperature. For example, proton matrix ENDOR associated with the tritolylmethyl radical is not seen in glassy toluene at 143°K but is quite prominent in polycrystalline toluene at the same temperature.

The amount of molecular motion appears to be significantly greater in the glassy phase. Qualitatively, matrix ENDOR will be observed when the tumbling frequency of the radical or matrix molecules becomes comparable with the ENDOR linewidth. Typical linewidths are 1 MHz, which correspond to molecular motions of 10^{-6} sec.

The occurrence of matrix ENDOR also depends upon whether there are magnetic nuclei in the local environment of the unpaired spin, typically within about 6 Å. For example, proton matrix ENDOR is seen for stable radicals in toluene, but the proton matrix ENDOR line disappears in deuterated toluene. If the radical site is located in a large complex molecule like a protein, contributions to the proton matrix ENDOR line will arise from protons in the protein itself as well as from the solvent molecules. When the proton matrix ENDOR intensity does not change when the solvent is deuterated, it suggests that the radical site is inaccessible to solvent molecules. This is a very useful qualitative application of matrix ENDOR that has been used in a number of biological systems.[8]

Matrix ENDOR seems to be rather generally observed for protons. Fluorine matrix ENDOR has also been observed in polymeric systems,[9] but little work has been reported. When both protons and fluorines occur in the matrix, resolution is sufficiently good that both matrix ENDOR lines may be distinguished, as shown in Figure 4.[10]

To understand how various physical parameters determine the matrix ENDOR quantitatively as a probe of the unpaired electron delocalization of radicals over nearby matrix molecules and the weak interactions of stable radicals with their molecular surroundings, it is necessary to have a lineshape model for matrix ENDOR. A simple line-shape model[11-13] has been initially applied that appears to lead to semiquantitative information. This model and its application to unpaired electron distributions and to motional

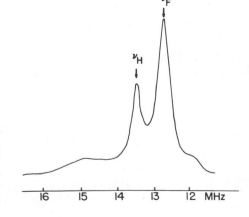

Figure 4. ENDOR spectrum of di-*tert*-butylnitroxide in $1H,1H$-heptafluor-1-butanol glass at 48°K showing resolved proton and fluorine matrix ENDOR lines. Taken from R. N. Schwartz, M. K. Bowman, and L. Kevan, unpublished work.

processes will be discussed in Sections 1.2.1–1.2.3. Then we describe an improved line-shape model that is currently under development in Section 1.2.4.

1.2.1. Simple Line-Shape Model

Matrix ENDOR lines contain information about the magnetic nuclear environment of a radical. For specificity, we will consider that these magnetic nuclei are protons. In the simple line-shape model, it is convenient to divide the protons into two groups—near protons on which there is appreciable spin density from the unpaired electron wave function and matrix protons, which are farther away and on which there is negligible unpaired electron spin density. This somewhat artificial distinction will be removed in the improved line-shape model. The electron-spin–nuclear-spin interaction involving the near protons, which "see" the electron wave function, will involve an isotropic part (Fermi contact interaction) and an anisotropic part (dipolar interaction). The Fermi contact interaction with these near protons gives rise to ENDOR lines at $v_p \pm A_p/2$ unless broadened beyond detection by the dipolar interaction or by some other mechanism, where v_p is the free proton resonance frequency in the magnetic field used and is typically 13–14 MHz at X-band fields of 3300 G. The electron-spin–nuclear-spin interaction with the matrix protons consists only of the dipolar interaction, since there is little or no electron spin density on these protons. This dipolar interaction produces a matrix ENDOR line centered at v_p.

We will write the general line shape of a matrix ENDOR signal as

$$f(v) = k \int_{\bar{a}}^{r} \int_{0}^{\pi} \int_{0}^{2\pi} R(r, \theta, \phi)g(v - v_0(r, \theta, \phi))r^2 \sin \theta \, dr \, d\theta \, d\phi \qquad (2)$$

where k is a constant that affects the amplitude but not the shape of the line, R is a function of various relaxation times that determine the ENDOR intensity, and $g(v - v_0)$ is the nuclear spin packet line shape centered at the resonant frequency v_0. The lower limit of the radial integral $\bar{a}$ represents an effective distance at which the "matrix" nuclei begin for which only a dipolar interaction is considered. The ϕ dependence of R and $g(v - v_0)$ can be eliminated by considering the magnetic moments to be aligned along the external magnetic field.

If the frequency dependence of $g(v - v_0)$ is determined mainly by the electron–nuclear dipolar interaction, as has been assumed, the matrix ENDOR line shape will depend on the nuclear relaxation processes that determine the form of R. In this simple line-shape model, it is assumed that nuclear relaxation occurs only by nuclear spin–lattice relaxation characterized by a time T_{1n} and that cross-relaxation is negligible. Then, if the nuclear transitions are assumed to be saturated by the rf field, the ENDOR intensity is proportional to T_{1n}^{-1}.

For the simplest case, T_{1n} may be considered to be independent of angle. An example of this is nuclear relaxation by time-dependent fluctuations in the isotropic hyperfine interaction. The angular dependence of the ENDOR line shape then depends only on the electron–nuclear dipolar interaction that determines the frequency dependence of $g(v - v_0)$. Simulations show that the matrix ENDOR line shape is relatively insensitive to Gaussian or Lorentzian functional forms for $g(v - v_0)$. In general, a doublet matrix line shape is predicted; however, most experimental observations of matrix ENDOR show a singlet line shape. This leads to the consideration of alternative nuclear relaxation mechanisms.

The electron–nuclear dipolar relaxation mechanism (END) is usually the dominant ENDOR active mechanism for aromatic radicals in solution. (See Chapters 3 and 4 for further discussion.) In disordered solids, where the unpaired electron and the nuclei are fixed in position, nuclear relaxation may occur by (a) the fluctuating magnetic field at the nucleus that is caused by relaxation of the electron spin or (b) changes in the magnitude of the electron–nuclear dipolar interaction caused by unusual motion that modulates the radial interaction distance.[11] In general, the END interaction can be written in terms of the spherical tensor components. The corresponding nuclear relaxation rates have different angular dependencies: $(1 - 3\cos^2\theta)$, $\sin^2\theta\cos^2\theta$, and $\sin^4\theta$. For mechanism (a), the $\sin^2\theta\cos^2\theta$ term dominates and a singlet ENDOR line is simulated by equation (2) that is similar to that observed experimentally. The introduction of angular dependence for T_{1n} changes the simulated doublet to a simulated singlet.

The nuclear relaxation is dominated by motional processes when either the radical or the matrix nuclei undergo random motion in a characteristic time τ_c much less than the electron spin–lattice relaxation time T_{1e}. Nuclear relaxation via motional processes is then faster than via the fluctuating field of the END interaction. Often, all of the angular dependencies of T_{1n} become comparable and must be summed. This leads to a matrix ENDOR line shape with a maximum at the free proton frequency and one or two sets of shoulders on both sides. Trapped hydrogen atoms in zeolite matrices exhibit this type of matrix ENDOR line shape.[12]

1.2.2. *Analysis of Unpaired-Electron Distribution*

Kevan and co-workers have used matrix ENDOR to determine average sizes of unpaired-electron wave functions.[9,13–20] They have used the line-shape model of equation (2) with the assumption that the ENDOR intensity is dominated by T_{1n} processes in which T_{1n}^{-1} is dependent on $\sin^2\theta\cos^2\theta$. The nuclear spin packet line shape $g(v)$ is taken as Lorentzian with a half-width at half-height of α. The matrix ENDOR line shape is essentially insensitive to the choice of either Gaussian or Lorentzian shapes for $g(v)$. After the above substitutions, the line-shape expression becomes

$$f(v) = N \int_0^\pi \int_{\bar{a}}^\infty \frac{\cos^2 \theta \sin^3 \theta}{r^4} \times \left\{ \frac{1}{\alpha^2 + [v - (q/r^3)]^2} + \frac{1}{\alpha^2 + [v + (q/r^3)]^2} \right\} d\theta\, dr$$

$$(3)$$

where $q = (4\pi)^{-1} \gamma_e \gamma_n h (3 \cos^2 \theta - 1)$. The upper limit $r = \infty$ on the radial integral yields results comparable to a finite upper limit, because the very distant protons make small dipolar contributions. Empirically, it is found that good fits with the experimental data are obtained for $r \sim 10$ Å as an upper limit. The integration of r implies that the matrix nuclei are in a uniform continuous distribution.

Equation (3) for the line shape contains two parameters, α and $\bar{a}$. By comparing various experimental spectra to simulated ones from numerical integration of equation (3), it was found that the shape fit is much more sensitive to α than to $\bar{a}$ and that the best fit is usually obtained with $\alpha \sim 80$ kHz. The ENDOR linewidth, full width at half-height, can then be simulated with a particular value of $\bar{a}$, the lower limit of the r integral. The $\bar{a}$ value can be considered to represent the average size of the unpaired electron wave function.

This type of analysis has been applied to a variety of different systems. For example, the average spherical sizes of the wave functions of excess electrons trapped in glassy matrices of varying polarity have been measured.[13] Significant effects of matrix polarity were observed that are consistent with independent theoretical calculations that are based on an assumed trapping potential that is dependent upon matrix properties. Matrix polarity effects on silver atoms and nitroxide radicals have also been studied.[15]

It is also possible to use specifically deuterated matrix molecules to infer how the wave function of a trapped electron overlaps its first solvation shell molecules.[16] The molecular details of solvation can also be studied by matrix ENDOR when the solvation process can be arrested at different stages.[18,19] The change in matrix ENDOR linewidth between partially and completely solvated states gives an indication of the distance changes to the matrix protons as the solvation process proceeds. In certain systems, matrix ENDOR may be of aid in radical identifications, as has been demonstrated in some polymeric systems.[9]

1.2.3. *Analysis of Motional Processes*

Matrix ENDOR signals can yield information about molecular or lattice motion if the motion has a cycle time shorter than electron relaxation times. In the simplest case, one can consider "lattice" vibrations modulating the dipolar interaction to weakly coupled protons. This may then give rise to a temperature-dependent ENDOR linewidth. To date, almost no cases have been reported or analyzed. One example occurs when partially deuterated polycrystalline sucrose is γ-irradiated at room temperature to produce an intense matrix ENDOR signal that is characterized by a strongly temperature-dependent ENDOR linewidth between 100 and 300°K.[21] This is striking because the ESR linewidth does not vary in this temperature range. The matrix ENDOR line was interpreted as a superposition of broad and narrow components due to different groups of matrix protons. The linewidth and the temperature dependence were related to the average distances and oscillation frequencies of the two proton groups. Sucrose is composed of pyranose and furanose rings, and reasonable assignments could be made for the two proton groups to the protons on these two rings.

Molecular motion may also manifest itself in matrix ENDOR by affecting the nuclear relaxation mechanism to give rise to additional structure in the matrix ENDOR line.[12] The ENDOR line shape from trapped hydrogen atoms in Y-type zeolite powders has been attributed to such motional modulation.[12,22] These zeolites are composed of SiO_4 and AlO_4 tetrahedra in the ratio of 2.4 to 1. These tetrahedra are arranged to form cubo octahedra sodalite cages of diameter 6–7 Å. The sodalite cages are joined together via their hexagonal faces, forming small hexagonal prismatic cages. ESR studies of trapped hydrogen atoms produced by γ-irradiation of zeolites activated to ~ 673°K indicate that the hydrogen atoms are trapped in the larger sodalite cages; this is based on an analysis of the matrix effects on the hydrogen atom hyperfine constants and g factors. The matrix ENDOR spectrum shows considerable structure and it is quite temperature dependent. This spectrum appears to be best explained in terms of a motional modulation mechanism based on a two-jump model of the trapped hydrogen atom between two positions in the large sodalite cage.

1.2.4. Improved Line-Shape Model

In the previous section, a simple line-shape model was presented to describe the matrix ENDOR line shape. This simple model, though quite useful for a preliminary understanding of the origin of matrix ENDOR lines, does not include all spin relaxation paths. The simple model also assumes that the nuclear transitions are very strongly saturated by the nuclear radio-frequency power, and consequently, the line-shape function does not include any dependence on rf power. However, we know experimentally that matrix ENDOR can be observed even when the nuclear transitions are mildly saturated and that the ENDOR line intensity and width depend on the strength of the nuclear rf power. Since ENDOR arises because of the competition between various relaxation paths, a realistic model should include the various relaxation paths explicitly as well as the nuclear rf and microwave magnetic fields. Such a model was developed by Hochmann, Zevin, and Shanina[23] using a density matrix formalism (see also Chapter 3). We have successfully adapted this model to describe the matrix ENDOR line shape.[24]

We shall consider a spin system characterized by $S = \frac{1}{2}, I = \frac{1}{2}$. The static Hamiltonian for this system with axial symmetry and an isotropic g value can be written as[25]

$$\mathcal{H} = hv_e S_z - hv_p I_z + hAS_z I_z + hBS_z I_x \tag{4}$$

where

$$hv_e = g\beta H_0 \qquad\qquad hv_p = g_n \beta_n H_0$$

$$hA = \frac{gg_n\beta\beta_n}{r^3}(3\cos^2\theta - 1) - h\mathbf{a} \qquad hB = \frac{gg_n\beta\beta_n}{r^3}3\cos\theta\sin\theta$$

and it is assumed that the z axis coincides with the direction of the magnetic field H_0 and that the nucleus lies in the xz plane. In the above equations v_e and v_p are the electron and nuclear Larmor frequencies, respectively; g and g_n are the electron and nuclear g factors; β and β_n are the electron and nuclear Bohr magnetons; r is the distance between the electron and the nucleus; θ is the angle the electron–nuclear vector makes with H_0; and $\mathbf{a}$ is the isotropic coupling in hertz. The above Hamiltonian can be transformed in a representation in which S_z and I_z are diagonal. The eigenvalues of this transformed Hamiltonian are given by[26]

$$E(m_s, m_I) = hv_e m_s + hKm_I \tag{5}$$

where

$$K = [(Am_s - v_p)^2 + B^2 m_s^2]^{1/2} \tag{5a}$$

and the corresponding eigenvectors are

$$\Psi(-\tfrac{1}{2}, -\tfrac{1}{2}) = \cos\frac{\phi^-}{2}\,|-\tfrac{1}{2}, -\tfrac{1}{2}\rangle - \sin\frac{\phi^-}{2}\,|-\tfrac{1}{2}, \tfrac{1}{2}\rangle$$

$$\Psi(-\tfrac{1}{2}, \tfrac{1}{2}) = \cos\frac{\phi^-}{2}\,|-\tfrac{1}{2}, \tfrac{1}{2}\rangle + \sin\frac{\phi^-}{2}\,|-\tfrac{1}{2}, -\tfrac{1}{2}\rangle$$

$$\Psi(\tfrac{1}{2}, -\tfrac{1}{2}) = \cos\frac{\phi^+}{2}\,|\tfrac{1}{2}, -\tfrac{1}{2}\rangle - \sin\frac{\phi^+}{2}\,|\tfrac{1}{2}, \tfrac{1}{2}\rangle \tag{5b}$$

$$\Psi(\tfrac{1}{2}, \tfrac{1}{2}) = \cos\frac{\phi^+}{2}\,|\tfrac{1}{2}, \tfrac{1}{2}\rangle + \sin\frac{\phi^+}{2}\,|\tfrac{1}{2}, -\tfrac{1}{2}\rangle$$

with

$$\tan\phi^- = \tfrac{1}{2}B/(\tfrac{1}{2}A + v_p) \qquad \tan\phi^+ = \tfrac{1}{2}B/(\tfrac{1}{2}A - v_p) \tag{5c}$$

K can take two values, K^+ and K^- for $m_s = \frac{1}{2}$ and $-\frac{1}{2}$, respectively. It is clear that K^+ and K^- are the nuclear resonance frequencies associated with $m_s = \frac{1}{2}$ and $-\frac{1}{2}$. The eigenvectors defined by equation 5b are slightly different from those given by Hochmann et al.[23] in that they consider only mixing due to isotropic coupling, while in the present formulation, mixing due to dipolar coupling is also considered.

The ENDOR effect can be described by using a density matrix formalism. The equations of motion for the density matrix elements are given by

$$\frac{i\,d\rho_{\alpha\beta}}{dt} = \left(\omega_{\alpha\beta} - \frac{i}{T_{\alpha\beta}}\right)\rho_{\alpha\beta} + [V, \rho]_{\alpha\beta}, \qquad \alpha \neq \beta \tag{6a}$$

$$\frac{i\,d\rho_{\alpha\alpha}}{dt} = i\sum_{\lambda} (W_{\lambda\alpha}\rho_{\lambda\lambda} - W_{\alpha\lambda}\rho_{\alpha\alpha}) + [V, \rho]_{\alpha\alpha} \tag{6b}$$

where the W are the transition probabilities shown in Figure 5 for which $W_{\alpha\lambda} = W_{\lambda\alpha}$ in the high-temperature approximation; the $T_{\alpha\beta}$ are the relaxation times; $\omega_{\alpha\beta}$ is the frequency separation between levels α and β; and V is the interaction of the spin system with both the microwave and the radiofrequency fields. (See also Chapter 3.) V can be written as

$$V = 2V_1 \cos \omega_1 t + 2V_2 \cos \omega_2 t \tag{7}$$

where ω_1 and ω_2 are the frequencies of the microwave and radiofrequency fields, respectively. The components V_1 and V_2 are given by

$$V_1 = g\beta H_1 S_x \tag{8a}$$

$$V_2 = -g_N \beta_N H_2 I_y \tag{8b}$$

where $2H_1$ and $2H_2$ are the linearly polarized microwave and radiofrequency fields, respectively.

The term V_1 denotes the coupling of the electron spin to the microwave field, and the term V_2 denotes the coupling of the radiofrequency field with the nuclear spin. The interaction of the microwave field with the nuclear spin has no significant effect on the resonant behavior on the spin system and so is neglected. We can solve equations (6a) and (6b) for the equilibrium spin population differences and define an equivalent Bloch susceptibility χ.

Following Seidel[27] we define the ENDOR signal $\delta_\pm$ as

$$\delta_\pm = \frac{[\chi'' - \chi''(H_2 = 0)]H_1}{[\chi''(H_2 = 0)H_1]_{max}} \tag{9}$$

where $\chi''(H_2 = 0)H_1$ simply corresponds to the magnetization measured by ESR. It is beyond the scope of this chapter to give the detailed derivation for

χ'' here (but see Chapter 3). We therefore write the final expression for the
ENDOR effect for an inhomogeneously broadened line[23]:

$$\delta_\pm = \left(\frac{X^2}{1+X^2}\right)^{1/2}\left\{\left[\frac{1+X^{-2}}{1+X^{-2}-(1-\alpha_\pm)(1+Y_\mp^{-2})}\right]^{1/2}-1\right\} \quad (10)$$

where $X^2 = 2P_{++}(\text{ESR})T_{1e}^{\text{eff}} = 2P_{--}(\text{ESR})T_{1e}^{\text{eff}}$ and $Y_\pm^2 = 2P_\pm(\text{ENDOR})T_{1n}^{\text{eff}}$. In
the above expressions for X^2 and Y^2, we define $P_{++}(\text{ESR})$ as the ESR transi-
tion probability induced by the microwave field between the levels $|-\frac{1}{2}, \frac{1}{2}\rangle$
and $|\frac{1}{2}, \frac{1}{2}\rangle$, while $P_{--}(\text{ESR})$ is the corresponding transition probability be-
tween the levels $|-\frac{1}{2}, -\frac{1}{2}\rangle$ and $|\frac{1}{2}, -\frac{1}{2}\rangle$. The ENDOR transition probability
$P_+(\text{ENDOR})$ denotes the nuclear transition probability induced by the radio-
frequency field between the levels $|\frac{1}{2}, -\frac{1}{2}\rangle$ and $|\frac{1}{2}, \frac{1}{2}\rangle$, while $P_-(\text{ENDOR})$ is
the corresponding transition between the levels $|-\frac{1}{2}, -\frac{1}{2}\rangle$ and $|\frac{1}{2}, \frac{1}{2}\rangle$. It
should be noted that the quantity i_{mn} defined by Hochmann et al.[23] is equal
to P_{mn} in our formulation. The transition probabilities can be evaluated by
Fermi's golden rule using the eigenvectors given in equation 5b. They are
given for a system with isotropic g by

$$P_{++}(\text{ESR}) = \frac{2\pi}{\hbar^2}\left(\frac{g\beta H_1}{2}\right)^2\cos^2\frac{\phi^+}{2}\cos^2\frac{\phi^-}{2}h(\omega) \quad (11)$$

$$P_\pm(\text{ENDOR}) = \frac{2\pi}{\hbar^2}\left(\frac{g_n\beta_n H_2}{2}\right)^2\cos^2\phi^\pm g(\omega) \quad (12)$$

where $h(\omega)$ and $g(\omega)$ are the corresponding line-shape functions, and actually, $g(\omega) = g(\omega - 2\pi K^\pm)$. We assume that $g(\omega)$ and $h(\omega)$ have a
Lorentzian distribution and are given by

$$g(\omega) = \left(\frac{T_{2n}}{\pi}\right)\frac{1}{1+(\omega-2\pi K_\pm)^2 T_{2n}^2} \quad (13)$$

$$h(\omega) = \left(\frac{T_{2e}}{\pi}\right)\frac{1}{1+(\omega-\omega_0)^2 T_{2e}^2} \quad (14)$$

where T_{2n} and T_{2e} are the nuclear and electronic transverse relaxation times
and ω_0 is the center frequency of the ESR line.

In the above expressions the "longitudinal" electron relaxation time
T_{1e}^{eff} is given by

$$T_{1e}^{\text{eff}} = \frac{1}{2(W_S+W_N)}\frac{2W_S W_N + (W_S+W_N)(W_X+W_{X'}) + W_X W_{X'} - W_N^2}{2W_S W_N + (W_S+W_N)(W_X+W_{X'}) + 2W_X W_{X'}} \quad (15)$$

and the "longitudinal" nuclear relaxation time T_{1n}^{eff} is given by

$$T_{1n}^{\text{eff}} = \frac{1}{2(W_S+W_N)}\frac{2W_S W_N + (W_S+W_N)(W_X+W_{X'}) + W_X W_{X'} + W_S^2}{2W_S W_N + (W_S+W_N)(W_X+W_{X'}) + 2W_X W_{X'}} \quad (16)$$

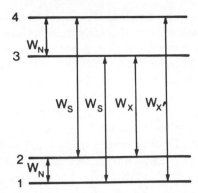

Figure 5. Energy level diagram for $S = \frac{1}{2}$, $I = \frac{1}{2}$ spin system. The W refer to the various relaxation transition probabilities. The levels 1 to 4 refer to the spin states $|-\frac{1}{2}, -\frac{1}{2}\rangle$, $|-\frac{1}{2}, \frac{1}{2}\rangle$, $|\frac{1}{2}, -\frac{1}{2}\rangle$, and $|\frac{1}{2}, \frac{1}{2}\rangle$, respectively.

The parameter α in equation (10) can take two values, α_+ and α_-, depending on the transitions we are considering

$$\alpha_+ = \alpha_{13,43} = \alpha_{42,21}$$

$$= \frac{W_S + W_N + W_X}{4T_{1e}^{\text{eff}} T_{1n}^{\text{eff}}(W_S + W_N)[2W_S W_N + (W_S + W_N)(W_X + W_{X'}) + 2W_X W_{X'}]} \tag{17}$$

$$\alpha_- = \alpha_{13,21} = \alpha_{42,43}$$

$$= \frac{W_S + W_N + W_X}{4T_{1e}^{\text{eff}} T_{1n}^{\text{eff}}(W_S + W_N)[2W_S W_N + (W_S + W_N)(W_X + W_{X'}) + 2W_X W_{X'}]} \tag{18}$$

The first two subscripts on α refer to the ESR transition and the second two subscripts refer to the nuclear transition. The various relaxation probabilities are shown in Figure 5.

When we consider ENDOR in a polycrystalline matrix, we have to average $\delta_\pm$ over all orientations. The ENDOR response is given by

$$\delta(r_i) = \langle \delta_+(r_i, \theta, \phi) + \delta_-(r_i, \theta, \phi) \rangle_\theta \tag{19}$$

for a single value of the distance r_i. Since the contribution to matrix ENDOR arises from a number of nuclei located at various distances, we must add the contributions from nuclei located at different distances to obtain the final ENDOR spectrum. We compute the quantity

$$\delta = \sum_{r_{\text{min}}}^{\infty} \delta(r_i) \approx \int_{r_{\text{min}}}^{\infty} \delta(r_i) \, dn \tag{20}$$

where r_{min} is the distance at which the closest nuclei are located. If we assume a spherical distribution, the number of nuclei between r and $r + dr$ is given by

$$dn = r^2 \, dr \tag{21}$$

and equation (10) predicts an infinite matrix ENDOR response. However, we know experimentally that this is not true. This weakness in the theory can be removed by incorporation of an overlap correction first introduced by Allendoerfer and Maki[28] to explain the relative intensities of various liquid phase ENDOR lines from a given radical. The way the correction works can be understood by referring to Figure 5. If we apply a strong microwave field to induce an ESR transition between levels 2 and 4, we essentially increase the spin temperature. Then when we apply a radio-frequency field to induce a transition between levels 3 and 4, we essentially reduce the spin temperature, and this results in an ENDOR enhancement. This is true because of the large population difference between levels 3 and 4 as a result of selectively saturating the $2 \leftrightarrow 4$ transition. Then we are implicitly assuming that the transitions $1 \leftrightarrow 3$ and $2 \leftrightarrow 4$ are well separated. However, this is not true if we are considering nuclei at large distances corresponding to small hyperfine couplings. In this case, it is not possible to selectively saturate one ESR transition. Consequently, the population difference between the nuclear levels approaches zero and so the radiofrequency field cannot induce any net transitions. This results in a decrease in the ENDOR enhancement. The correction term then becomes

$$C = \frac{\Delta M_z(\Delta\omega)}{\Delta M_z(\Delta\omega \to \infty)} \tag{22}$$

where ΔM_z is the difference in the z component of magnetization of the irradiated and unirradiated ESR lines that are to be coupled by radio-frequency fields. For solids, the magnetization is given by[29]

$$M_z = M_0 \left| \frac{\Delta\omega^2}{\Delta\omega^2 + \gamma^2[H_1^2 + 2(\delta H)^2]} \right| \tag{23}$$

Therefore,

$$C = \frac{\Delta\omega^2}{\Delta\omega^2 + 3\gamma^2 H_1^2} \tag{24}$$

where $\Delta\omega$ is the separation between the ESR transitions. It is assumed that $H_1 \sim \delta H$, the spin packet half-width.

The ENDOR enhancement with this correction becomes

$$\delta_\pm = \left(\frac{X^2}{1+X^2} \right)^{1/2} \left\{ \left[\frac{1+X^{-2}}{1+X^{-2} - (1+\alpha_\pm)C(1+Y_\pm^{-2})^{-1}} \right]^{1/2} - 1 \right\} \tag{25}$$

It is this expression we program to calculate the matrix ENDOR line shape. Notice that for large distances, $\Delta\omega \sim 0$ and $C = 0$. In this case, $\delta_\pm = 0$, consistent with experimental observation.

This formulation is clearly more realistic than the simple line-shape model described in Section 1.2.1. No assumptions about the strength of the radiofrequency and microwave magnetic fields are made. Consequently, we can predict the variation of ENDOR intensity both with H_1 and H_2 and compare with experimental studies. This provides additional experimental variables with which to obtain structural parameters from the matrix ENDOR line shape. Furthermore, since the expression for the ENDOR enhancement explicitly includes the various relaxation parameters, it is in principle possible to determine specific relaxation mechanisms that are operative in the system.

We have applied this improved line-shape model [equation (23)] to a disordered system well studied by other magnetic resonance techniques, namely, to trapped electrons in 2-methyltetrahydrofuran glass at 77°K. Good structural information about the molecular geometry around the trapped electron exists for this system from electron spin echo modulation data and second-moment ESR line-shape analysis.[30] Geometrical data have also been deduced for this system from simplified models of the matrix ENDOR response,[13,31] so we can compare the information obtained from matrix ENDOR using different line-shape models.

To calculate the ENDOR line shape we need r_{min}, the distance at which nuclei closest to the trapped electron are located; the small isotropic coupling $\mathbf{a}$; the values of H_1 and H_2 in gauss; the various relaxation parameters—W_S, W_N, W_X, and $W_{X'}$; and the transverse relaxation times T_{2e} and T_{2n}, which appear in the line-shape functions $h(\omega)$ and $g(\omega)$, respectively. The rotating magnetic fields H_1 and H_2 can be measured experimentally, while r_{min} and $\mathbf{a}$ have to be varied as parameters to obtain the best fit to the experimental spectrum. Some of the relaxation parameters can be measured experimentally and others can be eliminated as independent parameters for certain relaxation mechanisms.

The electron spin–lattice relaxation time T_{1e} can be directly measured by saturation recovery using a pulsed ESR method. For trapped electrons in 2-methyltetrahydrofuran at 77°K, T_{1e} has a value of 300 μsec.[32] The transition probabilities are defined in Figure 5 so that they refer to transitions between unmixed spin levels. However, since the actual levels contain a small mixture of other states due to the hyperfine coupling, the electron transition probability W_S is obtained by taking the appropriate matrix elements of the relaxation operator between the eigenstates defined by equation (5a). One has

$$W_S = |\langle \Psi(+, +)| \mathscr{H}_{SL} |\Psi(-, +)\rangle|^2 \tag{26}$$

$$= |\langle + + | \mathscr{H}_{SL} | - + \rangle|^2 \cos^2\left(\frac{\phi^+}{2} - \frac{\phi^-}{2}\right) \tag{26a}$$

$$= \frac{1}{T_{1e}} \cos^2\left(\frac{\phi^+}{2} - \frac{\phi^-}{2}\right) \tag{26b}$$

where $\mathscr{H}_{SL}$ is the relaxation operator that acts on the spin coordinates only. ϕ^+ and ϕ^- describe the mixing coefficients of the wave function; they are functions of the dipolar coupling and are dependent on angle. Therefore, the parameter W_S depends on angle θ. Thus, from the measured value of T_{1e}, W_S can be calculated using equation (26b).

The transverse electron relaxation time T_{2e} can be obtained from the decay of the two-pulse spin echo envelope. The decay of the envelope is characterized by phase memory time T_M. If we assume that T_M is limited by electron spin–spin flips,[33] we can approximate $T_M \approx T_{2e}$. The value of T_M for trapped electrons in 2-methyltetrahydrofuran glass at 77°K is ~ 0.5 μsec.

The nuclear relaxation probability W_N can be decomposed into two parts. The first part is due to the intrinsic nuclear relaxation and is independent of the proximity of a nucleus to the unpaired electron. This is denoted by W_N^{intr}, which can be measured using standard NMR techniques in the corresponding diamagnetic system. In the present case, the diamagnetic system is 2-methyltetrahydrofuran. Since we are not aware of any such measurements, we treated W_N^{intra} as a limited parameter to be varied. The relaxation of nuclei close to an unpaired electron is dominated by the dipolar interaction between the electron and the nucleus. The pseudosecular term of the dipolar interaction ($\propto S_z I_\pm$) provides the main relaxation path. This relaxation rate W^{para} depends on the electron relaxation time T_{1e}, because the fluctuating magnetic field at the nucleus due to the relaxing electron is responsible for the nuclear relaxation. This relaxation rate calculated by neglecting the nonsecular terms of the dipolar interaction is given by

$$W_N^{para} = \tfrac{9}{8}T_{1e}^{-1}\gamma_e^2\gamma_n^2\hbar^2 \sin^2\theta \cos^2\theta (2\pi v_p)^{-2} r^{-6} \tag{27}$$

The effective nuclear relaxation rate therefore becomes

$$W_N = W_N^{intr} + W_N^{para} \tag{28}$$

Due to the strong radial dependence of W_N^{para}, it is clear that W_N^{para} mainly affects the wings of the ENDOR line, because its largest contributions are from the closest nuclei, while W_N^{intr} affects mainly the center of the ENDOR line that has its largest contributions from the distant nuclei.

The transverse nuclear relaxation time T_{2N} can be measured from the NMR spectrum. However, in the present case, this is treated as a parameter to be varied.

The cross-relaxation probabilities W_X and $W_{X'}$ can be eliminated as independent parameters if we assume a relaxation mechanism for them. In the present case, the cross-relaxation arises due to the hyperfine coupling. Neglecting the nonsecular part of the hyperfine coupling, we have $W_X = W_{X'}$. The cross-relaxation probability can be calculated by evaluating

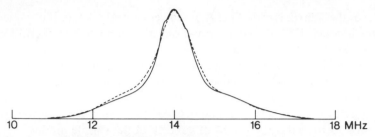

Figure 6. Simulated (solid curve) and experimental (dashed curve) matrix ENDOR line for the trapped electron in 2-methyltetrahydrofuran glass at 77°K. The parameters used for the simulation are $r_{min} = 3$ Å, $\mathbf{a} = 1.2$ MHz, $H_1 = 0.06$ G, $H_2 = 10$ G, $T_{1e} = 300$ μsec, $W_N^{intr} = 0.05$ sec^{-1}, and $T_{2n} = 40$ μsec. Note especially the good agreement between the simulated and experimental spectra in the wings.

the matrix element

$$W_X = |\langle \Psi(-,+)|\mathscr{H}_{SL}|\Psi(+,-)\rangle|^2 = \frac{1}{T_{1e}}\sin^2\left(\frac{\phi^+}{2} - \frac{\phi^-}{2}\right) \quad (29)$$

and so is dependent on angle.

Using the various measured parameters, we simulated ENDOR spectra for various values—r_{min}, $\mathbf{a}$, W_N^{intr}, and T_{2N}. The best visual fit with the experimental spectrum is obtained for $r_{min} = 3.0$ Å, $\mathbf{a} = 1.2$ MHz, $W_N^{intr} = 0.05$ sec^{-1}, and $T_{2N} = 40$ μsec. The simulated (solid) and the experimental (dashed) spectra are shown in Figure 6. It is interesting to compare these results with independent data on the same system obtained from the analysis of electron spin echo modulation patterns. The spin echo data[30] gives $r_{min} = 3.1$ Å to the nearest protons and $\mathbf{a} = 1.2$ MHz, which appear to be in excellent agreement with the matrix ENDOR line-shape analysis. Since we have considerable confidence in the spin echo data, we believe that the present model for the matrix ENDOR lineshape is also capable of giving reliable geometrical data. Although there are four parameters to be varied in simulating the ENDOR line, these parameters each predominantly affect a different portion of the ENDOR line, so it is possible to obtain a nearly unique set of parameters. For example a larger value of r_{min} mainly decreases the wingspan, while a smaller value of r_{min} mainly increases the wingspan. The isotropic coupling mainly controls the position and amplitude of the weak shoulders observed in the matrix ENDOR line. A larger hyperfine coupling increases the amplitude of the shoulders and pushes the shoulders away from the center. Also, the shape of these weak shoulders very strongly depends on the sign of $\mathbf{a}$.

The parameter T_{2N} mainly controls the width of the ENDOR line. A larger value of T_{2N} results in a narrower ENDOR line. This is understandable, since T_{2N} is inversely proportional to the width of the nuclear spin packet. Finally, as mentioned before W_N^{intr} mainly affects the shape of the central portion of the matrix ENDOR line. A narrower central portion is obtained for smaller values of W_N^{intr}. These observations are found to be general for the range of parameters we varied.

The simple matrix ENDOR line-shape model discussed in Section 1.2.1 gives $r_{min} = 2.8$ Å. This agrees fairly well with the improved lineshape model, although no information on the isotropic hyperfine coupling is obtained. More comparisons need to be made between the simple and improved line-shape models to assess the extent of validity of the simple model.

An even simpler matrix ENDOR line-shape model has been proposed by Iwasaki *et al.*[31] They assume that a matrix ENDOR line can be well represented by a Pake doublet composed of two powder ESR line shapes. For the closest protons, the matrix ENDOR linewidth between points of maximum slope is interpreted as $|A_\perp|$, and $A_\parallel$ is given by the total line span above the baseline. For $A_\perp < 0$, this analysis for trapped electrons in 2-methyltetrahydrofuran glass gives $r_{min} = 3.2 \pm 0.2$ Å and $a = 1.5 \pm 0.4$ MHz. These values are slightly different than originally published,[31] because $A_\parallel$ was not taken at the extreme ends of the wings of the matrix ENDOR line (M. Iwasaki, private communication). The agreement of this simple model with both the improved matrix ENDOR line-shape model and with the electron spin echo results is striking.

In conclusion, we feel that the improved line shape can sensitively represent experimental matrix ENDOR line shapes and will prove useful for extracting geometrical information therefrom.

2. ELDOR in Disordered Matrices

ELDOR has not been applied to disordered matrices as extensively as ENDOR. ELDOR has been most useful, thus far, for probing magnetic relaxation mechanisms, both within an inhomogeneously broadened line[34] and between two resonance lines of two different radicals.[35] It has also been recognized that ELDOR can reveal forbidden ESR transitions corresponding to simultaneous electron and proton spin flips.[36,37] Hyperfine coupling information can also be obtained by ELDOR for disordered systems.[38] In principle, the resolution obtainable should be somewhat better than by ESR, but this promise has largely yet to be realized.

2.1. Powder ELDOR—Hyperfine Information

The ELDOR signal intensity in a disordered system is a function of the powder spectrum at both the detecting and pumping frequencies. This intensity at a given magnetic field can formally be given by

$$I(H) = \iint\limits_{\theta\phi} [f_{M_1 \cdots M_k \cdots}(v_d, \theta, \phi)][f_{M_1 \cdots (M-1)_k \cdots}(v_p, \theta, \phi)]\sin\theta \, d\theta \, d\phi \qquad (30)$$

where the function f is a measure of whether a spin configuration with nuclear quantum numbers $M_1 \cdots M_k \cdots M_n$ has a resonance at the magnetic field H over the surface element $\sin \theta \, d\theta \, d\phi$. For simplicity, it has been assumed that the relaxation processes are angularly independent.

The most detailed study of powder ELDOR is that of Hyde and co-workers on a frozen solution of DPPH (diphenylpicrylhydrazyl) in toluene.[38] The interest in this system is whether any difference in the two nitrogen coupling constants can be discerned. ELDOR studies in liquid toluene did reveal a small difference; the ratio of the smaller to the larger isotropic coupling is 0.815.[38] The spectrum is an anisotropic five-line spectrum arising from the interaction of the unpaired electron with two nearly equivalent nitrogens. The extreme low- and high-field ESR components arise from nuclear configurations denoted by $[(+1, +1)_{\parallel}]$ and $[(-1, -1)_{\parallel}]$, where the numbers refer to the nuclear quantum number of the α and β nitrogen in DPPH and the $\parallel$ sign indicates that the magnetic field is parallel to the p-orbital of the two nitrogens and perpendicular to the molecular plane (the two p-orbitals are assumed parallel). The sign indicates that the field is perpendicular to the nitrogen p-orbitals. In DPPH, however, the largest nitrogen splitting occurs for the parallel direction. The intense central ESR line arises from several nuclear configurations but the largest is $[(0, 0)_{\perp, \parallel}]$.

When the detecting frequency is set at the central ESR line $[0, 0]$ and the pumping frequency is swept, a buildup of ELDOR intensity occurs at those frequencies where $v_p - v_d$ corresponds to a principal frequency of the hyperfine tensor components of the α or β nitrogens. The result is given in the top spectrum in Figure 7, where a single broad ELDOR line at 56.5 MHz occurs as a result of the contributions from both $[(-1, 0)_{\parallel}]$ and $[(0, -1)_{\parallel}]$;

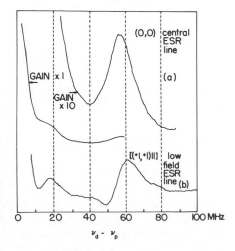

Figure 7. The frequency-swept ELDOR spectra of DPPH in solid solutions of toluene for $v_p < v_d$. Spectrum (a) is obtained at 153°K with the observing resonant condition set on the crossover of the center line, while spectrum (b) is obtained at 123°K while observing the peak of the low-field turning point. Adapted from J. S. Hyde, R. C. Sneed, Jr., and G. H. Rist, *J. Chem. Phys.* **51**, 1404 (1969).

the perpendicular components are much smaller. It is concluded that 56.5 MHz is the average of the anistropic α and β nitrogen hyperfine tensors for the parallel direction. If this ELDOR line were split, one could then distinguish the hyperfine components of the two nitrogens. When the detecting frequency is set on the low-field ESR line $[(+1, +1)_{\parallel}]$, the selection of configurations is more limited and one might expect the resolution to be better. However, the ELDOR spectrum at the bottom of Figure 7 still only shows a single line at 56 MHz.

Comparison of the ELDOR and ESR spectra in frozen solutions indicates that some improvement in resolution is achieved by ELDOR, but it is not sufficient to resolve the different splitting of the two nitrogens in DPPH. However, the asymmetry of the ELDOR lines suggests that an exact computer analysis of the powder spectra may enable a determination to be made of the principal direction of the hyperfine tensors of each nitrogen.

2.2. *Spin Packet Model and Spin Diffusion in Inhomogeneous Lines*

Field and frequency sweep ELDOR spectra can be observed from an inhomogeneous ESR line. It is most convenient to interpret these spectra in terms of the noninteracting spin packet model originally introduced by Portis.[39] Sets of electron spins that interact with different nuclear spin configurations form different spin packets, all of which superimpose to form the observed inhomogeneously broadened line. The word "noninteracting" implies no transfer of magnetic energy between spin packets. When this restriction is relaxed, we will speak of "spin diffusion" as a general term meaning transfer of saturation between different spin packets of an inhomogeneous line without implying any specific mechanism. Yoshida, Feng, and Kevan[34] find that the spin packet width and the degree, if any, of spin diffusion between spin packets can be estimated by simulating the ELDOR line.

Field swept ELDOR spectra from the one-line ESR spectrum of a trapped electron in 10 M NaOH aqueous glass are shown in Figure 8 on the left-hand side.[34] The spectra are shown as a function of pump power at a pumping frequency 20 MHz higher than the detecting frequency. In Figure 8a, with effectively no pump power applied, the normal ESR line is seen. As the pump power increases, the spectra become asymmetric. This asymmetry can be readily simulated by a spin packet model for the ELDOR line as shown by the right-hand side of Figure 8.

To simulate the ELDOR response as a function of pump power and Δv, the difference in the pumping (v_p) and detecting (v_d) microwave frequencies, the spin packet shape function is taken as Lorentzian with a half-width at half-height of $\Delta \omega_L T_{2e}^{-1}$ and a Gaussian distribution of spin packets. The half-width at half-height of the Gaussian line is taken as $\Delta \omega_G$ and the centers

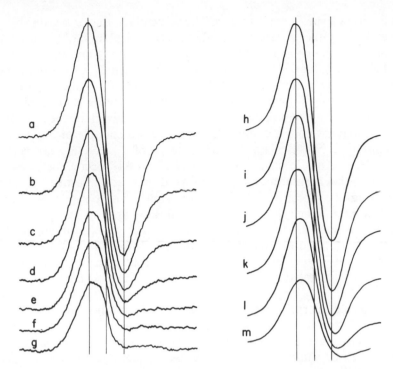

Figure 8. Field-swept ELDOR spectra of trapped electrons in γ-irradiated 10 M NaOH glass at 115°K with $v_p - v_d = 20$ MHz. Observed spectra are shown on the left at various pumping powers: (a) -39 dB, (b) -25 dB, (c) -20 dB, (d) -15 dB, (e) -10 dB, (f) -5 dB, and (g) 0 dB. Simulated spectra are shown on the right for the saturation parameter S equal to (h) 0, (i) 0.1, (j) 0.3, (k) 1.0, (l) 3.0, and (m) 10.0 with the inhomogeneity parameter $a = 0.37$. Adapted from H. Yoshida, D. F. Feng, and L. Kevan, *J. Chem. Phys.* **58**, 3411 (1973).

of the spin packets (ω') form a Gaussian distribution about ω_0, the center of the observed line. When the spin packets are assumed not to interact, the pumping and the detecting frequencies must be applied to the same spin packet to have any effect. After explicit forms of the line-shape functions are substituted into the equation for the imaginary part of the magnetic susceptibility χ' at the detecting frequency, χ'' can be calculated as a function of an inhomogeneity parameter a, where $a = \Delta\omega_L/\Delta\omega_G$, and a saturation parameter S, where $S = \gamma^2 H_{1p}^2 T_{1e} T_{2e}$ and H_{1p} is the microwave magnetic field of the pumping frequency. The observed ELDOR line shape depends on whether the field modulation cycle time is greater or less than T_{1e}. For the example in Figure 8, T_{1e} is longer than the field modulation time, and the derivative of χ'' with respect to the observing frequency ω_o ($d\chi''/d\omega_o$) was calculated. The theoretical field swept ELDOR spectra given in the right-hand side of Figure 8 were obtained by using the parameters S and a that gave the best fit to the experimental spectra. The value of parameter a gives the spin packet width. Other parameters needed for the calculation of χ'', such as ω_p (the pump frequency), ω_o, and $\Delta\omega_G$ are experimentally known.

From the values of S and a, the values of T_{1e} and T_{2e} can be determined as a function of pump power.[34] If a unique set of T_{1e} and T_{2e} values is calculated that is not dependent on pump power, then it is concluded that the noninteracting spin packet model is a correct model for the description of the inhomogeneous ESR line. However, for the example of trapped electrons in 10 M NaOH glass, a nonunique set of T_{1e} and T_{2e} values is obtained. This arises because spin diffusion actually occurs between spin packets, so that desaturation of the observed spin packet occurs. This causes the experimental spectra to saturate more slowly with power than the simulated spectra, which assume no spin diffusion. As the pump power increases, the spin diffusion rate increases, which results in an apparent decrease in T_{1e}. In other words, spin diffusion provides an additional relaxation path that is equivalent to spin–lattice relaxation. Analysis of this effect shows that the apparent T_{2e} is approximately independent of pumping power.

Frequency swept ELDOR spectra associated with the single ESR line of a trapped electron in 10 M NaOH aqueous glass are shown in Figure 9.[34] Such spectra have also been termed $\Delta M_I = 0$ spectra.[37] From the spin packet, with or without spin diffusion, the frequency swept ELDOR spectra are expected theoretically to be smoothly decreasing curves as the frequency difference increases. This general behavior is observed in Figure 9, but definite structure is also observed, as indicated by the arrows. This structure is more prominent at higher pumping power. Although this structure was originally interpreted as due to isotropic hyperfine interaction,[34] further work has made it clear that it is to be assigned to forbidden ESR transitions involving simultaneous spin flips of electrons and neighboring nuclei and occurring at a multiple of the free nuclear frequency away from the main ESR line.[40] This interpretation is in agreement with earlier work on trapped electrons in KCl single crystals[41] and with recent work on nitroxides in amorphous polymer matrices.[37] In the polymer matrices, the resolution of ELDOR spin-flip lines is rather sensitive to temperature.

For the trapped electron in 10 M NaOH aqueous glass, no spin-flip lines are observable in the 9 GHz ESR spectra, because they are buried in the linewidth. However, at 35 GHz, one set of proton spin-flip lines becomes detectable.[42] The remarkable thing about the ELDOR spectra is that four sets of proton spin-flip lines are observed spaced at v_H, $2v_H$, $3v_H$, and $4v_H$ from the center of the ESR line. These lines correspond to the simultaneous flip of the electron and one, two, three, and four proton spins. Consequently, there must be at least four nearest-neighbor protons around the trapped electron. Recent electron spin echo experiments on this same system have

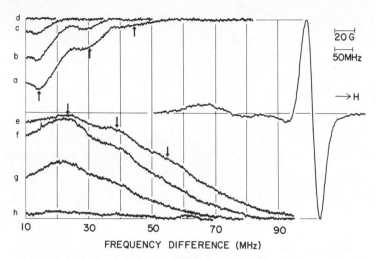

Figure 9. Frequency-swept ELDOR spectra of trapped electrons in 10 M NaOH glass recorded at 115°K as a function of the frequency difference $v_p - v_d$ with $v_p > v_d$. The spectra were recorded at a pumping power level of (a) 0 dB, (b) −10 dB, (c) −20 dB, and (d) −39 dB when detecting at the low-field peak of the derivative ESR spectrum and at a pump power level of (e) 0 dB, (f) −10 dB, (g) −20 dB, and (h) −39 dB while observing at the high-field peak of the ESR spectrum given at the right-hand side of the figure. Structure in the ELDOR spectrum is depicted by the arrows. Adapted from H. Yoshida, D. F. Feng, and L. Kevan, *J. Chem. Phys.* **58**, 3411 (1973).

shown that there are six nearest-neighbor proton around the electron,[43] which is consistent with the ELDOR results. If the spin-flip ELDOR frequencies can be measured accurately enough, it should be possible to calculate the average distance from the unpaired electron to the nearest matrix protons.[44]

2.3. Cross-Relaxation Studies

When two different spin systems are present in a solid, magnetic energy transfer may occur between them via cross-relaxation.[45] Radiolysis of disordered solids often produces more than one kind of trapped radical, and in general, it is possible that significant magnetic energy transfer may occur between different spin systems. Yoshida *et al.*[35] reported an example of such magnetic energy transfer between trapped electrons and the matrix radicals produced in the γ radiolysis of 2-methyltetrahydrofuran (MTHF) glass at 77°K by directly observing cross-saturation effects. Direct observation of cross-saturation is made possible by ELDOR. One microwave frequency pumps an ESR line of one radical, while the second microwave frequency detects the effect of this pumping on an ESR line of a second radical. For steady-state ELDOR, net cross-saturation can be observed in one

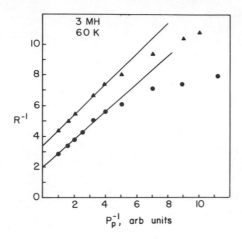

Figure 10. ELDOR saturation curves. Dependence of reciprocal reduction factor on reciprocal pumping microwave power for trapped electrons in γ-irradiated 3-methylhexane glass at 60°K and $\Delta f = 33$ MHz for doses of 0.1 Mrad (▲) and 1.4 Mrad (●).

direction if the relaxation time of one species is significantly less than that of the other species. In a number of γ-irradiated organic glassy matrices, both electrons and radicals are trapped with relatively high efficiency. In general, the electrons have longer relaxation times than the radicals, so such systems are good models for studying magnetic energy transfer by ELDOR.

In addition to electrons and radicals in methyltetrahydrofuran glass, cross saturation is observed between trapped electrons and radicals in γ-irradiated 3-methylhexane glass.[46] The ESR spectrum of irradiated 3-methylhexane glass at 77°K shows a sharp line in the center $(\sim 4\text{ G})$ due to a trapped electron (e_t^-) and a six-line background spectrum with a splitting of ~ 20 G due to a 3-methylhexane radical in which the radical site is located on carbon number 2. Field-swept ELDOR spectra can be obtained for which the pumping microwave frequency is applied at a position in the radical spectrum when the detecting microwave frequency is centered on the trapped electron line. As the pumping power is increased, the intensity of the trapped electron line decreases while the intensity of the radical lines remains nearly unaffected. This is the phenomenon of cross-saturation from the radical to the electron spin system. The ELDOR response of the trapped electron line is expressed by a reduction factor R, which is defined as (signal with pump on)/(signal with pump off). The dependence of R^{-1} on the reciprocal pump power is shown in Figure 10 for two doses. If there were no ELDOR responses, R^{-1} would be infinite. A linear relationship is expected theoretically according to

$$R^{-1} = 1 + \frac{T_{pd}}{T_{1d}} + (2W_{sp}T_{1p})^{-1}\left(1 + \frac{T_{pd}}{T_{1d}} + \frac{T_{pd}T_{1p}}{T_{1d}T_{dp}}\right) \tag{31}$$

where T_{1d} and T_{1p} are the spin–lattice relaxation times of the detected and pumped spins, respectively; T_{pd} and T_{dp} are the cross-relaxation times for the pumped-to-detected and the detected-to-pumped transitions, respectively; and W_{sp} is the stimulated transition probability for the pumped transition that is proportional to the pumping microwave power.

In general, equation (31) is followed at high pumping powers, but there is a deviation from the linear plot at low pumping powers. The intercept corresponding to infinite pump power gives $(1 + T_{pd}/T_{1d})$, which determines either the ratio of cross-relaxation time to the spin–lattice relaxation time of the detected spin or the cross-relaxation time itself if the spin–lattice relaxation time of the detected spin is measured independently.

It has been previously suggested[35] that the occurrence of saturation of the trapped electron line by pumping the radical line is due to a dipolar cross-relaxation mechanism.[45] In this mechanism, the direct spin–spin interaction between the two spin systems (electron and radical) causes simultaneous spin flips in both spin systems with any excess energy conserved by dipolar interactions as dependent on a function of the ESR line shapes of the two types of spins. In this formulation, the cross-relaxation line-shape function g_{12} is given by

$$g_{12} = \int\limits_{-\infty}^{\infty}\!\!\!\int g_1(\omega)g_2(\omega')\,\delta(\omega + \omega')\,d\omega\,d\omega' \tag{32}$$

where g_1 and g_2 are the line shapes of spins of type 1 and 2, respectively.[45] This formulation applies only to concentrated spin systems such as paramagnetic salts. In dilute spin systems, with $S = \frac{1}{2}$, such as we are considering, the second moment for cross-relaxation can be far greater than the sum of the second moments of the individual lines of the two spin species.[47] Kiel has shown that cross-relaxation can be significant even in cases where the separation between the interacting resonances is many times (20) the sum of the ESR linewidths.[47]

One can envision single-step and multistep cross-relaxation mechanisms. In the multistep mechanism, cross-relaxation occurs between the pumped radical line and an adjacent radical line stepwise until a central radical line is reached, which then cross relaxes to the electron line. In this mechanism, the two line shapes involved in an individual cross-relaxation process overlap. In the single-step mechanism, direct cross relaxation occurs between the pumped radical line and the electron line even though the two ESR lines involved may not overlap. These two mechanisms may be distinguished as follows. The single-step mechanism is the classic dipolar cross-relaxation process for which the associated cross-relaxation time is expected to be temperature independent and to vary with $(\Delta f)^2$. The multistep mechanism involves cross-relaxation between different nuclear substates of the same spin system; this is expected to occur via a temperature-dependent electron nuclear dipolar relaxation. In addition, the multistep process is expected to vary quite strongly with Δf [perhaps as $(\Delta f)^{2m}$ where m is an effective number of steps] as the number of steps involved increases.

The experimental data for trapped electron–radical systems shows that the cross-relaxation time T_{pd} is nearly temperature independent and depends on $(\Delta f)^n$ with $n < 2$.[46] Therefore, it is concluded that the single-step mechanism applies to these systems.

In order to relate the experimental cross-relaxation time to a correlation distance between the two cross-relaxing species, we consider a dilute spin system involving two different $S = \frac{1}{2}$ species. From time-dependent perturbation theory, the probability per unit time for the process that radical p flips down and radical d flips up with any energy discrepancy being taken up by the electron-spin–electron-spin dipolar interaction is

$$W_{pd} = 2\pi\hbar^{-2}|\mathscr{H}_{pd}|^2 g_{pd}(\omega = 0) \tag{33}$$

where $\mathscr{H}_{pd}$ is the dipolar interaction and g_{pd} is the frequency distribution function over which the dipolar interaction can absorb the excess Zeeman energy. The distribution function g_{pd} is also called the cross-relaxation line shape.

In the high-field approximation and when the g factors of the two species have the same sign (in this case positive) the $(S_{+p}S_{-d} + S_{-p}S_{+d})$ term in the dipolar interaction dominates. Then the square of the dipolar matrix element is given by[45]

$$|\mathscr{H}_{pd}|^2 = (16)^{-1}g^4\beta^4(1 - 3\cos\theta_{pd})^2 r_{pd}^{-6} \tag{34}$$

where we have assumed $g_p = g_d$ and β is the Bohr magneton. We must now average over all possible radical sites so that

$$\sum_d' |\mathscr{H}_{pd}|^2 = f \sum_d^n |\mathscr{H}_{pd}|^2 \tag{35}$$

where the prime on the summation indicates exclusion of terms with $p = d$ and where f is the fraction of sites occupied by radicals. The summation on the left-hand side goes over all sites occupied by a radical p or a radical d, while the summation on the right-hand side goes over all molecular sites n. In a disordered glassy system with randomly oriented radicals, the average of $(1 - 3\cos^2\theta_{pd})^2 = \frac{4}{5}$ and $\sum_d r_{pd}^{-6} = zd^{-6}$, where z is the effective number of nearest-neighbor sites and d is the average correlation distance.

Now, from the definition of cross-relaxation time,[45]

$$(2T_{pd})^{-1} = \sum_d W_{pd}P_d = \tfrac{1}{2}\sum_d W_{pd} \tag{36}$$

where P_d is the probability for spin d to be in the correct magnetic energy state form a transition to occur, we obtain

$$(T_{pd})^{-1} = \sum W_{pd} = \pi(10\hbar^2)^{-1}g^4\beta^4 fzd^{-6}g_{pd}(\omega = 0) \tag{37}$$

The crux of evaluating equation (37) is the form of the cross-relaxation line shape. For dilute spin systems, Kiel[47] has pointed out that the appropriate cross-relaxation lineshape can be computed in terms of its second moment. From Kiel's expression Lin *et al.*[46] obtain, after some approximations,

$$\hbar^2 \langle \omega_{pd}^2 \rangle = (\tfrac{6}{5}) g^4 \beta^4 f z d^{-6} \tag{38}$$

The question now arises as to what shape function to use for the cross-relaxation interaction. Lin *et al.*[46] conclude that a cut-off Lorentzian shape is perhaps the best approximation to the cross-relaxation shape in dilute $S = \tfrac{1}{2}$ systems. The equation finally obtained to relate the cross-relaxation time to the correlation distance between two cross-relaxing radicals is

$$T_{pd}^L = \frac{6.86 \hbar d^3}{g^2 \beta^2 (fz)^{1/2}} \left(1 + \frac{(2.12 \times 10^{12}) \Delta \omega^2 d^6 \hbar^2}{g^4 \beta^4 f z} \right) \tag{39}$$

where all quantities are in cgs units. To evaluate the correlation distance d, $z = 10$ may be chosen for a disordered system that is intermediate to $z = 8.5$ for a simple cubic lattice and $z = 14.5$ for a face-centered cubic lattice.[48]

Evaluation shows that the average electron–radical correlation distance is 9 Å in methyltetrahydrofuran glass and 11 Å in 3-methylhexane glass.[46] These appear to be reasonable distances. When equation (39) is evaluated, the second term in the parentheses is much larger than unity, so the dependence of the cross-relaxation time on the various parameters is given by

$$T_{pd}^L \propto d_L^9 (\Delta \omega)^2 f^{-3/2} \tag{40}$$

It appears that ELDOR is well suited for cross-relaxation measurements and that such measurements can lead to correlation distances between different types of radicals. This should be of importance in evaluating inhomogeneous radical decay kinetics in solid disordered systems.

ACKNOWLEDGMENTS

The assessment of ENDOR work was supported by the U.S. Army Research Office, and the assessment of ELDOR work was supported by the U.S. Energy Research and Development Administration under contract E(11-1)-2086.

References

1. M. Decaillot and J. Uebersfeld, *C. R. Acad. Sci. Paris* **265**, B155 (1967).
2. J. S. Hyde, G. H. Rist, and L. E. G. Eriksson, *J. Phys. Chem.* **72**, 4269 (1968).
3. L. Kevan and L. D. Kispert, *Electron Spin Double Resonance Spectroscopy*. Chapter 1, Wiley Interscience, New York (1976).

4. R. D. Allendoerfer, *Chem. Phys. Lett.* **17**, 172 (1972).
5. G. H. Rist and J. S. Hyde, *J. Chem. Phys.* **52**, 4633 (1970).
6. L. R. Dalton and A. L. Kwiram, *J. Chem. Phys.* **57**, 1132 (1972).
7. D. Becker and A. L. Kwiram, *Chem. Phys. Lett.* **39**, 180 (1976).
8. L. Kevan and L. D. Kispert, *Electron Spin Double Resonance Spectroscopy*, Chapter 7, Wiley-Interscience, New York (1976).
9. J. N. Helbert, B. E. Wagner, E. H. Poindexter, and L. Kevan, *J. Polym. Sci. (Phys.)* **13**, 825 (1975).
10. R. N. Schwartz, M. K. Bowman, and L. Kevan, unpublished work.
11. D. S. Leniart, J. S. Hyde, and J. C. Vedrine, *J. Phys. Chem.* **76**, 2079 (1972).
12. J. C. Vedrine, J. S. Hyde, and D. S. Leniart, *J. Phys. Chem.* **76**, 2087 (1972).
13. J. Helbert, L. Kevan, and B. L. Bales, *J. Chem. Phys.* **57**, 723 (1972).
14. J. Helbert and L. Kevan, *J. Chem. Phys.* **58**, 1205 (1973).
15. B. L. Bales, R. N. Schwartz, and L. Kevan, *Chem. Phys. Lett.* **22**, 13 (1973).
16. R. N. Schwartz, M. K. Bowman, and L. Kevan, *J. Chem. Phys.* **60**, 1690 (1974).
17. B. L. Bales, R. N. Schwartz, and L. Kevan, *Ber. Bunsenges Phys. Chem.* **78**, 194 (1974).
18. H. Hase, F. Q. H. Ngo, and L. Kevan, *J. Chem. Phys.* **62**, 985 (1975).
19. D. P. Lin and L. Kevan, *Chem. Phys. Lett.* **40**, 517 (1976).
20. F. Q. H. Ngo, S. Noda, and L. Kevan, in *Proc. of 4th International Symposium on Radiation Chemistry, Kesthely, Hungary, 1976* (P. Heddig and P. Schiller, eds.), pp. 951–962, Akademiai Kaido, Budapest, Hungary (1977).
21. E. G. Derouane and J. C. Vedrine, *Chem. Phys. Lett.* **29**, 222 (1974).
22. J. C. Vedrine, D. S. Leniart, and J. S. Hyde, *Ind. Chim. Belg.* **38**, 397 (1973).
23. V. L. Hochmann, V. Ya. Zevin, and B. D. Shanina, *Fiz. Tver. Tela* **10**, 337 (1968); English trans.: *Sov. Phys. Solid State* **10**, 269 (1968).
24. P. A. Narayana, R. N. Schwartz, M. Bowman, D. Becker, and L. Kevan, *J. Chem. Phys.* **67**, 1990 (1977).
25. W. Low, in *Solid State Physics* (F. Seitz and D. Turnbull, eds.), Suppl. 2, p. 52, Academic Press, New York (1960).
26. L. G. Rowan, E. L. Hahn, and W. B. Mims, *Phys. Rev.* **1374**, 61 (1965).
27. H. Seidel, *Z. Phys.* **165**, 239 (1961).
28. R. D. Allendoerfer and A. H. Maki, *J. Mag. Res.* **3**, 396 (1970).
29. A. G. Redfield, *Phys. Rev.* **98**, 1787 (1955).
30. L. Kevan, M. K. Bowman, P. A. Narayana, R. K. Boeckman, V. F. Yudanov, and Yu. D. Tsvetkov, *J. Chem. Phys.* **63**, 409 (1975).
31. M. Iwasaki, H. Muto, B. Eda, and K. Nunome, *J. Chem. Phys.* **56**, 3166 (1972).
32. M. Bowman, Ph.D. Thesis, Wayne State University (1975).
33. W. B. Mims, in *Electron Paramagnetic Resonance* (S. Geschwind, ed.), p. 263, Plenum Press, New York (1972).
34. H. Yoshida, D. F. Feng, and L. Kevan, *J. Chem. Phys.* **58**, 3411 (1973).
35. H. Yoshida, D. F. Feng, and L. Kevan, *J. Chem. Phys.* **58**, 4924 (1973).
36. J. S. Hyde, M. D. Smigel, L. R. Dalton, and L. A. Dalton, *J. Chem. Phys.* **62**, 1655 (1975).
37. M. M. Dorio and J. C. W. Chien, *J. Magnet. Resonance* **21**, 491 (1976).
38. J. S. Hyde, R. C. Sneed, and G. H. Rist, *J. Chem. Phys.* **51**, 1404 (1969).
39. A. M. Portis, *Phys. Rev.* **91**, 1071 (1953).
40. L. Kevan, D. F. Feng, and F. Ngo, unpublished work.
41. P. R. Moran, *Phys. Rev.* **135**, 247 (1964).
42. M. Bowman, L. Kevan, R. N. Schwartz, and B. L. Bales, *Chem. Phys. Lett.* **22**, 19 (1973).
43. P. A. Narayana, M. K. Bowman, L. Kevan, V. F. Yudanov, and Yu. D. Tsvetkov, *J. Chem. Phys.* **63**, 3365 (1975).
44. M. Bowman, L. Kevan, and R. N. Schwartz, *Chem. Phys. Lett.* **30**, 208 (1975).

45. N. Bloembergen, S. Shapiro, P. S. Pershan, and J. O. Artman, *Phys. Rev.* **114**, 445 (1959).
46. D. P. Lin, D. F. Feng, F. Q. H. Ngo, and L. Kevan, *J. Chem. Phys.* **65**, 3994 (1976).
47. A. Kiel, *Phys. Rev.* **120**, 137 (1960).
48. M. Goldman, *Spin Temperature and Nuclear Magnetic Resonance in Solids*, p. 226, Clarendon Press, Oxford (1970).

7

Crystalline Systems

Lowell D. Kispert

1. Introduction

For the past two decades or so, an extensive number of EPR studies of free radicals and triplet states in crystalline systems have been reported.[1-3] These studies have demonstrated the high specificity for detection, identification, and monitoring of the decay and formation of the reactive paramagnetic intermediates as a function of various experimental conditions. However, a fundamental limitation of EPR is its resolution. Since the electron magnetic double resonance methods can increase the effective resolution of EPR spectra, they are being used almost routinely for studies of radicals in solids.

In this chapter, we will review the use of two double resonance methods—electron–nuclear double resonance (ENDOR) and electron–electron double resonance (ELDOR)—in the study of free radicals in solids. It will be shown that these two techniques can complement each other, as well as play separate roles in the study of free radicals. A wide variety of chemical systems and a wide variety of nuclei have been studied. The reason for this is that one can always vary the temperature so that the relaxation times that make ENDOR and ELDOR possible can be short-circuited by the available microwave and rf power. For instance, if only low-power rf (1 W or less) is available, an ENDOR experiment can be carried out by simply lowering the temperature, thus increasing nuclear spin–lattice relaxation time, making it possible to saturate the nuclear transitions. In the case of ELDOR studies, the nuclear spin–lattice relaxation may have to be shortened in order to provide an ELDOR active relaxation pathway, and thus, higher temperature than for

---

Lowell D. Kispert • Department of Chemistry, The University of Alabama, Tuscaloosa, Alabama

ENDOR may be required. Because of this ability to change the relaxation times by simply varying the temperature, ENDOR studies have been carried out[4] in (1) defect centers in ionic crystals, (2) transition metal ions in ionic crystals, (3) polycrystalline and amorphous solids, and (4) molecular crystals (primarily organic).

Due in part to the lack of commercial ELDOR equipment, not as extensive a list of studies have been carried out, although a number of studies have been reported for radicals in molecular crystals (both $S = \frac{1}{2}$ and $S = 1$ cases) and polycrystalline and amorphous solids.[4] Very little work has been carried out for transition metal ions, as it appears that T_{1e} is too short in most cases to observe ELDOR. In this chapter, only ENDOR and ELDOR studies of radicals in organic crystals will be described. Typically, in organic crystals, $T_{1e} < T_{x1} \ll T_{1n}$. However, in irradiated crystals, $T_{1e} < T_{x1} \leq T_{1n}$ over the temperature range 77–300°K. At 300°K, $T_{1e} \simeq 10^{-6}$ sec.

In this chapter we describe how crystalline-phase ENDOR is used to determine small hyperfine splittings, quadrupole couplings, and reaction mechanisms of radical formation and how crystalline-phase ELDOR is used to determine large hyperfine splittings, to identify radicals with large quadrupole moments, and to study spin exchange processes. In the later part of the chapter, we will deal with the complementary role played by ENDOR and ELDOR spectroscopy in the separation of overlapping EPR spectra, in the study of proton–deuterium exchange, in the study of methyl groups undergoing tunneling rotation, and in the determination of the rates of intramolecular motion. It will not be possible to consider the effect of modulation frequency dealt with in Chapter 5, as little quantitative data exist. Most ELDOR spectra considered have been measured using 100-kHz field modulation.

Before it is possible to discuss these aspects of radicals in crystals, it will be necessary to explain how one measures the anisotropic hyperfine splittings that are exhibited by crystals. It is from such data that a large amount of the structural information of radicals is obtained.

2. Determination of Anisotropic Hyperfine Splitting Tensors

2.1. ENDOR

Most organic radicals in single crystals that are composed of carbon, hydrogen, nitrogen, oxygen, fluorine, chlorine, or deuterium are found to exhibit g values that possess little anisotropy and typically range from 2.0023 to 2.0080. However, the hyperfine splitting values are generally quite anisotropic and may vary as much as 200 G as in the case of α-fluorines. We can write the spin Hamiltonian for such a system as

$$\mathscr{H} = g\beta \mathbf{H} \cdot \mathbf{S} - g_n \beta_n \mathbf{H} \cdot \mathbf{I} + \mathbf{S} \cdot \mathbf{A} \cdot \mathbf{I} \tag{1}$$

where **A** is a tensor consisting of nine Cartesian components that is a measure of the coupling between two three-component vectors **S** and **I** while g is assumed isotropic. The first two terms represent the electronic and nuclear Zeeman interactions, respectively. Typically, an EPR experiment is carried out at 3300 G, and thus, for all radicals that we will deal with in this chapter, the first term is greater than the sum of the last two terms.

To solve for the hyperfine splitting tensor **A**, we can choose a set of laboratory axes x', y', z' and take the magnetic field to be in the z' direction. Assuming **A** to be small enough so that the electronic Zeeman interaction dominates causes $S_{z'}$ to be aligned along the magnetic field direction H_0 so all the $S_{x'}$ and $S_{y'}$ terms in the hyperfine interaction drop out. In this particular example we have

$$\mathscr{H} = g\beta H_{z'}S_{z'} - g_n\beta_n H_{z'}I_{z'} + h(S_{z'}A_{z'x'}I_{x'} + S_{z'}A_{z'y'}I_{y'} + S_{z'}A_{z'z'}I_{z'}) \quad (2)$$

Then for $S = \tfrac{1}{2}$, one can show[4] that the first-order ENDOR frequencies are given by

$$v_{\pm} = |v_n \pm R/2| \quad (3)$$

where

$$R = (A_{z'x'}^2 + A_{z'y'}^2 + A_{z'z'}^2)^{1/2} \quad (4)$$

If the hyperfine splitting tensor is diagonal in the laboratory x', y', z' axes, then $R = A_{z'z'}$. R is an effective hyperfine splitting at a given orientation of the magnetic field with respect to the laboratory axes. However, to determine all the components of the hyperfine splitting tensor, equation (1) must be solved using a general set of orthogonal axes.

Normally, one chooses three orthogonal crystal axes, x, y, z. It is convenient to choose one of these axes to lie parallel to a crystallographic axis of the crystal. This is usually easy to do, as one of the predominant external edges of a crystal will generally lie parallel to a crystallographic axis. For a triclinic crystal, it is only possible to use one of the crystallographic axis as a reference axis, as the three triclinic axes are nonorthogonal to each other. However, a large number of organic crystals crystallize as monoclinic crystals where the crystallographic a and c axes are orthogonal to the crystallographic b axis; however, a is not orthogonal to c. Nevertheless, in this case, the x and y axes can be chosen to lie parallel to the a and b crystal axes, while z lies parallel to the c^* axis, c^* being an axis perpendicular to a and b that lies near the c crystal axis. Usually, the predominant external edges of the crystal will be parallel or perpendicular to the crystallographic edges. However, in general, the x, y, z reference axes will not be coincident with the symmetry axes of the radical in the crystal. In this case, the direction of the external magnetic field relative to the x, y, z axes is given by the direction cosines l_x, l_y, l_z. If equation (1) is solved using general x, y, z axes, then R will be given by

$$R = [l_x^2(A_{xx}^2 + A_{xy}^2 + A_{xz}^2) + l_y^2(A_{yx}^2 + A_{yy}^2 + A_{yz}^2) + l_z^2(A_{zx}^2 + A_{zy}^2 + A_{zz}^2)$$
$$+ 2l_x l_y(A_{xx}A_{xy} + A_{xy}A_{yy} + A_{zx}A_{zy}) + 2l_x l_z(A_{xx}A_{xz} + A_{xy}A_{zy} + A_{xz}A_{zz})$$
$$+ 2l_y l_z(A_{xy}A_{xz} + A_{yy}A_{yz} + A_{yz}A_{zz})]^{1/2} \tag{5}$$

which can be written in an alternate form as

$$R = [l_x^2 T_{xx} + l_y^2 T_{yy} + l_z^2 T_{zz} + 2l_x l_y T_{xy} + 2l_x l_z T_{xz} + 2l_y l_z T_{yz}]^{1/2} \tag{6}$$

where T_{ij} are the elements of a symmetric tensor that is the square of the hyperfine tensor **A**. This can be proven by performing the matrix multiplication **A** times **A**.

The T_{ij} elements are experimentally obtained by collecting ENDOR data in the xy, yz, and zx planes. This is accomplished in the following manner. A crystal is mounted so that z is perpendicular to H_0 and H_0 is parallel to the xy plane. ENDOR data are collected at angles θ, where θ is defined as the angle between the x axis and H_0 in the xy plane. For all orientations in this plane, the direction cosines (l_x, l_y, l_z) of H_0 equal $(\cos\theta, \sin\theta, 0)$, so equation (6) becomes

$$R(\theta) = (\cos^2\theta T_{xx} + \sin^2\theta T_{yy} + 2\sin\theta\cos\theta T_{xy})^{1/2} \tag{7}$$

Values for T_{xx}, T_{yy}, and T_{xy} can be obtained by noting that

$$T_{xx}^{1/2} = R\,(0°), \qquad T_{yy}^{1/2} = R\,(90°)$$

and by fitting T_{xy} to the data taken at angles other than 0° and 90°. Similarly, rotation about the x and y axes, where the direction cosines equal $(0, \cos\theta, \sin\theta)$ and $(\sin\theta, 0, \cos\theta)$, respectively, gives the other values of T_{ij}. Note that in the yz plane, the angle $\theta = 0°$ occurs parallel to the y axis, while in the zx plane, the angle $\theta = 0°$ occurs parallel to the z axis.

The resulting 3×3 matrix must then be rotated to a new coordinate system X, Y, Z in which it is diagonal. This is always possible, as the **T** matrix is symmetric, since $T_{xy} = T_{yx}$, etc. To determine the diagonal values, one must solve for the roots λ in the equation $\det|\mathbf{T} - \lambda\mathbf{1}| = 0$ where **1** is the unit matrix. Expansion gives a cubic equation in λ, the three roots of which are the principal values of the diagonal matrix in the X, Y, Z coordinate system. In the diagonal form, the values $T_{XX} = A_{XX}^2$, $T_{YY} = A_{YY}^2$, and $T_{ZZ} = A_{ZZ}^2$. Thus, by taking the square root of T_{XX}, T_{YY}, and T_{ZZ}, one can determine the principal values of A_{XX}, A_{YY}, and A_{ZZ} except for an ambiguity in sign. Since the hyperfine splittings for a number of different radicals have been determined, the correct sign can be assigned by analogy to those that have been determined. Exceptions do occur when small isotropic couplings exist.

The principal hyperfine splittings can be separated into its isotropic (A_{iso}) and anisotropic (A_i) components. The isotropic splitting is given by

$$A_{\text{iso}} = (A_{XX} + A_{YY} + A_{ZZ})/3 \qquad (8)$$

and the anisotropic components are given by

$$B_{XX} = A_{XX} - A_{\text{iso}}$$
$$B_{YY} = A_{YY} - A_{\text{iso}} \qquad (9)$$
$$B_{ZZ} = A_{ZZ} - A_{\text{iso}}$$

Inspection of equations (8) and (9) shows that the anisotropic components add to zero. This means that the dipolar interaction goes to zero when averaged over all directions with respect to the magnetic field as observed for radicals in liquids.

Normally, the principal values of the diagonal $\mathbf{T}$ matrix are determined by computer routine that computes a transformation matrix that describes the rotation necessary to transform the $\mathbf{T}$ matrix in the x, y, z coordinates to that in the X, Y, Z coordinates. This transformation matrix consists of the direction cosines of the principal X, Y, Z axes relative to the laboratory x, y, z axes. This can be stated mathematically for the kth principal value axis, where T_k is the kth principal value as

$$T_{ij} l_{kj} = T_k l_{ki} \qquad \text{for } i = x, y, z \text{ and } k = X, Y, Z \qquad (10)$$

where l_{kj} for $j = x, y, z$ are the direction cosines. The mathematical operation that is performed by the computer routine is given in equation (11):

$$
\begin{vmatrix} A_{xx} & & \\ & A_{yy} & \\ & & A_{zz} \end{vmatrix} =
\begin{vmatrix} l_{Xx} & l_{Xy} & l_{Xz} \\ l_{Yx} & l_{Yy} & l_{Yz} \\ l_{Zx} & l_{Zy} & l_{Zz} \end{vmatrix}
\begin{vmatrix} A_{xx} & A_{xy} & A_{xz} \\ A_{yx} & A_{yy} & A_{yz} \\ A_{zx} & A_{zy} & A_{zz} \end{vmatrix}
\begin{vmatrix} l_{xX} & l_{xY} & l_{xZ} \\ l_{yX} & l_{yY} & l_{yZ} \\ l_{zX} & l_{zY} & l_{zZ} \end{vmatrix} \qquad (11)
$$

where the principal values A_{XX}, A_{YY}, and A_{ZZ} appear as the diagonal elements of the matrix that results when the similarity transformation $\mathbf{L} A_{\text{und}} \mathbf{L}^{-1}$ is performed on the undiagonal $\mathbf{A}$ matrix. The elements of the $\mathbf{L}$ matrix are the direction cosines l_{ij} between axis i and j. Experimentally, this operation transforms the $\mathbf{A}$ matrix in the x, y, z coordinate system to that in the X, Y, Z coordinate system.

The direction cosines l_{ij} can be quite valuable, for they give the orientation of the principal axis of the radical relative to the external reference axes x, y, z. Since $x, y,$ and z are parallel or at least related to one of the crystallographic a, b, c^* axes, one can locate the position of the radical center relative to the molecule in the crystal if the crystal structure is known. This can be used to locate the particular nucleus in the crystal that enables a structure determination to be made of the radical.

2.2. ELDOR

As long as the hyperfine splittings are greater than 30 MHz but less than approximately 600 MHz, the experimental procedure for obtaining the principal hyperfine splitting tensor components for each nucleus from an ELDOR spectrum is identical to the method outline in Section 2.1 for ENDOR. In particular, the allowed–allowed ELDOR line position[4] at the apparent hyperfine splitting for each nucleus is plotted as a function of angle for each plane. In the ij plane, the values of T_{ii}, T_{jj}, and T_{ij} are evaluated according to equation (7) for the data in each plane. The resultant $\mathbf{T}$ matrix is diagonalized according to the methods outlined above, resulting in values for the principal hyperfine splittings and direction cosines. Keep in mind that such a procedure assumes that a first-order treatment is adequate for the radical under study. This is not always the case, as small second-order corrections can usually be detected for splittings larger than 30 MHz. Hence, if the hyperfine splittings are greater than approximately 600 MHz or smaller than 30 MHz, the first-order theory will not be adequate. The corrections that must be applied for large splittings are described in the next section.

When the hyperfine splittings for radicals are less than approximately 30 MHz, additional lines than those predicted in first order will be observed, since the EPR transitions that are forbidden according to a first-order treatment ($\Delta M_s = \pm 1$, $\Delta M = \pm 1$) become partially allowed.[5] This occurs because the magnitude of $2v_n$, where v_n is the nuclear Zeeman frequency at 3300 G, is approximately 27 MHz and cannot be neglected whenever the hyperfine splitting is of similar magnitude. Because of this, ELDOR spectra at X-band frequencies of radicals containing splittings less than approximately 30 MHz consist of both allowed–allowed and forbidden–allowed lines[4] (or combination lines, as they are sometimes referred to), which depend on both A and v_n. This results in complex spectra. In such cases, it is necessary to carry out an exact calculation such as was recently reported[5] for the HOOCCHĊHCOO⁻ in irradiated single crystals of potassium hydrogen maleate, as well as being examined in some detail in Chapter 6 of Kevan and Kispert.[4]

Even though complex ELDOR spectra are always observed for radicals in crystals where $A \simeq 2v_n$, forbidden–allowed ELDOR lines are also observed in some systems when $A > 2v_n$. In order to predict the positions of such lines without resorting to an exact calculation, it is practical to have an approximate expression that can be used to quickly assign the various allowed and forbidden ELDOR transitions when the hyperfine splittings are larger than 30 MHz. Such an expression is given by equation (12), which approximately simulates the ELDOR spectral lines of a radical in a crystal predicted from exact expressions derived in Kevan and Kispert[4] and Iwasaki et al.[5] This

Table 1

First-Order ELDOR Transitions for Two Unequal
Couplings $(I = \frac{1}{2})^a$

Forbidden–allowed	Allowed–allowed
$\frac{1}{2}A_1 \pm v_1$	
$\frac{1}{2}A_1 + \frac{1}{2}A_2 \pm v_1 \mp v_2$	A_1
$\frac{1}{2}A_1 + \frac{1}{2}A_2 \pm v_1 \pm v_2$	$A_1 + A_2$
$A_2 + \frac{1}{2}A_1 \pm v_1$	A_2
$\frac{1}{2}A_2 \pm v_2$	
$A_1 + \frac{1}{2}A_2 \mp v_2$	

^{a} Assuming at a given crystal orientation,

$E(m_s, m_I^1, m_I^2)$

$$= g\beta H m_s + A_1 m_s m_I^1 + A_2 m_s m_I^2 - v_1 m_I^1 - {}_2 m_I^2$$

this table was constructed assuming the observing position is set at the high field line and $v_p > v_o$. If $v_p < v_o$, then the observing position is set at the low-field position. For more than two unequal nuclei, the ELDOR transitions can be calculated for any observing position from equation (12).

expression is

$$E(m_s, m_i^1, m_i^2, \ldots, m_i^n) = v_e m_s + \sum_{i=1}^{n} A_i(\text{eff})m_s m_I^i - \sum_{i=1}^{n} v_n^i m_I^i \qquad (12)$$

where $v_e = gH/h$, $v_n^i = g_n \beta_n H/h$ for the ith nucleus, $A_i(\text{eff})$ is the hyperfine splitting observed at any given angle, and m_s and m_I^i have been defined previously. As an example, the approximate ELDOR transitions for two non-equivalent nuclei with $I = \frac{1}{2}$ are given in Table 1. Note that all forbidden–allowed transitions are dependent on v_n^i. However, keep in mind that accurate hyperfine tensors can not be calculated using this equation; rather, an approach like that given in Iwasaki *et al.*[5] should be used.

2.3. Second-Order Effects

The second-order effects on the hyperfine couplings measured by ENDOR vary as $A^2/4v_e$. Because of this, they can generally only be detected for $A \geq 30$ MHz. In liquids, splittings of this magnitude are seldom investigated by double resonance, as they are usually resolvable from the EPR spectrum. However, in solids, it is important to obtain accurate dipolar tensors from which dipole–dipole distances can be obtained. For instance, the relative orientation of the β protons with respect to the p_z orbital containing the unpaired electron in alkyl-type radicals can be deduced from the dipole–dipole tensors. However neglect of the second-order contributions can lead to errors as large as 1 MHz,[6] resulting in an error in the calculated dipolar distance. The inclusion of second-order effects has also been used to correctly identify some ENDOR transitions[7] and to explain the reason for the splitting of some ENDOR lines.[8–9] For example, the ENDOR lines arising from the α proton in the methyl radical produced in x-irradiated lithium acetate dihydrate crystals[9] show doublet splittings due to second-order effects when the magnetic field is positioned on the $m_I = \pm\frac{3}{2}$ lines. The exact expression for the methyl radical ENDOR frequencies can be found in Table III of Toriyama et al.[9]

So far, in determining hyperfine splittings from ELDOR spectra, we have ignored the second-order corrections to the ELDOR line positions. Generally, when a line position fit is made to equation (7), it is assumed for all practical purposes that such corrections are small. Indeed, in most cases of protonated radicals, a first-order perturbation treatment is adequate to interpret the frequency-swept ELDOR spectra, since the ELDOR linewidths are from 1 to 3 MHz and the second-order shift is of the order of 0.3 MHz. In these cases, the allowed–allowed ELDOR transitions at a given angle will be observed at $A_i(M_i^p - M_i^o)$, where M_i^p is the ith quantum number for the pumped line and M_i^o is the ith quantum number for the observed line at a position A_i. Then the use of equation (7) is appropriate. However, in some cases, second-order theory will be needed. For example, the hyperfine splitting constant for the six methyl protons of $(CH_3)_2\dot{C}COOH$ in irradiated α-aminoisobutyric acid is nearly isotropic and equal to 64 MHz.[10,11] To a good approximation, the second-order downfield shift for each EPR line from that of a first-order spectrum is $(A^2/2H_0)[I(I + 1) - M_I^2]$, where M_I is the nuclear quantum number for each line and I is the total quantum number. Thus, an ELDOR line will occur at $A + 5A^2/2H_0$ when the observing field is set on the resonance position of the high-field $M_I^0 = -3$ EPR line and the $M_I^p = -2$ line is pumped. For $A = 64$ MHz, this shift from the first-order-splitting A equals 1.3 MHz, a small but measurable quantity. Such a correction varies depending on the observed and pumped lines and has been given in Table 6-2 of Kevan and Kispert.[4] It should be noted that the first-order expression for $R(\theta)$, equation (7), can still be used for ELDOR spectra showing significant second-order effects, providing that the ELDOR line positions are initially shifted to a first-order position.

There are further difficulties in determining a hyperfine tensor correct to second order if a large hyperfine anisotropy is observed such as that found for fluorine in $\dot{C}FHCONH_2$. When this occurs, the second-order shift is also angularly dependent.[1] The procedure for handling such situations has been recently outlined[4] and will not be repeated here.

2.4. *Signs of Hyperfine Constants*

Only the absolute magnitudes are usually determined by either EPR, ENDOR, or ELDOR. Yet, the sign of the hyperfine splitting is important, since it is a measure of the sign of the unpaired electron spin density and ultimately the electronic structure of the radical.

Currently, the relative signs of different hyperfine constants can be determined by a double ENDOR experiment that involves two simultaneous rf fields. In the double ENDOR experiment, one electron transition and one nuclear transition are saturated. When a second rf field is applied to the sample, there will either be an enhancement or reduction of the observed ENDOR transition depending on whether the ENDOR transition that has been scanned is related to a coupling with the same sign or opposite. If the spectrum consists of two couplings with the same sign, then the resulting double resonant spectrum will consist of two positive peaks, while if the two couplings are of opposite signs, the spectrum will consist of one positive and one negative peak. Although this method only gives relative signs, absolute signs can be obtained by taking into account other structural evidence of the radical. A complete description of this method has been given elsewhere.[4]

A determination of the absolute signs is possible providing that the second-order shifts can be experimentally measured. In such cases, one carries out a calculation of the ENDOR frequencies using a Hamiltonian that includes not only the electronic and nuclear Zeeman terms as well as the hyperfine terms, but also includes the nuclear–nuclear dipolar terms. It turns out that the second-order splitting that is calculated in the absence of the nuclear dipolar term is shifted from that which results when the nuclear dipolar is included. This shift in the second-order splitting is dependent on the sign of the hyperfine coupling. Since the sign of the nuclear dipolar term is known, the absolute sign of the hyperfine coupling can be determined. This method is somewhat tedious to apply; however, it has been used to determine the absolute sign of the α- and β-proton splittings in the $CH_3\dot{C}HCOOH$ radical[12] as well as the absolute signs of the α-proton splittings for the $\dot{C}H_3$ radical.[9]

3. *Unique Role Played by ENDOR*

Because the ENDOR linewidth is limited by the dipole–dipole interaction between neighboring nuclei in crystals, typical ENDOR linewidths of 100 kHz and as narrow as 20 kHz are observed.[4,13] This represents a 100–1000-fold increase in the spectral resolution over that of the single-crystal EPR spectrum. For instance, most EPR lines of radicals in crystals are broadened by unresolved hyperfine interaction with surrounding magnetic nuclei. This produces an inhomogeneously broadened line whose linewidth is generally of the order of 10 MHz or more. As a result of the narrow ENDOR lines and increase in resolution, hyperfine splittings can be determined more precisely than from EPR measurements alone, and small splittings (< 5 MHz) can be detected. This has proven invaluable in identifying the reaction mechanism of radical formation, since the radical structures can be identified in detail, and has made the measurement of small quadrupole couplings possible.

3.1. *Determination of Small Splittings*

In Section 2.1, it was pointed out that an ENDOR spectrum consists of lines at $v_{\pm} = |v_n \pm R/2|$ for each interacting nuclei. Since v_n equals 14.4 MHz for protons in a magnetic field of 3390 G, the proton ENDOR spectrum will consist of lines centered about v_n, each with a linewidth typically 0.1 MHz. Providing that the ENDOR spectrum does not consist of several overlapping lines, hyperfine couplings (R) can be measured of the order of 0.1 MHz. In some special cases, proton–proton interactions as small as 20 kHz can be measured.[14]

In most instances, ENDOR measurements are restricted to splittings that occur between 15 and 0.1 MHz, as larger hyperfine splittings appear as resolved splittings in the EPR spectrum. The hyperfine splittings that occur between 15 and 0.1 MHz are generally weakly coupled protons or lattice protons that are dipolar coupled to the free radical. From these couplings, accurate dipole–dipole distance can be obtained from which the structure of the radical and the location of the radical in the host lattice can be deduced. A recent ENDOR study[9] of the methyl radicals in irradiated single crystals of lithium acetate dihydrate demonstrates the importance of accurately determining the small dipolar splittings. In that study, seven different proton couplings were observed, and the principal hyperfine splittings and direction cosines were deduced for each. The estimated error was less than ± 0.14 MHz and $\pm 2°$ for the hyperfine splittings and direction cosines, respectively. From these data, accurate estimates of the distances and directions from the methyl carbon to the neighboring protons were calculated and compared to the corresponding distances and directions obtained from the crystal structure.[9] From this comparison, a measure of the spin delocalization to the neighboring methyl protons could be deduced, as well as the location of the methyl radical in the lithium acetate dihydrate lattice. It was found that, within an error of ± 0.05 Å in distance and $2°$ in direction, the methyl radical retained nearly the same position as the methyl group in the undamaged crystal. In fact, it was estimated that the recoil displacement of the methyl group was less than 0.3 Å. In other ENDOR studies, location of lattice protons have enabled one to determine the reaction mechanisms of radical formation.[4] This will be dealt with in more detail in Section 3.2.

Although considerable attention has been given to the measurement of proton ENDOR because of its potential importance in biological problems and is thus the most highly developed, a difficulty can arise if ENDOR measurements are attempted for nuclei other than protons or fluorine-19. A basic requirement for successful ENDOR is that nuclear saturation be achieved with the available rf fields. Such a requirement is met if

$$\gamma_n^2 H_2^2 T_{1n} T_{2n} \geq 1$$

where H_2 is the amplitude of the rf magnetic field in the rotating frame, γ_n is the gyromagnetic ratio of the nucleus, T_{1n} is the nuclear spin–lattice relaxation time, and T_{2n} is the spin–spin lattice relaxation time. Since H_2 is usually limited to about 10 G, nuclear saturation may be hard to achieve for those nuclei with small γ_n. However, it is fortunate that the rf field at the nucleus is enhanced[3] if the hyperfine splitting is nonzero and of the order of 10 G or larger. This is because the isotropic hyperfine interaction for a system of electron spins and nuclear spins creates an electronic magnetic field at the nucleus due to the electronic magnetic moment. Thus, for nuclei with small magnetic moments (^{14}N, ^{13}C, ^{23}Na, ^{87}Rb, and ^2D), intense ENDOR spectra will be observed for those nuclei with isotropic hyperfine splittings around 10 G or more.

Nevertheless, splittings for nuclei with small γ_n and essentially no isotropic values can be observed by simply lowering the temperature to lengthen T_{1n} and thus increase the ability to saturate the spin system. For instance, the free nuclear frequency for ^{39}K for $H_0 = 3347$ G is 0.66 MHz. Pairs of ^{39}K ENDOR lines were observed in the range of 0.5–0.85 MHz for an F center in $K^{81}Br$ at 90°K. This was due to an isotropic splitting of 0.27 MHz and an anisotropic splitting of (0.022, −0.011, −0.011 MHz) for ^{39}K nuclei in the third shell surrounding the electron center. Somewhat similar splittings were also observed for ^{81}Br. In fact, the good ENDOR resolution permitted an assignment of all ENDOR lines to nuclei in the first eight shells surrounding an excess electron F center in an irradiated alkali halide.[15]

3.2. Determination of Quadrupole Couplings

Quadrupole interaction occurs in radicals with nuclei with $I > \frac{1}{2}$, such as chlorine, rubidium, nitrogen, sodium, and deuterium. If the quadrupole interaction is much smaller than the hyperfine splitting, the quadrupole interaction will not affect the EPR spectrum in first order. However, it will lead to additional splitting in the ENDOR spectra and enable the quadrupole interaction to be measured. If the quadrupole interaction is of the order of the hyperfine interaction, a second-order description is necessary to describe the EPR spectrum. In such a situation, EPR lines, forbidden in first order, become partially allowed, and the quadrupole interaction can be determined by comparing the spectra calculated using a hamiltonian that includes the quadrupole interaction Q. This is usually tedious[16–18] and can require a great deal of computer time, as all transitions (forbidden and allowed in first order) must be calculated and then compared to the observed EPR spectrum. In many cases, the poorly resolved EPR spectrum prevents a unique solution for the quadrupole tensor from being derived. A much more straightforward measure of Q is by ENDOR, where only a few highly resolved lines are observed.

For illustrative purposes, we examine the simple cases where $I = 1$ (usually nitrogen and deuterium), the hyperfine (**A**) and quadrupole (**Q**) tensors have the same principal axes, $A \gg Q$ and the magnetic field is aligned along a principal axis (A). Then the four ENDOR frequencies observed while positioned on the $m_I = 0$ EPR line are given by[4]

$$v(\text{ENDOR}) = \tfrac{1}{2}A_{zz} \pm v_n \pm Q_{zz} \tag{13}$$

On the other hand, only two lines,

$$v_1 = |\tfrac{1}{2}A_{zz} + v_n + Q_{zz}| \quad \text{and} \quad v_4 = |\tfrac{1}{2}A_{zz} - v_n - Q_{zz}| \tag{14}$$

are observed for the $m_I = -1$ EPR line, and two lines,

$$v_2 = |\tfrac{1}{2}A_{zz} + v_n - Q_{zz}| \quad \text{and} \quad v_3 = |\tfrac{1}{2}A_{zz} - v_n + Q_{zz}| \tag{15}$$

are observed for the $m_I = +1$ EPR line. The reason for this is shown in Figure 1. When either the high- or low-field EPR line is observed, only transitions v_1 and v_4 or v_2 and v_3 have a level in common with observed EPR lines and thus are detected. However, all four ENDOR transitions have a level in common with the $m_I = 0$ EPR transition. Figure 1 also shows that if there is negligible quadrupole interaction, only two ENDOR transitions are observed for each line.

From inspection of the expressions for the ENDOR lines, it is seen that the average of the ENDOR frequencies associated with the $m_I = -1$ or the $m_I = +1$ lines depends only on the hyperfine coupling. This provides a convenient and handy way to obtain the hyperfine tensor.[19] However, this is true only if the quadrupole and hyperfine tensors have the same principal axes. Unfortunately, this is generally not true,[17–18] although it is often assumed to be. General methods for obtaining quadrupole and hyperfine tensors independent of any assumption regarding the direction of the principal axes have been described[17,18] for the case where the quadrupole interaction is smaller than the hyperfine interaction. Typically, the frequency position of each ENDOR line is recorded at different angles of rotation of the crystal in each of three orthogonal planes. The resulting data is then fit to

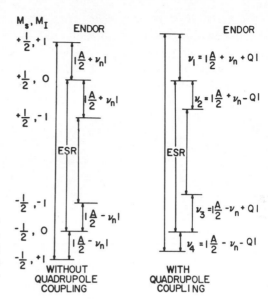

Figure 1. Energy level scheme for $S = \frac{1}{2}$, $I = 1$ system showing the ENDOR and EPR transitions in the absence and presence of quadrupole interactions. (From Kevan and Kispert.[4])

equations previously described,[17] which are functions of **g**, **A**, and **Q** tensors. Computational analyses give rise to the diagonal values of the **g**, **A**, and **Q** tensors.

Numerous examples have been reported for nitrogen quadrupole coupling tensors such as nitrogen atoms in diamond,[20] copper-8-hydroxyquinolate substituted into single crystals of phthalimide, and of 8-hydroxyquinoline[21] and NO_2^{2-} and NO_2.[19] The Cl quadrupole coupling ($I = \frac{3}{2}$) has been detected in KCl crystals[22] and the Na quadrupole coupling ($I = \frac{3}{2}$) with CO_2^- in sodium formate crystals.[16]

3.3. Reaction Mechanism of Radical Formation

The high resolution of ENDOR can be used to directly study radical formation and radical transformation mechanisms. In a large number of EPR studies of irradiated carboxylic acids and amino acid crystals,[23] the lower resolution of EPR restricts studies to radicals with resolvable splittings, such as exhibited by α and β protons. However, little is learned about the presence or absence of protons that exhibit small splittings (≤ 10 MHz), such as the OH proton on a COOH group. Unfortunately, the absence or presence of such protons is of utmost importance in elucidating the mechanism of radical formation, as proton transfer plays one of the most important roles in such formation.[24–28]

As an example, an elegant ENDOR study by Muto and Iwasaki[24] on irradiated L-alanine has shown that the particular proton transferred from a neighboring molecule can be identified. The crystal structure of L-alanine shows that the two oxygen atoms of the carboxyl ion form three different hydrogen bonds with the neighboring NH_3^+ protons. In the undamaged crystal, the protons H_2' and H_3' depicted in structure I, lie near the COO^- plane, while H_1' forms a hydrogen bond in a direction not in the COO^- plane.

$$CH_3-\underset{\underset{H_3}{\overset{+}{N}}-H_2}{\overset{\overset{H_1}{|}}{\overset{|}{C}}}-C\overset{O\cdots H_1'}{\underset{O\cdots H_2'}{\overset{-}{\diagdown}}}\quad H_3'$$

(I)

After irradiation at 77°K, the structure of the predominant radical can be indentified by determining the superhyperfine EPR couplings of all the hydrogen-bonded protons in the neighboring NH_3^+ group. Its structure has been deduced from ENDOR measurements to be

$$CH_3-\underset{\underset{H_3}{\overset{+}{N}}-H_2}{\overset{\overset{H_1}{|}}{\overset{|}{C_2}}}-\dot{C_1}\overset{O-H_1'}{\underset{\underset{H_3'}{O^-}}{\diagdown}}\quad H_2'$$

(II)

where the H_1' proton has been transferred to the radical anion across the hydrogen bridge and trapped in a position that is not coplanar with the COO^- plane. The protons H_2' and H_3' remain in the neighboring NH_3^+ group.

The details of this structure elucidation have been previously reviewed[4] and will not be repeated here. However, it serves as an excellent example of how the reaction mechanisms of radical formation can be deduced from ENDOR measurements. Motivated by the success of such an experiment, ENDOR studies have also been used to show that specific hydrogen-bonded protons are transferred to form neutral radicals in other systems, including L-valine·HCl,[26] glycine,[27] α-aminoisobutyric acid,[27] succinic acid,[25] and imidazole.[28] In most cases, the proton transfer is stereospecific, except where the transferred proton can occupy either of two geometrical positions in the neutral radical as was shown to occur for irradiated imidazole.[28]

3.4. *Distant ENDOR*

It is possible to carry out NMR experiments by ENDOR spectroscopy on the neutral molecules in a crystal with a greater sensitivity (typically 10^2 to 10^3) than is possible by direct detection using NMR spectroscopy. To carry out such an experiment, paramagnetic centers are introduced into a single crystal by either doping or exposure to high-energy radiation. Then the electron spin system is saturated with the available microwave power, which leads to spin polarization of the surrounding nuclei that have only weak isotropic and/or anisotropic hyperfine interaction with the electrons. The mechanism by which this polarization occurs will not be described here, but the reader is referred to a recent text[4] that adequately describes the mechanism. This nuclear polarization can then be transferred by nuclear spin diffusion to noninteracting "distant" nuclei in a time of order T_{1n} or somewhat shorter. If a saturating rf field such as employed in an ENDOR experiment is applied, the "distant" nuclei absorb their NMR frequencies and depolarize. This depolarization is transmitted back via a reverse path to the electron spins, where it causes a change in the electron spin level populations, which in turn results in an ENDOR response.

Both deuteron and proton distant ENDOR have been detected in partially deuterated 1,1-cyclobutanedicarboxylic acid, $(CH_2)_3C(COOD)_2$,[29] and in deuterated malonic acid, $CD_2(COOD)_2$.[30] From such studies, the location of the hydrogen-bonded carboxylic acid bridges and the conformation of the $(CH_2)_3C(COOD)_2$ molecules were determined. The deuteron spectra are the most interesting, as the deuteron distant ENDOR linewidth is as narrow as 2.5 kHz. This enables the quadrupole coupling tensors to be determined for different deuterons. An example of such spectra is given in Figure 2 for the $(CH_2)_3C(COOD)_2$ crystals. The spectra are centered about

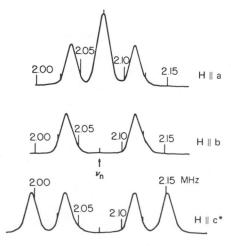

Figure 2. Deuteron distant ENDOR spectra at 4.2°K in single crystals of $(CH_2)_3$-$C(COOD)_2$ with the magnetic field parallel to orthogonal crystal axes c^*, b, and a. ν_n designates the free deuteron frequency of 2.07 MHz. (Adapted from Dalton and Kwiram.[29])

the free deuteron frequency of 2.07 MHz and show only nuclear quadrupole splitting, there being no hyperfine splitting, as these spectra are effectively pure NMR spectra. The deuteron quadrupole coupling is angularly dependent so two pairs of lines corresponding to the two different quadrupole splitting for $H_0 \| c^*$ changes to two equivalent splittings for the $H_0 \| b$ orientation. Poorer resolution is observed for proton distant ENDOR, which occurs near the free proton frequency. From this data, one can determine the proton–proton dipole tensor and the distance between two protons. One difficulty that does arise with detecting "distant" ENDOR is that the signal response time is of the order of the nuclear spin–lattice relaxation time. At 400°K, this appears typically to be ≥ 1 sec. Because of this, magnetic field and rf modulation normally employed in commercial ENDOR spectrometers cannot be used.

4. *Unique Role Played by ELDOR*

Spin exchange between different spin systems is one important relaxation mechanism that can drastically affect both the ENDOR and ELDOR responses. In fact, strong spin exchange (which occurs at high radical concentration) can reduce the ENDOR signal intensity to zero in the absence of strong T_{x1} processes,[4] while on the other hand, the ELDOR intensity increases with radical concentration. The fact that strong spin exchange results in a favorable increase in ELDOR spectral intensity permits the study of radical inhomogeneity (cluster formation) in irradiated crystals.

Intense ELDOR spectra can also occur for radicals with small T_{1n} ($< 10^{-4}$ sec). Thus, ELDOR studies are favored for radicals with large hyperfine splittings, as T_{1n} appears to decrease as the hyperfine splitting increases.[31] As T_{1n} decreases, it becomes more difficult to obtain an ENDOR spectrum. In addition, most ENDOR spectrometers do not operate above 100–200 MHz. Thus, ELDOR is probably the method of choice for studying spin exchange that occurs for triplet excitons and radicals containing large anisotropic hyperfine splittings such as α-fluorines. In addition, radicals that possess large quadrupole interactions, such as those that contain α-chloro or α-bromo nuclei, are ideally suited for ELDOR study, as the quadrupole relaxation can be orders of magnitude faster than the relaxation arising from magnetic dipole–dipole interactions[32,33] or T_{1n} processes. Most ELDOR spectra considered have been recorded at 100 kHz modulation.

4.1. *Spin Exchange Processes*

The potential of using ELDOR to study spin exchange processes in solids has recently been demonstrated in a study of methyl radical production at

$77°K$ as a function of metal cation[34] in irradiated acetate crystals and powders. During an ELDOR study of the rapidly reorienting methyl radical in various irradiated acetates, it was observed that instead of the expected enhanced lines,[10] surprisingly intense reduced-field-swept ELDOR lines were observed that varied from $R = 100\%$ for irradiated magnesium acetate tetrahydrate to near $R = 0\%$ for irradiated calcium acetate monohydrate.

By measuring the characteristic relaxation time $(T_{1e} T_{2e})^{1/2}$ for the methyl radical in the various acetates, it was found that the radiation-produced methyl radicals are trapped in clusters at low radiation dose (forming a nonuniform spatial distribution) to a greater degree in irradiated magnesium acetate tetrahydrate than in any of the other acetates under study. Assuming that the intense field-swept ELDOR reduction factors were due to an intermolecular relaxation process between the radicals in each cluster, it was concluded that the tendency for radicals to be trapped in clusters in irradiated acetates is dependent on the cation and decreases in the approximate order $Mg^{2+} > K^+ > Na^+ > Ca^{2+}$. Furthermore, this feature suggests that the radical formation mechanism is ionic in nature and that the metal cation plays an important role in the early stages of the radical formation process. It appears from this example alone that ELDOR measurements can be used extensively in the study of interradical processes in solids.

4.2. Determination of Large Splittings

Successful ELDOR studies have been carried out[35] for the $\dot{C}FHCONH_2$ and $\dot{C}F_2CONH_2$ radicals in irradiated fluoroacetamide and trifluoroacetamide single crystals where the fluorine hyperfine splitting varies from 0 to 560 MHz with angle between the principal axis and magnetic field. Such studies have shown that the relaxation times are highly dependent on the hyperfine anisotropy and that ELDOR measurements should permit an examination of the dominant relaxation mechanism. On the other hand, the large ELDOR intensity variation with fluorine hyperfine splitting increases the sensitivity of the ELDOR intensity to the presence of molecular motion. Because of this, it was possible to detect the libration of the CF_2 group as well as the torsional oscillation of the CF_2 group about the $C-C$ bond for $\dot{C}F_2CONH_2$.

ELDOR studies have also been carried out for triplet excitons,[36] where it was possible to detect triplet–triplet collisions at lower frequencies and thus at lower temperatures than by EPR. It was also possible to measure the exchange frequency as a function of hyperfine splitting. This suggests that ELDOR studies of systems that show large hyperfine anisotropy can give important information on anisotropic processes.

4.3. *Identification of Radicals with Large Quadrupole Interactions*

An ELDOR study of $\dot{C}ClFCONH_2$ has shown that the chlorine quadrupole-induced relaxation is dominant over relaxations arising from magnetic dipole–dipole interactions, hyperfine splitting anisotropy, and intramolecular rotational motion.[32] Furthermore, it was also shown that ELDOR transitions not containing the quadrupole term are ELDOR active and that no ELDOR lines are observed due to hyperfine interactions of fluorine, nitrogen, or hydrogen for $\dot{C}ClF_2CONH_2$. In fact, from studies so far reported, nuclei such as nitrogen-14 or deuterium, which exhibit smaller quadrupole interactions, do not exhibit such dominant effects as that of chlorine,[4] and so ^{14}N or D ELDOR lines are rather difficult to observe. However, since the chlorine nucleus does exhibit these dominant effects, it has been possible to identify the presence of $\dot{C}Cl_3$ and $\dot{C}Cl_2CONH_2$ in irradiated trichloroacetamide.[37] Also, since one of the ELDOR lines occurs at $A(Cl)$ it has been possible to obtain one of the chlorine hyperfine splittings directly[32] without recourse to extensive computer calculation to simulate the complex EPR pattern, which consists of forbidden and allowed lines from both isotopes of chlorine.

5. *Complementary Role Played by ENDOR and ELDOR*

In a number of areas, ENDOR and ELDOR measurements can complement each other. As has been stressed in this and earlier chapters, ENDOR measurements are easier to perform if T_{1n} is long ($\geq 10^{-4}$ sec), while ELDOR measurements require T_{1n}, T_{1x}, or T_{ss} times to be shorter ($< 10^{-4}$ sec). Thus, there are temperature ranges over which ELDOR signals can be observed for which no ENDOR signals have been recorded. In addition, the two techniques rely on somewhat different relaxation processes. To emphasize this point, selected examples will be discussed in four different areas. They are the application of ENDOR and ELDOR to the separation of overlapping EPR spectra, the study of methyl groups undergoing tunneling rotation, the study of proton–deuteron exchange, and the determination of intramolecular motion.

5.1. *Separation of Overlapping Spectra*

The appearance of overlapping spectra is rather common in single crystals. First, for a given orientation of the crystal in a magnetic field, there will generally be two or more magnetically inequivalent radical orientations of two chemically identical radicals. Second, if the EPR study concerns the study of radicals produced in irradiated crystals, there will be at least two different primary radicals formed. These correspond to the cation (oxidized form) and the anion (reduced form), providing the undamaged molecule was neutral. There are a number of different approaches to separating EPR spectra. Although no hard and fast rules apply in each case, examples will be given that demonstrate the different approaches that might be used.

5.1.1. *ENDOR of Nonoverlapped Line*

One of the most usual ways of separating overlapping EPR spectra is to rotate the crystal so that some of the high-field EPR lines do not overlap one another. An ENDOR spectrum is recorded for each line. As long as this EPR line is not made up of a combination of lines, the ENDOR spectrum will correspond to just one radical. Identification of other radicals can be carried out by positioning the magnetic field on other nonoverlapped EPR lines. For example, this has been used successfully in the separation of the overlapping EPR spectra in irradiated crystals of *N*-methylcytosine,[38] as well as being able to identify the four different radicals in x-irradiated succinic acid,[39] in addition to determining the direction cosines and the hyperfine tensors for each.

5.1.2. *ENDOR-Induced EPR*

In some cases, the EPR spectrum is so complex that there are no non-overlapped lines of one of the radicals. In such situations, another approach, commonly called ENDOR-induced EPR, can be useful in separating the spectral components of each radical. In this method, one first records the ENDOR spectrum for an overlapped EPR line. Next, the intensity of one of these ENDOR lines is recorded by setting the rf oscillator to that ENDOR frequency while the magnetic field is swept. For each ENDOR line monitored, an ENDOR-induced EPR spectrum will be observed. If all the ENDOR lines belong to one radical, then the same ENDOR-induced EPR spectrum occurs for all monitored ENDOR frequencies. On the other hand, if all the ENDOR frequencies do not belong to the same radical, then a different ENDOR-induced EPR spectrum will result. Such a technique has been used very successfully to separate the overlapping spectra of radicals in solution.[40] In crystals, the method is not only important for separating overlapping spectra due to chemically different radicals, but also in assigning the ENDOR frequencies that belong to different magnetically inequivalent sites of a radical. Such a technique was used quite successfully to assign the ENDOR lines corresponding to magnetically inequivalent sites for nitrogen atoms in diamond crystals.[20,41]

5.1.3. ENDOR-Induced EPR Spectra for Radicals Containing Quadrupole Moments

When radicals contain a quadrupole moment, the ENDOR-induced EPR spectrum recorded will only consist of a portion of the EPR spectral lines instead of the full EPR spectrum. This is because, in the absence of a quadrupole interaction, each ENDOR transition ($|\frac{1}{2}A + v_n|$ or $|\frac{1}{2}A - v_n|$) has a level in common with each EPR transition, as shown in Figure 1 for an $S = \frac{1}{2}$ and $I = 1$ system. Thus, the ENDOR-induced EPR spectrum recorded will appear similar to the spectrum of one radical. However, when a quadrupole interaction is present, this is not the case, as four ENDOR frequencies (labeled v_1, v_2, v_3, and v_4) instead of two are observed. For instance, if ENDOR frequency v_1 for one radical is monitored, then only the $m_I = +1$ and 0 EPR transitions will be observed and not all three EPR transitions, which might have been expected. However, it should be noted that if ENDOR frequency v_2 (lower frequency transition) for the same radical site is pumped, the $m_I = -1$ and 0 EPR transitions will be observed, i.e., the central EPR line and a line at high field will appear. Such a spectrum is given in Figure 3 for one of the four magnetically nonequivalent sites of nitrogen in diamond[41] where two low-field lines (Figure 3b) and two high-field lines (Figure 3c) are observed as two different ENDOR frequencies are monitored. To carry out an ENDOR-induced EPR measurement, it is useful to overmodulate the EPR spectrum when recording the initial ENDOR spectrum so that contributions from all the radicals appear with their maximum intensity in the ENDOR spectrum.

5.1.4. Double ENDOR

The ENDOR-induced EPR effect can also be used in a somewhat different manner to assign the EPR lines of an overlapping spectrum. This is to perform a double ENDOR experiment where one monitors one ENDOR transition with one rf frequency and then pumps the other ENDOR transitions with a second rf frequency. If all the pumped ENDOR frequencies cause a change in the monitored ENDOR frequency, then the ENDOR transitions all arise from the same radical. Hampton and Alexander[42] used this method to study the radicals produced in x-irradiated cytidine single crystal at room temperature. They found three proton coupling constants and showed by double ENDOR that they all have the same sign.

One does need to be cautioned about trying to carry out double ENDOR experiments, as they do not always work. Basically, there needs to be the presence of a transient ENDOR signal that can occur for radicals with long relaxation times. Further details have been previously given.[4]

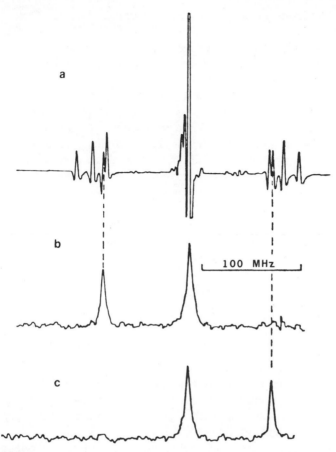

Figure 3. (a) The EPR spectrum of the four resolved nitrogen sites in diamond. (b) The ENDOR-induced EPR spectrum with the rf equal to 44.948 MHz. (c) The ENDOR-induced EPR spectrum with the rf equal to 46.665 MHz. Q is negative and thus opposite in sign to that assumed in Figure 1. (From Cook.[41])

5.1.5. ELDOR

Although overlapping EPR spectra can be separated by various ENDOR techniques, it is generally difficult to separate EPR spectra when intermolecular spin–spin exchange is a dominant relaxation mechanism or when a radical substituent undergoes intramolecular motion with a correlation time that competes with the nuclear Zeeman frequency. The presence of such intramolecular motion tends to shorten T_{1n}, making it more difficult to saturate the nuclear transition with the available rf power. Yet, shortening T_{1n} gives rise to more intense ELDOR spectra, especially if an END-type relaxation mechanism (see Chapter 3) is a significant mode of relaxation as the ELDOR signal intensity is proportional to T_{1e}/T_{1n}. In addition, the presence of significant intermolecular spin–spin relaxation tends to reduce the ENDOR signal while it increases the ELDOR signal intensity. This was demonstrated in Section 4.1.

Further complications can also arise whenever a radical substituent undergoes intramolecular motion. As the temperature is raised, nonequivalent nuclei become equivalent. When this occurs, some of the rigid-lattice low-temperature EPR lines broaden and disappear, being replaced by narrow lines at higher temperatures. It has been difficult to obtain ENDOR over a temperature range where radicals exhibit such behavior (see Chapter 4). However, this is not the case for ELDOR spectra. In fact, the spectral broadening can be used to an advantage.

A recent ELDOR study of the $\dot{C}H_2COO$ radical in irradiated zinc acetate has shown[43] that no ELDOR spectral lines will occur from the central two broad lines when the observing position is set on the high-field line and the following condition is met:

$$\Delta\omega\tau_c \simeq 2 \tag{16}$$

where τ_c is the correlation time of the motion which the CH_2 group undergoes and $\Delta\omega$ is the frequency difference between the central lines of the rigid lattice.[44–45] Techniques have been described[43] in which τ_c can be estimated as a function of temperature from ELDOR data taken at the slow exchange limit (when two central EPR lines are present) or from ELDOR data taken in the rapidly rotating limit. Thus, for most practical purposes, a good estimate of τ_c as a function of temperature is obtainable for any radical undergoing molecular motion.

This feature can be used to separate completely overlapped EPR spectra[43] by taking advantage of the probable fact that neither the low temperature separation of its inner lines nor its correlation time will be identical for both radicals. Thus, it only requires that an orientation be selected for one radical so that the condition $\Delta\omega\tau_c \simeq 2$ occurs at a temperature where narrow inner lines are observed for the second radical. In this case, only ELDOR lines from the radical with the narrow lines (i.e., for that one for which $\Delta\omega\tau_c = 2$) will be observed. This method has been used successfully to separate the overlapped EPR spectra of $\dot{C}H_2COO^-$ and $NH_3^+\dot{C}HCOO^-$ in irradiated glycine single crystals[43] under conditions of significant spin–spin relaxation, the absence of nonoverlapped lines, and over a range of temperatures where large spectral changes occur. Experiments of this type suggest that an ENDOR experiment might be done at low temperatures to identify the structure of radicals where the relaxation conditions favor ENDOR observation and an ELDOR experiment at higher temperatures to study the dynamic processes involved. An example of such a study is given in Section 5.3.

5.2. Methyl Groups Undergoing Tunneling Rotation

In numerous examples,[46-49] methyl groups attached to radicals of the type $CH_3-\dot{C}R_1R_2$ undergo quantum mechanical tunneling rotation at low temperature instead of a freezing out of the classic rotational motion of the methyl group. Tunneling rotation occurs when two successive proton pair exchanges correspond to rotation of the methyl group through $2\pi/3$. Tunneling rotation, unlike classic rotation, is spin dependent because it is correlated with exchanges of proton spin states. Thus, the tunneling rotational motion of the methyl group affects the energies of the spin states and appears as an additional term in the spin Hamiltonian.[47a,b] The energy diagram that results (Figure 4) is composed of the usual diagram to which has been added the E and A rotational energy states split by the tunneling splitting $3J$. Furthermore, this results in a eight-line EPR spectrum, as shown in the right-hand portion of Figure 4.

Although the study of the energy levels and the resulting EPR spectrum as a function of the magnitude of the tunneling splitting has been the purpose of numerous articles, it is necessary to note here that one of the important contributions of ENDOR measurements to studies of this type has been the ability to measure the tunneling frequency when the tunneling splitting $3J$ is larger than the proton hyperfine splitting. For example, the four high-frequency $\Delta m = \pm 1$ ENDOR transitions v_1, v_2, v_3, and v_4 shown in Figure 4 are equal to first order.

However, the second-order shift is significant and lowers the center two levels in the lowest quartet of energy levels to make $v_1 = v_2 = v_3$. Note that the EPR frequencies and the ENDOR frequencies v_2 and v_4 are not affected by the second-order shift. However, ENDOR frequencies v_1 and v_3 contain information about J in second order according to the relationship

$$v_1 = v_n + \frac{1}{6} A_{zz}^A + \frac{|A_{zz}^E|^2}{216J}$$

where the tensor components A_{zz} are determined independently, and thus, $3J$, the tunneling magnitude, can be measured.[46]

However, when the tunneling and hyperfine splittings are nearly equal, the EPR spectra may give information on $3J$. An example of this occurred for the $CH_3\dot{C}(COOH)_2$ radical in irradiated methylmalonic acid,[47b] where $3J = 192$ MHz and $A \simeq 65$ MHz. Calculations for this system showed that when $3J \simeq A$ (the hyperfine splitting), weak satellite lines appear symmetrically displaced about the central allowed EPR lines at lower and higher magnetic fields than the seven main lines of the tunneling EPR spectrum. Unfortunately, these satellite lines are obscured by the spectrum of a more intense second radical, $CH_3\dot{C}HCOOH$ (indicated by letter I in the lower portion of Figure 5), occurring as a result of the irradiation of methylma-

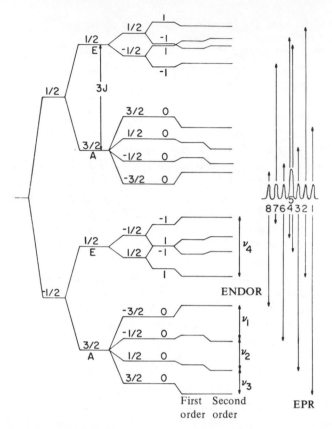

Figure 4. Schematic energy level diagram for a methyl group undergoing tunneling rotation where the classic rotation is frozen out. The energy diagram corresponds from left to right to the electron Zeeman splitting, the tunneling splitting $3J$, the hyperfine splitting of the A rotational state, the hyperfine splitting of the E rotational state, the nuclear Zeeman shift, and the second-order shift. The ENDOR transitions are given by ν_1, ν_2, ν_3, and ν_4. The EPR spectrum shown to the right consists of transitions 1, 3, 6, and 8 associated with the A rotational state and the others with the E state. (From Kevan and Kispert.[4])

Ionic acid. Annealing techniques only partially resolved these satellite lines (labeled a, b, c, and d) shown under high gain in the upper portion of Figure 5. In this case, ELDOR studies[50] played a complementary role to the ENDOR studies in enabling a resolution of these satellite lines by methods outlined in Section 5.1. These studies also enabled a qualitative measure of the frequency of the classic rotation and the quantum mechanical hopping rate as well as leading to a confirmation of the assignment of the satellite lines. Since this example shows the potential importance of ELDOR measurements to the study of radicals in irradiated material, we will outline the details of that study here.

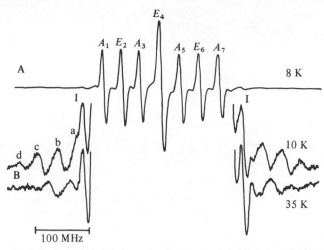

Figure 5. The first-derivative EPR absorption spectrum at (A) low gain and (B) high gain of irradiated methylmalonic acid crystal mounted with the magnetic field parallel to one of the crystal axes. The main lines are labeled according to symmetry and line position. The weak sidebands a, b, c, and d are forbidden transitions. 100-kHz field modulation was used throughout. (From Mottley et al.[50])

For a four-level system, if the relaxation processes are dominated by nuclear spin-flip probabilities W_n, the ELDOR intensity (ignoring the effects complicated by the field modulation frequency for the time being, see Chapter 5) varies proportionally to $W_n/(2W_e + W_n)$ when an allowed EPR line is observed and an allowed EPR line is pumped. On the other hand, if significant W_{x1} and W_{x2} processes occur such as expected when the reorientational rate of the methyl group is competitive with the electron Zeeman frequency and simultaneous electron and nuclear flips are possible, then the ELDOR intensity varies proportionally to $-W_{x1}W_{x2}/[W_e(W_{x1} + W_{x2}) + W_{x1}W_{x2}]$, and thus, enhanced ELDOR spectra will result. For the case of the methyl group undergoing tunneling rotation, the ELDOR intensity varies according to $W_n/(W_e + W_n)$ when an allowed line is observed and a satellite line is pumped. Accurate values for W_e, W_n, W_{x1}, and W_{x2} are not known. However, for the sake of discussion here, we will set the methyl group reorientational rate equal to W_n except at large reorientational rates ($> 10^8$), where it will equal W_{x1}, and $6W_{x1}$ will equal W_{x2}. Since $W_e = 1/T_{1e}$, we will assume T_{1e} data measured for the radicals in irradiated carboxylic acids[51] to be approximately correct. The resulting estimated data is plotted in Figure 6 as a function of $\ln(1/\tau)$ versus $1/T$, where τ is the correlation time of the methyl group reorientation and T is the absolute temperature. Within the inner rectangle, the reorientation correlation time is accurately known from line-shape measurements. In the outer rectangle, the classic reorientational correlation time and the $A \to E$ and $E' \to E''$ transition times have

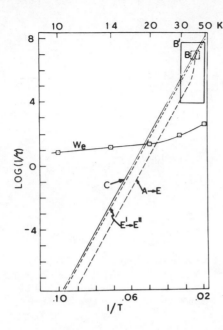

Figure 6. A plot of $\log(1/\tau)$ versus $1/T$, where τ is the correlation time for C, classic rotation; $E' \to E''$, tunneling rotation from symmetry state E' to E'' and $A \to E$, the tunneling rotation from state A to E. T is the absolute temperature. W_e is the spin–lattice relaxation probability and $1/\tau = W$, the relaxation probabilities for the processes mentioned.

been measured by methods other than line-shape measurements.[46–49] Outside the rectangular area, the respective correlation times are assumed to have the same exponential dependence as found within the rectangular areas.

There are certain features of Figure 6 that should be noted. First, at 50°K or higher, enhanced ELDOR spectra should be observed, as the reorientational rate is competitive with the electron Zeeman frequency. In fact, the ELDOR spectrum obtained with the observing condition set at the crossing point of line $A_7(-\frac{3}{2}A)$ and pumping the low-field portion of the EPR (Figure 7) shows enhanced ELDOR lines at 68 MHz. Furthermore, as the temperature

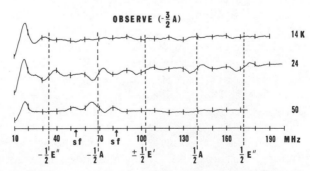

Figure 7. ELDOR spectra obtained as a function of temperature with the observing condition set on the crossing point of line $A_7(-\frac{3}{2}A)$ of Figure 5 and pumping the low-field part of the EPR spectrum. Spin-flip line positions (sf) are indicated by arrows. (From Mottley et al.[50])

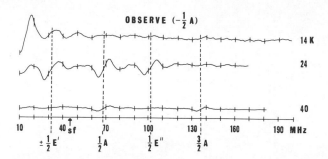

Figure 8. ELDOR spectra obtained as a function of temperature with the observing conditions set on the crossing point of line $A_5(-\frac{1}{2}A)$ of Figure 5 and pumping the low-field part of the EPR spectrum. Spin-flip line positions (sf) are indicated by arrows. (From Mottley *et al.*[50])

is lowered, the phase of the line at 68 MHz changes to that of a reduced ELDOR spectrum as is predicted if W_n decreases. Note also the near lack of ELDOR intensity at 14°K. This is predicted from Figure 6 as the reorientation rate, for $A \rightarrow E$, $W_n \ll W_e$ and thus the ELDOR intensity approaches zero, since $W_n/(2W_e + W_n) \rightarrow 0$.

Other features of the relaxation probabilities given in Figure 6 are also demonstrated by setting the observer at the crossing point of line $E_6(-\frac{1}{2}E'')$ (Figure 8), and line $A_5(-\frac{1}{2}A)$ (Figure 9) and pumping the low-field part of the spectrum. In contrast to the absence of ELDOR lines at 14°K in Figure 6, ELDOR lines at 68 MHz do appear at 14°K in Figure 8 when the pumped and observed transitions have the same value of m. This is because the $E' \rightarrow E''$ rate is of an order of magnitude greater than the $A \rightarrow E$ rates at 14°K, enabling ELDOR to be observed while observing EPR lines of E symmetry. In addition, only $m = \frac{1}{2}$ to $m = -\frac{1}{2}$ ELDOR transitions are observed at 24°K, while at 40°K, transitions occur for all values of m. Also not shown is the fact that ELDOR spectra are unobservable below 12°K. This variation in the

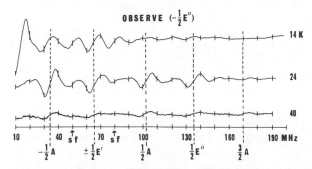

Figure 9. ELDOR spectra obtained as a function of temperature with the observing condition set on line $E_6(-\frac{1}{2}E'')$ of Figure 5 and pumping the low-field part of the EPR spectrum. Spin-flip line positions (sf) are indicated by arrows. (From Mottley *et al.*[50])

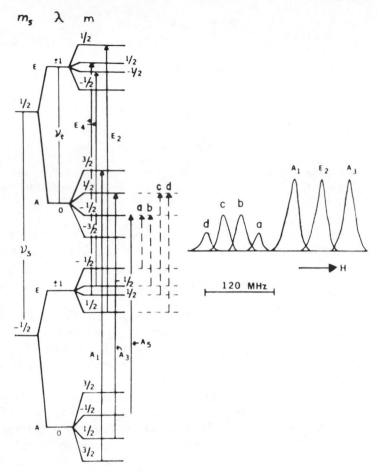

Figure 10. The energy-level diagram given in Figure 4 with the main EPR transitions A_1, E_2, E_4, A_3, and A_5 indicated by solid lines and the sideband transitions a, b, c, and d indicated by dotted lines. To the right of the energy diagram is given the low-field half of the absorption EPR spectrum for $CH_3\dot{C}(COOH)_2$ at $8°K$, where the sideband intensity is greatly enhanced relative to the main lines for clarity.

observed ELDOR intensity appears to reflect the fact that the reorientational rate W_n varies as a function of temperature. In other words, below $12°K$, $W_n \ll W_e$ for all transitions; at $14°K$, $W_n < W_e$; at $24°K$, $W_n \simeq W_e$; at $40°K$, $W_n \simeq$ hyperfine splittings; and at $50°K$. $W_n \simeq$ electron Larmor rate.

Another interesting ELDOR measurement is the assignment of the weak satellite lines shown in Figure 5. An energy diagram for $CH_3\dot{C}(COOH)_2$ undergoing tunneling rotation is given in Figure 10, where the main EPR lines as well as the weak satellite lines have been indicated by the solid and

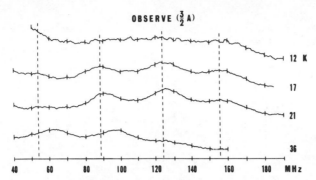

Figure 11. ELDOR spectra obtained by observing the bottom of the derivative $A_1(\frac{3}{2}A)$ peak of Figure 5 and pumping the low-field sideband regions. The dashed vertical lines represent line positions of the tunneling sidebands at 17°K. The observing power was 0.5 μW, while the pump power was attenuated 5 dB from 150 mW maximum. (From Mottley *et al.*[50])

dotted lines, respectively. It is noted from Figure 10 that the satellite lines have levels in common with only certain of the allowed transitions. For instance, transitions c and d have a level in common with transition A_3 and transition d has a level in common with E_2, while the allowed transition A_1 has no level in common with any of the satellite lines a, b, c, and d. As long as no relaxation exists that is significant enough to connect the energy levels of four different satellite lines, it should be possible to pump a satellite line and to observe a reduction in the EPR line intensity for those lines with a level in common. In addition, a measure of any significant relaxation that might connect various levels can be detected by observing the A_1 line. The results are shown in Figures 11–13. From Figure 11, we note that significant relaxation probabilities that might connect various levels are absent below 12°K, as also suggested from the ELDOR results given in Figures 7–9 below 12°K. Thus, when the E_2 line is observed and the satellite lines are pumped (Figure 12), it is noted that only line d shows an effect at 11°K, as one would predict from the energy level diagram in Figure 10. Likewise, lines c and d show an effect when line A_3 is observed (Figure 13). This, then, confirms the assignment of the satellite transitions given in Figure 10.

On the other hand, significant W_n relaxation probabilities are required to observe the four satellite lines that are obscured in the EPR spectrum by an impurity radical. Such relaxation processes are present at 24°K, as shown by the ELDOR results given in Figure 11. Thus, we are able to resolve the four satellite lines at 24°K that were obscured in the EPR spectrum (Figures 12 and 13). In addition, if intermolecular relaxation (T_{ss} processes) were significant at this temperature, the satellite lines would not have been resolved, as all spectral lines would have been connected. Thus, intermolecular relaxation does not appear significant at 24°K.

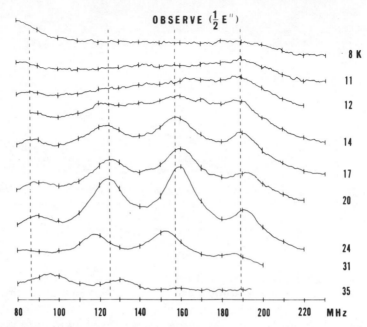

Figure 12. ELDOR spectra obtained by observing the bottom of the derivative $E_2(\frac{1}{2}E'')$ peak of Figure 5 and pumping the low-field sideband region. The applied power and scale are the same as in Figure 11. (From Mottley *et al.*[50])

It is interesting to note that, because of the presence of a small W_n below 10°K, intense ENDOR is observed,[46] especially at 4°K. However, with increasing temperature and increasing W_n (decreasing T_{1n}), ENDOR spectra become difficult to record, yet ELDOR spectra become easier to observe.

5.3. *Proton–Deuteron Exchange*

Both ENDOR and ELDOR studies can complement each other in the study of H/D exchange in irradiated crystals. The high resolution of ENDOR permits the detection of protons as well as deuterons that are present in a complex EPR spectrum, since the free proton frequency occurs at 14.4 MHz in a field of 3390 G, while the free deuteron frequency occurs at 2.21 MHz. Recently, Teslenko[52] was able to use this feature to resolve a controversy over whether or not H/D exchange does occur in irradiated glycine-d_3 at room temperature. He used the fact that in the absence of H/D exchange, the ENDOR spectrum will consist of the deuterium ENDOR from the ND_3^+ deuterons as well as the proton ENDOR spectrum from the CH_2- protons. However, the ENDOR showed that the protons at all positions of the radical were replaced by deuterons. This was verified by comparing the deuterium

ENDOR frequencies for CD_2— calculated from the relationship $v_D = 0.153512v_H$ to those additional deuterium ENDOR peaks observed experimentally, where v_D is the deuterium ENDOR frequency and v_H the ENDOR proton frequency.

One possible disadvantage of the ENDOR method is the requirement that W_n be long enough so that the nuclear levels can be saturated by the available rf power. This in many cases may necessitate an observing temperature for ENDOR that does not correspond to that at which the H/D exchange takes place. For instance, Teslenko[52] carried out his ENDOR experiments at 77°K and not at room temperature, where the H/D exchange processes occurred.

ELDOR, on the other hand, can be useful in following the H/D exchange process over the temperature range where it occurs near room temperature, since W_n becomes shorter, due in part to radical substituents undergoing intramolecular motion. In fact, ELDOR measurements proved especially useful in following the rates of H/D exchange processes that occur between 193°K and room temperature in irradiated glycine crystals containing a mixture of 45% $NHD_2^+CH_2COO^-$, 45% $ND_3^+CH_2COO^-$, and 10%

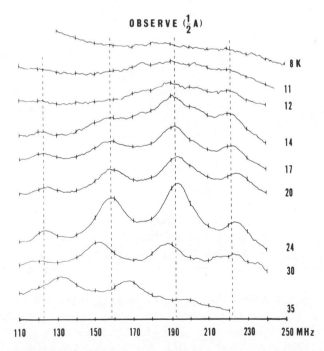

Figure 13. ELDOR spectra obtained by observing the bottom of the derivative $A_3(\frac{1}{2}A)$ and pumping the low-field sideband region. The applied powers and scale are the same as in Figure 11. (From Mottley *et al.*[50])

$NH_2D^+COO^-$. It was observed[53] that both the $NHD_2^+\dot{C}HCOO^-$ and $ND_3^+\dot{C}HCOO^-$ radicals are both formed at approximately 140°K. However, an increase in the spectral intensity of $ND_3^+\dot{C}HCOO^-$ and $NHD_2^+\dot{C}HCOO^-$ occurs at 215°K and could be followed by ELDOR. A subsequent H/D exchange also occurs upon further warming, to form $NHD_2^+\dot{C}DCOO^-$ as well as $ND_3^+\dot{C}DCOO^-$ radicals near room temperature.[54] It is important to point out that ELDOR spectra are obtained over this entire temperature range; however, the spectra contain information only about the proton hyperfine splittings, there being no splittings from deuterium. This apparently is due to a competition between the W_n relaxation processes for proton and deuterium nuclei. Since $W_n \propto \gamma_n^2$, the ratio of $W_D/W_H = \gamma_D^2/\gamma_H^2 = \frac{1}{42}$, where γ is the gyro-magnetic ratio for either the deuterons or protons. Thus, W_n processes for protons appear to be the dominant relaxation process; the quadrupole relaxation of the deuteron apparently does not contribute sufficiently for deuteron ELDOR to be observed. The assignment of the deuterium splittings was obtained by studying the H/D exchange along crystal directions where the proton hyperfine splittings were known from ENDOR, converting these splittings to deuterium splittings, and using the resulting combination of proton and deuteron splittings to simulate the EPR spectrum observed.

5.4. Intramolecular Motion

The resolution possible by ENDOR spectroscopy permits the study of molecular motions whenever freezing out the motion is revealed by the inequivalence of small hyperfine couplings. For instance, the cyclobutyl-1-carboxylic acid radical

$$H_2C-\dot{C}-COOH$$
$$\mid \quad \mid$$
$$H_2C-CH_2$$

is formed by x-irradiation of the dicarboxylic acid at room temperature. At 4°K, the ENDOR spectra of this radical show two sets of β-proton axial and equatorial splittings,[29] indicating that the ring exists in a static pucker relative to the hyperfine frequency. Yet, at 298°K, this radical shows[55] equivalent β-proton splittings, suggesting rapid inversion relative to the hyperfine frequency. The precision with which hyperfine splittings can be determined also reveals the presence of hindered motion, even though no freezing out of the motion exists. For instance, an ENDOR study of CH_3 radicals irradiated in $CH_3COOLi \cdot 2H_2O$ crystals revealed that the methyl radical undergoes hindered oscillation about a second axis in the radical plane in addition to rapid reorientation about the C_3 axis. It is apparent from these examples that ENDOR studies make it possible to identify the type of molecular motion that a particular radical undergoes.

Although the ELDOR resolution is significantly lower, the ELDOR relaxation mechanism is extremely sensitive to the correlation time of molecular reorientation. The application of ELDOR as a monitor of the molecular reorientation rate is quite new, and extensive studies have not been carried out as exact calculations in which the modulation frequency as discussed in Chapter 5 must be included for a quantitative understanding of the correlation times. However, a number of preliminary studies suggest the possibilities of such studies. Both enhanced and reduced ELDOR spectra have been observed for the $(CH_3)_2\dot{C}COOH$ radical in irradiated α-aminoisobutyric acid as a function of temperature.[10] Qualitatively, a fit to the spectral variation was found to be due to a variation with temperature below 10°C of the methyl group reorientation rate followed at higher temperatures by the exchange of the two methyl groups.[10] Other studies[10] of $CH_3\dot{C}HCOOD$ in irradiated L-alanine, of $NH_3^+\dot{C}HCOO^-$ and $\dot{C}H_2COO^-$ in irradiated glycine,[43] of $\dot{C}F_2CONH_2$[35] in irradiated trifluoroacetamide, of $\dot{C}Cl_3$ in irradiated trichloroacetamide, and of the cyclobutyl-1-carboxylic acid radical[55] have also shown the ELDOR intensity to be sensitive to the reorientation rates. However, in order to extract correlation times of the motion, exact time-dependent calculations like those described in Chapter 5 must be performed.

ACKNOWLEDGMENTS

The author expresses his appreciation to Deborah Green for her expert typing of the manuscript and to Dr. T. C. S. Chen for his criticism of the manuscript. This work was partially supported by United States Energy Research and Development Administration under contract no. E-(40-1)-4062 and partially under NATO research grant no. 825. This is ERDA document ORO-4062-30.

References

1. J. E. Wertz and J. R. Bolton, *Electron Spin Resonance*, McGraw-Hill Book Co., New York (1972).
2. N. M. Atherton, *Electron Spin Resonance*, Halsted Press-Wiley, New York (1973).
3. A. Abragram and B. Bleaney, *Electron Paramagnetic Resonance of Transition Ions*, Oxford University Press, London (1970).
4. L. Kevan and L. D. Kispert, *Electron Spin Double Resonance Spectroscopy*, John Wiley and Sons, New York (1976).
5. M. Iwasaki, K. Toriyama, and K. Nunome, *J. Chem. Phys.* **61**, 106 (1974).
6. A. L. Kwiram, *J. Chem. Phys.* **55**, 2484 (1971).
7. P. Gloux, *Mol. Phys.* **21**, 829 (1971).
8. S. F. J. Read and D. H. Whiffen, *Molec. Phys.* **12**, 159 (1967).

9. K. Toriyama, K. Numone, and M. Iwasaki, *J. Chem. Phys.* **64**, 2020 (1976).
10. L. D. Kispert, K. Chang, and C. M. Bogan, *J. Chem. Phys.* **58**, 2164 (1973), and references cited therein.
11. L. D. Kispert, K. Chang, and C. M. Bogan, *J. Phys. Chem.* **77**, 629 (1973).
12. R. J. Cook and D. H. Whiffen, *J. Chem. Phys.* **43**, 2908 (1965).
13. G. Feher, *Phys. Rev.* **105**, 1122 (1957).
14. C. A. Hutchinson, Jr., personal communication.
15. H. Seidel, *Z. Physik* **165**, 239 (1961).
16. R. J. Cook and D. H. Whiffen, *J. Phys. Chem.* **71**, 93 (1967).
17. K. A. Thomas and A. Lund, *J. Magnet. Resonance* **18**, 12 (1975).
18. M. Iwasaki, *J. Magnet. Resonance* **16**, 417 (1974).
19. S. N. Rustgi and H. C. Box, *J. Chem. Phys.* **59**, 4763 (1973).
20. R. J. Cook and D. H. Whiffen, *Proc. Roy. Soc. London* **295A**, 99 (1966).
21. G. H. Rist and J. S. Hyde, *J. Chem. Phys.* **50**, 4532 (1969).
22. J. M. Spaeth and M. Sturn, *Phys. Stat. Sol.* **42**, 739 (1970).
23. M. Iwasaki, *MTP Int. Rev. Sci., Phys. Chem.*, Ser. 1, Vol. 4, p. 327 (C. A. McDowell, ed.), Butterworths' Publications, London (1972).
24. H. Muto and M. Iwasaki, *J. Chem. Phys.* **59**, 4821 (1973).
25. H. Muto, K. Nunome, and M. Iwasaki, *J. Chem. Phys.* **61**, 1075 (1974).
26. H. Muto, K. Nunome, and M. Iwasaki, *J. Chem. Phys.* **61**, 5311 (1974).
27. H. Muto and M. Iwasaki, *J. Chem. Phys.* **61**, 5315 (1974).
28. B. Lamotte and P. Gloux, *J. Chem. Phys.* **59**, 3365 (1973).
29. L. R. Dalton and A. L. Kwiram, *J. Amer. Chem. Soc.* **94**, 6930 (1972).
30. R. C. McCalley and A. L. Kwiram, *Phys. Rev. Lett.* **24**, 1279 (1970).
31. J. S. Hyde, R. C. Sneed, Jr., and G. H. Rist, *J. Chem. Phys.* **51**, 1404 (1969).
32. L. D. Kispert, K. Chang, and C. M. Bogan, *Chem. Phys. Lett.* **17**, 592 (1972).
33. W. T. Huntress, Jr., *Adv. Mag. Res.* **4**, 1 (1965).
34. C. Mottley, L. D. Kispert, and P. S. Wang, *J. Phys. Chem.* **80**, 1885 (1976).
35. L. Kispert and K. Chang, *J. Mag. Res.* **10**, 162 (1973).
36. V. A. Benderskii, L. A. Blumenfeld, P. A. Stunzas, and E. A. Sokolov, *Nature* **220**, 365 (1968); P. A. Stunzhas, V. A. Benderskii, L. A. Blumenfeld, and E. A. Sokolov, *Opt. Spektrosk.* **28**, 278 (1970) [English trans.: *Opt. Spectrosc.* **28**, 150 (1970)]; P. A. Stunzhas, V. A. Benderskii, and E. A. Sokolov, *Opt. Spektrosk.* **28**, 487 (1970) [English trans.: *Opt. Spectrosc.* **28**, 261 (1970)]; P. A. Stunzhas and V. A. Benderskii, *Opt. Spektrosk.* **30**, 1041 (1971) [English trans.: *Opt. Spectrosc.* **30**, 559 (1971)].
37. L. D. Kispert and M. T. Rogers, *J. Chem. Phys.* **58**, 2065 (1973).
38. H. C. Box, *J. Mag. Res.* **14**, 323 (1974).
39. E. E. Budzinski and H. C. Box, *J. Chem. Phys.* **63**, 4927 (1975).
40. N. M. Atherton and A. J. Blackhurst, *J. Chem. Soc. Faraday Trans. II* **68**, 470 (1972).
41. R. J. Cook, *J. Sci. Instr.* **43**, 548 (1966).
42. D. A. Hampton and C. Alexander, *J. Chem. Phys.* **58**, 4891 (1973).
43. C. Mottley, K. Chang, and L. Kispert, *J. Mag. Res.* **19**, 130 (1975).
44. W. M. Tollis, L. P. Crawford, and J. L. Valenti, *J. Chem. Phys.* **49**, 4745 (1968).
45. H. Ohigashi and Y. Kurita, *Bull. Chem. Soc. Japan* **41**, 275 (1968).
46. S. Clough and F. Poldy, *J. Chem. Phys.* **51**, 2076 (1969).
47a. J. H. Freed, *J. Chem. Phys.* **43**, 1710 (1965).
47b. S. Clough, J. Hill, and F. Poldy, *J. Phys. C: Solid State Phys.* **5**, 518 (1972).
48. S. Clough and F. Poldy, *J. Phys. C: Solid State Phys.* **6**, 2357 (1973).
49. For example, W. L. Gamble, I. Miyagawa, and R. L. Hartman, *Phys. Rev. Lett.* **20**, 415 (1968); P. S. Allen and S. Clough, *Phys. Rev. Lett.* **22**, 1351 (1969); S. Clough, *J. Phys. C: Solid State Phys.* **4** 280 (1971); F. Apaydin and S. Clough, *J. Phys. C: Solid State Phys.* **1**, 932 (1968); C. Mottley, T. B. Cobb, and C. S. Johnson, Jr., *J. Chem. Phys.* **55**, 5823 (1971);

R. Ikeda and C. A. McDowell, *Molec. Phys.* **25**, 1217 (1973); S. Clough, M. Starr, and N. D. McMillan, *Phys. Rev. Lett.* **25**, 839 (1974); P. S. Allen, *J. Phys. C: Solid State Phys.* **7**, L22 (1974); S. Clough and J. R. Hill, *J. Phys. C: Solid State Phys.* **7**, L20 (1974); and references cited therein.

50. C. Mottley, L. Kispert, and S. Clough, *J. Chem. Phys.* **63**, 4405 (1975).
51. L. R. Dalton, A. L. Kwiram, and J. A. Cowen, *Chem. Phys. Lett.* **17**, 495 (1972).
52. V. V. Teslenko, *Chem. Phys. Lett.* **32**, 332 (1975).
53. L. Kispert and K. Chang, unpublished work.
54. L. L. Gautney, Jr. and I. Miyagawa, *Radiat. Res.* **62**, 12 (1975).
55. W. L. Gamble, L. A. Dalton, L. R. Dalton, and A. L. Kwiram, *Chem. Phys. Lett.* **34**, 565 (1975).

8

ENDOR on Hemes and Hemoproteins

Charles P. Scholes

1. Introduction

Heme compounds within proteins are crucial in many biological areas; for example, in carrying oxygen (hemoglobin), in storing oxygen (myoglobin), and in transporting electrons (cytochromes). A better knowledge of the electronic structure of the heme and its immediate protein environs should lead to a more complete understanding of how heme functions in its biological roles. The technique of ENDOR allows precise measurement of the hyperfine interaction of paramagnetic heme electrons near the Fe^{3+} site of heme with many nearby nuclei. The strength of hyperfine interactions gives detailed information on the electronic distribution at heme and its neighbors and thus gives insight into the details of the electronic wave function of heme. Since heme represents the active center of the protein, such detailed information should lead to a more complete, even quantum mechanical, understanding of the relationship between heme's electronic structure and its function in proteins.

1.1. Biological Role of Hemoproteins

1.1.1. Myoglobin and Hemoglobin

Myoglobin is a relatively simple and well-understood monomeric heme protein of molecular weight about 17,800. Its purpose is to receive oxygen from hemoglobin and to store it for later use in biological oxidations. Because myoglobin is a monomeric protein with only one heme per molecule,

Charles P. Scholes • Department of Physics, State University of New York at Albany, Albany, New York it exhibits simple, noncooperative binding of oxygen, and the oxygenation of myoglobin is well understood in a phenomenological way.[1] However, the intimate changes in electronic structure that accompany and perhaps promote the reversible binding of oxygen are still not clear. Sperm whale myoglobin was the first protein to have its three-dimensional crystallographic structure determined.[2] Compared to other proteins, its atomic coordinates are known to high accuracy. Starting with the work of Ingram *et al.*,[3a-c] its EPR spectrum has been studied by many groups. The initial ENDOR work on heme proteins was done with ferric myoglobin.

Hemoglobin is a tetrameric molecule, consisting of two pairs of myoglobinlike subunits, called the α and the β subunits, and its overall molecular weight is about 64,500. Hemoglobin is more complex than myoglobin, because its subunit chains communicate with each other so that one subunit can signal the state of its oxygenation to its neighbors. Although hemoglobin has been the subject of extensive study, there is still controversy about the intimate structural changes (called conformational changes) that occur upon binding of oxygen. Various other spectroscopic techniques (optical, circular dichroism, NMR) have shown that changes associated with heme do occur upon binding of oxygen and upon change of the hemoglobin molecular conformation.[4a-c] However, to understand the fundamental electronic details behind these changes, it is important to know the intimate electronic structure at the oxygen-binding sites of the molecule, i.e., at the hemes. It is here that ENDOR can make a significant contribution.

The heme molecule is shown in Figure 1 as it is found in myoglobin and hemoglobin. For ENDOR, it will be important to know which nuclei will most strongly interact with the electrons centered on the iron. As indicated in Figure 1, the strongest interactions are expected to be with the four nitrogens of heme, with the proximal histidine nitrogen, and with the iron nucleus itself. There may also be strong interactions with the sixth ligand, which can be O_2, CO, H_2O, or any of a number of anions. The heme group in hemoglobin and myoglobin, which is called protoheme, is not covalently linked by the side chains of its outer pyrrole rings to the protein; rather, it sits in a hydrophobic pocket within the protein. Ferric myoglobins or hemoglobins, which we study by ENDOR, are commonly called metmyoglobins or methemoglobins.

1.1.2. Cytochromes

Cytochromes occur in organisms ranging all the way from one-celled bacteria to man. In higher organisms, they often occur in organelles called mitochondria. This general class of proteins serves to transport electrons from oxidized foodstuffs to the point (cytochrome *c* oxidase) where electrons finally reduce oxygen. The best understood cytochromes are the *c*-type cytochromes, which are easily extracted from the mitochondria and which

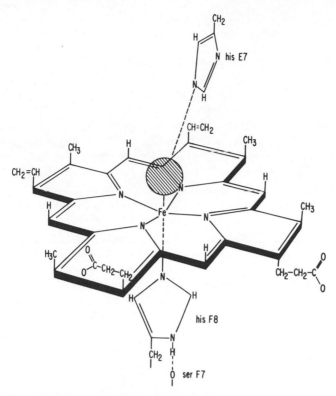

Figure 1. Heme is the large planar molecule in the center of this figure, and it is depicted here as it is found in myoglobin and hemoglobin. Note the four heme nitrogens attached to the heme iron. A proximal histidine (F8) nitrogen is also attached to the heme iron. The shaded area represents the binding site for the sixth ligand. Distal histidine (E7) interacts with the sixth ligand but not with the heme iron. The histidines F8 and E7 are amino acids connected to the protein and are not integrally part of the heme. If the heme is removed from the protein, then the histidines in this figure disappear, but the shaded area can still be occupied by an anion. (Reprinted with permission from Feher *et al.*[5] Copyright by the New York Academy of Sciences.)

have been crystallized and studied by x-ray diffraction.[6a–c] Cytochrome *c* is a heme protein of molecular weight about 12,400, and its fifth and sixth ligands are known to be histidine nitrogen and methionine sulfur. The heme group in cytochrome *c* has its vinyl groups substituted and is covalently linked through these substituted groups to the rest of the protein by thioether bonds. Many of the other cytochromes that occur in the mitochondria are less well understood than cytochrome *c*. This lack of understanding is due to the difficulty in extracting them from the lipid membrane and to the fact that many of them have several subunits and are much more complex than cytochrome *c*.

1.2. Paramagnetism of Heme Systems

1.2.1. Hyperfine Interactions Measured in Heme

Magnetic studies of heme systems allow one to gain information on how electrons distribute themselves at the various atoms of heme and its fifth and sixth ligands. In heme, the central iron nucleus and the nuclei of the immediate iron ligands have a large amount of unpaired electron density. Considerable electron spin finds its way into s orbitals, and the important interaction is the Fermi contact interaction. In the case of heme nitrogens, which have unpaired electron density in their valence 2s orbitals, one can straightforwardly relate [see equation (6)] observed hyperfine structure to the percentage of unpaired electron in a nitrogen 2s orbital. (2p contributions for nitrogen hyperfine structure have also been considered by Scholes.[7]) For ^{57}Fe, there is little s character to the valence orbitals of iron; the Fermi interaction with the ^{57}Fe nucleus arises in a more complicated fashion by core polarization of inner-shell iron s electrons.[8] Small contact interactions occur with outer heme ring protons and have been related to unpaired electron density on adjacent carbons by semiempirical McConnell relations.[9,10a,b]

Distant hydrogens can interact with the electron spin at the iron by direct dipolar interaction. The dipolar first-order coupling in such a case is

$$A_{dipole} = \frac{g_n \beta_n g_e \beta_e}{r^3} (3 \cos^2 \alpha - 1) \tag{1}$$

where g_n is the nuclear g value, β_n is the nuclear Bohr magneton, g_e is the particular electronic g value along the applied magnetic field, and β_e is the electronic Bohr magneton. r is the iron-to-proton distance and α is the angle between the vector joining the iron–proton coordinates and the external magnetic field.

The quadrupole moments of nuclei such as ^{14}N or ^{35}Cl interact with nearby electric field gradients (EFGs). The quadrupole interaction is related to the net nonspherical electron population about a particular nucleus. Das and co-workers have performed calculations that include both valence and nonlocal contributions to the EFG of heme iron[11a] and heme nitrogens.[11b]

Also important is the direct nuclear Zeeman interaction of a nucleus with the external magnetic field. This interaction, which ENDOR in particular measures, is proportional, via well-known and tabulated nuclear g values, to the magnetic field strength. The resolution of nuclear Zeeman interactions by ENDOR gives no immediate new information on electronic structure but does serve the purpose of identifying the nucleus being observed.

1.2.2. Spin Hamiltonian of Heme Systems

The interactions of electron spins and nuclear spins with each other and with internal and external fields are written in terms of spin operators. The form of the resultant spin Hamiltonian is determined by symmetry, and the constants that indicate the strength of various interactions are generally left to be determined by experiment.

1.2.2.1. Electronic Spin Hamiltonian. High-spin ferric heme has five unpaired electrons $(S = \frac{5}{2})$ centered on the heme iron; metmyoglobin with H_2O as the sixth ligand is a high-spin ferric heme compound. In high-spin ferric heme, the fourfold symmetric heme electronic environment serves to split the $S = \frac{5}{2}$ electronic ground state into three doublets; the splitting between doublets is called zero-field splitting. Zero-field splitting arises in spin Hamiltonian formalism from a term of the form, $D[S_z^2 - \frac{1}{3}S(S + 1)]$, where D is the zero-field splitting constant and S is the electron spin operator. The $S_z = \pm\frac{1}{2}$ doublet, on which EPR and ENDOR are done, lies lowest in energy. The $\pm\frac{3}{2}$ doublet is at an energy $2D$ above the ground doublet, and the $\pm\frac{5}{2}$ doublet is $6D$ above the ground doublet.

The energy separation between the lowest doublet $(S_z = \pm\frac{1}{2})$ and the higher doublets of high spin ferric heme is large compared to the electronic Zeeman energy.[12,13a] Thus, we can treat the lowest electronic doublet as an effective electronic spin of $\frac{1}{2}$, and the electronic Zeeman spin Hamiltonian has the form $\mathscr{H} = g_{e\perp}\beta_e(H_xS_x + H_yS_y) + g_{e\parallel}\beta_e H_zS_z$, where $g_{e\perp} \sim 6.0$ and $g_{e\parallel} = 2.00$. The direction corresponding to $g_{e\parallel} = 2.00$ is normal to the heme plane, and the directions corresponding to $g_{e\perp} = 6.0$ are in the heme plane.[3a]

In low-spin ferric heme, the single unpaired electron resides primarily in an iron orbital of d_π symmetry. The local heme electronic environment here is distorted away from fourfold symmetry, and first-order effects of unquenched angular momentum are important.[13b] The upshot for low-spin ferric heme is a highly rhombic g tensor. In cytochrome c one has $g_z = 3.06$, $g_y = 2.25$, and $g_x = 1.25$.[14a] In both cytochrome c and metmyoglobin cyanide, the g_z axis lies near the heme normal.[14a,b]

1.2.2.2. Nuclear Spin Hamiltonian. The magnetic hyperfine, quadrupole, and nuclear Zeeman interactions are next incorporated into the spin Hamiltonian. For heme systems, the nuclear interactions, as well as the electronic g values, are highly anisotropic, and the complete spin hamiltonian contains a large number of terms.[15a] However, in most of the systems we have studied by ENDOR, the magnetic field has been chosen to point along the heme normal (i.e., along a direction from the fifth ligand through the iron atom to the sixth ligand as in Figure 1). In such a case, the spin Hamiltonian becomes a simpler, more manageable axial spin Hamiltonian. For a nucleus of spin I and an electron with effective spin $\frac{1}{2}$, the first-order axial spin

Hamiltonian (electronic Zeeman + nuclear terms) is

$$\mathscr{H} = g_{e\parallel}\beta_e S_z H + A_{zz}I_z S_z + P_{zz}[I_z^2 - \tfrac{1}{3}I(I+1)] - g_n\beta_n H I_z \qquad (2)$$

A_{zz} is the z component of the magnetic hyperfine interaction, and P_{zz} is the z component of the quadrupole interaction. The last term in the spin Hamiltonian is the direct nuclear Zeeman interaction. By diagonalizing the above spin Hamiltonian, one obtains for the nucleus in question the expressions v_{ENDOR} for the ENDOR transition frequencies.

For an $I = \tfrac{1}{2}$ nucleus $(P_{zz} = 0)$, the two ENDOR transition energies are

$$hv_{\text{ENDOR}} = |\tfrac{1}{2}|A_{zz}| \pm g_n\beta_n H| \qquad (3)$$

For an $I = 1$ nucleus, the four ENDOR transition energies are

$$hv_{\text{ENDOR}} = |\tfrac{1}{2}|A_{zz}| \pm |P_{zz}| \pm g_n\beta_n H| \qquad (4)$$

For an $I = \tfrac{3}{2}$ nucleus, the six ENDOR transition energies are

$$hv_{\text{ENDOR}} = |\tfrac{1}{2}|A_{zz}| \pm g_n\beta_n H| \qquad (5a)$$

$$= |\tfrac{1}{2}|A_{zz}| + 2|P_{zz}| \pm g_n\beta_n H| \qquad (5b)$$

$$= |\tfrac{1}{2}|A_{zz}| - 2|P_{zz}| \pm g_n\beta_n H| \qquad (5c)$$

The $I = 1$ nitrogen nucleus is the nucleus that we have most often studied, and equation 4 predicts four ENDOR lines for a set of equivalent nitrogen nuclei. The ^{14}N ENDOR lines occur in two pairs, the lines in each pair being separated by twice the nuclear Zeeman energy. The left side of Figure 2 shows the resultant electron and nuclear spin energy levels for one set of equivalent ^{14}N nuclei, while the expected pattern of the ENDOR spectrum is shown on the right-hand side of Figure 2. An actual four-line ENDOR pattern from heme nitrogens alone is shown in Figure 6b.

1.2.3. *Electron Paramagnetic Resonance*

Ingram and co-workers[3a–c] used the orientation-dependent EPR of myoglobin single crystals to establish the heme plane orientation prior to the determination of this orientation by x-ray diffraction. Unfortunately, direct EPR on heme proteins either in solution or in single crystals cannot resolve electron–nuclear hyperfine couplings with the ^{14}N, ^{1}H, or ^{57}Fe. The numerous interactions from these nuclei merely contribute to the EPR linewidth. In the protein crystals, there is additional broadening due to slight misalignment of one heme with respect to the next.[3c] Spin–spin interaction between hemes (~ 25 Å apart) may also broaden the EPR line. However, prior to the ENDOR work on heme proteins, a heme model system was developed in which heme was doped in a magnetically dilute fashion into the

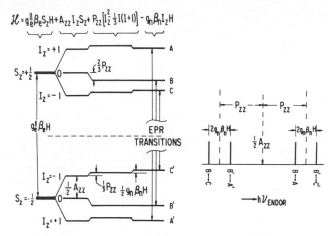

Figure 2. Energy levels (*left*) and ENDOR frequencies (*right*) of a system with an effective electron spin of $\frac{1}{2}$ and a nuclear spin of 1 (e.g., ^{14}N), with the magnetic field along the heme axial symmetry axis. The contributions of electron Zeeman, electron–nuclear hyperfine, nuclear quadrupole, and nuclear Zeeman interactions are shown. In ENDOR, the EPR transitions are monitored while the nuclear transitions are induced. The energies of the four nuclear transitions are found to occur in pairs separated by $2g_n\beta_n H$. The energy levels (left) are not drawn to scale; the EPR frequency is approximately three orders of magnitude larger than the ENDOR frequencies. (Reprinted with permission from Feher *et al.*[5] Copyright by the New York Academy of Sciences.)

organic crystal perylene.[7] The heme was sufficiently well oriented and magnetically dilute that hyperfine interactions with the heme nitrogens were resolved directly by EPR, as shown by the nine-line EPR pattern of Figure 3. $|A_{zz}|$ for hemin in perylene was 8.18 ± 0.34 MHz, and the Fermi contribution for the heme nitrogen was 8.68 ± 0.16 MHz. Unpaired electron spin density in a nitrogen 2s orbital is related to A_{Fermi} by[16]

$$A_{\text{Fermi}} = \frac{8\pi g_n \beta_n \beta_e}{3S} |\psi_0|^2 f_s \tag{6}$$

where $S = \frac{5}{2}$, $|\psi_0|^2$ is the square of nitrogen 2s wave function at the nitrogen nucleus, and f_s is the fraction of unpaired electron spin in the nitrogen 2s orbital. With $|\psi_0|^2 = 33.4 \times 10^{24}$ cm^{-3},[17] f_s was found to be 2.7% of an unpaired s electron on each heme nitrogen.

1.2.4. Electron Spin–Lattice Relaxation (T_1) Measurements

The amplitude of ENDOR signals depends critically on different spin relaxation processes, many of which are temperature dependent. The electron spin–lattice relaxation T_1 was obtained by monitoring the exponential recovery of the EPR signal after the application of a pulse of saturating

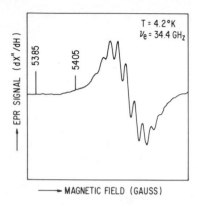

Figure 3. EPR signal from nitrogens of heme doped into perylene. The nine-line signal is the EPR signal that is characteristic of four equivalent $I = 1$ heme ^{14}N's. (Note: Only four ENDOR lines would be seen from any number of equivalent nitrogens.) (Reprinted with permission from Scholes.[7] Copyright by the American Institute of Physics.)

microwave power. As shown in Figure 4, the electron spin relaxation rate in high-spin ferric metmyoglobin varied by more than three orders of magnitude in the temperature range from 2° to 4°K. (ENDOR signals were later obtained from these samples near 2°K, where T_1 are in the millisecond regime.) Besides indicating where ENDOR measurements should be tried, the T_1 experiments led to the determination of the zero-field splitting parameter D.

The mechanism that accounts for the rapid decrease in spin–lattice relaxation rate in high-spin ferric heme between 4° and 2°K is an electron spin relaxation mechanism explained by Orbach.[18] In the Orbach process, lattice vibrations (phonons) with energies approximately equal to the energy difference $2D$ between the ground $S_z = \pm\frac{1}{2}$ state and first excited $S_z = \pm\frac{3}{2}$ state are absorbed and emitted by the electron spin system. The relaxation rate $(1/T_1)$ is given by

$$1/T_1 = A[\exp(2D/kT) - 1]^{-1} \approx Ae^{-2D/kT} \tag{7}$$

where k is the Boltzmann constant, T the absolute temperature, and A the coefficient of spin–phonon coupling. By measuring the exponential temperature dependence of this rate between 2° and 4°K, one measured the zero-field splittings. Values of D were found in frozen solutions to be 9.26 ± 0.11 cm^{-1} for aquometmyoglobin and 6.21 ± 0.12 cm^{-1} for metmyoglobin fluoride.[19] These values agreed well with values on the same compounds measured by far infrared spectroscopy[12] and magnetic susceptibility,[20] and the T_1 method required much smaller samples than the other techniques.

Spin–lattice relaxation in single crystals of low-spin ferric cytochrome c was measured by Mailer and Taylor,[21] who related the electronic T_1 to the phase lag of the field-modulated dispersion EPR signal. In the regime from 4° to 8°K, T_1, as measured by Mailer's technique, was approximately constant at 4 μsec. Between 8° and 20°K the spin–lattice relaxation rate $(1/T_1)$ went as T^7.

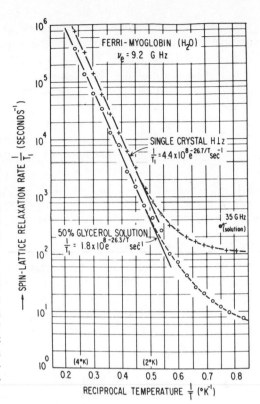

Figure 4. Spin–lattice relaxation rate near $g_{e\perp} = 6.0$ versus the reciprocal temperature for aquometmyoglobin in a single crystal and in a frozen solution. Above 2°K, the Orbach process predominates; below 2°K, a competing mechanism takes over. The zero-field splitting $2D$ was obtained from the slope of the line [equation (7)] with appropriate corrections for the low-temperature relaxation mechanism. A few points were taken with EPR at 35 GHz, and these showed the Orbach behavior above 2°K, but below 2°K the relaxation mechanism was found to be strongly field dependent, as shown by the 35 GHz point on the graph. (Reprinted with permission from Scholes *et al.*[19] Copyright by Elsevier North-Holland Biochemical Press.)

1.3. *Rationale for Using ENDOR to Study Heme*

Often an unpaired electron spin will experience hyperfine interactions with the magnetic moments of nearby nuclei; from the strength of such interactions, we learn of the distribution of electrons at particular nuclei. If only a few nuclei should be coupled to an unpaired electron, then one may often resolve hyperfine interactions directly from the splitting of the EPR signal. But all too often (especially in proteins), there are too many interacting nuclei, and as a result, one sees an EPR absorption that is an unresolved blur of hyperfine information. One might consider using NMR on nuclei near the unpaired electrons, but in the instances of ^{14}N, ^{35}Cl, or ^{57}Fe, for example, the NMR sensitivity of such nearby nuclei is far too low for them to be seen. At this point, ENDOR comes to our aid in resolving hyperfine structural information.

Proton NMR of liquid samples may give higher resolution of contact-shifted protons than ENDOR of the same protons in frozen solution. This is particularly true in the case of low-spin ferric hemes, where very rapid room temperature electron spin relaxation is favorable to narrow, highly resolved

NMR lines. On the other hand, the NMR spectroscopist is faced with more complexity in separating contact and dipolar ("pseudocontact" in NMR terminology) interactions than is the ENDOR spectroscopist.[22] ENDOR of frozen solutions may also be a better technique for resolving hyperfine interaction from species that are complicated by rapid chemical exchange in liquid.

2. ENDOR Technique Applied to Heme Systems

2.1. Factors Affecting the ENDOR Signal

In an ENDOR experiment, nuclear transitions are induced by an auxiliary radiofrequency field and monitored by observing the electron spin resonance signal. In ENDOR on heme systems, two conditions hold: (1) the EPR line is inhomogeneously broadened by the electron–nuclear hyperfine structure that we would like to resolve; (2) the EPR line is at least partially saturated during the experiment. The detailed mechanisms giving rise to ENDOR signals are complicated.[15b] However, it can generally be said that the ENDOR signals result from the rf-induced flipping of nuclear spins; this opens new relaxation pathways and changes the degree of saturation of the EPR signal.

Detailed mechanisms and explanations for the ENDOR process in solids depend on several different spin relaxation processes, and many of these relaxation processes (e.g., electron spin–lattice relaxation) are temperature dependent.[15b] Thus, temperature is a most important variable in optimizing the ENDOR signal. Spin–spin interactions between paramagnetic centers also enter into the ENDOR process.[23] A high concentration of paramagnetic centers, as may be found in pure single crystals of metmyoglobin or in hemin aggregates, can reduce or altogether eliminate ENDOR signals by opening competing spin–spin relaxation pathways. So the concentration of heme must be controlled and be fairly dilute for optimum ENDOR signals.

Another possible ENDOR mechanism for heme follows from the fact that application of microwave power to the electron spins can lead to dynamic nuclear polarization of the nearby nuclei.[15c] The nuclei that are so polarized act back on the electron spin and slightly shift the EPR line.[24] The application of an ENDOR rf field depolarizes the nuclei and causes a change in the EPR signal by shifting the EPR line back toward its unsaturated position. The idea for this mechanism originated in the course of ENDOR experiments on metmyoglobin, where shifts of the EPR line were noted upon saturating heme EPR signals. This depolarization ENDOR scheme was subsequently shown to work even in the case of homogeneously broadened EPR lines in Li–LiF.[24]

Most of our ENDOR work is carried out with 100-kHz field modulation, which can pass through the spin packets of the heme EPR line so quickly that the energy absorption during the time of any one modulation cycle may be small. Nevertheless, after many passes of the modulation field, a nonequilibrium steady state is set up whose EPR signal (both χ' and χ'') depends in a complex fashion on microwave-induced transitions, spin relaxation, and the modulation amplitude and frequency. With heme systems, EPR lines that show such passage effects have been good candidates for ENDOR. Apparently, the change in effective relaxation rates brought on by ENDOR-induced transitions strongly affects the rapid-passage signal intensity. We have found that our dispersion (denoted $d\chi'/dH$ in figures) rapid-passage ENDOR signals are generally ENDOR-induced *decreases* in the rapid passage EPR signals. On the other hand, our absorption (noted $d\chi''/dH$ in figures) ENDOR signals are generally ENDOR-induced *increases* in the absorption EPR signal. We are mainly interested in ENDOR frequencies in this article, and the labeling of vertical axes on the ENDOR spectra is not meant to imply the sign of the ENDOR-induced change.

Because we are dealing with a complex bioinorganic spin system, and because we often do not do ENDOR under slow-passage conditions,[15d,e] the mechanisms that give rise to our ENDOR signals are, at best, empirically understood. Besides temperature and sample concentration, other variables of a technical nature have their effect on the ENDOR signals and are optimized in the course of our experiments. These include the saturating microwave power, magnetic field setting, rf power, rf frequency sweep rate, magnetic field modulation amplitude and frequency, and tuning to the in-phase (χ') or the out-of-phase (χ'') part of the EPR signals. It is not critical for this work to understand the exact mechanism giving rise to ENDOR amplitudes, because we use ENDOR as a spectroscopic tool whose main information is contained in the frequencies and not the amplitude or the sign of the signals.

2.2. The ENDOR Spectrometer

For ENDOR at the relatively low microwave powers needed (< 1 mW) for heme, a superheterodyne spectrometer similar to that in reference 25 was again used in Feher's laboratory. At Albany, a homodyne system is used that is based on the Varian V-4500 bridge, which we have modified for low-power operation by addition of a circulator and bypass reference arm. Field modulation of 100 kHz has generally been used in both labs for ENDOR on hemes; 100 kHz field modulation effectively discriminates against low-frequency microphonic noise near the cavity, and such a high field modulation frequency is not a limitation on doing rapid sweeps of the ENDOR rf frequency.

A set of standard Pyrex double Dewars is used to attain liquid helium temperatures. Of considerable importance for this work is the Gordon coupler,[26] which is used to match the EPR cavity immersed in liquid helium to the waveguide. This spring-loaded coupler is specially designed to prevent mechanical vibration from modulating the coupling and producing noise.[27]

ENDOR on heme systems has been done with lightly silvered TE_{102} cavities that have provision for easy insertion and removal of large frozen sample tubes. In the Feher laboratory, R. A. Isaacson has constructed the cavities from quartz, and at Albany, they have been made from the plastic Hysol.[28]

The rf for doing ENDOR is provided by a generator (Hewlett-Packard 8601A), whose frequency is swept by an external voltage from the signal averager. After amplification (10-W ENI amplifier), the rf is carried by coaxial transmission line to the vicinity of the ENDOR cavity, where it is transferred to coils wrapped about the ENDOR cavity. The rf frequency is swept rapidly (typically at 10 MHz/sec) over a wide frequency range without tuned circuits. By application of appropriate resistive loads in series with the ENI, the rf system is made broad band in its frequency response. The magnitude of the rf field has been measured to be about 1 G peak-to-peak at the sample.

The ENDOR frequency is repetitively swept over the desired frequency range, and the resultant ENDOR signals are stored in the memory of the signal averager in synchrony with the ENDOR frequency sweep.

3. ENDOR Measurements on Hemes and Hemoproteins

3.1. ENDOR from High-Spin $(S = \frac{5}{2})$ Ferric Heme Proteins

3.1.1. ENDOR from Nitrogen in Frozen Solutions of Metmyoglobin and Methemoglobin

Prior to 1971, attempts had been made to observe ENDOR signals from single crystals of pure ferric myoglobin.[29,30] Apparently, the close proximity of the hemes to one another in these crystals produced strong spin–spin interactions that shorten some of the relaxation processes important in the ENDOR mechanism. No nitrogen ENDOR was reported. A large proton distant ENDOR signal was reported from the single crystals.[30] Additional proton ENDOR was seen both in crystals[30] and solutions[31a,b] of metmyoglobin azide, apparently from heme protons.

ENDOR from the nitrogens in heme and in heme proteins was initially obtained from frozen solutions in which the heme separation and concentration could easily be varied.[32] Because of the large electronic g anisotropy of high-spin ferric heme,[3a] the EPR spectrum of a frozen solution extends over

Table 1

ENDOR Transitions from Frozen Solutions of Metmyoglobin[a]

ENDOR frequency (MHz)	Assignment	$|A_{zz}|$ (MHz)	$|P_{zz}|$ (MHz)
2.39 ± 0.03	Heme	7.60 ± 0.02	0.44 ± 0.01
3.02 ± 0.02	Histidine[b]		
3.27 ± 0.01	Heme[b]		
4.33 ± 0.01	Heme		
4.95 ± 0.01	Histidine[b]		
5.22 ± 0.01	Heme[b]		
6.48 ± 0.03	Histidine		
8.47 ± 0.01	Histidine	11.46 ± 0.03	1.75 ± 0.02

Preliminary values of the X and Y components of the ^{14}N hyperfine and quadrupole tensor of the heme nitrogens from dilute single crystals of myoglobin[c]	$|A_{xx}| = 20.5 \pm 1.5$ MHz $|P_{xx}| = 1.55 \pm 0.20$ MHz $|A_{yy}| = 30.0 \pm 2.5$ MHz $|P_{yy}| = 1.10 \pm 0.20$ MHz

[a] $T = 2.1°$K, $H = 3196$ G, $v_e = 8.936$ GHz. Metmyoglobin chromatographically purified, 6 mM in 50% (v/v) glycerol, 0.1 M potassium phosphate buffer, pH 6.0. ENDOR frequencies connected by arrows are separated to first order by twice the nuclear Zeeman energy. Assignments are based on a comparison of the metmyoglobin ENDOR spectrum with the ENDOR spectrum from a hemin that had no histidine. Values of $|A_{zz}|$ and $|P_{zz}|$ were based on these assignments. The rms error was obtained from the results of several runs in which alternately increasing and decreasing frequency sweeps were used. Data and parameters from Scholes et al.[32]

[b] The assignment of the lower-frequency pair of ENDOR lines to the histidine nitrogens is not as clear-cut as the assignment of the pair at 6.48 and 8.47 MHz, since hemin compounds with no histidine do not have ENDOR transitions from nitrogen in the 6–9 MHz region but do have transitions below 5 MHz. In view of the relative signal amplitudes, we have tentatively assigned the 3.0 and 4.9 MHz lines to the histidine nitrogen.

[c] As discussed in the text, some of the ENDOR lines were split; their average frequencies were taken in determining these values. The X and Y axes are in the heme plane, with the Y axis probably along the iron–nitrogen bond and the X axis perpendicular to this bond. Parameters from Feher et al.[5]

a large magnetic field range corresponding to $2 \le g_e \le 6$. However, Hyde et al.[33a,b] had previously recognized that information-rich, single-crystal-like ENDOR spectra could be obtained by doing ENDOR at the g value extrema of frozen or polycrystalline samples. In our case, the magnetic field was set at the $g_{e\parallel} = 2.00$ extremum, where one selects only those heme molecules that have their normals parallel to the magnetic field. The ENDOR spectra in such a case are more easily interpretable since the four high-spin ferric heme nitrogens are equivalent.

The experimental results from the nitrogens in aquometmyoglobin are shown in Table 1 and in Figure 5. The assignment of various peaks in Figure 5 to heme nitrogens or to the proximal histidine nitrogen was based on a comparison of the myoglobin ENDOR spectrum, which has peaks for both heme and histidine, with the four-line ENDOR pattern from a simple heme, which has no histidine ligands.[32]

To extract electronic information from a spectrum such as that of Figure 5, we used the spin Hamiltonian of equation (2) and the expression for ^{14}N ENDOR frequencies of equation (4). Salient points on the information gained thereby are:

1. Four ENDOR lines from the heme nitrogens and four ENDOR lines from the histidine nitrogen were predicted, and a total of eight nitrogen lines was seen.

2. Hyperfine couplings to the unpaired electrons were determined for both heme and histidine nitrogens. The heme nitrogen value of $|A_{zz}| = 7.60 \pm 0.02$ MHz for metmyoglobin was reasonably consistent with the value of 8.18 ± 0.34 MHz from EPR studies of heme in perylene.[7] The theoretical work (extended Hückel, self-consistent charge method) of Das and co-workers[34] recently predicted A_{zz} values for heme and histidine nitrogens in metmyoglobin that are about 60% of the experimentally measured values. Das believes that his method may be better at predicting trends and ratios of coupling constants than their absolute values. Thus, it is interesting to note that his theoretically predicted ratio of heme's A_{zz} to histidine's A_{zz} is 1.87, while the experimental ratio is 1.51.

3. Quadrupole interaction of the nitrogen nuclei with surrounding EFGs was seen. The quadrupole coupling of the histidine was roughly consistent with quadrupole couplings measured by application of quadrupole resonance techniques to the nitrogens of imidazole.[35]

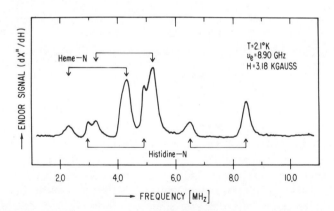

Figure 5. ENDOR spectrum from 6mM aquometmyoglobin in 1 : 1 (v/v) glycerol, 0.1 M potassium phosphate buffer, pH 6.0. This spectrum reflects the interaction of the electron spin with the nitrogen ligand nuclei. Arrows labeled "Heme-N" show the peaks assigned to the heme nitrogens. Arrows labeled "Histidine-N" show peaks from the histidine nitrogen. The magnetic field is perpendicular to the plane of the heme. Experimental conditions: sample volume $\approx$ 1 ml; ENDOR rf $\approx$ 0.5 G p.t.p.; ENDOR sweep rate $\approx$ 10 MHz/sec; 100 kHz modulation amplitude $\approx$ 5 G p.t.p.; microwave power $\approx$ 10 μW; trace taken with $\approx$ 1 hr of signal averaging; ENDOR represents a positive 2% increment of EPR signal. (After Scholes et al.[32])

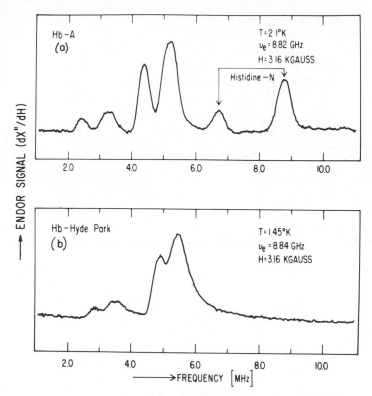

Figure 6. Nitrogen ENDOR spectra from hemoglobins. (a) Normal methemoglobin A, concentration ≈ 6 mM in 0.1 M phosphate, pH 7. Note similarity to metmyoglobin spectrum in Figure 5. (b) ENDOR from the histidine-less β chains of hemoglobin M$_{Hyde\ Park}$. Note lack of the high-frequency peak assigned to proximal histidine. (Reprinted with permission from Feher *et al.*[5] Copyright by the New York Academy of Sciences.)

Our ^{14}N results on metmyoglobin have been independently reproduced in the thesis work of M. R. Pattison.[36] The main thrust of Pattison's work, however, has been to study the electron spin dynamics of several metmyoglobin compounds at low temperatures and to investigate in detail the behavior of the distant proton ENDOR signal from these systems.

After nitrogen ENDOR from the simple monomeric protein metmyoglobin had been obtained, ENDOR from the more complex tetramer, hemoglobin, was studied. The ENDOR spectrum of high-spin ferric methemoglobin A in Figure 6a closely resembles that of metmyoglobin. The nitrogen frequencies are quite close to those of metmyoglobin; the main difference is that the frequencies of the histidine nitrogens are about 0.1 MHz higher than those in the metmyoglobin and those of the heme nitrogens are about 0.1 MHz lower. Thus, in methemoglobin A, the low-frequency histidine peaks and adjacent heme nitrogens have merged, and not so many lines are resolved as in metmyoglobin.

A number of human hemoglobins have been found that have genetic mutations where one of the amino acids has been altered. In particular, the replacement of either a proximal or a distal histidine in a particular subunit may cause the heme to be stabilized in the high-spin ferric form, which will not bind oxygen but which is paramagnetic and gives an ENDOR signal. Mutant hemoglobins where either the α or the β subunit is stabilized in the ferric form have been given the name hemoglobin M. One such mutant is hemoglobin $M_{Hyde\ Park}$ (i.e., $\alpha_2 \beta_2^{92His \to Tyr}$), where the proximal histidine of the β chain has been replaced by a tyrosine.[37] Tyrosine lacks a nitrogen for liganding to the heme iron. Figure 6b shows the ENDOR spectrum from the mutant β chains of Hyde Park; the normal α chains are in ferrous form and do not give an ENDOR signal. The most obvious feature of this ENDOR spectrum is the absence of the high-frequency histidine nitrogen ENDOR lines. It had been speculated that the distal histidine of the β chain could have taken over the role of the proximal histidine and was binding to heme.[37] The lack of histidine hyperfine structure in Figure 6b indicates that the distal histidine is not appreciably involved in binding to the iron. A counterpart to hemoglobin $M_{Hyde\ Park}$ is hemoglobin M_{Iwate}, where the α subunits have their proximal histidine replaced by tyrosine and the β chains are normal (i.e., $\alpha_2^{87His \to Tyr} \beta_2$). The ENDOR results from Iwate were essentially identical to those from Hyde Park.

3.1.2. *ENDOR Used to Study Cooperative Oxygenation in Hemoglobins*

When one or more subunits in a hemoglobin tetramer binds oxygen, the entire hemoglobin molecule may undergo a conformational change that alters the affinity of the hemoglobin molecule for oxygen. The heme groups where the oxygen binds are several tens of angstroms away from other hemes, and yet the binding of oxygen to one heme does affect the affinity of its neighbors for oxygen. ENDOR is particularly useful to determine whether the binding of oxygen to heme in some of the subunits of hemoglobin leads to specific electronic structural changes at the hemes of neighboring subunits that are, atomically speaking, far away.

Investigation by ENDOR requires that at least some of the subunits of hemoglobin be in the paramagnetic ferric form. Such mixed ferrous–ferric forms of hemoglobin are called *valency hybrids*. Hemoglobins $M_{Hyde\ Park}$ and M_{Iwate} are naturally occurring valency hybrids, but x-ray crystallographic studies on these[38] have shown that the genetic defects in these two proteins have greatly altered the protein structure away from its normal conformation. For the work on cooperative oxygen binding, another mutant, hemoglobin $M_{Milwaukee}$ ($\alpha_2 \beta_2^{67Val \to Glu}$), was selected because two of its four subunits

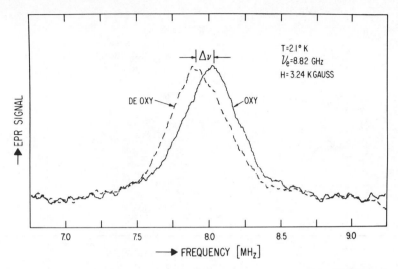

Figure 7. Effect of oxygenation of hemoglobin M$_{Milwaukee}$ on one of the histidine nitrogen ENDOR lines. The ferrous subunits were oxygenated while the ENDOR from the β ferric peaks was monitored. Washed whole cells were used. Vertical axis is the ENDOR signal. (Reprinted with permission from Feher et al.[5] Copyright by the New York Academy of Sciences.)

were ferric and would give ENDOR signals and because its structure determined by x-ray diffraction in both oxy and deoxy forms[39] was isomorphous with the respective oxy and deoxy forms of normal hemoglobin A. As a function of oxygen binding to the two normal α ferrous subunits within hemoglobin M$_{Milwaukee}$, the ENDOR spectrum of the other two β ferric subunits was followed. The mutant ferric β chains of this hemoglobin still have their proximal histidine, and changes in the histidine electronic structure from these ferric subunits were seen by observing the ENDOR signal from the β histidine nitrogen. As shown in Figure 7, there was a difference in the ENDOR frequency of the β histidine nitrogen as a function of oxygenating or deoxygenating the normal α chains. Similarly, interpretable changes from protons near the β hemes of Milwaukee were seen by NMR workers.[40] Perutz has suggested that when hemoglobin switches from its high oxygen affinity form to its low oxygen affinity form there is a lengthening of the iron–nitrogen bonds.[4] It may be that the decrease in ENDOR frequency upon going from oxy to deoxy form of hemoglobin M$_{Milwaukee}$ reflects this bond lengthening.

3.1.3. ENDOR from Protons and ^{57}Fe

After the initial study of heme nitrogens, the ENDOR from several other nuclei was observed.[5] Because they are relatively far from the heme iron, the protons are only weakly coupled to the heme iron. Weak coupling means

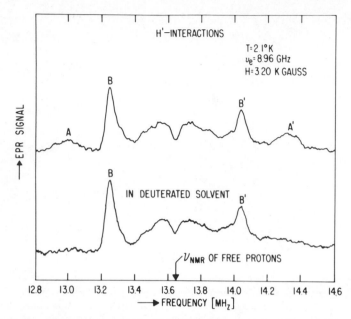

Figure 8. Proton ENDOR from metmyoglobin in deuterated and nondeuterated solvent and protein. Samples same as in Figure 5. Note absence of peaks AA′ in deuterated sample. Peaks BB′ are assigned to heme meso protons. Vertical axis is the ENDOR signal. (Reprinted with permission from Feher *et al.*[5] Copyright by the New York Academy of Sciences.)

that these protons interact more strongly with the applied magnetic field H by the direct Zeeman interaction than they do by hyperfine interaction with the electron spin; i.e., $g_n \beta_n H \gg |A_{zz}|$. In this case, there are two ENDOR lines per proton, equally spaced about the proton Larmor frequency, with frequencies given by equation (3). As shown in Figure 8, the most pronounced lines correspond to couplings $|A_{zz}|$ of 0.80 and 1.30 MHz. The latter two lines $AA′$ must belong to protons that exchange with the solvent, because they disappeared when the solvent and the protein were deuterated. Such exchangeable protons may be the protons of the aquo ligand or the proton that hydrogen binds to the more distant nitrogen of the proximal histidine. From a knowledge of the heme coordinates,[2] the dipolar contribution [equation (1)] was calculated for many of the heme protons. The largest calculated dipolar contribution was for the meso protons, starred in Figure 10, and this contribution was about -0.85 MHz. Estimates from NMR on high-spin ferric heme compounds show the direct contact contribution from the meso protons to be in the range from $+0.10$ MHz[41a] to -0.20 MHz.[41b] Thus, it seemed likely that the protons lines $BB′$ were from meso protons.

Initial ENDOR studies were performed on ^{57}Fe in metmyoglobin and methemoglobin with ^{57}Fe in 90% isotopic enrichment. The $I = \frac{1}{2}$ ^{57}Fe nucleus is strongly coupled to the paramagnetic electrons, so $|A_{zz}| \gg g_n \beta_n H$. Equation (3) holds for the two ENDOR frequencies, but they are centered at $\frac{1}{2}|A_{zz}|$ and split from each other by $2g_n \beta_n H$. (See Scholes *et al.*[42] for small second-order corrections to ^{57}Fe Zeeman splitting.)

Near $g_{e\parallel} = 2.00$, we were able to observe two ^{57}Fe ENDOR lines, from which we obtained A_{zz} for the iron. However, the ^{57}Fe ENDOR near $g_{e\parallel} = 2.00$ was complicated by an interesting phenomenon; when the Larmor frequency of the bulk free protons overlaps with one of the ^{57}Fe transitions, the amplitude of the corresponding ^{57}Fe line is reduced to zero, apparently by a ^{57}Fe-proton nuclear cross-relaxation process. Figure 9 shows several ^{57}Fe spectra that were taken at $g_{e\parallel} = 2.00$ but with different EPR–ENDOR cavities, and at different EPR frequencies ν_{EPR}. Consequently, the magnetic field and the free proton frequencies vary from one spectrum to the next.

Note that when the free proton frequency is near where one of the ^{57}Fe lines ought to be, the intensity of that ^{57}Fe line is reduced. Away from $g_{e\parallel} = 2.00$, the overlap of ^{57}Fe and proton frequencies was not a problem, and by tuning to the dispersion mode (χ'), ^{57}Fe ENDOR at all fields between $g_{e\parallel} = 2.00$ and $g_{e\perp} = 6.0$ was obtained. The intrinsic hyperfine parameters for the ^{57}Fe in both types of samples were found to be quite isotropic, but a

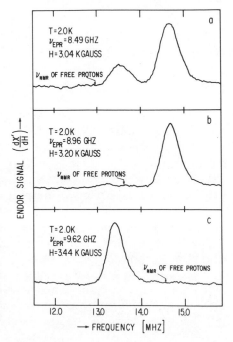

Figure 9. ^{57}Fe ENDOR spectra of 90% isotopically enriched metmyoglobin sample. Sample is 2 mM in heme and is dissolved in 50% (v/v) glycerol, 0.1 M phosphate buffer (pH 7.0). Spectra were taken at different EPR frequencies, ν_{EPR}, and magnetic field strengths H, all corresponding to $g_e = 2.00$. When the Larmor frequency of the free protons overlaps with one of the ^{57}Fe transitions, note how the amplitude of the corresponding ENDOR line is reduced to zero. (a) Larmor frequency of protons less than either ENDOR transition. (b) Larmor frequency of protons falls close to low-frequency ENDOR line. (c) Larmor frequency of protons is identical to the high-frequency ENDOR line. (Reprinted with permission from Scholes *et al.*[42] Copyright by the Elsevier/North-Holland Biomedical Press.)

Table 2

^{57}Fe Hyperfine Parameters for Metmyoglobin and Methemoglobin[a]

Sample	A_{iso} (MHz)[b]	Δ (MHz)[c]	$A_{\parallel}$ (MHz) = $A_{iso} + 2\Delta$	$A_{\perp}$ (MHz) = $A_{iso} - \Delta$
Metmyoglobin	-27.29 ± 0.10[d]	-0.24 ± 0.10	-27.77 ± 0.20	-27.05 ± 0.10
Methemoglobin	-26.85 ± 0.10[d]	-0.15 ± 0.10	-27.15 ± 0.20	-26.70 ± 0.10

[a] Parameters derived from data and formulas of Scholes *et al.*[42]
[b] A_{iso} is the same as the Fermi contact interaction for the ^{57}Fe nucleus.
[c] Δ is a small anisotropy parameter.
[d] The sign of the hyperfine coupling can be determined from the sign of the observed second-order pseudonuclear shift to the ^{57}Fe nuclear g value.[42]

small anisotropic part was found that caused $A_{\perp}$ and $A_{\parallel}$ to differ by about 2%, as tabulated in Table 2. This difference may reflect an axial distortion in the density of unpaired electron spin at the iron nucleus. Mössbauer spectroscopy has been reported insensitive to differences of less than 10% between $A_{\perp}$ and $A_{\parallel}$ for both high-spin ferric myoglobin[43a] and hemoglobin.[43b] Having found where the ENDOR signals occurred from isotopically enriched samples, we were able to obtain ^{57}Fe ENDOR near $g_{e\perp} = 6.0$ from samples containing ^{57}Fe in only 2.2% natural abundance.[5] As indicated in Table 2 and as seen in both enriched and nonenriched samples, small but definite differences were seen between the myoglobin and hemoglobin ^{57}Fe couplings, with the hemoglobin couplings slightly smaller than those of the myoglobin. The implication is that there is an electronic difference at the heme iron between myoglobin and hemoglobin.

3.1.4. Single-Crystal ENDOR

In order to obtain a complete picture of the heme wave function, one would like to resolve completely all the anisotropic components of the magnetic hyperfine and quadrupolar tensors. This is not easily done with frozen solutions, because for g values not equal to 2.00, the high-spin ferric heme ENDOR signal will correspond to a broad powder average[33a,b] from many orientations of the heme with respect to the magnetic field. To obtain the desired complete resolution, one prefers to do ENDOR on single crystals. With the information from the single-crystal study, we then simulate the powder ENDOR spectra with the hope of obtaining more complete information from the powder ENDOR spectra.

The problem with early single-crystal studies was that the pure metmyoglobin crystals were magnetically too concentrated. However, orientation-dependent ENDOR could be obtained from mixed myoglobin crystals that were grown to contain about 90% diamagnetic, CO-liganded ferrous myoglobin and only about 10% high-spin ferric metmyoglobin.[5] Such spectra were obtained with the magnetic field making various angles with respect to the heme. The spectrum with the magnetic field along the heme normal was identical to the spectrum measured in frozen solution at $g_{e\parallel} = 2.00$. The spectra measured with the magnetic field in the heme plane, however, gave considerable new information. The general features of the ENDOR orientation pattern for heme were fit for all orientations by the hyperfine and quadrupole parameters of Table 1. The Y and X directions are directions parallel and perpendicular to heme nitrogen–nitrogen diagonals, with Y most likely along the iron–nitrogen bond direction and X perpendicular to that bond direction.

There are some as yet unexplained complexities, and thus, the X and Y parameters in Table 1 should be considered as preliminary estimates. In the heme plane, some of the heme nitrogen ENDOR lines were unexpectedly split (by up to 0.4 MHz), and their average frequencies were used in calculating the X and Y parameters in Table 1. These unexpected additional splittings were not the result of crystal misorientation, and all the observed split ENDOR lines showed the orientation behavior with respect to heme axes that one expects for heme (as opposed to histidine) nitrogen. Another complexity was that in the heme plane it was not possible to distinguish any nitrogen ENDOR as clearly coming from the histidine. To clear up these problems, work is in progress to prepare myoglobin crystals with heme that contains ^{15}N in 90% + abundance.[44] Nitrogen-15 ($I = \frac{1}{2}$) heme will give only two ENDOR lines per nitrogen instead of four. The ^{15}N heme should give a simpler heme ENDOR pattern in the heme plane and may reveal ENDOR from ^{14}N histidine as well.

3.2. ENDOR Studies on High-Spin Ferric Hemin Compounds

The preceding sections show a substantial body of ENDOR data on myoglobin and hemoglobin. ENDOR data should, however, be understood in more than the present empirical fashion, and heme should be studied in a less complex environment than a protein. Systematic studies are called for on factors, such as ligand-induced or medium-induced factors, that affect heme's electronic structure.

An obvious model system to study is the heme group free of protein. Ferric heme compounds free of protein are called hemins, and the structure of some of these compounds that we are studying is shown in Figure 10. Several high-spin ferric hemins have been shown by x-ray work on crystalline material to form anion complexes with just one (fifth) axial ligand.[45] Of relevance to the ENDOR work is a group of experiments by far-infrared,[12] Mössbauer,[46] proton NMR,[47] and near-infrared[48] spectroscopies. These experiments were done on hemin compounds with the following series of

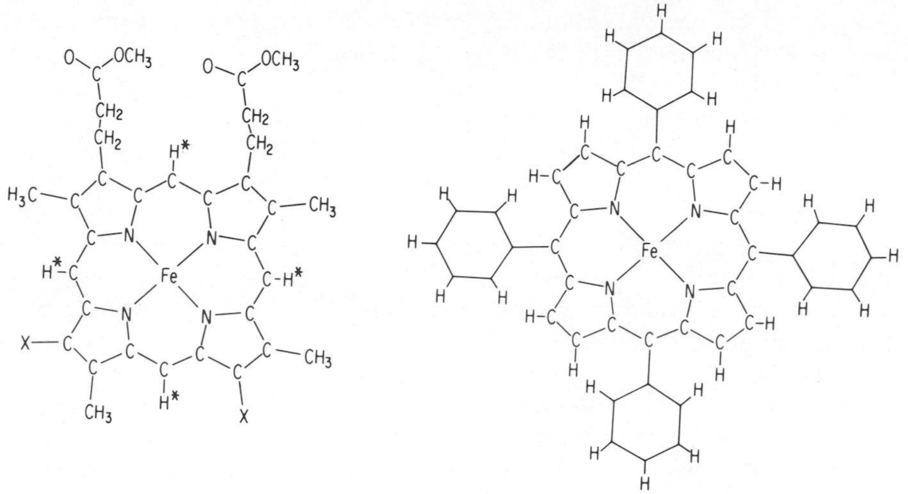

Figure 10. The structure of several hemin compounds {*left*: Fe(III) [proto/deutero] hemin dimethyl ester; *right*: ferric tetraphenylporphine} that we have used for ENDOR study. Axial anion is not shown in this projection. Heme protons are shown; note meso protons for the proto- and deuterohemes and lack of these in the ferric tetraphenylporphine. X = HC=CH$_2$, protohemin; X = H, deuterohemin; the asterisks denote meso protons.

axial, anionic ligands: methoxide, phenoxide, fluoride, acetate, azide, chloride, bromide, and iodide. The zero-field splittings from the far infrared work, the ^{57}Fe quadrupole interactions measured by Mössbauer, the heme proton contact shifts from NMR, and the optical wavelengths from near infrared all *increased* as one went through the series from methoxide to iodide. Caughey *et al.*[48] noted that the ligand order that these spectral properties followed was an order corresponding to the inverse of the spectrochemical series. It was suggested that these spectral properties could be explained in terms of a stronger interaction or bonding between iron and the axial anion, resulting in a weaker interaction between porphyrin nitrogens and iron. We were thus motivated to do the present study because ENDOR is a good tool for determining directly and quantitatively just how the electronic interactions and overlap with the heme nitrogen, and other atoms as well, may vary as we change the axial anion on the hemin.

Work is now in progress to observe variation in ENDOR frequencies from hemin nitrogen brought on by changing the axial anion on hemin. A problem with work of this kind is that the study has to be done in frozen solution, and hemin systems tend to aggregate in solution. This aggregation causes spin–spin interactions between paramagnetic hemin centers, which may altogether eliminate ENDOR signals. Since the solute (i.e., hemin) may segregate upon freezing of the solvent, a solvent system that forms a glassy matrix is more likely to accommodate the hemin in nonaggregated fashion in the solid phase. Some solvents may keep the hemin monomeric in solution but may replace the desired axial anion. With the three aims of providing a nonaggregating solvent, of providing a glassy matrix, and of not replacing the desired axial anion, we first developed the solvent of 1 : 1 (v/v) tetrahydrofuran (THF) and chloroform ($CHCl_3$). To assure that the proper halide ligand stayed on the hemin, we added in our initial experiments a tenfold molar excess of appropriate halide ligand in the form of tetrabutylammonium halide[49] to the THF–$CHCl_3$ solution. We have recently found that providing this additional halide is an unnecessary precaution. The difference between the nitrogen hyperfine parameters of chloro and bromo hemes occurs with or without added excess halide ligand.

With the magnetic field along the normal to the heme, we observed magnetic hyperfine and quadrupolar interactions with the heme nitrogens of protohemin dimethyl ester chloride and bromide.[50] Bromide hyperfine structure was seen by EPR, and chloride hyperfine structure was seen by ENDOR, showing that the solvent had not replaced the axial ligand of either compound. As shown in Table 3, the magnetic hyperfine coupling $|A_{zz}|$ for the nitrogens of protohemin chloride is 7.55 ± 0.05 MHz; the coupling for the heme nitrogens of the corresponding bromide compound is 7.28 ± 0.05 MHz. Figure 11 shows a comparison of the nitrogen ENDOR spectra from the chloro and bromo derivatives of protohemin, reflecting about a

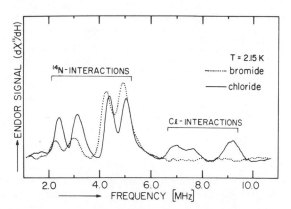

Figure 11. ENDOR spectra from protohemin chloride and protohemin bromide dimethyl esters near $g = 2.00$ at $T = 2.1°K$, primarily reflecting the interaction of the electron spin with the heme nitrogen nuclei. Sample conditions were as noted in Table 3, footnote *a*. Note the difference in nitrogen hyperfine couplings between the two compounds, shown by the shift of the entire nitrogen pattern from the bromide compound to lower frequency than the pattern from the chloride compound. Both protohemin chloride and protohemin bromide dimethyl esters were in a 1 : 1 (v/v) THF–$CHCl_3$ solution, which was also 30 mM in tetrabutylammonium chloride or bromide, respectively. (Reprinted with permission from Van Camp.[50] Copyright by the American Chemical Society.)

Table 3

^{14}N ENDOR Results and ^{35}Cl and ^{37}Cl ENDOR Transitions in Protohemin Halide Dimethyl Esters[a]

Compound	^{14}N			^{35}Cl ENDOR frequency (MHz)	^{37}Cl ENDOR frequency (MHz)
	ENDOR frequency (MHz)	$|A_{zz}|$ (MHz)	$|P_{zz}|$ (MHz)		
Protohemin	2.44 ± 0.03	7.55 ± 0.05	0.35 ± 0.03	1.6 (shoulder)	5.7 ± 0.2^b
chloride	3.17 ± 0.03			6.9 ± 0.1	7.6 ± 0.1
dimethyl	4.40 ± 0.02			9.18 ± 0.04	11.7 ± 0.2
ester	5.08 ± 0.02			14.7 ± 0.1	c
				17.4 ± 0.1	
Protohemin	2.29 ± 0.04	7.28 ± 0.05	0.36 ± 0.03		
bromide	3.06 ± 0.03				
dimethyl	4.27 ± 0.03				
ester	4.94 ± 0.03				

[a] Both protohemin chloride and protohemin bromide dimethyl esters were 3 mM in a 1 : 1 (v/v) THF–CHCl$_3$ solution, which was also 30 mM in the appropriate tetrabutylammonium halide. The ENDOR data were taken at v_e = 9.02 GHz, H = 3.22 kG for the chloride and v_e = 9.02 GHz, H = 3.26 kG for the bromide. The coupling constants and ENDOR frequencies quoted here were taken for the chloride compound at 5 G above the g = 2.00 EPR extremum and for the bromide compound at 5 G above the highest field peak of the resolved, four-line bromide hyperfine pattern. These field positions were chosen to provide (commensurate with good EPR and ENDOR sensitivity) well-oriented subsets of heme molecules with the magnetic field along their normals. The rms error in ENDOR frequencies and the coupling constants was obtained from the results of several separate runs on each compound in which alternately increasing and decreasing frequency sweeps were used. Data and parameters from Van Camp et al.[50]
[b] Better resolution of this peak was obtained with decreasing ENDOR sweeps from 20 to 1 MHz or with ^{37}Cl-enriched hemin. Figure 12a was taken with increasing ENDOR sweeps from 1 to 20 MHz.
[c] There should be a ^{37}Cl peak at about 14 MHz; such a peak is predicted to occur at twice the ^{37}Cl nuclear Zeeman energy above the 11.7 MHz peak. This predicted peak may suffer interference from near-lying proton peaks. It was recently seen by using ^{37}Cl-enriched hemin.

3.5% difference between the heme nitrogen couplings for the two compounds and possibly a slightly higher electron density at the nitrogen of the chloride compound. The quadrupole couplings for both compounds stayed at about 0.35 MHz.

For the hemin chloride, nonnitrogen ENDOR lines were seen in the region from 0.5 to 20.5 MHz, as shown in Figure 12 and Table 3. Such lines have *only* been seen from the chlorohemins, and their pattern is most definitely not the pattern of protons. We assigned the four most intense of these lines to ^{35}Cl and several less intense lines were apparently from ^{37}Cl, whose abundance is one-third that of ^{35}Cl. We fitted the observed ^{35}Cl data to an axial spin Hamiltonian [equation (2)], which gives for the $I = \frac{3}{2}$ ^{35}Cl nucleus the expressions of equation (5) for ENDOR transition frequencies. The detailed argument for our fit of the chloride data is given in Ref. 50. The ^{35}Cl ENDOR lines at 6.9 and 9.18 MHz were fit to equation (5a), and the ^{35}Cl lines at 14.7 and 17.4 MHz were fit to equation (5b). The results of this fit gave for ^{35}Cl $|A_{zz}| = 16.1 \pm 0.1$ MHz and $|P_{zz}| = 4.0 \pm 0.1$ MHz. In equation (5c), these values of $|A_{zz}|$ and $|P_{zz}|$ almost exactly cancel one another and predict a low-energy transition of energy about equal to $g_n \beta_n H \approx 1.4$ MHz. Such a low-energy transition was indeed found, as shown in Figure 12b. Back calculation of the ^{37}Cl ENDOR frequencies gave good agreement with the frequencies of ENDOR lines that we thought to be from ^{37}Cl. By use of the chloride magnetic hyperfine couplings determined by ENDOR, the change of EPR linewidth (peak width at half height) at the $g_{e\parallel} = 2.00$ derivative extremum from a width of 14 G without chloride to a width of 22.5 G with chloride was well explained, and the change in EPR line shape was computer-simulated very well. By studying hemin chloride prepared with 90% ^{37}Cl, we have recently checked our values for chloride's magnetic hyperfine and quadrupole parameters and confirmed our assignment of ^{37}Cl lines.

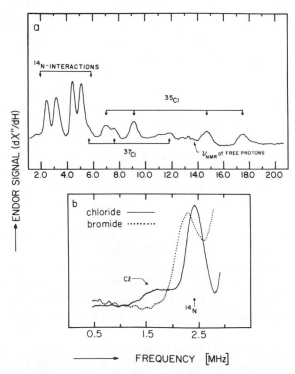

Figure 12. (a) ENDOR spectrum for protohemin chloride dimethyl ester taken near $g = 2.00$ at $T = 2.1°$K, reflecting the interaction of the electron spin with both chloride and nitrogen ligands. Sample conditions were as noted in Table 3, footnote *a*. (b) ENDOR spectrum from the protohemin chloride compared with the ENDOR spectrum from protohemin bromide dimethyl ester in the region from 0.5 to 2.5 MHz. The purpose of this figure is to demonstrate in the 1.5-MHz region the existence of a low-energy chloride ENDOR transition. (Reprinted with permission from Van Camp.[50] Copyright by the American Chemical Society.)

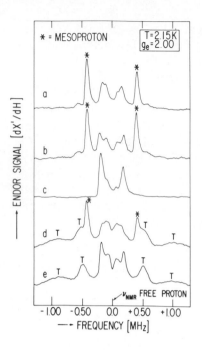

Figure 13. Proton ENDOR from the protons of several hemin compounds. Spectra are centered at the free proton frequency which was near 14 MHz for all samples. The EPR cavity frequency ν_{EPR} will vary slightly from one sample to the next. Thus, the resonance field at $g = 2.00$ differs slightly from one sample to the next. So that we would not have to label separately the frequency axis for each spectrum and so that we could easily compare the proton hyperfine splittings between samples, we have referred all spectra to the free proton frequency, arbitrarily taken as the zero of frequency. Spectra a, b, and c are in fully deuterated THF-d_8 and CDCl$_3$ (no tetrabutyl-ammonium salts used). Spectra d and e were done with CDCl$_3$ and protonated THF. Note additional new outlying peaks, marked T (to denote the THF protons). ENDOR (a) from protohemin chloride dimethyl ester—*has* meso protons; (b) from deuterohemin chloride dimethyl ester—*has* meso protons; (c) from ferric tetraphenylporphine chloride—*has no* meso protons. (d) from proto-hemin dimethyl ester; (e) from ferric tetraphenylporphine chloride.

Chloride and nitrogen ENDOR signals were obtained with the absorption EPR mode, fairly large field modulation (≈ 5 G peak to peak), and fast ENDOR frequency sweeps (≈ 10 MHz/sec). Proton ENDOR could be brought out by using the dispersion mode, small field modulation (≈ 0.1 G p.t.p.), and slow ENDOR frequency sweeps (≈ 0.1 MHz/sec). To date, we have done proton ENDOR on hemin compounds to assign proton resonances to various heme protons and to determine if solvent protons (as from THF) were giving hyperfine interaction. As shown in Figure 13a–c, we first compared heme proton ENDOR spectra from protohemin, deuterohemin, and ferric tetra-phenylporphine chloride dissolved in fully deuterated THF-d_8–CDCl$_3$. The most striking feature is the pair of sharp peaks from the proto- and deutero-hemes with separation of about ± 0.40 MHz from the free proton frequency (i.e., $|A_{zz}| \approx 0.8$ MHz). This feature is similar to previously observed proton splittings in Figure 8 that we assigned to meso protons.[5] Figure 13c shows that the peaks with splittings of ± 0.40 MHz are absent from the proton ENDOR spectrum of ferric tetraphenylporphine, which lacks meso protons. Small differences between the weakly coupled inner proton peaks of the three compounds were seen; small peaks were seen lying outside the meso protons of proto- and deuterohemes. We believe these peaks reflect the differing outer pyrrole substituents of the three compounds. We hope that future work on ferric octaethylporphyrin, with only ethyl groups on its outer pyrrole rings and which does have meso protons, will clear up more proton assignments. Next, the mixed solvent deuterated $CDCl_3$ and protonated THF was used, and several new proton ENDOR lines appeared with separations of ± 0.95 and ± 0.55 MHz, labeled T in Figure 13d and e. These latter two proton spectra indicate that ferric heme systems can certainly interact with THF. Perhaps THF acts as a sixth ligand, although it definitely does not replace the desired axial chloride or bromide.

Since proton ENDOR has shown that THF interacts with the heme, we have been concerned lest some of our ENDOR results, such as the values of the nitrogen hyperfine couplings, might depend on the solvent system used. Thus we have developed another solvent system in which ferric protohemes or deuterohemes do not aggregate with themselves. This system is a glass-forming 1 : 1 (v/v) mixture of chloroform and methylene chloride that contains diamagnetic, metal-free mesoporphyrin ester added in a fivefold molar ratio to the paramagnetic hemes. Mesoporphyrin is needed in this system, because in its absence, the protoheme or deuteroheme will aggregate. In this new chloroform–methylene chloride–mesoporphyrin system, the heme nitrogen hyperfine couplings rose appreciably from their previous values in THF–$CHCl_3$, which were 7.55 ± 0.05 MHz for the protohemin chloride and 7.28 ± 0.05 MHz for the protohemin bromide. In this new solvent system, the nitrogen hyperfine coupling $|A_{zz}|$ has been found to be about 8.10 ± 0.05 MHz for the protohemin chloride and about 7.85 ± 0.05 MHz for the protohemin bromide. Thus, there is approximately the same *difference* in nitrogen couplings between chloro- and bromohemes, as was seen in the THF–$CHCl_3$ solvent. At present, we are looking at nitrogen and proton ENDOR in both our solvent systems from protohemin and deuterohemin with fluoride, formate, acetate, azide, and iodide ligands. Recent work has shown that we do not need as elaborate solvent systems to prevent ferric tetraphenylporphine aggregation, and work is underway on this heme model compound as well.

3.3. *ENDOR from Low-Spin* $(S = \frac{1}{2})$ *Ferric Hemes and Hemoproteins*

The relaxation behavior of low-spin ferric heme compounds differs considerably from that of high-spin ferric heme.[21] The conditions of rapid passage that have been used for EPR study of cytochrome c[14a] were good conditions for doing ENDOR on these systems. By using the dispersion mode, a fairly small 0.5-G p.t.p. modulation amplitude, and relatively slow ENDOR frequency sweeps (≈ 1 MHz/sec), we obtained ENDOR from a number of low-spin ferric heme systems.[51] Detailed in Table 4 and Figure 14 is ^{14}N information from three low-spin ferric heme compounds: cytochrome c, metmyoglobin cyanide, and protohemin mercaptide. Each of these spectra was taken from a frozen solution near the g value (g_{max}) that was shown by

324

Charles P. Scholes

Table 4

ENDOR Frequencies from ^{14}N in Low-Spin Ferric Heme Compounds[a]

Sample and conditions	Frequencies (MHz)[b]	Hyperfine parameters (MHz)
Hemin mercaptide $T = = 3.6°K$ $H = 2.90$ kG $g_z = 2.32$ $2g_n\beta_n H = 1.78$ MHz	1.47 ± 0.06 2.25 ± 0.1 3.22 ± 0.02 3.89 ± 0.03	$\|A_{zz}\| = 5.42 \pm 0.06$ $\|P_{zz}\| = 0.36 \pm 0.03$ (heme nitrogen)
Metmyoglobin cyanide $T = 2.1°K$ $H = 1.95$ kG $g_z = 3.42$ $2g_n\beta_n H = 1.20$ MHz	1.44 ± 0.1 2.90 ± 0.01 3.24 ± 0.02 4.11 ± 0.01 4.38 ± 0.02	$\|A_{zz}\| = 7.32 \pm 0.02^c$ $\|P_{zz}\| = 0.15 \pm 0.01^c$ (heme or histidine nitrogen)
Cytochrome c $T = 2.1°K$ $H = 2.22$ kG $g_z = 3.01$ $2g_n\beta_n H = 1.38$ MHz	1.40 ± 0.07 2.48 ± 0.04 (shoulder) 2.70 ± 0.01 4.02 ± 0.01	(heme or histidine nitrogen)

[a] Data and parameters from Scholes and Van Camp.[51]
[b] Arrows indicate Zeeman pairs.
[c] Tentative values.

single-crystal studies of the first two compounds[14a,b] and linear electric field effect on the last[52] to be near the heme normal. ENDOR on the mercaptide complex was done because this complex (suggested by J. Peisach) lacks nitrogeneous fifth and sixth ligands, so the nitrogen ENDOR from it must be from heme nitrogens. The nitrogen hyperfine couplings for the mercaptide complex are $\|A_{zz}\| = 5.42 \pm 0.06$ MHz and $\|P_{zz}\| = 0.36 \pm 0.03$ MHz. The ENDOR spectrum from the metmyoglobin cyanide also showed two pairs of Zeeman-split nitrogen lines, which, if they originate from equivalent nitrogens, have $\|A_{zz}\| = 7.32 \pm 0.02$ MHz and $\|P_{zz}\| = 0.15 \pm 0.01$ MHz. Cytochrome c showed one sharp pair of nitrogen lines, with the possibility of another pair as shoulders on the low-frequency side of these sharp lines. We noted also an ENDOR resonance of low intensity near 1.5 MHz in both the cytochrome c and the metmyoglobin cyanide, and we speculate that this line may be part of a weaker nitrogen pattern that is partly obscured by the other more intense lines.

The nitrogen ENDOR lines from the protohemin mercaptide do occur in the same general region as the lines from the metmyoglobin cyanide and the cytochrome c, but the spectra are sufficiently different that we are as yet unable to assign nitrogen lines in the two proteins to heme or to histidine nitrogen. We hope to resolve the heme versus histidine assignment problem by looking at samples of methemoglobin or metmyoglobin cyanide that have been isotopically enriched with ^{15}N heme. Although the ENDOR work on low-spin ferric heme compounds is just starting, it is clear that we can obtain ENDOR from low-spin ferric hemes and that there exists a substantial difference between the nitrogen ENDOR from the three low-spin ferric heme systems observed.

Shown in Figure 15 is proton ENDOR from protohemin cyanide and metmyoglobin cyanide. Protohemin cyanide, as prepared with dimethylform-amide (which is an aprotic solvent) and with D_2O to solubilize the KCN, was designed to give ENDOR only from those protons that are nonexchange-able heme protons. The proton spectrum from metmyoglobin cyanide shows a pattern differing from the simpler protohemin cyanide, probably reflecting the additional interactions with nonheme protons. Both of these spectra were taken at g values where the magnetic field is near the heme normal. On the basis of equation (1), the electron–proton dipolar interaction was determined. For protochemin cyanide, the value of the dipolar interaction calculated with $g_z = 3.63$ was -1.58 MHz for the meso protons (at 4.5 Å the closest of any protons to the heme iron) and -0.53 MHz for the

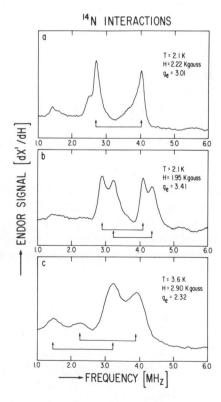

Figure 14. Part of the ENDOR spectra of (a) cytochrome c, (b) metmyoglobin cyanide, and (c) protohemin mercaptide, reflecting the interaction of the electron spin with ligand nitrogen nuclei. Each spectrum was taken at a g value where the magnetic field is near the heme normal. The protein samples were 6 mM in 1 : 1 (v/v) glycerol, pH 7 phosphate buffer. Protohemin mercaptide prepared from proto-hemin chloride dimethyl ester was put in 1 : 1 DMF–mercaptoethanol. Specific ENDOR fre-quencies are given in Table 4. Lines denoted with arrows appear in pairs where the frequency difference between lines within a pair corres-ponds to approximately twice the ^{14}N nuclear Zeeman energy at the respective magnetic field. (Reprinted with permission from Scholes and Van Camp.[51] Copyright by Elsevier/North-Holland Biomedical Press.)

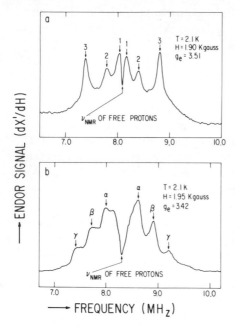

Figure 15. Part of the ENDOR spectrum of (a) protohemin cyanide dimethyl ester and (b) metmyoglobin cyanide reflecting the interaction of the electron spin with nearby protons. Each spectrum was taken at a g value where the magnetic field is along the heme normal. Lines denoted with the same number in (a) and with the same Greek letter in (b) occur at equal spacings from the free proton frequency and therefore are due to the same kind of proton. Splitting of the peaks marked 3 is ± 0.72 MHz from the free proton frequency and of peaks marked β is ± 0.60 MHz from the free proton frequency. (Reprinted with permission from Scholes and Van Camp.[51] Copyright by Elsevier/North-Holland Biomedical Press.)

ring methyl protons (~ 6.5 Å from the iron). NMR results of Shulman *et al.*[10a] and Wüthrich and Baumann[10b] showed only a small contact contribution for the meso protons ($\sim +0.15$ MHz) and a larger contribution ($\sim +0.5$ MHz) for the ring methyls. We thus assigned the peaks labeled 3 in Figure 15a, which have $|A_{zz}| = 1.43 \pm 0.01$ MHz, to the meso protons. The peaks labeled 1 and 2 are apparently from the outer pyrrole ring substituents.

The dipolar contribution calculated for the meso protons of the metmyoglobin cyanide, taking $g_z = 3.44$, is -1.50 MHz, so that dipolar plus contact interaction gives a net -1.35 MHz. This suggests that the peaks labeled β in Figure 15b, which have $|A_{zz}| = 1.19 \pm 0.03$ MHz, are from meso protons, provided that some reduction of the dipolar contribution from covalent effects is introduced.[10a] To assign more proton resonances, future work will be done with heme compounds with replacements on outer pyrrole rings and with deuterated solvents.

4. Summary

In this chapter, we have reviewed the present information on ENDOR of hemes and hemoproteins. The initial ENDOR experiments in Feher's laboratory on high-spin ferric metmyoglobin are recalled, and information from ^{14}N, ^{1}H, and ^{57}Fe is summarized. The initial work was extended to methemoglobin and small differences in ^{14}N and ^{57}Fe couplings with respect to metmyoglobin were seen. The application of ENDOR to problems of biological interest was shown by experiments on several mutant hemoglobins. ^{14}N ENDOR has been seen from naturally occurring mutant hemoglobins that have two subunits in the ferric form and two subunits in the ferrous form. In hemoglobin $M_{Hyde\ Park}$ and hemoglobin M_{Iwate}, where the mutant ferric chains have no proximal histidincs, wc showcd that the distal histidine on the mutant ferric chains does not appreciably bind to the ferric iron. In hemoglobin $M_{Milwaukee}$ it was shown by ENDOR that the primary event of oxygen binding to one set of hemoglobin subunits produces electronic structural change at a specific site (the histidine nitrogen) of another heme in a neighboring subunit. ENDOR signals from magnetically dilute single crystals of metmyoglobin have been observed, and preliminary information from these observations is given.

More recently, we have observed ENDOR from heme nitrogen, heme protons, and chloride ligand in axially liganded high-spin ferric hemin compounds. With the hemin compounds, we have measured the change in ENDOR frequencies from changing axial ligand and solvent. We have assigncd protons from the hemin ENDOR spectra to heme meso protons and to solvent protons. ENDOR signals from heme-linked nitrogen and protons in low-spin ferric hemes and hemoproteins have recently been seen. We expect to continue to use ENDOR in the future as a probe for biological changes, as an analytical tool, and as a means to discover the heme wave function.

ACKNOWLEDGMENTS

Dr. H. L. Van Camp has been a most important worker in building the ENDOR spectrometer at Albany and in performing many of these experiments. Mr. R. A. Isaacson has given us considerable technical advice, and he took a very important part in the construction of the Feher ENDOR apparatus and in many of the ENDOR measurements on metmyoglobin and hemoglobins. The author is grateful to Prof. G. Feher, in whose laboratory much of this work was performed, and from whom the author learned the details of the ENDOR technique. Prof. W. S. Caughey has kindly provided us with deuterohemin esters. Since 1971, this work has been supported by the following grants: NIH grant no. AM-17884 (C.P.S.), NIH grant no. GM 13191 (G.F.), NSF grant no. GH-36615X (G.F.), and NSF grant no. DMR-24361 (G.F.). C. P. Scholes is the recipient of NIH Research Career Development Award No. 1 K04 AM00274.

References

1. E. Antonini and M. Brunori, *Hemoglobin and Myoglobin in Their Reactions with Ligands*, Chap. 9, North-Holland Publishing Co., Amsterdam, The Netherlands (1971).
2. H. C. Watson, in *Progress in Stereochemistry* (B. J. Aylett and M. M. Harris, ed.), Vol. 4, pp. 299–333, Butterworth, London (1969).
3. (a) J. E. Bennett, J. F. Gibson, and D. J. E. Ingram, *Proc. Roy. Soc. (London)* **A240**, 67–82 (1957). (b) J. E. Bennett, J. F. Gibson, D. J. E. Ingram, T. M. Haughton, G. A. Kerkut, and K. A. Munday, *Proc. Roy. Soc. (London)* **A262**, 395–408 (1961). (c) G. A. Helcké, D. J. E. Ingram, and E. F. Slade, *Proc. Roy. Soc. (London)* **B169**, 275–288 (1968).
4. (a) M. F. Perutz, A. R. Fersht, S. R. Simon, and G. C. K. Roberts, *Biochemistry* **13**, 2174–2186 (1974). (b) M. F. Perutz, E. J. Heidner, J. E. Ladner, J. G. Beetlestone, C. Ho, and E. F. Slade, *Biochemistry* **13**, 2187–2200 (1974). (c) M. F. Perutz, J. E. Ladner, S. R. Simon, and C. Ho, *Biochemistry* **13**, 2163–2173 (1974).
5. G. Feher, R. A. Isaacson, C. P. Scholes, and R. L. Nagel, *Ann. N.Y. Acad. Sci.* **222**, 86–101 (1973).
6. (a) R. E. Dickerson, T. Takano, D. Eisenberg, O. B. Kallai, L. Samson, A. Cooper, and E. Margoliash, *J. Biol. Chem.* **246**, 1511–1535 (1971). (b) T. Takano, O. B. Kallai, R. Swanson, and R. E. Dickerson, *J. Biol. Chem.* **248**, 5234–5255 (1973). (c) F. R. Salemme, J. Kraut, and M. D. Kamen, *J. Biol. Chem.* **248**, 7701–7716 (1973).
7. C. P. Scholes, *J. Chem. Phys.* **52**, 4890–4895 (1970).
8. M. F. Rettig, P. S. Han, and T. P. Das, *Theoret. Chim. Acta (Berlin)* **12**, 178–182 (1968).
9. H. M. McConnell, *J. Chem. Phys.* **24**, 764–766 (1956).
10. (a) R. G. Shulman, S. H. Glarum, and M. Karplus, *J. Molec. Biol.* **57**, 93–115 (1971). (b) K. Wüthrich and R. Baumann, *Helv. Chim. Acta* **57**, 336–350 (1974).
11. (a) J. C. Chang, T. P. Das, and D. Ikenberry, *Theoret. Chim. Acta (Berlin)* **35**, 361–368 (1974). (b) P. S. Han, T. P. Das, and M. F. Rettig, *Theoret. Chim. Acta (Berlin)* **16**, 1–21 (1970).
12. G. C. Bracket, P. L. Richards, and W. S. Caughey, *J. Chem. Phys.* **54**, 4383–4401 (1971).
13. J. S. Griffith, *The Theory of Transition Metal Ions*, Cambridge University Press, Cambridge (1971). (a) Chapter 12, Section 12.4.12 and 12.4.5. (b) Chapter 12, Section 12.4.10.
14. (a) C. Mailer and C. P. S. Taylor, *Can. J. Biochem.* **50**, 1048–1055 (1972). (b) H. Hori, *Biochim. Biophys. Acta* **251**, 227–235 (1971).
15. (a) A. Abragam and B. Bleaney, *Electron Paramagnetic Resonance of Transition Metal Ions*, Clarendon Press, Oxford (1970). (a) Chapter 3. (b) Chapter 4. (c) Section 1.12. (d) Section 2.8. (e) Section 4.6.
16. J. Owen and J. H. M. Thornley, *Rept. Progr. Phys.* **29**, 675–728 (1966).
17. D. R. Hartree and W. Hartree, *Proc. Roy. Soc. (London)* **A193**, 299–304 (1948).
18. R. Orbach, *Proc. Roy. Soc. (London)* **A264**, 458–495 (1961).
19. C. P. Scholes, R. A. Isaacson, and G. Feher, *Biochim. Biophys. Acta* **244**, 206–210 (1971).
20. H. Uenoyama, T. Iizuka, H. Morimoto, and M. Kotani, *Biochim. Biophys. Acta* **160**, 159–166 (1968).
21. C. Mailer and C. P. S. Taylor, *Biochim. Biophys. Acta* **322**, 195–203 (1973).
22. K. Wüthrich, *Struct. Bonding* **8** (in particular pp. 86–89) (1970).
23. G. Feher and E. A. Gere, *Phys. Rev.* **114**, 1245–1256 (1959).
24. G. Feher and R. A. Isaacson, *J. Mag. Res.* **7**, 111–114 (1972).
25. G. Feher, *Bell System Tech. J.* **36**, 449–484 (1957).
26. J. P. Gordon, *Rev. Sci. Instr.* **32**, 658–661 (1961).
27. R. A. Isaacson, *Rev. Sci. Instr.* **47**, 973–974 (1976).
28. H. L. Van Camp, C. P. Scholes, and R. A. Isaacson, *Rev. Sci. Instr.* **47**, 516–517 (1976).
29. G. Feher, *Electron Paramagnetic Resonance with Applications to Selected Problems in Biology*, Chap. 2, Gordon and Breach Science Publishers, New York (1970).

30. P. Eisenberger and P. S. Pershan, *J. Chem. Phys.* **47**, 3327–3333 (1967).
31. (a) J. S. Hyde and G. H. Rist, "ENDOR of Proteins," presented at the *Third International Conference* on *Magnetic Resonance in Biological Systems*, Airlie House, Warrenton, Virginia (1968). (b) J. S. Hyde, in *Ann. Rev. Phys. Chem.* **25**, 407–435 (see in particular p. 418) (1974).
32. C. P. Scholes, R. A. Isaacson, and G. Feher, *Biochim. Biophys. Acta* **263**, 448–452 (1972).
33. (a) J. S. Hyde, in *Magnetic Resonance in Biological Systems* (A. Ehrenberg, G. B. Malmström, and T. Vänngard, eds.), pp. 63–85, Wenner-Gren Symposium Series, Pergamon Press, Oxford (1967). (b) G. H. Rist and J. S. Hyde, *J. Chem. Phys.* **52**, 4633–4643 (1970).
34. S. K. Mun, J. C. Chang, and T. P. Das, *Bull. Amer. Phys. Soc. Ser.* 2, **21**, 317 (1976).
35. J. Koo and Y. N. Hsieh, *Chem. Phys. Lett.* **9**, 238–241 (1971).
36. M. R. Pattison, Ph.D. Thesis, Wayne State University, Detroit, Michigan (1974).
37. P. Heller, *Amer. J. Med.* **41**, 799–814 (1966).
38. J. Greer, *J. Molec. Biol.* **59**, 107–126 (1971).
39. M. F. Perutz, P. D. Pulsinelli, and H. M. Ranney, *Nature New Biol.* **237**, 259–263 (1972).
40. T. R. Lindstrom, C. Ho, and A. V. Pisciotta, *Nature New Biol.* **237**, 263–264 (1972).
41. (a) R. J. Kurland, R. G. Little, D. G. Davis, and C. Ho, *Biochemistry* **10**, 2237–2246 (1971). (b) F. A. Walker and G. N. La Mar, *Ann. N.Y. Acad. Sci.* **206**, 328–348 (1973).
42. C. P. Scholes, R. A. Isaacson, T. Yonetani, and G. Feher, *Biochim. Biophys. Acta* **322**, 457–462 (1973).
43. (a) G. Lang, *Quart. Rev. Biophys.* **3**, 1–60 (1970). (b) M. R. C. Winter, C. E. Johnson, G. Lang, and R. J. P. Williams, *Biochim. Biophys. Acta* **263**, 515–534 (1972).
44. A. Lapidot, private communication.
45. (a) D. F. Koenig, *Acta Cryst.* **18**, 663–673 (1965). (b) J. L. Hoard, M. J. Hamor, T. A. Hamor, and W. S. Caughey, *J. Amer. Chem. Soc.* **87**, 2312–2319 (1965).
46. T. H. Moss, A. J. Bearden, and W. S. Caughey, *J. Chem. Phys.* **51**, 2624–2631 (1969).
47. W. S. Caughey and L. F. Johnson, *J. Chem. Soc. D* **1969**, 1362–1363 (1969).
48. W. S. Caughey, H. Eberspaecher, W. H. Fuchsman, S. McCoy, and J. O. Alben, *Ann. N.Y. Acad. Sci.* **153**, 722–737 (1968).
49. G. N. La Mar, *J. Amer. Chem. Soc.* **95**, 1662–1663 (1973).
50. H. L. Van Camp, C. P. Scholes, and C. F. Mulks, *J. Amer. Chem. Soc.* **98**, 4094–4098 (1976).
51. C. P. Scholes and H. L. Van Camp, *Biochim. Biophys. Acta* **434**, 290–296 (1976).
52. (a) J. Peisach and W. B. Mims, *Proc. Natl. Acad. Sci. U.S.* **70**, 2979–2982 (1973). (b) W. B. Mims and J. Peisach, *Biochemistry* **13**, 3346–3349 (1974).

9

ENDOR and ELDOR on Iron–Sulfur Proteins

Richard H. Sands

1. Introduction

It is fitting that a book of this nature should include a chapter on the application of double resonance techniques to the study of iron–sulfur proteins. Magnetic resonance techniques have permitted a detailed study of the nature of the active sites of many proteins that had not been possible previously. Among these is the whole class of iron–sulfur proteins. These proteins have been implicated, primarily as electron carriers, in the mediation of a wide variety of biochemical reactions in plants, bacteria, and mammals. They are involved in photosynthesis, oxidative phosphorylation, nitrogen fixation, and many hydroxylations, to name but a few important functions.

Iron–sulfur proteins, according to the IUPAC–IUB Commission and Biochemical Nomenclature, are defined as those proteins containing iron coordinated to sulfur—either cysteine or inorganic sulfur—and in spite of a few exceptions may be divided into four categories: (i) Rubredoxins, having no acid-labile sulfur and a negative midpoint potential at pH 7. The iron is coordinated to cysteinyl sulfur, and the oxidized state exhibits a $g = 4.3$ EPR spectrum. (ii) Ferredoxins, having equal numbers of iron and labile sulfur atoms and a negative midpoint redox potential at pH 7. They are characterized by EPR spectra appearing on reduction and having $\bar{g}$ values less than 2. Such proteins are ubiquitous in nature, appearing in plants, bacteria and

Richard H. Sands • Biophysics Research Division, Institute of Science and Technology, and Department of Physics, University of Michigan, Ann Arbor, Michigan

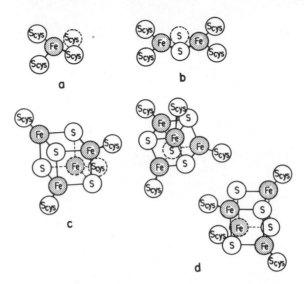

Figure 1. Representations of the structures of the active sites of the iron–sulfur proteins: (a) a rubredoxin from *C. pasteurianum* as determined by x-ray analysis, (b) a two-iron ferredoxin as determined by spectroscopy and model compound incorporation, (c) a four-iron cluster from the high-potential iron protein from *C. vinosum D* as determined by x-ray analysis and a four-iron ferredoxin as determined by spectroscopy and model compound incorporation, and (d) the two four-iron clusters in the eight-iron ferredoxin from *M. aerogenes* as determined by x-ray analysis.

animals. (iii) High-potential iron–sulfur proteins, having a midpoint potential that is positive at pH 7 and characterized by an EPR spectrum in the oxidized state that has $\bar{g}$ slightly greater than 2.* (iv) Conjugated iron–sulfur proteins, comprising proteins that contain any of the above redox centers plus additional prosthetic groups.

There are several excellent review articles[1–3] and a series of books edited by W. Lovenberg[4] that describe the state of knowledge about these proteins in considerable detail. Some of the properties of the iron–sulfur proteins are summarized below. The emphasis here will be on the physical properties, and the reader is referred to the literature via the above review articles for the chemistry.

Let us begin by summarizing the known structural properties of the active sites of these proteins. The basic structure (Figure 1a) is that evidenced in the single-iron proteins, the rubredoxins, and is that of an iron atom liganded to four sulfur atoms (cysteine sulfurs here) in a distorted tetrahedral arrangement. The importance of this structure is evident when one sees (Figure 1b,c,d) that the multi-iron proteins have active sites, which are clusters of these tetrahedra joined at edges (two-iron ferredoxins) or

---

* $\bar{g} \equiv (g_x + g_y + g_z)/3$.

faces (four- and eight-iron proteins). Knowledge of the structure of the center in the rubredoxins comes from x-ray crystallographic data.[4] in addition, these proteins have been studied extensively by chemical, spectroscopic, extended x-ray absorption fine structure techniques,[5] and magnetic susceptibility.[4] The iron is high-spin ferric in the oxidized state and high-spin ferrous in the reduced state.

The structure of the active site of the two-iron ferredoxins[6] is shown in Figure 1b; it consists of two iron atoms, each of which is in a distorted tetrahedron of sulfur atoms. The bridging sulfurs are acid-labile, whereas the remaining ligands are cysteinyl sulfur atoms. Until recently, the elucidation of this structure came from purely spectroscopic evidence,[6] much of which will be presented in this chapter. Now, however, extrusion and reconstitution experiments based upon model compound syntheses, the latter stimulated by the previous spectroscopic evidence, have removed all doubts. Figure 1c displays the structure of the center of a four-iron ferredoxin and also that of the four-iron high-potential iron protein. These consist of distorted cubes with alternate iron and labile sulfur atoms at the corners, and in addition, each iron atom is liganded to a single mercaptide sulfur from cysteine. The evidence in the case of the four-iron ferredoxins was spectroscopic until, more recently, extrusion and reconstitution experiments based upon model compounds and competing ligand exchange have confirmed the conclusion. The structure of the center in the high-potential iron protein from *Chromatium vinosum D* has been determined by x-ray crystallography.[7] The earlier evidence for the four-iron ferredoxins was the similarity between the chemical data and optical and EPR spectra for these proteins with those of an eight-iron protein, which consists of two such four-iron clusters as determined by x-ray crystallography as described next. The structure of the active site of the eight-iron ferredoxin from *M. aerogenes*[8] is shown in Figure 1d and consists of two such four-iron clusters, as shown in Figure 1c, separated by a mean distance of 11.5 Å.

Thus the structure of the active site of a high-potential iron protein (a strong oxidizing agent), redox potential +0.35 V, has been shown to be the same as that of the four-iron ferredoxin and one of the clusters of an eight-iron ferredoxin (strong reducing agents), redox potential −0.38 V. This is an interesting puzzle for the biophysicist, because apparently the difference in these redox potentials must be accounted for on the basis of the formal valence states of the iron atoms involved (or, to be more precise, on the basis of molecular orbital theory) and the electrostatic charges of the surrounding ligands and residues.

In the case of the two-iron ferredoxins it has been possible apparently to determine the formal charge states of the iron atoms by spectroscopic means, as will be shown. Because we know that charge must be delocalized it is interesting to discover that spin, at least, is localized. In the studies on the four- and eight-iron clusters, the conclusion that spin is localized is further supported. Is charge also localized in these systems? Studies on the two-iron ferredoxins appeared to indicate that it was. On the other hand, similar studies on the four- and eight-iron clusters are being interpreted as supporting charge delocalization. ENDOR, ELDOR, and Mössbauer studies are essential for an understanding of these questions, and it is the purpose of this chapter to demonstrate the application of double resonance spectroscopies to this task.

2. Theory

This section is intended to remind the reader of the principles involved in the *electron paramagnetic resonance* (EPR), *electron nuclear double resonance* (ENDOR), and the *electron–electron double resonance* (ELDOR) spectrometries of amorphous (frozen aqueous solution) protein samples. For a comprehensive discussion of the EPR and ENDOR spectrometries of transition element ions the reader is referred to the text by Abragam and Bleaney.[9]

2.1. Fundamental Interactions in EPR

As has already been discussed in the previous theory chapters, the shapes of electronic spectral absorption lines are determined by the various interactions that the electrons have with their surroundings. For iron-group ions, the magnetic electrons are exposed to the electrostatic fields from surrounding atoms and the resulting interaction is great enough to quench the orbital angular momentum. The only orbital moment arises because of the admixture caused by the spin–orbit coupling. The combined effect of these interactions in general is to produce an anisotropic g tensor and a zero-field splitting of the different spin projection states as indicated by the following spin Hamiltonian:

$$\mathscr{H} = -\mathbf{S} \cdot \tilde{g} \cdot \mathbf{H} + D[S_z^2 - \tfrac{1}{3}S(S+1)] + E(S_x^2 - S_y^2) \tag{1}$$

In order to understand the origin of each of these terms, one must consider a specific system. This was done for the heme proteins in previous chapters, where the rhombic distortion (E) was small; however, for the single-iron iron–sulfur proteins (rubredoxins) in the oxidized state, the iron is high-spin ferric with large rhombic distortion, E nearly equal to $\frac{1}{3}D$, its maximum value.[10] The Kramers doublet states are equally separated in energy (Figure 2), and when the external applied magnetic field splitting is small, each of the doublets yields an EPR spectrum that is usually described in the literature by a set of three effective g values (although these so-called g values change with field)—at X-band the lowest doublet gives an EPR spec-

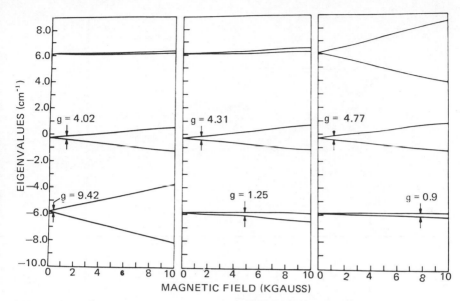

Figure 2. Energy levels with the magnetic field applied along the three principal axis directions of the g tensor for *P. oleovorans* rubredoxin assuming $D = 1.76$ cm^{-1} and $E = 0.495$ cm^{-1}. The arrows indicate energy separations corresponding to X-band EPR quanta and the measured effective g values at X band. (From Peisach et al.[10])

trum starting at $g_{\text{eff}} \simeq 9$ and extending to $g_{\text{eff}} \simeq 0.6$; the middle doublet gives absorptions near $g_{\text{eff}} \simeq 4.3$, and the highest doublet gives an absorption similar to that of the lowest doublet. The resulting EPR spectrum is a superposition of all three such absorptions with temperature-dependent relative intensities (Figure 3).

For the multi-iron proteins, the EPR spectra bear no resemblance to that in Figure 3, even though in most instances there are high-spin ferric atoms present in the cluster. The reason for this is that there are electrostatic exchange interactions (discussed below) between the iron atoms that result in the total electronic spin of the ground state being 0 or $\frac{1}{2}$, depending upon the oxidation state of the complex. This means that at low temperatures (below 77°K), the Hamiltonian describing the system no longer contains the D or E terms of equation (1), but only the first (Zeeman) term. The g values are all near $g = 2$, and the EPR spectra are similar to those shown in Figure 9.

In Chapter 6, Kevan and Narayana discussed the synthesis of EPR spectra for amorphous samples, and these are detailed by several authors.[11–14] A brief review follows. For spin $\frac{1}{2}$ (or pseudospin $\frac{1}{2}$) systems, the resonance condition may be expressed as

$$h\nu_0 = g\beta H \tag{2}$$

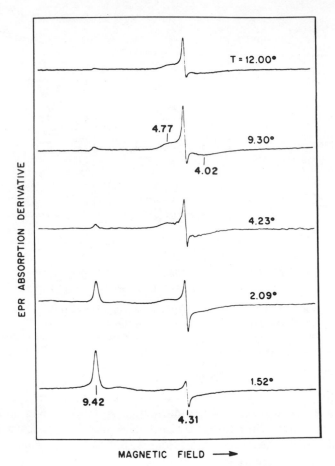

Figure 3. *P. oleovorans* rubredoxin EPR spectra at various temperatures (relative intensities are arbitrary). Effective *g* values are indicated on the spectra. (From Peisach *et al.*[10])

where ν_0 represents the applied microwave frequency, β is the Bohr magneton, H is the applied magnetic field and for a magnetic field applied at the Euler angles θ, ϕ with respect to the molecular *g*-tensor principal axes,

$$g = (g_x^2 \sin^2 \theta \cos^2 \phi + g_y^2 \sin^2 \theta \sin^2 \phi + g_z^2 \cos^2 \theta)^{1/2} \qquad (3)$$

Under the assumption that all molecular orientations are equally likely, the number of molecules having a magnetic field oriented at angles between θ and $\theta + d\theta$, ϕ and $\phi + d\phi$ is proportional to the solid angle at these orientations; i.e.,

$$dN = N_0 \sin \theta \, d\theta \, d\phi \qquad (4)$$

From this, the number of molecules undergoing EPR at magnetic field values

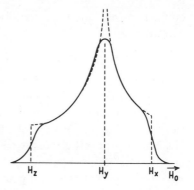

Figure 4. A synthesized EPR absorption spectrum versus applied magnetic field for an amorphous sample with $g_z \geq g_y \geq g_x$. The dashed curve represents dN/dH, the number of molecules resonant at applied fields between H and $H + dH$. The solid curve is the result of including a Gaussian line shape for each molecule. In practice, first-derivative spectra are almost always determined (e.g., Figure 9), not the absorption spectrum depicted here. (Adapted from Fritz et al.[15])

between H and $H + dH$ is given by dN/dH. From equations (2) and (3), one may obtain an analytic expression relating $\sin \theta \, d\theta \, d\phi$ to H and dH. These relations are given by several authors.[11–14] The dotted lines in Figure 4 represent dN/dH versus H for a specific set of parameters. To obtain the actual absorption spectrum, it is necessary to multiply the number of absorbing molecules by the transition probability and absorption line shape at each orientation. This is usually done by computer,[15] particularly when hyperfine interactions are also present. The result is displayed as the solid line in Figure 4.

A discussion of this absorption is useful for what will follow. $H_x = h v_0 / g_x \beta$ is the magnetic field value at which molecules having their g-tensor x axes collinear with the applied field will undergo EPR. $H_z = h v_0 / g_z \beta$ is the magnetic field value at which molecules with their g-tensor z axes collinear with the applied field will absorb, but because $H_z < H_y < H_x$ by assumption, $H_y = h v_0 / g_y \beta$ is the magnetic field where not only molecules with their y axes aligned collinear with the field will absorb but also all molecules where the magnetic field lies anywhere in the plane containing the y axis and the direction in the xz plane where $g = g_y$ will absorb. This is the explanation for the much greater absorption at H_y.

Pertinent to the above is an effect that contributes to the line shapes owing to a distribution of structures or environments in the sample of interest; that is, one protein molecule is not the same as another. Individual molecules may exist either in a range of conformations or in a distribution of electrostatic charges which are hard to demonstrate by other means but which may yield g values, which are distributed about mean values. Nothing is known a priori about the expected shape of such distributions, often called "g strain," but this constitutes a definite source of linewidth, as will be demonstrated in the discussion of the EPR spectra to follow.

The magnetic hyperfine interactions including the magnetic dipole–dipole interaction and the Fermi contact interaction have been discussed at length in previous chapters. Suffice it to remark here that the dipolar and contact terms may be combined using tensor notation to yield a Hamiltonian

$$\mathscr{H} = -\beta \mathbf{S} \cdot \tilde{g} \cdot \mathbf{H} + \mathbf{I} \cdot \tilde{A} \cdot \mathbf{S} - g_n \beta_n \mathbf{I} \cdot \mathbf{H} + \text{higher-order terms} \quad (5)$$

where $\tilde{g}$ is the g tensor and $\tilde{A}$ is a hyperfine tensor, and for a spin $\frac{1}{2}$ electronic system coupled to a spin $\frac{1}{2}$ nuclear system such as ^1H or ^{57}Fe, there are two electronic transitions, corresponding to the two different nuclear orientations, which occur for the same frequency but at two different applied fields. If such absorptions are resolved in EPR, they can be used to identify nuclear constituents of the paramagnetic center, but when they are unresolved, which is usually the case for proteins, they simply contribute to the linewidth to such an extent that most of the EPR signal shapes encountered in biological systems are the sums of large numbers of narrower unresolved hyperfine resonance absorptions. Under such circumstances, ENDOR spectrometry is particularly useful, because it permits the identification of the nuclear constituents and the quantitative measurement of the hyperfine interactions. This will be discussed in detail later.

As has been mentioned previously, spin-exchange interactions have been demonstrated to exist among the iron atoms in several of the multi-iron proteins and between the two four-iron clusters in the eight-iron ferredoxins. The presence of these interactions determines in large measure magnetic resonance spectra as well as the magnetic susceptibilities of these proteins; in addition, because of this interaction it has been possible in the two-iron proteins, to assign ligands to individual iron atoms as will be discussed later in this chapter.

Because sulfur produces weak ligand fields, *if the electrons are localized* (which must be established experimentally), the iron atoms can be expected to be in the high-spin state in all cases; yet, due to an antiferromagnetic spin exchange interaction, the ground spin states of the multi-iron complexes are 0 or $\frac{1}{2}$ only, depending upon the particular protein and its oxidation state. An example is offered by the two-iron ferredoxins where at low temperatures the total spin $S = S_1 + S_2 = 0$ in the oxidized state with the individual iron spins given by $S_1 = S_2 = \frac{5}{2}$ and total spin $S = S_1 + S_2 = \frac{1}{2}$ with $S_1 = \frac{5}{2}$ and $S_2 = 2$ in the reduced state, as shown by a combination of spectroscopies, including Mössbauer in particular.[6]

The total spin state of a four-iron protein is also 0 or $\frac{1}{2}$ at low temperatures depending on oxidation state, and Mössbauer spectroscopy indicates that the four-iron atoms are nearly equivalent to each other in the diamagnetic state. This implies either a distributed electron model for these proteins in this oxidation state, dynamic Jahn–Teller distortions, or rapid "hopping." These studies are continuing in hopes of clarifying this matter, but for the moment, it is a matter for conjecture as to which model is correct, although present thinking seems to lean toward a distributed electron model. In the other oxidation states (paramagnetic), there is evidence that the electron spins are localized and that the iron atoms may again be in the high-spin states.

The combined effect of an electrostatic exchange interaction and spin–orbit coupling yields a spin–spin coupling that can be anisotropic. In addition, there can be electronic magnetic dipole interactions at a distance. Several authors have discussed this in detail,[16,17] and these anisotropic electron-spin–electron-spin interactions play a major role in determining the EPR spectra of the eight-iron ferredoxins.

Necessary to the observation of EPR is the difference in the populations of the various states. The relaxation mechanisms by which this difference population is established and maintained are transitions induced by coupling to the lattice. These transitions, limit the lifetime of a spin in a given state and therewith broaden the energy levels of these spins so that at high temperatures the spin-state lifetimes are very short ($\sim 10^{-11}$ sec) and the levels so correspondingly broad that it is no longer possible to observe EPR in many cases, or the structure of interest is totally obscured. Thus, it is necessary to freeze the protein samples and detect EPR at low temperatures. It has been found empirically that the freezing of the samples does not denature the protein, at least not those proteins studied here; however, the rate of freezing can be important because of clustering. Such effects have been seen for the iron–sulfur proteins and the EPR spectra of the high-potential iron protein from *Chromatium vinosum* are particularly sensitive in this regard.

2.2. The Phenomenon of ENDOR

Recently, *electron nuclear double resonance* (ENDOR) spectrometry has been applied to the study of proteins. The method of ENDOR and the theory behind it have been explained in previous chapters. Suffice it to say that it is possible to use the anisotropy in the electron g factor to select molecules of a given set of orientations to undergo EPR and hence ENDOR; a brief discussion follows.

As was discussed for EPR spectral synthesis above, setting the magnetic field at some value between H_x and H_z permits only those molecules having a specific set of orientations to be studied; e.g., if the magnetic field is set below H_z (assuming $H_x > H_y > H_z$), then primarily those molecules whose z axes are aligned along the field will be undergoing EPR. It is possible to simulate the ENDOR spectrum at any given applied field, and by comparison with the experimental spectrum to obtain the components of the A tensor. The application of ENDOR spectrometry to proteins offers the opportunity for the determination of the spin states of transition element ions and for the determination of the bonding ligands in these paramagnetic sites by the detection and determination of the hyperfine couplings.

2.3. The Phenomenon of ELDOR

The method of electron–electron double resonance (ELDOR) spectrometry and the theory behind it have been explained in previous chapters. Once again, it is possible to use the anisotropy in the electron g factor to select molecules of a given set of orientations to undergo EPR and hence ELDOR. One may use ELDOR to investigate the same set of energy levels and hyperfine coupling parameters as was described for ENDOR, and in fact, that is useful whenever the hyperfine interaction exceeds the radiofrequency range of the ENDOR instrument (~ 60 MHz). Such is the case for cupric ions, for example. However, that is not the purpose to which we will apply it here. As has been indicated, the individual paramagnetic centers in proteins are coupled by two interactions, the electrostatic exchange interaction, $-2J\mathbf{S}_1 \cdot \mathbf{S}_2$, and the magnetic dipole–dipole interaction. We have been using ELDOR to measure these interactions when less than 1.5 GHz, or rather to confirm the determinations from EPR spectral syntheses. The specific application to be described here will be to the two coupled iron–sulfur centers in an eight-iron ferredoxin; however, it may be stated with confidence that spin–spin interactions may be detected between nearly all paramagnetic centers in conjugated proteins. When J is small, the magnetic dipole–dipole interactions provide a means for determining r^{-3} and hence the distance between centers; but when J is large, the spin–orbit coupling will introduce anisotropic spin–spin (pseudodipolar) couplings that mask the true dipolar interactions, and then the latter may not be determined.

In order to understand the application here most easily, we will ignore the dipolar interaction and consider two electron spin $\frac{1}{2}$ centers coupled by an isotropic spin-exchange interaction. The Hamiltonian is then

$$\mathscr{H} = -\beta \mathbf{S}_1 : \tilde{g}_1 : \mathbf{B} - \beta \mathbf{S}_2 : \tilde{g}_2 : \mathbf{B} - 2J\mathbf{S}_1 \cdot \mathbf{S}_2 \qquad (6)$$

which is reminiscent of the Hamiltonian describing type AB spectra in high-resolution NMR; in fact, the analogy is exact. The reader is referred to any text on high-resolution NMR[18] for a discussion of the quantum mechanics. Figure 5a shows the energy levels at fixed magnetic field B, and Figure 5b shows the allowed transitions at fixed field. The separations are given as $2C$ and $2J$, the latter replacing J in NMR because of the factor of 2 appearing in the expression for the spin–spin interaction as customarily defined in EPR, and

$$C = \tfrac{1}{2}[(g_{1_{\text{eff}}} - g_{2_{\text{eff}}})^2 \beta^2 H^2 + 4J^2]^{1/2} \qquad (7)$$

from analogy with NMR. If one now imagines a double resonance experiment where one observes at a microwave frequency corresponding to the 4–2 transition and sweeps the frequency of a second saturating microwave field through the other transitions, one has the ELDOR experiment to be described

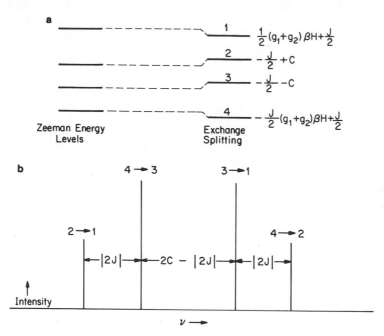

Figure 5. Energy level diagram (a) and stick spectrum (b) for an exchange coupled electron $(S_1 = S_2 = \frac{1}{2})$ system with J negative. See text for explanation of J and C.

later in this chapter. Clearly, it is possible by such an experiment to detect the presence of these other transitions and hence to measure $2J$ and $2C$; i.e., one may measure both the strength of the spin coupling between the two centers and the effective g values of the coupled species. The latter may be used to determine the relative orientations of the two g tensors[19] and the former has a direct bearing on the electron transfer rate within the protein.

If there also exists a magnetic dipole–dipole interaction between the centers, it should be possible to measure it as well and to get thereby the distance between the centers. Such is the case for many coupled centers in proteins.[19]

3. Instrumentation

3.1. ENDOR Instrumentation

There were two ENDOR instruments used for the studies described here. The X-band system was built by Dr. John Fritz and is described in detail in his doctoral thesis[20] and briefly in a published paper.[15]

The basic ENDOR instrument is nearly the same as that described by Hyde and Maki[21] and Hyde,[22] and the reader is referred to these papers for details. The one major difference is that the magnetic field modulation (at 47 Hz) is approximately square wave rather than sinusoidal. This was an essential change that results in an ENDOR spectrum from only one magnetic field setting (the other being chosen off the EPR resonance absorption when possible). When the EPR signal is isotropic as in the case for free radicals in solution as studied by Hyde and Maki[21] and Hyde,[22] the sinusoidal modulation presents no problem; however, when the EPR spectrum is anisotropic, the sinusoidal modulation yields ENDOR spectra from a continuum of magnetic field values and hence molecular orientations and thereby greatly complicates the spectral interpretation.

The second ENDOR instrument operated at K_u band and was constructed by Drs. Joseph Reid and W. Richard Dunham. It is described in detail in Reid's doctoral thesis.[23] The major difference was that the radio-frequency field was applied by means of four wires internal to the cylindrical cavity. In other respects it was similar to the above.

3.2. ELDOR Instrumentation

The ELDOR instrument used for the studies described here was designed and built by D. Palaith and M. Fung and is described in detail elsewhere.[19]

The basic ELDOR instrument is nearly the same as the commercial Varian system except for three features: (i) the available pump power is 3 W, provided by a traveling wave tube amplifier; (ii) the field modulation frequency is 6 kHz; and (iii) the pump cavity has its resonance frequency changed by mechanically inserting sapphire rods into it at the strong electric field positions. The latter permits continuous tuning of the pump frequency over $v_p - v_o = 0$ to 1500 MHz at X band.

The cavity is a crossed dual-mode (TE_{104} pump and TE_{102} observing) system similar to that described by Hyde *et al.*[24] The low-temperature system is a modified Air Products Heli-Tran system that uses a continuous flow of cold helium gas to cool the sample. The pump power is square-wave modulated at 47 Hz. The resulting signal is phase-detected at 6 kHz and 47 Hz.

4. Iron–Sulfur Proteins—ENDOR and ELDOR

4.1. Two-Iron Ferredoxins

In collaboration with many other colleagues, we have studied the two-iron ferredoxins from plants, algae, bacteria, and mammals. These are: mammalian—beef and pig adrenal cortex; plant—cotton, spinach and parsley; algal—*Synechococcus lividus* and *Chlorella vulgaris*; bacterial—*Pseudomonas putida, Azotobacter vinelandii,* and *Clostridium pasteurianum.*

In addition to these studies, Hall *et al.* have studied nearly all of the above plus the two iron ferredoxins from *P. aminovorans*,[25] *A. tumefaciens*,[26] *Chloropseudomonas ethylica*,[27] and *Halobacterium halobium*,[28] and the algal ferredoxin from *Euglena*[29]; Tsibris *et al.*[30] and Munck *et al.*[31] have made detailed studies of the bacterial ferredoxin from *P. putida*; Kerscher *et al.*[28] have studied the ferredoxin from *H. halobium*; and Orme-Johnson and Beinert[32] have studied the ferredoxin from adrenal cortex. The similarities between most of the bacterial and mammalian two-iron ferredoxins on the one hand and the plant and algal on the other are striking; however, there are distinct differences between the classes that should be noted.[31] *Azotobacter vinelandii* has a two-iron ferredoxin in each class, and that from *C. pasteurianum* and *C. ethylica* looks like the plant and algal variety, so it is not strictly proper to divide them as simply as is indicated.

4.1.1. *Common Properties*

The biochemistry that established these proteins as a class characterized by their having two iron and two acid-labile sulfur atoms per molecule[32] took place concomitantly with the physical studies described here.

These proteins have the following common properties. In addition to the amino acid sequence, there are two atoms of iron and two atoms of labile sulfur and nothing else. The proteins will accept only one electron upon reduction with a redox potential near that of the hydrogen electrode, ~ -420 mV, and they exhibit a characteristic EPR spectrum in the reduced state that quantitates to give $S = \frac{1}{2}$ and g values close to the free electron value; upon close examination, two of the g-tensor principal values are below the free electron value and one is above. Table 1 lists these properties for a few of the proteins, including the molecular weights. The amino acid sequences for these two-iron proteins exhibit strong similarities[4] and at least four cysteine residues, which have been implicated in the iron–sulfur complex of these proteins.

It is appropriate at this point to summarize the results of the many optical, magnetic resonance, and Mössbauer experiments in terms of a model[6] for the active center of these proteins (Figure 6). This model cannot be fully justified in this chapter, because it is based in large measure on the results of Mössbauer spectroscopy. The latter could not have been properly interpreted at low temperature without the results of ENDOR to be discussed here. The reader is referred to the review paper by Sands and Dunham[33] for a complete discussion of all of these techniques. These proteins are believed to contain two iron atoms, each of which are tetrahedrally coordinated by four sulfur atoms. Two of the sulfur atoms (labile sulfides) are bridging

Table 1

Some Properties of Two-Iron Ferredoxins

Protein	Molecular weight	Fe	S^{2-}	Number of electrons	E' (mV)	g value	Optical peaks wavelengths (nm)
Azotobacter I	21,000	2	2	1	~ −350	1.93, 1.94, 2.01	331, 419, 460
Azotobacter II		2	2	1	~ −350	1.91, 1.96, 2.04	344, 418, 460
Parsley ferredoxin	10,600	2	2	1	−413	1.90, 1.96, 2.06	330, 422, 463
Adrenodoxin	12,500	2	2	1	−274	1.93, 1.93, 2.02	330, 415, 453
Spinach ferredoxin	10,600	2	2	1	−420	1.89, 1.95, 2.05	325, 420, 465
Putidaredoxin	12,500	2	2	1	−235	1.93, 1.93, 2.02	325, 415, 455
Cotton	10,950	2	2	1	−420	1.89, 1.95, 2.05	325, 419, 460
Clostridium pasteurianum	25,000	2	2	1	−403	1.92, 1.94, 2.01	333, 425, 463
H. halobium	14,800	2	2	1	−345	1.90, 1.97, 2.07	329, 421, 467
Eqisetum telmateia	10,500	2	2	1		1.896, 1.966, 2.057	325, 420, 465

ligands common to both iron atoms; the other four sulfur atoms are provided by cysteine residues. Please notice that the cysteine ligands for the ferric atom are only dashed in this picture; this is intended to indicate that their presence has not been established by any definitive spectroscopic experiment. Whether or not these two cysteine ligands were present in the protein remained to be established by these studies; however, the correctness of the above conclusions are attested to by model compound syntheses. Specifically, the studies by Mayerle et al.[34] on the $[FeS(SCH_2)_2C_6H_6]_2^{2-}$ anion have shown its structure to be that in Figure 7. Furthermore, the beautiful extrusion and reconstitution experiments of Que and Holm[35] have firmly established all aspects of the above model.

As has been indicated, the initial evidence for the model in Figure 6 was obtained from a combination of chemical data and the physical studies previously mentioned.[6,15,36–40] These studies made use of the magnetic properties of the reduced states of these proteins determined by isotopic substitutions for the atoms in this center. By observing the interaction between the electronic spins and these nuclear spins, one could deduce the

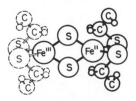

Figure 6. Schematic structure of the iron–sulfur complex in the two-iron ferredoxins as determined by spectroscopic and chemical studies. (Adapted from Dunham et al.[6])

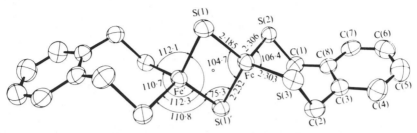

Figure 7. The structure of the centrosymmetric $[FeS(SCH_2)_2C_6H_4]_2^{2-}$ anion. Fifty percent probability ellipsoids of thermal vibration are shown; hydrogen atoms are omitted for the sake of clarity. Other important structural parameters: Fe $\cdots$ Fe′, 2.698(1); S(1) $\cdots$ S(1), 3.498(3); S(2) $\cdots$ S(3), 3.690(2); C(1) $\cdots$ C(2), 3.046(5)A; dihedral angle between the planes FeS(1)S(1)′ and FeS(2)S(3), 89.95(5)°. (From Mayerle et al.[34])

vicinal character of each atom in the active center. Thus, one might expect to see magnetic coupling to both iron atoms, both labile sulfides, all four cysteine sulfur atoms and the α-CH and β-CH$_2$ protons of the cysteine residues.[6,39,40] In addition, the presence of iron in the active center makes it possible to use Mössbauer spectrometry[41] on the ^{57}Fe to study the magnetic and electrostatic environment of the two iron atoms. Furthermore, the closeness of the two iron atoms indicates a possible electron-exchange interaction and the use of static magnetic susceptibility[38] to measure this interaction in both the oxidized and reduced states. These measurements show that in both oxidation states there is an antiferromagnetic exchange interaction between the electron spins of the two iron atoms. In the oxidized state, the two nearly equivalent iron atoms are both high-spin ferric ($S = \frac{5}{2}$), resulting in diamagnetism below 77°K ($J \simeq -200$ cm^{-1}) and paramagnetism that increases with temperature above 77°K. Upon single-electron reduction, the ferredoxin displays Curie $S = \frac{1}{2}$ paramagnetism below 77°K and paramagnetism greater than Curie above 77°K, resulting from the antiferromagnetic exchange interaction (-100 cm^{-1}) between a high-spin ferric and a high-spin ferrous iron. Using simple crystal field theory, the ferrous iron atoms may be described as being in a rhombically distorted tetrahedral ligand field (probably consisting of two labile and two cysteine sulfur atoms), with the reducing electron in a d_z^2 orbital. Whether the application of such a theory is appropriate is the subject of continuing study.

4.1.2. EPR Studies

Shetna et al.[42] provided the first demonstration that the paramagnetic center of these proteins contained iron by growing Azotobacter on a medium enriched in ^{57}Fe ($I = \frac{1}{2}$) and showing that the isolated iron–sulfur protein yielded a broadened EPR signal in comparison to that of the control, prepared from cells grown on a medium of low natural abundance ^{57}Fe. This broadening was attributed to the unresolved nuclear hyperfine interactions with iron. This work stimulated several authors to put forth hypothetical models to account for the observed EPR g values on the basis of iron being involved.[43–46] Additional experimental data was required to distinguish among these models and to prove conclusively the nature and number of the iron atoms involved.

The evidence that *two* vicinal iron atoms were involved came initially from the EPR studies of Beinert and Orme-Johnson[32] and of Tsibris *et al.*[30] on proteins in which the ^{56}Fe had been substituted by ^{57}Fe (see Figure 9B). The observed $1 : 2 : 1$ triplet hyperfine splitting at g_z from the ^{57}Fe protein indicated that since ^{57}Fe has a nuclear spin of $\frac{1}{2}$, there must be two iron atoms interacting comparably to produce a $1 : 2 : 1$ triplet as opposed to the doublet expected from just one iron.

Exchange experiments[47] where the labile sulfur had been substituted by ^{80}Se and ^{77}Se demonstrated that the selenium and hence the labile sulfur atoms were an integral part of the active center, and by virtue of a similarly observed $1 : 2 : 1$ EPR hyperfine triplet for proteins with ^{77}Se $(I = \frac{1}{2})$ atoms substituted, the conclusion that there were two labile sulfur (or selenium) atoms involved in the active center was established.

Der Vartanian *et al.*[48] showed by a ^{33}Se growth experiment on *P. putida* and a labile sulfide exchange, that cysteine sulfur was also a part of the active center, although the number of cysteines involved could not be established.

Before continuing to the main subject of this chapter, the ENDOR studies, it is appropriate to call attention to a striking feature of the EPR spectra of all of these iron–sulfur proteins, namely, the existence of a field-dependent inhomogeneous linewidth, often referred to as " g strain." Strong,[49] in his thesis studies, collected a great deal of data on this feature. Figure 8 shows his observed EPR linewidths at 2.8, 9.6, and 35 GHz for several reduced two-iron ferredoxins. Your attention is called to the fact that the linewidths are nearly linear with frequency, deviating only at the lowest frequency (due to unresolved hyperfine interactions), thus verifying that the sample exhibits a " distribution " of principal-axis g values. The cause of this distribution has not been established, although it is thought by many to arise from a statistical distribution of charges on the protein molecules or a statistical distribution of conformations, rendering one molecule different from another. The EPR linewidth is presently the most sensitive indicator of these differences. It is interesting to note that the great utility of the ENDOR technique to these proteins is due to the presence of this g strain, which tends to obscure any resolution of the superhyperfine interactions in the EPR spectra but permits such resolution by ENDOR.

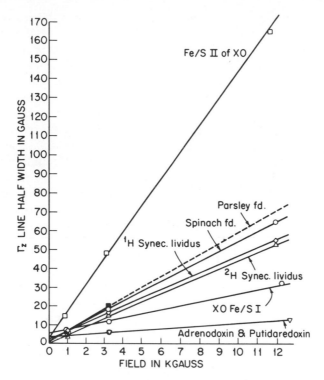

Figure 8. Magnetic field dependence of the EPR line half-width at half-maximum amplitude of the g_z feature for a variety of two-iron–two-sulfur proteins. (From Strong.[49])

4.1.3. Iron ENDOR Studies

It is possible by measuring the principal components of a hyperfine tensor to obtain knowledge about the electron spatial distribution. Because the electronic g tensor is anisotropic for the two-iron ferredoxins, one may use ENDOR not only to obtain the principal components of the hyperfine (A) tensor but to obtain the relative orientation of the principal axes of the A and g tensors (see Fritz et al.[15]).

The experimental EPR spectra for native and ^{57}Fe-enriched adrenodoxin are shown in Figure 9A and B. It is possible to pick out for study only those molecules whose z axes lie along the applied field direction by setting the field at H_z, hence only those molecules will be undergoing EPR and ENDOR. By setting the field at $H_x = H_y = H_\perp$ all molecules where the field is in the xy-plane are saturated and undergoing EPR and ENDOR. Figure 10 displays the resulting ENDOR spectra obtained at 19°K as the applied radio-frequency field was swept in frequency from 6 to 30 MHz. When the field was set at H_z, the top spectrum was obtained and when the field was set at $H_\perp$, the bottom spectrum resulted. Notice that there are several pairs of lines in the top spectrum centered at 13.5 MHz. These arise from protons (nuclear Zeeman frequency equals 13.5 MHz in this applied field) that are coupled to the paramagnetic center by hyperfine interactions. The asymmetric doublet at 21 MHz tailing to higher frequencies is associated with one of the ^{57}Fe atoms. That doublet is identified as arising from iron, because the separation is twice the nuclear Zeeman frequency for iron. The center of that doublet must therefore be $A_z/2$ for that iron. The ENDOR signal from the second iron is located under the proton signals as can be determined from the ^{57}Fe–^{56}Fe difference spectra.[15] Your attention is called to the fact that in the bottom spectrum the signal to the right here shows two doublets, one centered at 25 MHz and the other at 27.5 MHz. These can be identified as $A_y/2$ and $A_x/2$, respectively, for iron.[15] The ENDOR signal vanishes in this region for the ^{56}Fe sample.

Fritz *et al.*[15] display the ^{57}Fe–^{56}Fe difference spectra in the lower frequency region, where the proton ENDOR signals are dominant. The components of the hyperfine tensor for the second iron could be estimated from this data, but as long as the protons were present in the protein, we could not hope to get very accurate parameters in the x–y region. Similar results were obtained using putidaredoxin (see Fritz *et al.*[15]).

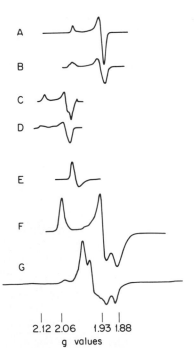

A

B

C

D

E

F

G

| | | | |
2.12 2.06   1.93 1.88
g values

Figure 9. Derivative EPR spectra for several iron-sulfur proteins: (A) reduced native two-iron adrenodoxin, (B) ^{57}Fe reconstituted adrenodoxin, (C) oxidized HiPIP from *R. tenue*, (D) oxidized HiPIP in 0.2 *M* NaCl from *C. vinosum D*, (E) oxidized HiPIP-type center S-3 from pigeon mitochondrial membrane, (F) reduced four-iron ferredoxin I from *B. polymyxa*, and (G) reduced eight-iron ferredoxin from *M. Lactilyticus*.

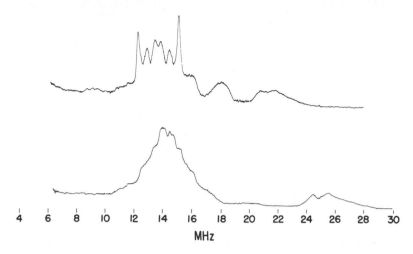

Figure 10. ENDOR spectrum recorded for reduced adrenodoxin in D_2O at 19°K and with the applied field (upper) set at H_z so as to select molecules with the z axes aligned along the magnetic field and (lower) set at $H = H_x = H_y$ so as to select molecules with orientations such that the magnetic field is somewhere in the xy plane of the molecular g-tensor principal axis system. (From Fritz *et al.*[15])

The protons proved to be even more of a problem in attempting to get the iron ENDOR from spinach and parsley ferredoxin. Figure 11 displays the ENDOR spectra obtained for ^{56}Fe and ^{57}Fe spinach ferredoxin. Spectra a and c were obtained with the magnetic field set at H_z. The difference spectra at 17.7 and 22 MHz represent the ENDOR of the two different ^{57}Fe nuclei; there is no discernable doublet structure in these lines. However, when the magnetic field was changed to $H_z + 31$ G, spectra b and d were obtained, and two features were noted—a doublet appeared at 21.5 MHz with the absorption tailing to higher frequency and the difference spectra in the low-frequency region had gotten weaker and moved to lower frequency. When the magnetic field was set at $H_z + 62$ G, spectrum e was obtained, and when the magnetic field was changed to H_x, spectrum f was obtained. This data made it possible to determine the principal axis values of the A tensor for the one iron (those ENDOR signals in the 21–25 MHz range), but the ENDOR for the second species showed only the A_z value, and the other components were lost under the proton ENDOR due to spin diffusion to the greater number of matrix protons. Similar behavior was observed for parsley ferredoxin.

From that data, it was clear that we needed a completely deuterated protein if we were to obtain the complete hyperfine tensors for spinach and parsley ferredoxin. Dr. Henry Crespi at Argonne National Laboratories had such a ferredoxin from the growth of an alga, *S. lividus*, on nearly pure D_2O and kindly provided ample amounts for our study. This algal ferredoxin is

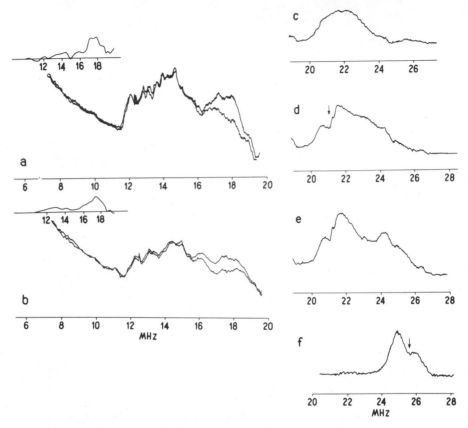

Figure 11. ENDOR spectra at 15°K and $\nu_0 = 9250$ MHz for reduced ^{57}Fe reconstituted (upper or single trace) or ^{56}Fe (lower trace) spinach ferredoxin in D_2O with the magnetic field set at (a and c) H_z, (b and d) $H_z + 31$ G, (e) $H_z + 62$ G, and (f) H_x. The difference spectra in (a) and (b) are plotted on a reduced scale in the upper left. (From Fritz *et al.*[15])

nearly identical to that from spinach in regard to molecular weight and EPR, ENDOR, and optical spectra. The proteins may not be distinguished by these criteria; however, because of the absence of proton ENDOR signals, it was possible to obtain the desired iron ENDOR signals for the second iron with the field set at H_x and H_y in addition to those with the field set at H_z.[50,51]

The algal ferredoxin was reconstituted with either ^{57}Fe or ^{56}Fe in the active site by standard procedures.[50,51] Figure 12 displays the ^{57}Fe ENDOR spectra in the region above 20 MHz, where no ^{56}Fe ENDOR could be detected. The resolved ENDOR doublet centered at 25.7 MHz observed with the field set at $H_x + 60$ G, where predominantly only a single protein orientation satisfies EPR and ENDOR resonant conditions indicated by accompanying computer simulations that $A'_x = 51.6$ MHz (the prime indicates values in the principal axis frame of the effective A tensor). Similarly, H_z selects a single protein orientation; however, the observation of a doublet at $H_z + 30$ G and not at H_z indicates[15] that the hyperfine tensor is rotated $30° \pm 5°$ about the $g_x(A_x)$ axis with respect to the g tensor. $A'_z = 42$ MHz from the computer simulations. Similarly, $A'_y = 50 \pm 2$ MHz.

The computer simulations in Figure 12 were obtained using the hyperfine values listed in Table 2 under the simplistic model that the relaxation times and transition probabilities were independent of orientation. Such a model should be good whenever only a small number of orientations are present, but when, as in Figure 12C, there are a large number of orientations undergoing EPR and ENDOR, the fit is not expected to be very good. No attempt has been made to refine the model, because we do not know how the relaxation times depend on orientation.

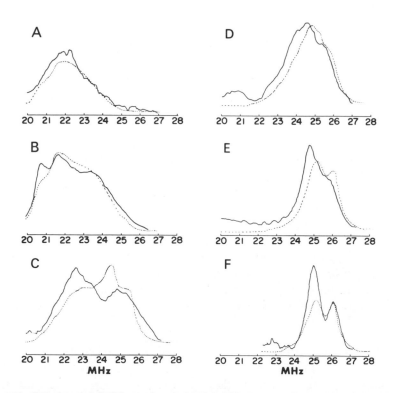

Figure 12. Experimental (solid curve) and simulated (dashed curve) ferric ENDOR spectra of reduced ^{57}Fe reconstituted algal ferredoxin from *Sy. lividus*. $T = 13°$K. The spectra were recorded with the magnetic field set at (A) H_z, (B) $H_z + 30$ G, (C) $H_z + 120$ G, (D) $H_x - 80$ G, (E) H_x, and (F) $H_x + 60$ G. (From Anderson *et al.*[51])

Table 2

Measured Principal-Axis Components of the Effective Hyperfine Tensors for $S = \frac{1}{2}$ as Determined from ENDOR Spectra[a]

Protein	Ferric site			Ferrous site		
	A_x	A_y	A_z	A_x	A_y	A_z
Spinach ferredoxin[b]	19.3 ± 0.4	$18.2 \, ^{+0.7}_{-2.6}$	14.6 ± 0.5	n.d.[c]	n.d.[c]	12.4 ± 0.7
Parsley ferredoxin[b]	19.2 ± 0.4	$18.2 \, ^{+0.7}_{-2.6}$	14.6 ± 0.7	n.d.[c]	n.d.[c]	12.1 ± 0.9
Adrenodoxin[b]	18.5 ± 0.5	$20.7 \, ^{+0.4}_{-1.1}$	$15.2 \, ^{+0.4}_{-0.7}$	6.3 ± 1.5	8.9 ± 1.5	12.4 ± 0.5
Putidaredoxin[b]	18.5 ± 0.5	$20.7 \, ^{+0.4}_{-1.1}$	$15.2 \, ^{+0.4}_{-0.7}$	6.3 ± 1.5	8.9 ± 1.5	12.4 ± 0.5
S. lividus ferredoxin[d]	$19.5 \, ^{+0.1}_{-0.2}$	18.2 ± 0.7	$14.6 \, ^{+0.1}_{-0.2}$	$4.9 \, ^{+0.1}_{-0.2}$	5.4 ± 0.4	12.7 ± 0.2

[a] In equivalent gauss at the net electron spin.
[b] From Fritz *et al.*[15]
[c] Not determined.
[d] From Anderson.[51]

The nearly isotropic hyperfine interaction can be assigned to a high-spin ferric atom that is undergoing spin-exchange with another iron atom.[36,51]

Figure 13 displays the $^{57}Fe-^{56}Fe$ ENDOR difference spectra in the low-frequency region, and just as for spinach ferredoxin with the field set at H_z, the difference spectrum occurs at 17.5 MHz. With the field set at H_x, the difference signal (Figure 13C and F) is centered at 6.5 MHz, establishing $A_x = 13.0 \, ^{+0.1}_{-0.5}$ MHz by computer simulation. This signal has never been observed in a protonated ferredoxin of this type.

With the field set at H_y, the $^{57}Fe-^{56}Fe$ difference ENDOR spectrum could not be scanned in a single rf sweep range, and three sweep ranges had to be used. The spectra were set to be identical at 3 MHz and at 19 MHz, assumptions that were justified by the A_x and A_z measured values. The observed difference spectrum was nearly flat from 7.6 to 14 MHz. Computer simulations[50,51] showed $A_y = 15 \pm 1$ MHz; however, Anderson *et al.*[51] were unable to detect any indication that the A-tensor principal axes were rotated with respect to the g-tensor axes. Table 2 summarizes the iron ENDOR data; the assignment of the respective tensors to a ferric and ferrous atom is made from the Mössbauer results.[51]

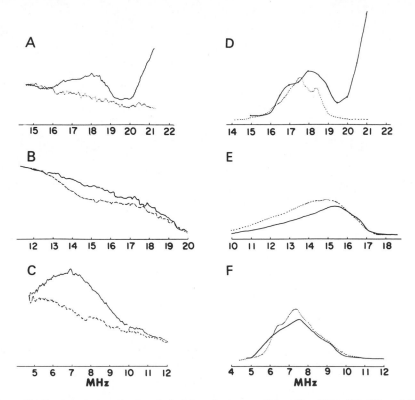

Figure 13. ENDOR spectra of reduced algal ferredoxin recorded at $T = 10°K$. (A), (B), and (C) are experimental ^{57}Fe (solid) and ^{56}Fe (dashed) reconstituted ferredoxin spectra recorded with the magnetic field set at H_z, $H_z + 60$ G, and H_x, respectively. (D), (E), and (F) are experimental differences (solid) and computer-simulated (dashed) spectra for the magnetic field set at H_z, $H_z + 60$ G, and H_x, respectively. (From Anderson et al.[51])

4.1.4. Proton ENDOR Studies

Figure 14 displays the proton ENDOR spectra obtained under various conditions for the algal ferredoxin from *S. lividus*. The ENDOR spectrum of the fully protonated protein in *tris*-chloride buffer with the field set at H_x is shown in Figure 14A. The 10–20 MHz segment shows six pairs of semi-resolved proton lines centered at the free proton frequency (14.65 MHz at H_x) and split by their hyperfine values. None of these resolved resonances disappears when the sample is freeze-dried and then redissolved in D_2O; however, the overall shape of the ENDOR changes, suggesting some unresolved protons are exchangeable in addition to the "matrix" water protons.

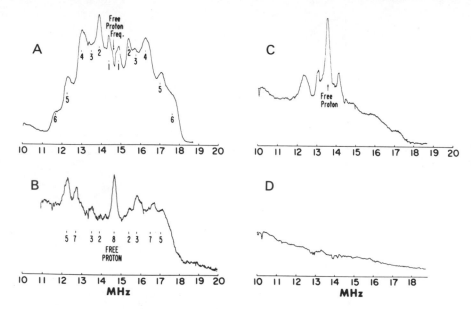

Figure 14. Proton ENDOR spectra of reduced algal ferredoxin recorded at $T = 20°K$. (A) Fully protonated with the magnetic field set at H_z. (B) and (C) Fully deuterated ferredoxin extracted in protonated *tris*-chloride buffers, twice freeze-dried, and soaked in D_2O for 1 week, at H_x and H_z, respectively. (D) Fully deuterated ferredoxin extracted in deuterated phosphate buffers. (From Anderson et al.[51])

Dr. Crespi normally isolated the deuterated proteins using protonated buffers. When the algae were grown on a fully deuterated medium, extracted in protonated phosphate buffers, transferred to *tris*-chloride buffer, twice freeze dried, and taken up with D_2O, the ENDOR spectrum in Figure 14B is obtained; the relative gain in Figure 14B is five times that for Figure 14A. A deuterated sample extracted in totally deuterated phosphate buffers showed only the sloping baseline at low frequency (Figure 14D); thus, the resonances seen in Figure 14B and C are due to protons that exchanged on during protein isolation but that do not totally reexchange even if placed in D_2O for 1 week. It was discovered, however, that these protons rapidly exchanged during the iron reconstitution procedure if performed in deuterated buffers.

The reader's attention is called to the fact that protons 2, 3, and 5 appear in both Figures 14A and B, whereas the weakly coupled protons, 8, exchange readily except for those due to the nonexchangeable *tris* protons. These *tris* protons are weakly coupled to the unpaired spin of the center and can be distinguished by the intensity of the signal being proportional to the *tris* concentration. There are no strongly coupled, readily exchangeable protons.

4.1.5. Discussion

These ENDOR studies demonstrated clearly that there were two distinctly different iron sites in each of the two-iron ferredoxins. For example, with the magnetic field set for resonance for those molecules with the z axes aligned along the field, Figures 11–13 show that there are ENDOR signals at 17.5 and 21.5 MHz that are associated with ^{57}Fe. These absorptions are separated by more than twice the nuclear Zeeman frequency for iron and therefore cannot arise from a single iron site (^{57}Fe in its ground state has nuclear spin one-half and therefore no quadrupole moment that could be responsible for such a splitting either). The hyperfine interaction of two iron atoms with the net electronic spin is the only possible interaction that could account for iron ENDOR signals at these frequencies; therefore, these signals yield direct measurements of the z components of the effective A tensors for the two iron atoms. Because the EPR spectra quantitate to a net electronic spin $S = \frac{1}{2}$, these ENDOR signals must be centered at $A_z/2$ for each atom yielding 35 MHz and 43 MHz for the respective A_z values for the two iron atoms. These are sufficiently close that they give the observed $1:2:1$ triplet splitting of the EPR spectra in the g_z region for adrenodoxin (Figure 9B) and putidaredoxin.

The studies also demonstrated the relative orientations of the effective g and A tensors for the sites. These data were essential in permitting unequivocal interpretation of the Mössbauer data on these proteins, and the latter data permitted us to assign these A tensors to two antiferromagnetically coupled high-spin ferrous and high-spin ferric atoms. That assignment provided a means to interpret the high-resolution proton NMR data and to unequivocally assign two cysteine ligands[6] to the ferrous atoms and pointed toward two cysteine ligands for the ferric atoms as well. This was a sufficiently detailed picture of the active site to point the way for the syntheses of model compounds.[34]

4.2. Four-Iron–Sulfur Proteins

4.2.1. Classifications and the Three-State Model

Four-iron–sulfur proteins may be divided into three classes:

1. High-potential iron proteins (HiPIP), which have been isolated from *Chromatium*,[52] *Rhodopseudomonas gelatinosa*,[53] *Rhodospirillum tenue*,[54] *Thiocapsa pfennigii*,[55] and other bacteria. These proteins are paramagnetic in the oxidized state and are characterized by a positive redox potential $\sim +350$ mV and EPR spectra with $\bar{g} = 2.06$.

2. High-potential iron-type proteins isolated from mitochondria[56-58] and nitrogen-fixing bacteria that exhibit an EPR spectrum with small g anisotropy and $\bar{g} = 2.01$ and have small positive to large negative (-350 mV) redox potentials. These proteins are paramagnetic in the oxidized state.

3. Four-iron ferredoxins, which have been isolated from *Bacillus polymyxa*,[59] *P. ovalis*,[60] *Spirochaeta aurantia*,[61] and *Bacillus stearothermophilus*.[62] These proteins are paramagnetic upon reduction and exhibit low redox potentials $\sim -350 \text{ mV}$ and EPR spectra with $\bar{g} = 1.96$.

X-ray crystallographic data on the HiPIP center in *C. vinosum D*[8] shows a cubanelike iron–sulfur center with 4Fe, $4S^{2-}$ and 4 cysteine residues (Figure 15B).

X-ray data on an eight-iron ferredoxin from *M. aerogenes*[7] shows two four-iron cubane structures (Figure 15C) of similar composition to that of HiPIP, from which it is concluded that this structure is a common structural element in the four- and eight-iron–sulfur proteins.

The redox and magnetic properties summarized above plus the fact that the optical spectra and magnetic properties of reduced HiPIP and oxidized ferredoxin were the same led Carter *et al.*[63] to postulate a three-state scheme for the redox properties of these centers:

$$\begin{array}{c} \text{Oxidized} \\ \text{HiPIP} \end{array} + e^- \rightleftarrows \begin{array}{c} \text{Reduced} \\ \text{HiPIP} \end{array} \equiv \begin{array}{c} \text{Oxidized} \\ \text{ferredoxin} \end{array} + e^- \rightleftarrows \begin{array}{c} \text{Reduced} \\ \text{ferredoxin} \end{array}$$

or

$$C^- + e^- \rightleftarrows \qquad C^{2-} \qquad + e^- \rightleftarrows \qquad C^{3-}$$

where C^{n-} corresponds to the formal valences of the centers: $(3Fe^{3+}, 1Fe^{2+}, 4S^{2-}$, and $4RS^-) \equiv C^-$; $(2Fe^{3+}, 2Fe^{2+}, 4S^{2-}$, and $4RS^-) \equiv C^{2-}$; and $(1Fe^{3+}, 3Fe^{2+}, 4S^{2-}$, and $4RS^-) \equiv C^{3-}$. (*Note:* Carter *et al.* used $+, 0,$ and $-$ for these labels, but the formal valences seem to be preferred by most workers in the field.)

Cammack[64] provided the first direct evidence to support the above three-state hypothesis when he showed that by treating HiPIP in 80% dimethyl sulfoxide, he could reduce it by dithionite to yield an EPR spectrum similar to that of a ferredoxin and reversing this process by addition of ferricyanide yielded the HiPIP spectrum again, with recovery of the protein.

The different properties of the various four-iron–sulfur proteins may be attributed to the different environments of the centers—the cluster in HiPIP is in a hydrophobic pocket[63] and in the bacterial ferredoxins it is close to the surface.

Model compounds have been synthesized by Holm and his colleagues.[65,66] X-ray data of the inner portion of the $[Fe_4S_4(SCH_2Ph)_4]$ anion is shown in Figure 15A. One of the surprising features of these model

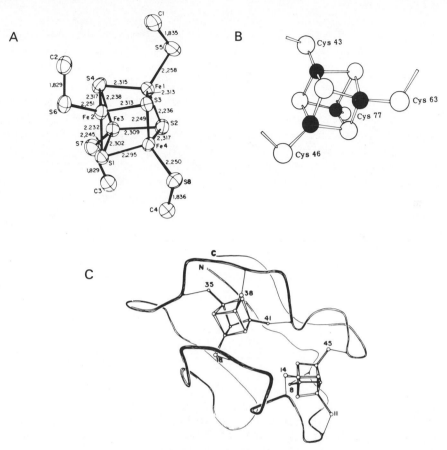

Figure 15. X-ray structures determined for (a) the inner portion of the $[Fe_4S_4(SCH_2Ph)_4]^{2-}$ anion with the hydrogen atoms omitted for the sake of clarity, (b) the HiPIP center from *C. vinosum*, and (c) the eight-iron ferredoxin from *M. aerogenes* showing only the carbon iron and sulfur positions. Fifty percent ellipsoids are shown. See text for references.

compounds is that the centers have structural stability unto themselves. The ligands may be exchanged[67] and the centers may be incorporated by competitive ligand substitution into the protein or extruded from them.[35] This has provided a potential means of distinguishing the two- and four-iron centers in many conjugated iron–sulfur proteins.

4.2.2. High-Potential Iron Proteins

The amino acid sequence of the HiPIP from *C. vinosum D* is known.[4] From x-ray data, the cysteine residues at positions 43, 46, 61, and 70 bind the iron–sulfur cluster. One portion of the molecule contains five hydrophobic

residues in a row (residues 34–38) and two other portions with four hydrophobic residues adjacent. In summary, this has been described loosely as an
" oil drop with an iron–sulfur cluster in the center." The x-ray structure of
the center is shown in Figure 15B.

   4.2.2.1. *EPR Studies.* The EPR spectrum (see Figure 9C) of the four-iron
HiPIP from *R. tenue* is describable by a single g tensor and quantitates to
$S = \frac{1}{2}$. The EPR spectrum obtained in frozen aqueous solution for the HiPIP
from *C. vinosum D* reveals a heterogeneity.[68] Space does not permit the
delineation of the full evidence[69] for the heterogeneity that comes from
titration with oxidant and subsequent freezing and recording of the EPR
spectrum. It is clear that in 0.2 M NaCl the EPR of fully oxidized protein
(Figure 9D) shows the presence of at least three resonances at g_z in addition
to the classical "bump" near $g = 2.09$ (3170 G). The three peaks near g_z do
not appear in the titration until more than 0.25 mole equivalent of oxidant
has been added, whereas the optical spectrum recorded at room temperature
shows a linear Nernst plot over the 5–95% oxidation range; thus, the heterogeneity revealed by EPR must be induced by the freezing process. EPR at
35 GHz demonstrates that the three peaks near g_z at X band must originate
from a field-independent spin–spin coupling of some sort, but the bump at
$g = 2.09$ does correspond to a real g value. The signal associated with the
bump represents approximately 15% of the total EPR intensity. This salt-
induced heterogeneity was not observed with *R. gelatinosa* HiPIP or with *R.
tenue* HiPIP (Figure 9C).

   4.2.2.2. *ENDOR Studies.* The electron nuclear double resonance
(ENDOR) at K_u band[23] of oxidized *C. vinosum* (^{57}Fe enriched) HiPIP shows
only two resolved doublets (see Figure 16). These are nearly isotropic and
correspond to ^{57}Fe A values of 22.6 and 32.5 MHz in agreement with the
work of Anderson *et al.*[51] Notice, however, that they had to do spectral
subtractions and had unresolved doublets. The K_u band field moves the
proton ENDOR signals out of the way. No additional iron ENDOR signals are
discernible for the $g = 2.09$ signal, indicating that the latter could be due to a
spin-coupled $(S = 1)$ system. Additional EPR rapid-freezing experiments[69]

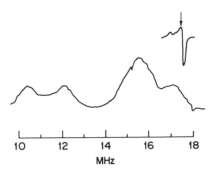

10   12   14   16   18
              MHz

Figure 16. ENDOR of oxidized ^{57}Fe-enriched
HiPIP from *C. vinosum* recorded at 16 GHz
so that the proton ENDOR signals are at higher
frequency. (From Reid.[23])

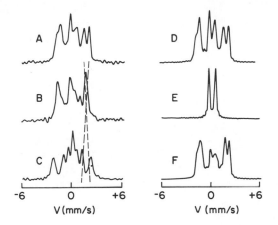

Figure 17. Mössbauer spectra and computer simulations for ^{57}Fe-enriched oxidized HiPIP from C. vinosum D. (A)–(C) Spectra recorded at 4.2°K with the applied magnetic field along the x-ray direction and set at 0.115 T, 3.0 T, and 6.0 T, respectively; (D) computer fit to (A) consisting of 16% of (E) and 84% of (F); (E) computer fit to the high-temperature spectrum of oxidized HiPIP to match the quadrupole splitting; (F) typical synthesized spectrum assuming the two distinctly different and opposite in sign A tensors obtained from ENDOR and quadrupole splittings adjusted to fit the experimental spectra.

have demonstrated that the 2.09 signal and the structure at g_z are induced by freezing and presumably are caused by protein association and the resulting spin–spin coupling.

The significance of the ENDOR results lies in the fact that only two different A tensors are discernible, though there are four iron atoms in the HiPIP center. This means that the iron atoms may be viewed as consisting of two pairs, each atom of the pair being equivalent. This was a surprise, because in the Carter three-state hypothesis discussed earlier, the formal valence states of the iron atoms would be three ferric and one ferrous. This will be discussed in more detail later, but the pairing is further supported by Mössbauer spectrometry.

Figure 17A,B,C displays the experimental Mössbauer spectra for fully oxidized *Chromatium* HiPIP[69] recorded with applied magnetic fields of 0.115, 3.0, and 6.0 Tesla, respectively. Notice how the outside lines cross over, indicating that the applied field and the smaller hyperfine field add in the ground state, whereas the larger hyperfine field is opposite in direction to the applied field; that is, the two hyperfine tensors identified in the ENDOR data are of opposite sign. These represent approximately 85% of the total iron. The fact that the sample is heterogeneous does not become obvious until one tries to synthesize the spectrum. Figure 17F is a Mössbauer synthesis consisting of two nonequivalent species, one with $A \simeq 22$ MHz and one with $A \simeq 32$ MHz and small quadrupole splittings (~ 0.7 mm/sec).

Table 3

Parameters for Mössbauer Syntheses for HiPIP from *Chromatium vinosum D*

	Composition[a] (%)	Isomer shift (mm/sec)	QS (mm/sec)	η[b]	$GQ\beta$[c]	A value[d]		
						A_x	A_y	A_z
Fe*1	16	0.15	−0.75	−1	—	—	—	—
Fe*2	42	0.08	−0.65	−1	0	8.14	8.14	7.57
Fe*3	42	0.15	−0.75	0	90	−11.42	−11.42	−10.63

[a] The percent composition is that used to synthesize the spectrum in Figure 17D.
[b] η is the asymmetry parameter for the electric field gradient.
[c] $GQ\beta$ is the Euler angle β in the Whittaker convention for the orientation of the electric field gradient tensor with respect to the g tensor. All other Euler angles are zero.
[d] A values (from ENDOR) are given in gauss at the electron.

Notice that the intensity in the middle is too low when compared to this experimental spectrum in Figure 17A. When one adds 15% of a species with the spectrum in Figure 17C, one gets Figure 17D, which begins to look more like the experimental spectrum (Table 3 lists the parameters and weightings used in Figure 17D); that is, the experimental spectrum contains, in addition to the species having a detectable hyperfine splitting, still another species of approximately 15% abundance that exhibits no hyperfine field. Thus, we see that both EPR and Mössbauer spectroscopies show a species in approximately 15% abundance that is not like the remainder of the centers. If we assume that the same species is responsible for both spectra, then it is paramagnetic with a g anisotropy approximately one-half of the anisotropy exhibited by the more abundant species and it shows no effective hyperfine field at the nuclei. Such a species could be a spin-coupled "dimer" of some sort, because the shorter T_1 would tend to average the hyperfine field to zero and the spin coupling could average the g values of the spin-coupled system to one-half of their previous anisotropies. There is one further attraction to the "dimer" assignment, and that is that the 15% species exhibits a quadrupole splitting that is the exact replica of the high-temperature spectrum for oxidized HiPIP (Figure 17C), which is what would be expected for a spin-coupled dimer with its shorter relaxation time T_1. However, we have no chemical evidence to support this conjecture in as much as no dimer is discerned by gel electrophoresis nor does the spectrum change with dilution; however, it is to be remembered that the EPR heterogeneity, at least, is a freezing artifact.

The Mössbauer spectrum of the oxidized HiPIP from *R. tenue* shows very little, if any, of the doublet spectrum in addition to the characteristic hyperfine coupled spectrum. This is in accord with the EPR that shows little or no $g = 2.09$ bump.

4.2.3. $g = 2.01$ High-Potential Iron-Type Proteins

These proteins are identified by EPR in mitochondria[56–58] and appear to be due to a four-iron–sulfur cluster.[56] They have a range of redox potentials ($+160$ to -350 mV) between those of the bacterial HiPIPs and those of the four-iron ferredoxins. They are paramagnetic in the oxidized state, analogous to the bacterial HiPIP, hence the label "HiPIP-type." This HiPIP-type system was found first[56] in the succinate-ubiquinone-reductase segment (complex II) of the mitochondrial membrane and was later labeled as center S-3 (see Figure 9C) by Ohnishi[70] and found still later[71] in the cytoplasmic side of the inner mitochondrial membrane and labeled center bc-3, indicating that it is physically (not necessarily chemically) associated with the cytochrome-b–cytochrome-c region of the respiratory chain. Proteins of this same type have been identified in a bacterium,[72] Nictobacter, and similar signals are seen in HB-8 membranes.[73]

Quite recently, Bothe and Yates[74] have isolated four-iron–sulfur proteins from Mycobacterium flavum and A. chroococcum that give $g = 2.01$ EPR spectra under oxidation and operate between the C^-, C^{2-} couple at very low potentials (Bothe and Yater called them ferredoxins). Sweeney et al.[75] have reported a $g = 2.01$ EPR signal in the oxidized state of a four-iron–sulfur protein called ferredoxin I from A. vinelandii. No ENDOR has been reported on these proteins.

4.2.4. Four-Iron Ferredoxins

These proteins are found in bacterial systems as previously noted; in particular, they have been found in B. polymyxa,[59] Desulfovibrio desulfuricans,[76] S. aurantia,[61] B. stearothermophilis,[62] and P. ovalis.[60] Redox potentials range from -330 to -420 mV. Molecular weights are ~ 6300–9500 daltons. No large single crystals have been reported for this protein and thus no x-ray structure exists. These proteins are diamagnetic in the oxidized state and accept one electron to become paramagnetic ($S = \frac{1}{2}$) in the reduced state with an EPR spectrum (for B. polymyxa I) shown in Figure 9f.[59]

No ENDOR spectra or magnetic susceptibility data for these proteins have been published. The evidence that these proteins have a four-iron cluster and not two plant-type two-iron clusters is that (a) they accept only one electron per four iron atoms, (b) the optical spectra are not at all like the two-iron systems but resemble the spectra for the eight-iron proteins instead, (c) the Mössbauer spectrum for B. stearothermophilus[62] looks like that of the eight-iron ferredoxins at high magnetic fields, and (d) extrusion and reconstitution take place via the four-iron model compounds and not with the two-iron systems.

4.2.5. Discussion

The ENDOR studies on the four-iron–sulfur proteins are incomplete. ENDOR spectra have been reported only for the high-potential iron proteins. These spectra have indicated inequivalent iron atoms; specifically, pairs of iron atoms in the four-iron cluster. Mössbauer studies have shown that the effective A tensors for the pairs are of opposite sign, i.e., each iron of a pair has an A tensor of the same sign and magnitude, but the opposite pair have equivalent A tensors to each other but differ from that for the other pair both in magnitude and direction. The isomer shifts of the members of each pair are the same. This implies electron localization in pairs but charge delocalization within a given pair. Whether one has spin localization and simultaneous charge delocalization is difficult to know. Clearly, more work needs to be done on these systems.

4.3. Eight-Iron Ferredoxins

4.3.1. Some Properties

These proteins contain two four-iron–sulfur clusters in close proximity. The x-ray structure of oxidized $M.$ *aerogenes* ferredoxin[7] is shown in Figure 15C. At this resolution, the two four-iron clusters appear to be identical but rotated with respect to one another. The center-to-center separation of the two cubes is 11.5 Å. The eight cysteine residues occur in clusters of four, one being Cys-8, -11, -14, and -18, and the other being Cys-35, -38, -41, and -45. An interesting feature of the structure is that these are not the respective ligands for the separate clusters; rather, one cluster has Cys-8, -11, -16, and -45 and the other has Cys-18, -35, -38, and -41. This apparently results in improved stability. Similar behavior occurs in $C.$ *acidi-urici.*[77] The region between the clusters has no amino acid side chains to block it and no water is apparent.

4.3.2. EPR Studies

The EPR spectra for these proteins shows a typical four-iron cluster EPR spectrum upon partial reduction that gradually shows spin–spin coupling (Figure 18) between the two clusters as the protein becomes fully reduced (one electron per four-iron cluster). This was first explained by Mathews *et al.*,[78] who demonstrated the nature of the splitting by recording the K-band EPR and the " $\Delta m = 2$ " signal at half-field at X band. Schepler[19] then wrote a computer program to synthesize such spin-coupled spectra, and his results are discussed below.

High-resolution PMR spectra at room temperature[79] show several contact-shifted proton resonances, but no quantitative explanation has been provided for the temperature dependencies observed as in the case of the two-iron ferredoxins.[6]

4.3.2.1. *Spin–Spin Coupling in Several Eight-Iron–Sulfur Proteins.* Spurred by the observations that there was extra structure appearing in the EPR spectra of several eight-iron ferredoxins whenever more than one four-iron cluster was reduced on each protein, Mathews grew several bacteria and isolated the ferredoxins from five bacteria, *M. lactilyticus, C. pasteurianum, C. butyricum, C. acidi-urici,* and *P. elsdenii.* The EPR spectra from the fully reduced forms of these eight-iron ferredoxins all showed spin–spin coupling.[80]

Mathews[80] made a detailed chemical and spectral study of the ferredoxin from *M. lactilyticus.* He concluded that it was an eight-iron ferredoxin exhibiting the following properties: (1) eight iron atoms per mole of protein, (2) eight labile sulfur atoms per mole of protein, (3) eight cysteines per mole of protein, (4) a molecular weight of approximately 6900 daltons, (5) an abundance of acidic and neutral amino acid residues and a small number of basic and aromatic residues, (6) an extinction coefficient of about 30,000 liter/mole cm at 390 nm, (7) accomodation of two electrons per mole of protein on full reduction with sodium dithionite.

The X-band spectra of the reduced proteins (all) were found to be generally intractable to analysis based on the assumption that they represented the superposition of independent spin one-half systems. Investigations at K_a band (35 GHz) indicated that the spectra of the reduced proteins contained field independent components. The interpretation of the EPR spectra thus recorded appeared to be most consistent with the view that these proteins when fully reduced contain two interacting paramagnetic sites and the spectra obtained therefore represent the total interacting system. Mathews searched for and detected the $\Delta m = 2$ (half-field resonance characteristic of spin-coupled spin $\frac{1}{2}$ systems with magnetic dipolar coupling. See Mathews *et al.*[78] for details. The parameters characteristic of the spectra, such as interaction constants and appropriate spin values of the sites, had to await a more thorough investigation performed by Schepler (see below).

4.3.2.2. *Computer Synthesis of Spin-Coupled EPR Spectra Observed in 8Fe Ferredoxins.* In order to understand the EPR spectra of 8Fe–8S* ferredoxins and to gain more information about the structure and function of these proteins, Schepler[19] wrote a computer program to simulate EPR spectra of interacting spin $\frac{1}{2}$ centers with anisotropic g tensors of arbitrary relative orientations. This program calculates cases when a dipole–dipole interaction and an exchange interaction are both present. In the case of the eight-iron ferredoxins, the program was used to test various models for the relative orientations and interactions of the two four-iron clusters against the experimental data. Common to all of the models tested was the concept of two interacting g tensors. The two 4Fe–4S* active sites are assumed to be described by identical g tensors. The success of the various spectral fits is such as to confirm this assumption. The g values used were obtained by fitting the partially reduced spectrum of *M. lactilyticus*.

Different combinations of the dipolar and exchange interactions were tested, both as to their relative size and as to the dipole orientation relative to the g-tensor principal axes. A significant finding was the strong dependence of spectral shape upon the relative orientation of the two g tensors. This was true not only of calculated spectra with large dipole contributions (which would be expected, since the dipole interaction itself is orientation dependent) but also for cases where exchange was considered to be the dominant spin–spin interaction mechanism. Just why this should be so is understood by the high-resolution NMR spectroscopists in terms of the relative sizes of chemical shift differences and the spin–spin coupling. In fact, the case of large exchange and small dipolar coupling is just the case of AB spectra in NMR. In fact, the EPR spectra for fully reduced *M. lactilyticus* ferredoxin are basically superpositions of AB spectra calculated for each orientation of the applied field relative to the two four-iron cluster system. This was confirmed after the fact by applying ELDOR spectroscopy to measure the frequency differences in such a four-line spectrum for a given

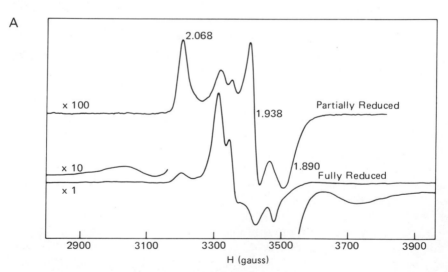

Figure 18. Experimental (A) X-band and (B) K_a-band EPR spectra of the eight-iron ferredoxin from *M. lactilyticus* 20% reduced (upper trace) and fully reduced (lower trace). (C) X-band EPR spectra of three fully reduced ferredoxins in the half-field ($\Delta m = 2$) region. The spectrometer gain is some 5000 times higher than used to record spectrum A in the $\Delta m = 1$ region. Arrows indicate the expected location for the $\Delta m = 2$ signals resulting from two inequivalent coupled spins having g average values of 2.00 or 1.94 (From Matthews et al.[78])

B

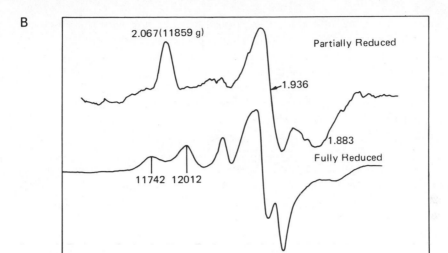

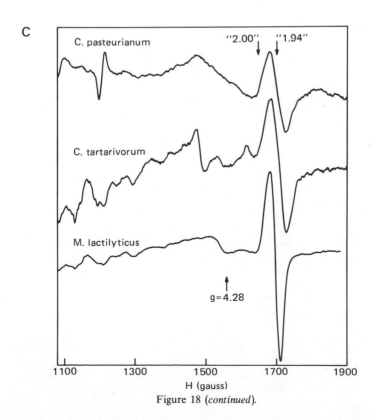

Figure 18 (*continued*).

orientation; i.e., by setting at an extremum line position with the observing frequency, one could select molecules of a particular orientation and sweep the pump frequency through the other lines in the four-line spectrum for that orientation. The transitions were shown in Figure 5B. To understand the situation in EPR, let us consider the energy dependencies of the lines in this spectrum and their intensities. From analogies with AB spectra in NMR (e.g., see Memory[81]) we could write

$$C = \tfrac{1}{2}[(g_1 - g_1)^2 \beta^2 H^2 + 4J^2]^{1/2} \tag{8}$$

where the spin–spin interaction is given by $-2J\mathbf{S}_1 \cdot \mathbf{S}_2$. Thus, the separation between the two central lines increases as the difference in the g values increases. Also, if $g_1 = g_2$, the separation goes to zero and the two inner lines become degenerate. The transition probabilities also have a g-value dependence. The most significant dependence comes from the manner in which the g-value dependence affects the angle η used in the intensity expressions shown in Table 4.

Table 4

EPR Simulation Parameters for the Spin-Coupled Spectra in the Fully Reduced Eight-Iron Ferredoxin from *M. lactilyticus*[a]

Parameter	X band	K band
Frequency	9.24 GHz	34.16 GHz
g_x	2.0700	2.0704
g_y	1.9439	1.9448
g_z	1.8915	1.8919
Half-width x	5 G	10 G
Half-width y	5 G	10 G
Half-width z	7 G	15 G

g-Tensor Orientations: The relative orientation of g tensor 2 to g tensor 1 in Euler angles in the Whittaker conventions: α, $0 \pm 5°$; β, $90 \pm 5°$; γ, $75 \pm 5°$.

Exchange Interaction: To approximate a Gaussian distribution of J values each simulated spectrum is a composite of three J values (300, 350, and 400 MHz) weighted in the ratio $0.6 : 1.0 : 0.6$.

Dipole Interaction: Radius, 8 ± 0.5 Å; radius vector is parallel to the g_x^1 direction.

[a] From Schepler.[19]

Since $\tan \eta \equiv 2J/[(g_2 - g_1)\beta H]$, the angle η is a trigonometric means of measuring the ratio of the exchange interaction relative to the difference in the Zeeman interactions. When $|g_2 - g_1| \ll g_1, g_2$ then the ratio of a middle line to an outer line is

$$\sim \frac{1 + \sin 2\eta}{1 - \sin 2\eta} \tag{9}$$

Thus, if $|2J| \ll |g_2 - g_1| \beta H$, the two lines will be nearly equal in intensity. As $|g_2 - g_1|$ becomes smaller, the outer line becomes smaller than the middle line, and when $g_1 = g_2$, the outer line intensity goes to zero.

This has several important consequences for the powder spectrum of g tensors interacting via an exchange coupling. Spatially equivalent g tensors (i.e., g values are the same and axes are parallel) will have unaltered EPR shapes due to spin–spin interactions. Spatially inequivalent g tensors will display, in general, quite complex spectra even when the principal g values are the same. Each powder orientation has its own distinctive pair of g values and four-line spectrum.

Another important characteristic of g tensor exchange interactions is the signal shape dependence on the EPR frequency band being used. This effect can again be traced through the definition of η. When the Zeeman interaction is the dominant interaction, the resonant magnetic field will be roughly proportional to the microwave frequency used; i.e., $h\nu \approx g_{\text{average}} \beta H$. Therefore a typical value of η at K band will be smaller than the corresponding X-band value and smaller yet than the S-band η value. Since η determines the relative size of the middle and outer EPR lines, increasing (decreasing) the microwave frequency will in turn decrease (increase) the middle-to-outer-line ratio. This is an important means of identifying an exchange interaction.

The experimental spectra and their computer simulations are shown in Figure 19. The detailed parameters of the fits are shown in Table 4, but a general description of the best simulation will be presented here. The relative orientation of the g tensors must be nearly the same as that shown in Figure 19C. The Euler rotation of g_2 relative to g_1 is R $(0°, 90°, 75°)$ (Whittaker convention). The dipolar radius vector is oriented along $g_x^{(1)}$ and the dipolar interaction is given by

$$\left\langle \frac{3Z_{12}^2 - r_{12}^2}{r_{12}^5} \right\rangle = \left(\frac{1}{8 \text{ Å}} \right)^3 \tag{10}$$

where the angle brackets mean integrated over the spin densities of the two four-iron cubes. This corresponds to a center-to-center distance of 10–12 Å, which is close to the value of 11.5 Å determined for M. aerogenes[7] by x-ray studies. The exchange coupling is $J = 350$ MHz, which is much larger than the dipole coupling $\mu/r^3 = 18$ G = 50 MHz.

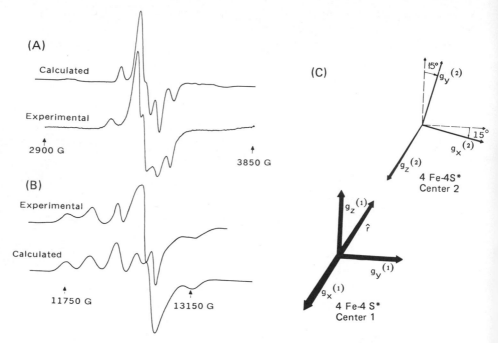

Figure 19. Experimental (A) X-band and (B) K_a-band ENDOR spectra and computer simulations of fully reduced eight-iron ferredoxin from *M. lactilyticus* assuming two spin-coupled $S = \frac{1}{2}$ centers corresponding to the two four-iron centers. (C) The relative orientations of the two four-iron center g tensors as determined by the computer simulation. (From Schepler.[19])

4.3.3. ELDOR Studies

As a means of testing this model, the *M. lactilyticus* sample was analyzed by Schepler[19] with ELDOR spectroscopy. Detection of ELDOR signals could confirm the presence of spin–spin interactions and could give a measure of their size.

Since the computer simulations indicate that the high-field and low-field wings are the weak outer resonances of the four-line spectra of certain crystal orientations, the first ELDOR experiment set the observing mode frequency (by adjusting the magnetic field) to correspond to the peak of the low-field wing. The pump mode frequency was then swept through the center of the EPR spectrum. Figure 20 shows the results of that sweep at three different temperatures. The spectra show two signals. Both are quite weak, having reduction factors less than 7%, and the signal at 930 MHz has a much more pronounced temperature dependence than the 730-MHz signal.

These ELDOR signals are readily interpreted in terms of a large exchange plus a small dipole model. Setting the observing mode at the low-field peak singles out a small set of molecular orientations that have resonances at this field and frequency. And since this position is the lowest field at which a resonance occurs, the orientations correspond to one of the active-site g tensors having its largest principal g value parallel to the static magnetic field. The other g value(s) involved will depend on the relative orientations of the two g tensors.

Ignoring any dipolar contributions for the time being, consider the crystal orientation resonances shown in Figure 20. Here the two ELDOR

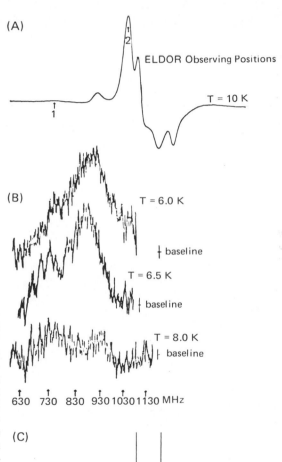

Figure 20. (A) The X-band EPR spectrum of fully reduced *M. lactilyticus* ferredoxin showing the observing location 1 for ELDOR. (B) The ELDOR spectra recorded at three different temperatures as a function of the difference between the swept pump frequency and the fixed observing frequency; the magnetic field is set so that the observing frequency is at location 1 as shown in (A). (C) The interpretation of the ELDOR signals in terms of a spin-coupled "AB" stick spectrum for the magnetic field oriented along the z axis of the g-tensor frame for one of the four-iron centers. (From Schepler.[19])

resonances at 730 MHz and 930 MHz are assumed to correspond to $2J$ and to $2C$. Substituting the resonance numbers into $2C$ yields a g value difference of 0.13. Using the low-field g value of 2.07, this predicts the other g value involved is 1.94, which is the g_y principal value. This prediction agrees with the simulation model, which has $g_y^{(1)}$ nearly parallel to $g_x^{(2)}$.

From Table 4, the relative orientations of the g tensors of the two clusters is determined within $\pm 5°$ and dipolar separation within ± 0.5 Å. That is not bad, by any standards.

4.3.4. ENDOR Studies

The only ENDOR studies reported on an eight-iron ferredoxin are those of Anderson *et al.*[82] on the ferrodoxin from *C. pasteurianum*. They obtained the spectrum from fully reduced protein that displays the spin-coupled EPR spectrum previously described. They interpreted their data in terms of all eight iron atoms having similar spin densities and A tensors, although their spectra were quite broad. They quote $A_x = 25 \pm 2$ MHz, $A_y = 29 \pm 5$ MHz, and $A_z = 33 \pm 1$ MHz.

The complication caused by the spin coupling is unfortunate, and very little else can be concluded except that the A tensors observed were all in this range. Clearly, much more work needs to be done with these proteins.

4.3.5. Discussion

The utility of the spin coupling in these proteins as a means for determining relative orientations and separations of the two clusters is obvious. The complexity of the ENDOR spectrum of the fully reduced proteins makes it difficult to interpret except in broad terms.

5. Summary

Double resonance spectrometries have provided essential data on spin–spin couplings, both electron–electron and electron–nuclear, throughout the entire class of iron–sulfur proteins. In combination with EPR and Mössbauer spectrometries, they have provided structural and chemical information not otherwise available. The ENDOR studies have been necessary for the detailed interpretation of the Mössbauer data, from which most of the information about the chemical and physical properties of the active sites of these proteins has come.

Inspired by the conclusions drawn from these spectrometries and x-ray crystallography, the inorganic chemists have synthesized model compounds for these active sites that have sufficient structural stability that they can be extruded or reconstituted in and out of the protein by competitive ligand exchange, thereby confirming the structural features of these sites.

ELDOR spectrometry coupled with careful EPR synthesis has provided information about the relative locations of the centers in the eight-iron ferredoxins, and the EPR synthesis has provided relative orientations as well.

ACKNOWLEDGMENTS

This chapter has been written in the form of a review. Most of the double resonance work described herein has been done by a relatively few people—John Fritz, Russell Anderson, Richard Dunham, and Kenneth Schepler, in particular. To them I extend my congratulations and whatever honor this review may bring them.

To my biochemistry colleagues, I can only express my appreciation for their many hours of hard work in elucidating the biochemical properties of these proteins. The names of Helmut Beinert, William H. Orme-Johnson, I. C. Gunsalus, David Hall, Y. I. Shetna, D. Der Vartanian, Graham Palmer, and John Tsibris are recorded repeatedly in the history of this subject. The truly great honor for this work is largely due them.

Finally, the names of George Feher and James Hyde are paramount in the history of the instrumentation. Others of us have simply followed in their footsteps making small improvements where we could.

References

1. J. C. M. Tsibris, and R. W. Woody, Structural Studies of Iron–Sulfur Proteins, *Coord. Chem. Revs.* **5**, 417 (1970).
2. D. O. Hall, R. Cammack, and K. K. Rao, in *Iron in Biochemistry and Medicine* (A. Jacob and M. Worwood, eds.), Chapter 8, p. 279, Academic Press, London (1974).
3. G. Palmer, in *The Enzymes* (P. D. Boyer, ed.), Vol. XII, 3rd ed., pp. 1–56, Academic Press, New York (1975).
4. W. Lovenberg (ed.), *Iron–Sulfur Proteins*, Vols. I–III, Academic Press, New York (1973).
5. R. G. Shulman, P. Eisenberger, W. E. Blumberg, and N. A. Stombaugh, *Proc. Nat. Acad. Sci. USA* **72**, 4003 (1975).
6. W. R. Dunham, G. Palmer, R. H. Sands, and A. J. Bearden, *Biochim. Biophys. Acta* **253**, 373 (1971).
7. E. T. Adman, L. C. Sieker, and L. H. Jensen, *J. Biol. Chem.* **251**, 3801 (1976).
8. C. W. Carter, Jr., S. T. Freer, Ng H. Xuong, R. A. Alden, and J. Kraut, *Cold Spr. Harb. Symp. Quant. Biol.* **36**, 381 (1971).
9. A. Abragam and B. Bleaney, *EPR of Transition Ions*, Oxford University Press (Clarendon), London, New York (1970).
10. J. Peisach, W. E. Blumberg, E. T. Lode, and M. J. Coon, *J. Biol. Chem.* **246**, 5877 (1971).
11. F. K. Kneubühl and B. Natterer, *Helv. Phys. Acta* **34**, 710 (1961).
12. R. Neiman and D. Kivelson, *J. Chem. Phys.* **35**, 156 (1961).

13. H. R. Gersmann and J. D. Swalen, *J. Chem. Phys.* **36**, 3221 (1962).
14. M. M. Malley, *J. Molec. Spectrosc.* **17**, 210 (1965).
15. J. Fritz, R. Anderson, J. Fee, G. Palmer, R. H. Sands, W. H. Orme-Johnson, H. Beinert, J. C. Tsibris, and I. C. Gunsalus, *Biochim. Biophys. Acta* **253**, 110 (1971).
16. T. Moriya, *Phys. Rev.* **120**, 91 (1960).
17. J. Owen and C. M. Harris, in *Electron Paramagnetic Resonance* (S. Geschwind, ed.), Chapter 6, Plenum Press, New York (1972).
18. See, for example, A. Carrington and A. C. McLachlan, *Introduction to Magnetic Resonance*, p. 46, Harper and Row, New York (1967).
19. K. Schepler, *Electron Spin–Spin Interactions in Three Biological Systems*, Ph.D. Thesis University of Michigan (1975).
20. J. W. Fritz *ENDOR Studies of Tetraphenylpyrryl and Four Ferredoxins*, Ph.D. Thesis, University of Michigan (1969).
21. J. S. Hyde and A. H. Maki, *J. Chem. Phys.* **40**, 3117 (1964).
22. J. S. Hyde, *J. Chem. Phys.* **43**, 1806 (1965).
23. J. A. Reid, *Electron–Nuclear Double Resonance Spectrometry on a High Potential Iron Protein*, Ph.D. Thesis, University of Michigan (1976).
24. J. S. Hyde, J. C. W. Chien, and J. H. Freed, *J. Chem. Phys.* **48**, 4211 (1968).
25. D. O. Hall, R. Cammack, and K. K. Rao, in *Iron in Biochemistry and Medicine* (A. Jacobs and M. Worwood, eds.), pp. 279–334, Academic Press, London, New York (1974).
26. J. Van Beenmen, J. deLey, D. O. Hall, R. Cammack, and K. K. Rao, *FEBS Lett.* **59**, 146 (1975).
27. M. C. W. Evans, R. V. Smith, A. Telfer, and R. Cammack, in *Proceedings of the 1st European Biophys. Congress* (E. Broda, A. Locker, and H. Springer-Lederer, eds.), Vol. 4, pp. 115–119, Vienna Medical Academy (1971).
28. L. Kerscher, D. Oesterhelt, R. Cammack, and D. O. Hall, *Eur. J. Biochem.* **71**, 101 (1976).
29. D. O. Hall and M. C. W. Evans, *Nature (London)* **223**, 1342 (1969).
30. J. C. M. Tsibris, R. L. Tsai, I. C. Gunsalus, W. H. Orme-Johnson, W. H. Hansen, and H. Beinert, *Proc. Nat. Acad. Sci. USA* **59**, 959 (1968).
31. E. Münck, P. G. Debrunner, J. C. M. Tsibris, and I. C. Gunsalus, *Biochemistry* **11**, 855 (1972).
32. W. H. Orme-Johnson and H. Beinert, *J. Biol. Chem.* **244**, 6143 (1969).
33. R. H. Sands and W. R. Dunham, *Quart. Revs. Biophys.* **7**, 443 (1975).
34. J. J. Mayerle, R. B. Frankel, R. H. Holm, J. A. Ibers, W. Phillips, and J. F. Weiher, *Proc. Nat. Acad. Sci. USA* **70**, 2429 (1973).
35. L. Que, Jr., R. H. Holm, and L. E. Mortenson, *J. Amer. Chem. Soc.* **97**, 463 (1975).
36. W. R. Dunham, A. Bearden, I. Salmeen, G. Palmer, R. H. Sands, W. H. Orme-Johnson, and H. Beinert, *Biochim. Biophys. Acta* **253**, 134 (1971).
37. W. A. Eaton, G. Palmer, J. A. Fee, T. Kimura, and W. Lovenberg, *Proc. Nat. Acad. Sci. USA* **68**, 3015 (1971).
38. G. Palmer, W. R. Dunham, J. A. Fee, R. H. Sands, T. Iizuka, and I. Yonetani, *Biochim. Biophys. Acta* **245**, 201 (1971).
39. M. Poe, W. D. Phillips, J. D. Gleckson, and A. San Pietro, *Proc. Nat. Acad. Sci. USA* **68**, 68 (1971).
40. I. T. Salmeen and G. Palmer, *Arch. Biochem. Biophys.* **150**, 767 (1972).
41. S. De Benedetti, F. De S. Barros, and G. R. Hoy, *Ann. Rev. Nucl. Sci.* **16**, 31 (1966).
42. Y. I. Shetna, P. W. Wilson, R. E. Hansen, and H. Beinert, *Proc. Nat. Acad. Sci. USA* **55**, 1263 (964).
43. H. Brintzinger, G. Palmer, and R. H. Sands, *Proc. Nat. Acad. Sci. USA* **55**, 397 (1966).
44. W. E. Blumberg and J. Peisach, in *Non-Heme Iron Proteins* (A. San Pietro, ed.), pp. 101–107, Antioch Press, Yellow Springs, Ohio (1965).

45. J. F. Gibson, D. O. Hall, J. H. M. Thornley, and F. R. Whattey, *Proc. Nat. Acad. Sci. USA* **56**, 987 (1966).
46. J. H. M. Thornley, J. F. Gibson, F. R. Whatley, and D. O. Hall, *Biochem. Biophys. Res. Commun.* **24**, 877 (1966).
47. W. H. Orme-Johnson, R. E. Hansen, H. Beinert, J. C. M. Tsibris, R. C. Bartholomaus, and I. C. Gunsalus, *Proc. Nat. Acad. Sci. USA* **60**, 368 (1968).
48. D. V. Der Vartanian, W. H. Orme-Johnson, R. E. Hansen, H. Beinert, R. L. Tsai, J. C. Tsibris, R. C. Bartholomaus, and I. C. Gunsalus, *Biochem. Biophys. Res. Commun.* **26**, 569 (1967).
49. L. H. Strong, *Biophysical Studies on Selected Iron–Sulfur Proteins*, Ph.D. Thesis, University of Michigan (1976).
50. R. E. Anderson, *Study of the Active Sites of Ferredoxin from Synechococcus lividus*, Ph.D. Thesis, Univ. Michigan (1972).
51. R. E. Anderson, W. R. Dunham, R. H. Sands, A. J. Bearden, and H. L. Crespi, *Biochem. Biophys. Acta* **408**, 306 (1975).
52. K. Das, H. Deklerk, K. Sletten, and R. G. Bartsch, *Biochim. Biophys. Acta* **140**, 291 (1967).
53. S. M. Tedro, T. E. Meyer, and M. D. Kamen, *J. Biol. Chem.* **251**, 129 (1976).
54. T. E. Meyer, S. M. Tedro, and M. D. Kamen, private communication.
55. S. M. Tedro, T. E. Meyer, and M. D. Kamen, *J. Biol. Chem.* **249**, 1182 (1974).
56. N. R. Orme-Johnson, W. H. Orme-Johnson, R. E. Hansen, H. Beinert, and Y. Hatefi, *Biochem. Biophys. Res. Commun.* **44**, 446 (1971).
57. T. Ohnishi, T. Asakura, T. Yanetani, and B. Chance, *J. Biol. Chem.* **246**, 5960 (1971).
58. F. J. Ruzicka, and H. Beinert, *Biochem. Biophys. Res. Commun.* **58**, 556 (1974).
59. N. A. Stombaugh, R. H. Burris, and W. H. Orme-Johnson, *J. Biol. Chem.* **248**, 7951 (1973).
60. T. Matsumoto, J. Tobari, K. Suzuki, T. Kimura, and T. T. Tchen, *J. Biochem.* **79**, 937 (1976).
61. P. W. Johnson, and E. Canale-Parola, *Arch. Mikrobiol.* **89**, 3363 (1973).
62. R. N. Mullinger, R. Cammack, K. K. Rao, D. O. Hall, D. P. E. Dickson, C. E. Johnson, J. D. Rush, and A. Simopoulos, *Biochem. J.* **151**, 75 (1975).
63. C. W. Carter, Jr., J. Krant, S. J. Freer, R. A. Alden, L. C. Sieker, E. Adman, and L. H. Jensen, *Proc. Nat. Acad. Sci. USA* **69**, 3526 (1972).
64. R. Cammack, *Biochem. Biophys. Res. Commun.* **54**, 548 (1973).
65. See, for example, R. Averill, T. Herskovitz, R. H. Holm, and J. A. Ibers, *J. Amer. Chem. Soc.* **95**, 3523 (1973).
66. T. Herskovitz, B. A. Averill, R. H. Holm, J. A. Ibers, W. D. Phillips, and J. F. Weiher, *Proc. Nat. Acad. Sci. USA* **69**, 2437 (1972).
67. L. Que, Jr., M. A. Bobrik, J. A. Ibers, and R. H. Holm, *J. Amer. Chem. Soc.* **96**, 4168 (1974).
68. B. Antanaitis and T. Moss, *Biochim. Biophys. Acta* **405**, 262 (1975).
69. M. S. Goldsmith, J. A. Fee, W. R. Dunham, and R. H. Sands, to be published.
70. T. Ohnishi, D. B. Winter, J. Lim, and T. E. King, *Biochem. Biophys. Res. Commun.* **60**, 1017 (1974).
71. T. Ohnishi, W. J. Ingledew, and S. Shiraishi, *Biochem. J.* **153**, 39 (1976).
72. W. J. Ingledew, and P. J. Halling, *FEBS Lett.* **67**, 90 (1976).
73. J. A. Fee, private communication.
74. H. Bothe, and M. G. Yates, private communication.
75. W. V. Sweeney, J. C. Rabinowitz, and D. C. Yoch, *J. Biol. Chem.* **250**, 7842 (1975).
76. J. A. Zubieta, R. Mason, and J. R. Postgate, *Biochem. J.* **133**, 851 (1974).
77. W. J. Orme-Johnson, *Biochem. Soc. Trans.* **1**, 30 (1973).
78. R. Mathews, S. Charlton, R. H. Sands, and G. Palmer, *J. Biol. Chem.* **249**, 4326 (1974).
79. W. D. Phillips, and M. Poe, in *Iron–Sulfur* Proteins (W. Lovenberg, ed.), Vol. II, p. 255, Academic Press, New York (1973).

80. R. A. Mathews, *Chemical and Physical Studies of a Bacterial Ferredoxin*, Ph.D. Thesis, University of Michigan (1973).
81. J. D. Memory, *Quantum Theory of Magnetic Resonance Parameters*, McGraw-Hill, New York (1968).
82. R. E. Anderson, G. Anger, L. Peterson, A. Ehrenberg, R. Cammack, D. O. Hall, R. Mullinger, and K. K. Rao, *Biochim. Biophys. Acta* **376**, 63 (1974).

10

Radiation Biophysics

Harold C. Box

1. Introduction

It would be difficult to mention an area of research that has more fundamental implications for mankind than does radiation biophysics. Mutagenesis, carcinogenesis, and cancer therapy all fall within the purview of radiation biophysics. These radiobiological phenomena are but manifestations of radiation effects occurring at the molecular level. The main objective of radiation biophysics is to explain the biological effects of radiation by discovering and interpreting the molecular effects. Numerous chemical and physical techniques have contributed to our understanding of the molecular effects of radiation; they including ESR spectroscopy, mass spectroscopy, chromatography, autoradiography, and pulse radiolysis, to name but a few. Our purpose in this chapter is to describe the contributions of another technique, namely ENDOR, that has only recently been deployed in radiation research. Our discussion of radiation biophysics will be highly selective, treating only those aspects of the subject where ENDOR has had notable impact.

ESR–ENDOR spectroscopy is uniquely suited to the study of radiation damage. The majority of biomolecules are composed of an even number of electrons whose spins are paired and are therefore diamagnetic and not amenable to study by ESR–ENDOR spectroscopy. When diamagnetic molecules are exposed to ionizing radiation, odd-electron free radical products are formed that can be studied by ESR and ENDOR spectroscopy. Consider how these paramagnetic products are formed. A beam of ionizing radiation passing through matter causes ionizations and excitations of the

Harold C. Box ● Biophysics Department, Roswell Park Memorial Institute, Buffalo, New York

constituent atoms and molecules. The primary chemical consequences of these events are the following: 1. Free radical oxidation products are generated when molecules are deprived of an electron in the ionization process. Depending upon the efficiency of intramolecular transfer of electron vacancies, one or more oxidation products are formed. 2. Electrons lost in the ionization process may be picked up by other undamaged molecules, generating reduction products that are also paramagnetic. 3. Primary odd-electron oxidation and reduction products are often unstable and may dissociate. One of the products formed in any such dissociation process must be a free radical. The radical formed by oxidation or reduction may interact with a diamagnetic molecule to form a secondary free radical. 4. Excited molecules may undergo homolytic dissociation, generating a pair of free radicals. In the solid state, recombination of the pair of radicals produced in a homolytic dissociation is likely unless one of the radicals is so small a fragment that it can diffuse away from its geminate partner. ESR–ENDOR spectroscopy is employed to detect and characterize the free radical products formed in solids in each of the aforementioned steps of the radiation damage process. Cryogenic techniques are useful for stabilizing the most primitive stages of the radiation damage process.

The contributions of ESR–ENDOR spectroscopy to radiation biophysics have come mainly from studies on model systems. Even in simple model systems, the ESR spectrum induced by ionizing radiation is often complicated. Difficulties arise whenever multiple free radical products are generated whose absorptions superimpose. Enter here the ENDOR method. The various components of a complex ESR absorption can often be distinguished by applying ENDOR technology. This possibility derives from the high resolution with which hyperfine couplings can be measured by the ENDOR method. Since not only hyperfine couplings but also the g value of the ESR absorption are anisotropic, ESR–ENDOR spectroscopy has been useful mainly for analyzing radiation damage in single crystals wherein molecules have like spatial orientations.

We shall focus on several problems of particular interest in radiation biophysics and mention the contributions that ENDOR has made in these areas.

2. Indirect Effects

Water is the main constituent of most living organisms. A knowledge of the effects of ionizing radiation on the aqueous portion of the cell is fundamental to understanding the biological effects of radiation. The deposition of energy in water by a beam of ionizing radiation has the following consequences:

$$H_2O \xrightarrow{\text{ionization}} H_2O^+ + e^- \tag{1}$$

$$H_2O \xrightarrow{\text{excitation}} H_2O^* \tag{2}$$

Three highly reactive species result from ionizations [equation (1)] and excitations [equation (2)], namely, solvated electrons, atomic hydrogen, and hydroxyl radicals, which are produced in the following reactions:

$$e^- \xrightarrow{\text{solvation}} e_{aq}^- \tag{3}$$

$$H_2O + e^- \longrightarrow H + OH^- \tag{4}$$

$$H_2O^+ \xrightarrow{H_2O} OH + H_3O^+ \tag{5}$$

$$H_2O^* \longrightarrow OH + H \tag{6}$$

Hydrogen atoms can be detected by ESR spectroscopy. The spectrum is characterized by an isotropic doublet pattern arising from the interaction between electron and proton. Solvated electrons are conveniently detected by optical spectroscopy. The solvated electron optical spectrum has a broad intense absorption centered at 700 nm. Hydroxyl radicals are not conveniently identified by either ESR or by optical spectroscopy. ENDOR has played a role in characterizing this latter species. Hydroxyl radicals are produced and stabilized in ice irradiated at 4°K (Figure 1). ENDOR measurements on

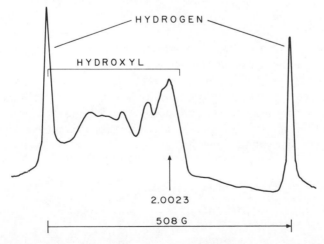

Figure 1. ESR absorption spectrum from a single crystal of ice irradiated at 4.2°K. The field was parallel to the *a* axis. Analysis of the OH absorption by ENDOR provided the results given in Table 1.

Table 1

Principal Values of the Hyperfine Coupling Tensors
(MHz) and g Tensors for OH Radicals Produced in
Ice

Set of characteristics	Hyperfine coupling tensor	g Tensor
1	-80.0	2.0028
	-124.4	2.0089
	9.2	2.0597
2	-81.2	2.0031
	-124.4	2.0089
	19.4	2.0571
3	-78.2	2.0027
	-126.2	2.0088
	15.6	2.0581

single crystals of ice x-irradiated at low temperature distinguish three OH radicals with slightly different hyperfine couplings. The anisotropy of the OH g value, the anisotropy of the hyperfine splitting, and the multiplicity of overlapping absorptions makes the OH portion of the ESR absorption in irradiated ice complex.[1,2] However, ENDOR provides a means for sorting out of the various components of the absorption so that not only the hyperfine coupling tensors but also the g tensors can be obtained (Table 1). The three slightly different sets of characteristics listed in Table 1 pertain to OH radicals produced in different crystalline environments.[3]

There is evidence that strongly suggests that hydroxyl radicals are more important than either solvated electrons or hydrogen atoms for instigating damage in biological systems. One indication of the importance of OH attack comes from studies of agents that are able to provide cells irradiated in culture with a measure of protection. For at least one class of protective agent, it has been shown that the capacity to protect cells correlates with the compound's reactivity with the hydroxyl radical.[4] Thus, removal of the OH radical before it can react with a biomolecule protects the organism.

3. Nucleic Acid Constituents

The prevailing view in cellular radiobiology is that damage to the DNA structure of cells is of overwhelming importance compared with such damage as may occur to other cellular components. This view is defensible,

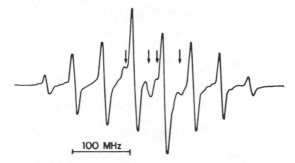

100 MHz

Figure 2. Derivative of ESR absorption from a single crystal of thymidine x-irradiated at room temperature. Magnetic field parallel to the *bc* plane, 10° from the *c* axis. Arrows indicate presence of a second free radical, which was identified from ENDOR measurements. [From J. N. Herak and C. A. McDowell, *J. Mag. Reson.* **16**, 434 (1974). Reproduced with permission.]

since it seems probable that catastrophic events such as mutagenesis, carcinogenesis, and lethality (judged from the cell's inability to undergo mitosis) have their molecular basis in modified DNA. The far-reaching implications of these radiobiological effects have fostered an intensive study of the nature of the chemical lesions produced in nucleic acid constituents by ionizing radiation. The ENDOR method has played a role in identifying the free radical products of irradiation in several DNA constituents.

Consider first the relatively stable free radicals that have been observed by ESR–ENDOR spectroscopy in single crystals irradiated at room temperature. Figure 2 shows the derivative spectrum of the ESR absorption from an irradiated single crystal of thymidine (I). The major component of the absorption, which has an eight-line hyperfine pattern, has long been recognized as arising from the radical (II). Another component in the ESR

Table 2

Principal Values of Proton Hyperfine Coupling Tensors Deduced from ENDOR
Measurements on Irradiated Single Crystals of Nucleic Acid Constituents

Compound	Radical	Proton	Principal values (MHz)		
Methyl cytosine[a]		C_5—H	134.0	123.3	122.2
		C_5—H	93.4	82.8	82.4
		C_6—H	−19.2	−46.2	−81.1
Methyl cytosine[a]		C_5—H	−22.9	−46.9	−79.4
		C_6—H	142.2	131.1	127.8
		C_6—H	150.2	141.2	139.7
Thymidine[b]		C_6—H	123.2	113.6	111.6
		C_6—H	117.8	107.4	106.0
Thymidine[c]		C_7—H	−67.0	−42.5	−22.0
		C_7—H	−65.8	−42.8	−21.0
		C_6—H	−14.3	−29.7	−42.2
Methyl uracil[d]		C_5—H	106.15	97.58	96.36
		C_5—H	104.94	96.34	92.12
		C_6—H	−24.74	−52.02	−86.17

[a] S. N. Rustgi and H. C. Box *J. Chem. Phys.* **60**, 3343 (1974).
[b] H. C. Box, E. E. Budzinski, and W. R. Potter *J. Chem. Phys.* **61**, 1137 (1974).
[c] J. N. Herak and C. A. McDowell *J. Mag. Res.* **16**, 434 (1974). Negative couplings assigned by reviewer.
[d] J. N. Herak and C. A. McDowell *J. Chem. Phys.* **61**, 1129 (1974).

absorption is suggested by those features indicated by the arrows in Figure 2. The radical responsible for this absorption could not be identified from the nebulous information provided by the ESR record. Application of the ENDOR technique provided a complete characterization of proton couplings. The principal values of the coupling tensors are included in Table 2. From these data, the minor component of the absorption could be assigned with certainty to the radical (III).

A similar situation occurs in single crystals of methyl cytosine (IV), where the major radical observed in crystals irradiated at room temperature could be identified from ESR studies as the radical (V). A second species (VI), produced in lesser concentration, could be identified only through the use of ENDOR. Complete hyperfine coupling tensors for both radical species as determined from ENDOR measurements are included in Table 2.

(IV)   (V)   (VI)

Even when ENDOR is not essential for identification purposes, it is useful to characterize radicals from ENDOR measurements. Suppose an irradiated crystal exhibits an ESR absorption arising from a single free radical species. The ESR spectrum may nevertheless be a superposition of several components due to the same free radical in different orientations. The phenomenon has come to be called "site splitting." Site splitting arises whenever the point symmetry relationships present in the crystal space group result in a given radical species being produced in two or more different spatial orientations that are related by a rotation or a reflection or both. Site splitting may not interfere with obtaining a reasonably good estimate of the isotropic part of the hyperfine splitting from ESR measurements if there are three orthogonal axes for which all symmetry-related radicals are equivalently oriented with respect to the applied field. However, the anisotropic part of the hyperfine coupling deduced from the ESR record may be substantially in error, since specification of this part of the coupling depends on measurements of hyperfine splittings in additional crystal orientations for which ESR may fail to resolve the individual components of the absorption generated by site splitting. The higher resolution afforded by the ENDOR method seldom fails to resolve differences in hyperfine couplings arising from site splittings. The usual consequence of this difference in resolving powers is that the overall anisotropy in hyperfine coupling estimated from ESR is significantly lower than the true anisotropy obtained from ENDOR measurements.

Table 2 is a compilation of proton couplings in free radicals formed by addition or abstraction of hydrogen to nucleic acid bases. These are relatively stable products observed in single crystals irradiated at room temperature and characterized from ENDOR measurements.

The ENDOR method is especially advantageous in the study of primary radiation products. Irradiations are performed at low temperature. If stabilization of primary radicals is successful, a minimum of two paramagnetic species must be observed, one from a primary reduction product and one from a primary oxidation product. ENDOR may be indispensible for sorting out the ESR absorption patterns that usually superimpose.

The most electrophilic constituents of nucleic acids are the base derivatives of pyrimidine and purine. It is not surprising, therefore, that the primary radiation-induced reduction products observed in single crystals of nucleosides and nucleotides are the result of electron addition to the base constituent. In these species, the unpaired electron occupies a delocalized π molecular orbital. Our attention will focus mainly on the hyperfine interactions of certain σ protons that are coplanar with the conjugated rings. The act of removing an electron or of adding an electron to a conjugated ring structure does not significantly perturb the structure especially if the experiment is carried out at very low temperature. Consequently, the orientation of σ proton hyperfine coupling tensors can be closely correlated with directions calculated from crystal structure. Thus, if substantial spin density occurs on the carbon atom in a $>C-H$ group, the principal axis corresponding to minimum negative coupling with the proton is along the $C-H$ bond axis; the intermediate principal value is in the direction normal to the plane of the conjugated molecule, and maximum negative coupling is orthogonal to these directions. Another characteristic of the hyperfine coupling of σ protons to be noted is the relative size of the anisotropic and isotropic parts of the coupling. For $>C-H$ protons, the ratio of the overall anisotropy to the isotropic part is about 1.2.

The numbering convention for pyrimidine and purine derivatives is shown in Table 3. In the pyrimidine base reduction products, the greatest concentration of spin density is on the C_6 carbon atom and the largest proton hyperfine interaction is with the C_6-H proton. The principal values of the C_6-H proton coupling tensors obtained from ENDOR measurements on irradiated single crystals of pyrimidine base derivatives are tabulated in Table 3. The crystals were x-irradiated at 4.2°K and maintained at this temperature throughout the measurement procedure. In every instance, the principal axes of the proton coupling tensor, as deduced from ENDOR measurements, correlated closely with directions calculated from crystal structure, so that there is little doubt the C_6-H proton is the origin of the interaction. Since the anion, but not the cation, is expected to have substantial spin density on C_6, the absorptions can be attributed with confidence to the reduced species.

Table 3

Principal Values of Proton Hyperfine Coupling Tensors in Reduction Products Produced by x-Irradiation in Nucleic Acid Constituents

Compound	Reduction product	Proton	Principal values (MHz)		
Thymidine[a]	(structure) deoxyribose	C_6-H	−13.0	−29.9	−55.5
Uridine 5'-monophosphate[b] (Ba salt)	(structure) ribose MP	C_6-H	−16.24	−32.92	−59.12
Cytidine 3'-monophosphate[b]	(structure) ribose MP⁻	C_6-H	−14.68	−33.98	−59.26
Cytosine hydrochloride[c]	(structure)	C_6-H	−25.8	−45.6	−66.6
Bromouridine[c]	(structure) ribose	C_6-H	−14.72	−34.86	−61.44
Adenine dihydrochloride[d]	(structure)	C_8-H	−10.74	−22.16	−40.64

[a] H. C. Box and E. E. Budzinski, *J. Chem. Phys.* **62**, 197 (1975).
[b] H. C. Box, W. R. Potter, and E. E. Budzinski, *J. Chem. Phys.* **62**, 3476 (1975).
[c] H. C. Box, *J. Chem. Soc., Faraday Disc. No. 63*, 264 (1978).
[d] H. C. Box and E. E. Budzinski, *J. Chem. Phys.* **64**, 1593 (1975).

Protonation of the pyrimidine base anions at higher temperatures is a mechanism for producing the hydrogen adducts to which we have already made reference. Hydroxyl-base adducts as well as hydrogen-base adducts are undoubtedly formed in irradiated cells. Recent work by Hariharan and Cerutti[5] has demonstrated that at least one type of base damage identified in model systems is produced *in vivo* also.

The radiation-induced reduction process in the presence of bromine-substituted uracil is of special interest. It is well known that 5-bromodeoxyuridine (VII) furnished as a nutrient to proliferating cells is incorporated into the nucleic acid structure of the cells. Incorporation occurs by way of substitution of (VII) for the normal thymidine nucleoside (VIII). The presence of the brominated base in the DNA enhances the sensitivity of cells to both ultraviolet and ionizing radiations. Hope persists that in favorable circumstances malignant growths such as brain tumors, which develop in a substantially nonproliferating host tissue, can be differentially sensitized, thereby making the malignancy more susceptible to treatment by radiation. The crucial step in the sensitization mechanism is probably the initial scavenging of electrons by bromouracil base. It is of interest that the anion so formed in x-irradiated bromouridine has been characterized by the ENDOR technique (Table 3.) The characteristics of the C_6—H proton coupling in this species is not markedly different from the couplings observed in other pyrimidine reduction products.

(VII)   (VIII)

In a purine base anion, the spin density on the C_8 carbon atom can be expected to generate a sizable hyperfine coupling with the C_8—H proton, which should serve to identify the species. Relatively little ENDOR data is available to date concerning purine base anions. Only the C_8—H proton coupling for the reduction product formed in single crystals of adenine dihydrochloride has been reported. The directions of the principal axes of the C_8—H proton coupling tensor take the expected orientation and serve to identify the reduction product unequivocally. The principal values of the coupling are given in Table 3. The ratio of the overall anisotropy of the coupling to the isotropic part is 1.2, consistent with the ratio observed for pyrimidine reduction products.

Information concerning radiation-induced oxidation products formed in nucleic acids is meager. If base cations are formed, the important hyperfine couplings are likely to involve the ^{14}N nuclei and the σ protons associated with N—H groups. ENDOR data are available only for the cation (IX) produced by irradiation of 5-nitro-6-methyluracil. The spin density on N_1 results in large couplings with the nitrogen nucleus and with the N_1—H proton. The principal values of the proton coupling are -1.86, -18.72, and -25.94 MHz, and the principal axes of the coupling tensor correspond with the directions of the N_1—H bond, the normal to the uracil ring, and orthogonal to these directions. The ratio of the overall anisotropy to the isotropic part of the coupling is 1.6, close to the ratio for the σ protons in another free radical containing $>N$—H groups reported by Ngo *et al.*[6] Since the N—H bond is polar, the proton is readily exchanged for deuterium by growing the crystal from heavy water solution. Unfortunately, attempts to measure the nitrogen hyperfine coupling by the ENDOR method proved unsuccessful.

$$
\begin{array}{c}
O \\
\parallel \\
C \\
HN^3 \quad {}^4 \quad {}^5C—NO_2 \\
\mid \quad \oplus \quad \parallel \\
O{=}C^2 \quad {}_1 \quad {}^6C—CH_3 \\
\diagdown N \diagup \\
\mid \\
H
\end{array}
$$

(IX)

Other types of damage to DNA must occur that have not yet been as well characterized as has base damage. One may confidently expect ENDOR to play a leading role in recognizing other types of radiation damage as may occur, for example, in the sugar component of DNA.

At least passing mention should be made of the fact that the proton couplings measured by ENDOR in various radiation products are of interest from a quantum chemical point of view also. Local concentrations of spin densities can be inferred from measurements of the isotropic as well as from the anisotropic parts of the hyperfine couplings. These densities may be compared with spin density distributions calculated from quantum chemical approximations, including simple Hückel theory, the INDO method, and various other approximations.

4. *Protein Constituents*

The typical protein consisting of scores of amino acid residues connected by amide linkages is an exceedingly complex molecule. A sensible approach to the study of radiation damage in protein begins with a study of

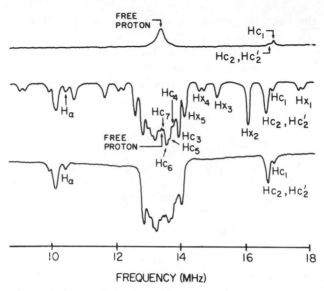

Figure 3. ENDOR spectra from a single crystal of acetylglycine x-irradiated at room temperature. Magnetic field parallel to the $a' = b \times c$ axis. Upper spectrum taken at nominal rf power level. Middle spectrum is negative ENDOR obtained with intense rf field. Lower spectrum is negative ENDOR from crystal in which exchangeable protons were replaced with deuterons. [From H. A. Helms, Jr., I. Suzuki, and I. Miyagawa, *J. Chem. Phys.* **59**, 5055 (1973). Reproduced with permission.]

simpler model compounds. One much-studied model compound is acetylglycine (X). A stable radical (XI) is produced in irradiated acetylglycine. Helms *et al.*[7] obtained the ENDOR spectra shown in Figure 3 from single crystals of acetylglycine irradiated at room temperature. These investigators utilized "negative ENDOR" (Figure 3) to obtain detailed information concerning the multitude of protons with which the unpaired electron in (XI) interacts. The ENDOR resonances were assigned to various protons within and in the vicinity of the radical as indicated in Figure 4.

$$CH_3-CONH-CH_2-COOH \qquad CH_3-CONH-\dot{C}H-COOH$$

$$\text{(X)} \qquad\qquad\qquad\qquad \text{(XI)}$$

In order to elucidate the primary events leading to the production of (XI), the precursor radicals formed in the primary steps of the radiation damage process must be stabilized and identified. The ESR absorption obtained from a single crystal of acetylglycine x-irradiated at 4.2°K is shown in Figure 5. This model compound serves as an excellent example illustrating how ENDOR can be used to analyze an otherwise uninterpretable ESR absorption. Using ENDOR, the reduction products (XII) and (XIII) and the primary oxidation product (XIV) can be identified. Reduction products analogous to

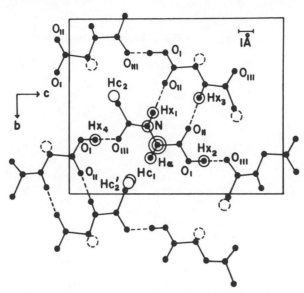

Figure 4. Attribution of various proton resonances shown in Figure 3 according to Helms *et al.*[7] [From H. A. Helms, Jr., I. Suzuki, and I. Miyagawa, *J. Chem. Phys.* **59**, 5055 (1973). Reproduced with permission.]

(XII) and (XIII) have been identified in a variety of amino acids and peptides irradiated at low temperature. The hyperfine tensors that characterize the proton couplings in these products have been determined from ENDOR measurements. The principal values of the coupling tensors for the nonexchangeable protons in reduction products of the types (XV), (XVI), and (XVII) are listed in Table 4. The carbon orbital carrying the unpaired electron in (XV)–(XVII) is known to be slightly admixed with *s* state; consequently, the bonding of the carbon is somewhat pyrimidal. One consequence of nonplanarity is that the range of β-proton couplings (Table 4) is much less than the range (approximately 3–123 MHz) observed for β-proton couplings in planar carbon-centered π-electron radicals. The reduction products (XV) and (XVI), formed in carboxylic acids as well as in amino acids and their salts, may protonate rapidly even at low temperature. Japanese workers have employed ENDOR in an extensive study of the protonation mechanism

Figure 5. ESR spectrum from a single crystal of acetylglycine x-irradiated at 4.2°K. Magnetic field parallel to the *b* axis. By isotropic substitution and by ENDOR analysis, the absorption can be shown to be a superposition of spectra arising from (XIII), (XIV), and two conformations of (XII).

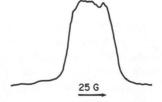

25 G

Table 4

Principal Values of Proton Coupling Tensors in Reduction Products Formed in Amino Acids and Peptides[a]

Compound	Principal values (MHz)		
Acetylglycine*	17.0	3.4	2.0
L-Tyrosine HCl	19.7	4.6	3.0
Glycylglycine HCl*	21.2	6.5	4.9
Glycine[b]	23.0	8.7	6.6
Acetylglycine	23.2	9.4	8.2
Acetylglycine*	25.4	11.4	10.6
DL-Aspartic acid HCl[c]	27.1	12.0	10.3
Acetylglycine	29.8	15.2	14.0
L-Valine HCl[d]	33.0	17.1	14.8
Glycylglycine HCl	37.1	21.3	18.7
L-Alanine[e]	61.4	50.8	49.0
L-Cysteine HCl[f]	68.2	58.3	56.2
DL-Serine	70.3	61.3	58.0
L-Histidine HCl	71.2	60.8	59.1
Glycylglycine HCl*	77.2	67.8	66.4
Acetylglycine	82.4	74.6	72.2
Glycine[b]	82.8	72.8	71.4
Aspartic acid HCl[c]	83.7	74.6	71.9
Acetylglycine	83.8	75.8	72.6
Aspartic acid HCl[c]	88.8	79.6	77.3
Glycylglycine HCl	91.6	81.8	78.4
Acetylglycine*	92.0	82.6	81.4

[a] Data refer to nonexchangeable proton couplings in radicals of the types

$$\begin{array}{ccc} \diagdown\!\!\!\!\raisebox{0pt}{}CH-\overset{\displaystyle\cdot}{C}\diagup^{O^-(H^+)}_{\diagdown O^-} , & \diagdown\!\!\!\!CH-\overset{\displaystyle\cdot}{C}\diagup^{O^-(H^+)}_{\diagdown OH} , & or \quad \diagdown\!\!\!\!C-\overset{\displaystyle\cdot}{C}\diagup^{O^-}_{\diagdown NH-} \end{array}$$

The last type, formed by reduction of the carbonyl oxygen of the peptide linkage, are marked by asterisks. Unless otherwise indicated, data are taken from H. C. Box, H. G. Freund, K. T. Lilga, and E. E. Budzinski, *J. Chem. Phys.* **63**, 2059 (1975), and references therein.
[b] M. Iwasaki and H. Muto, *J. Chem. Phys.* **61**, 5315 (1974).
[c] S. M. Adams, E. E. Budzinski, and H. C. Box, *J. Chem. Phys.* **65**, 998 (1976).
[d] H. Muto, K. Nunome and M. Iwasaki, *J. Chem. Phys.* **61**, 5311 (1974).
[e] H. Muto and M. Iwasaki, *J. Chem. Phys.* **59**, 4821 (1973).
[f] W. W. H. Kou and H. C. Box, *J. Chem. Phys.* **64**, 3060 (1976).

in single crystals. In L-alanine, the proton that neutralizes the reduction product is associated with a hydrogen bond between the carbonyl oxygen and nitrogen of a neighboring molecule. Upon reduction of the carbonyl oxygen, the proton associated with the hydrogen bond moves from the nitrogen side of the bond to the oxygen side.[8] In glycine and α-aminoisobutyric acid, a similar protonation of the anion occurs, but the movement of the proton is less pronounced.[9] In each case, it is a proton associated with a hydrogen bond out of the plane of the carboxylate group that transfers.

$$CH_3-CONH-CH_2-\overset{\displaystyle O^-}{\underset{\displaystyle OH}{\overset{|}{\underset{|}{C}}}}$$

(XII)

$$CH_3-\overset{\displaystyle O^-}{\overset{|}{C}}-NH-CH_2-COOH$$

(XIII)

$$CH_3-CONH-\dot{C}H_2$$

(XIV)

$$\underset{\diagup}{\overset{\diagdown}{C}}H-\overset{\displaystyle O^-}{\underset{\displaystyle O^-}{\overset{|}{\underset{|}{C}}}}$$

(XV)

$$\underset{\diagup}{\overset{\diagdown}{C}}H-\overset{\displaystyle O^-}{\underset{\displaystyle OH}{\overset{|}{\underset{|}{C}}}}$$

(XVI)

$$\underset{\diagup}{\overset{\diagdown}{C}}H-\overset{\displaystyle O^-}{\underset{\displaystyle NH}{\overset{|}{\underset{|}{C}}}}$$

(XVII)

The primary oxidation product observed in acetylglycine results from the dissociation of the oxidized molecule, yielding (XIV) and CO_2. Oxidation products formed in an analogous manner in various amino acids and peptides that have been characterized from ENDOR measurements are tabulated in Table 5.

In addition to decarboxylation, other types of oxidation effects are observed in irradiated amino acids containing conjugated rings. ENDOR studies have demonstrated that the conjugated rings in histidine hydrochloride and tyrosine hydrochloride are oxidized by ionizing radiation.[6,10] The interaction between unpaired electron and σ protons in these conjugated rings yield readily interpretable coupling tensors.

Entirely different types of primary oxidation and reduction products are observed in amino acids containing sulfur. In cystine hydrochloride (XVIII), the principal primary oxidation and reduction products observed in single crystals irradiated at low temperature are the cation (XIX) and the

Table 5

Principal Values of α-Proton Coupling Tensors in Free Radicals Formed by Decarboxylation in Amino Acids and Peptides[a]

Compound	Principal values (MHz)		
Acetylglycine	−80.2	−47.4	−22.6
	−72.4	−38.0	−11.2
Glycylglycine HCl	−80.9	−47.5	−22.7
	−78.1	−42.9	−19.9
β-Alanine	−91.6	−57.2	−21.2
	−91.3	−60.2	−20.3
L-Histidine HCl	−94.2	−54.3	−22.7
DL-Serine	−95.0	−56.2	−25.0
L-Cysteic acid	−104.0	−65.0	−29.6

[a] Delocalization of spin away from α carbon atom occurs most expeditiously in the peptide oxidation products resulting in smaller couplings. Data from H. C. Box, H. G. Freund, K. T. Lilga, and E. E. Budzinski, *J. Chem. Phys.* **63**, 2059 (1975), and references therein.

anion (XX). In both primary products, the unpaired electron is localized on the disulfide group. The couplings to the protons in the CH_2 groups adjacent to disulfide group have been obtained from ENDOR measurements. The results are given in Table 6. When a sulfhydryl group is present, it is a likely site for oxidation. Thus, in cysteine hydrochloride (XXI) x-irradiated at 4.2°K, the radical (XXII) is observed. Only one of the CH_2 proton couplings could be obtained in this study (Table 6).

$$HOOC-CH(NH_3Cl)-CH_2-(S-S)-CH_2-CH(NH_3Cl)COOH$$

(XVIII)

$$HOOC-CH(NH_3Cl)-CH_2-(S-S)^+-CH_2-CH(NH_3Cl)COOH$$

(XIX)

$$HOOC-CH(NH_3Cl)-CH_2-(S-S)^--CH_2-CH(NH_3Cl)COOH$$

(XX)

$$HOOC-CH(NH_3Cl)-CH_2-SH \qquad HOOC-CH(NH_3Cl)-CH_2-\dot{S}$$

(XXI)   (XXII)

Having discussed the types of damage that are observed in amino acids and peptides, it becomes appropriate to consider which of these lesions is likely to be important in irradiated proteins. If the protein contains cysteine residues linked by disulfide bridges, the results obtained with cystine indicate these residues are likely to dominate the reduction process. The disulfide group is a more powerful scavenger of electron than is the carbonyl

Table 6

Principal Values of CH_2 Proton Coupling Tensors in Amino Acid Oxidation and Reduction Products Containing Sulfur

Amino acid radical	Principal values (MHz)		
Cystine di-HCl[a]	21.2	17.3	16.4
$-CH_2-(S-S)^+-CH_2-$	14.3	5.47	4.12
Cystine di-HCl[a]	33.1	27.6	25.9
$-CH_2-(S-S)^--CH_2-$	24.0	18.9	17.3
Cysteine HCl[b]			
$-CH_2-\dot{S}$	107.8	98.8	96.4

[a] A. Naito, K. Akasaka, and H. Hatano, *J. Mag. Reson.* **24**, 53 (1976). Values quoted are average of two essentially equivalent couplings.
[b] W. H. Kou and H. C. Box, *J. Chem. Phys.* **64**, 3060 (1976).

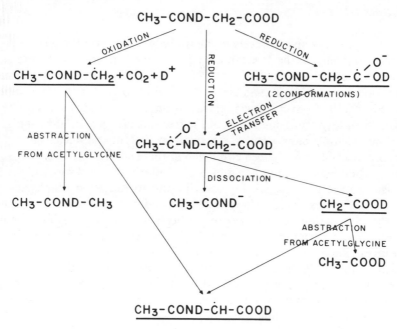

Figure 6. Radiation damage processes in acetylglycine according to Sinclair and Codella.[11]

group. When considering radiation-induced reduction processes, distinctions drawn between direct and indirect effects lose significance, since the origin of the electron effecting the reduction is inconsequential. However, for radiation-induced oxidation processes, we retain the distinction and refer only to direct oxidation effects. Just how specific the direct effects of radiation-induced oxidation in protein may be is not clear. It appears that specificity is determined by how effectively electron vacancies can transfer intramolecularily. Little definitive experimental data on this question is available.

Let us return now to our model compound, namely, acetylglycine (X). On warming acetylglycine, the primary products (XII), (XIII), and (XIV) are lost, and (XI) is formed. The intermediate steps in the overall radiation damage scheme set forth in Figure 6 were proposed by Sinclair and Codella.[11] The scheme is based on an ESR study of crystals that were irradiated at 77°K and subsequently warmed to intermediate temperatures. Radical identifications were reinforced by studies on ^{13}C and ^2H isotopically substituted crystals. ENDOR could profitably be employed to verify the correctness and the completeness of the radiation damage processes shown in Figure 6. This effort has not yet been made, nor has a thorough investigation of the entire radiation damage process been made in any other model system using the ENDOR method.

5. Conclusion

The role of ESR spectroscopy in the study of radiation damage processes has been to delineate the free radical steps in these processes. The ENDOR adjunct has greatly expanded the usefulness of ESR spectroscopy in this application by making it possible to analyze absorptions that are superimposed spectra from a variety of radiation products. ENDOR has been indispensible for identifying many of the free radicals produced by ionizing radiation in simple biomolecules, especially in protein and nucleic acid subunits. Single crystals have been employed in these studies in order to realize the important experimental advantage of similarly oriented molecules.

The potential of ENDOR for contributing to radiation biophysics remains largely unexplored. Many of the secondary processes that proceed with the elimination of primary free radicals and the generation of more stable secondary free radicals observed at room temperature have not as yet been investigated with the aid of ENDOR. The ENDOR method has only been applied to the study of radiation damage in rather simple biomolecules. It appears feasible to apply the method to molecules of considerably greater complexity, and it may be anticipated that deeper insights into radiation damage processes will come from these studies.

ACKNOWLEDGMENT

The support of this work through contract E(11-1)3212 with the Energy Research and Development Administration and grant no. FD00632 from the National Institutes of Health is gratefully acknowledged.

References

1. G. H. Dibdin, *Trans. Faraday Soc.* **63**, 2098–2111 (1967).
2. J. A. Brivati, M. C. R. Symons, D. J. A. Tinling, H. W. Wardale, and P. O. Williams, *Trans. Faraday Soc.* **63**, 2112–2116 (1967).
3. H. C. Box, E. E. Budzinski, K. T. Lilga, and H. G. Freund, *J. Chem. Phys.* **53**, 1059–1065 (1970).
4. I. Johansen and P. Howard-Flanders, *Radiat. Res.* **24**, 184–200 (1965).
5. P. V. Hariharan and P. A. Cerutti, *J. Mol. Biol.* **66**, 65–8; (1972).
6. F. Q. Ngo, E. E. Budzinski, and H. C. Box, *J. Chem. Phys.* **60**, 3373–3377 (1974).
7. H. A. Helms, Jr., I. Suzuki, and I. Miyagawa, *J. Chem. Phys.* **59**, 5055–5062 (1973).
8. H. Muto and M. Iwasaki, *J. Chem. Phys.* **59**, 4821–4829 (1973).
9. M. Iwasaki and H. Muto, *J. Chem. Phys.* **61**, 5215–5320 (1974).
10. H. C. Box, E. E. Budzinski, and H. G. Freund, *J. Chem. Phys.* **61**, 2222–2226 (1974).
11. J. Sinclair and P. Codella, *J. Chem. Phys.* **59**, 1569–1576 (1972).

11

Polymer Studies

Martin M. Dorio

1. Introduction

The physical and mechanical properties of solid polymers are dependent on the molecular motions as well as the inter- and intramolecular interactions. Several methods have been useful in the study of these motions: dielectric relaxation, dynamic mechanical relaxation,[1] and nuclear magnetic resonance.[2] EPR has also been used, of course, to probe molecular motions. The method has seen applications in both biological systems[3] and in polymers.[4–7] The latter were for polymeric systems in solution or for solid polymers in the vicinity of their melting transitions. This has been due primarily to the limited sensitivity of the EPR method in the slow-motion region. At these low temperatures, line-shape changes become small and subtle, thereby reducing the amount of information extractable from the spectra. Nevertheless, EPR studies in the solid state region [i.e., correlation times (τ_c) less than 10^{-7} sec] have been carried out[8–11] on a few systems.

In this chapter, I wish to discuss the applications of ENDOR and ELDOR to polymeric systems. It was only after the advent of these multiple resonance techniques that certain of the long-standing polymer problems lent themselves to solution. Other chapters of this book have already pointed out the variety of problems that multiple resonance has probed. The multiple resonance techniques appear, at this early stage, to offer the capability to probe polymer dynamical processes in a similar way.

Martin M. Dorio • Diamond Shamrock Corporation, T. R. Evans Research Center, Painesville, Ohio

2. Irradiated Polymers

As early as 1951, Schneider[12] had begun applying the EPR technique to irradiated polymers. Shortly thereafter, Libby[13] had found the five-line spectrum (Figure 1) in polyethylene, which he straightforwardly identified as due to the presence of the alkyl radical

$$-CH_2-\dot{C}H-CH_2-$$

Since that time, EPR has allowed study of radicals in many irradiated polymer systems.[14-20] Often times the classic work of Fessenden and Schuler[21] on alkyl radicals was drawn upon. A variety of such studies allowed a detailed comprehension of the mechanics of radical propagation, creation, and the kinetics of their reactions. This literature has been reviewed.[14-16]

As soon as complicated irradiated polymeric systems were investigated, however, it became clear that the multiplicity of radicals present led to severely overlapped spectra. Several controversies arose as a result of the myriad interpretations possible. Ayscough, for instance, presents such arguments in a paper in the *Journal of Polymer Science*[17] as they related to polypropylene. Throughout the years, a variety of clever experiments were necessary to elucidate the exact nature of these radicals. Charlesby and co-workers[18] resorted to use of model systems to help elucidate the nature of the radical signals observed. Oriented,[19] deuterated,[19] as well as photo-induced changes in the EPR[20] provided the data from which radical identifications were deduced. In general, a great many polymers were explored, and with each came a greater number of questions about radical dynamics within the polymer matrices. It is not the purpose of these remarks to provide a detailed review of the field; it is, however, to illustrate the complexity of the research area and to demonstrate the vastness of the methodology employed to address these problems.

Immediately after the first solution application of ELDOR,[22] it became clear[23] that the long-standing problems of polymer systems could be approached using the new ELDOR technique. As a model system, the overlapped spectra in irradiated malonic acid were separated.[24] Frequency-swept ELDOR were taken, before the advent of automatic sweep equipment, by manually stepping the frequency difference. As long as the dominant relaxation paths arose from intramolecular mechanisms, overlapped spectra could be separated. Irradiated polymethylmethacrylate (PMMA) was examined in a similar way. The data suggested that only one spin center was present.[23] There had been differing views as to the interpretation of the nine-line spectrum from PMMA.[25] The ELDOR, however, with deenhancements at each 33-MHz interval, provided convincing evidence that all lines arise from a single radical species.[26]

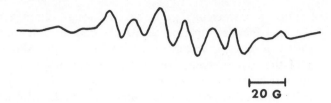

Figure 1. Typical five doublets of EPR spectrum of irradiated polyethylene. (After Libby.[13])

In recent work,[27] several attempts were made to apply the ELDOR technique to a variety of polyolefins. These experiments used relatively low-intensity ultraviolet as the radical initiator. In the ELDOR cavity, EPR signal intensity of the radicals formed was very low in all cases. This is probably attributable to the lowered cavity Q at the low temperatures at which the experiments were carried out. As was discussed in Chapter 2, cavity Q is a crucial parameter for observation of good double resonance spectra. The signal-to-noise ratio in the spectra so obtained was very low, resulting in poor interpretations at best.

3. Spin Probes within Polymeric Matrices

One technique to attempt to solve the problems of low signal intensity alluded to in Section 2 is to introduce a radical species of stronger intensity, for example, a molecule containing a radical.[28-38] The utility of this approach is that one is then afforded a reasonably high signal-to-noise ratio in the double resonance spectra. This allows the deduction and/or extraction of motional information on a molecular level from these spectra. In ELDOR, for example, the R factor, the measure of the double resonance effects of the EPR spectra, is defined by

$$R = \frac{\text{EPR(pump off)} - \text{EPR(pump on)}}{\text{EPR(pump off)}} \tag{1}$$

or

$$R = 1 - I/I_0 \tag{2}$$

where I_0 is the EPR intensity and I is the perturbed intensity with application of the pump power. In general, R is a function of the frequency separation $\Delta v = v_p - v_o$; H_p, the pump power; H_o, the observing power; and temperature T. The R factor has been derived in orientation space by Dalton *et al.*[39] As modulation amplitude is reduced to zero,

$$R(\theta, \phi, S_p) = \frac{F(\theta, \phi, S_p)}{4\pi} \sum_l \frac{(2l + 1)P_l(\cos \theta)}{l(l + 1)D + T_{1e}^{-1}} \tag{3}$$

where D is the diffusion coefficient, T_{1e} the effective longitudinal relaxation time, $P(\cos \theta)$ the lth Legendre function, and F the distribution function. R then may be related to $R(\Delta v)$ in frequency space. This relationship permits the correlation of the experimentally measured effects with the physical interpretation of molecular reorientation as reflected by the spectral diffusion. The difficulty with this approach, of course, is that the computations are rather complicated, and to date, theoretical studies of model systems[38–42] outnumber actual calculations of equation (3) from the measured $R(\Delta v)$. One hopes the future will see careful experimental studies balanced by detailed calculations to derive precise angular data for the probe molecules. In some early theoretical calculations for a simple nitroxide radical,[42] simulated line shapes permitted the identification of a 0.15-radian jump angle using a free diffusion model as a reasonable description of the molecular motion of the radical.

Work published so far has tended to the characterization of the nature of radicals interactions with the environment. Homogenization of the inhomogeneous line shapes has been elucidated in a series of experiments using nitroxide spin probes in relatively high molecular weight plastic matrices.[33] The presence of a single-line ELDOR response at ~ 14 MHz was found, interpreted as indicating the forbidden transitions arising from dipole–dipole coupling of the electronic spin to the cyclohexyl protons of the matrix. Such coupling is not thermally averaged due to the immobility imposed by the rigidity of the plastic matrix at $T = 230°$K. Care was taken experimentally by use of low modulation amplitudes and frequencies to avoid promoting homogenization of the relatively narrow 1.5-MHz spin packets. In further work,[35] it became clear that observation of discrete spin packages was possible as long as

$$\gamma_e H_1 > \frac{1}{2\pi T_{1e}} \geq V_n \tag{4}$$

where H_1 is the pumping microwave amplitude, T_{1e} the electronic relaxation time, V_n the rate of homogenization. Interestingly, in these experiments,[33–37] the pump frequency was fixed, as was the field, and the observing frequency was scanned through the resonance of interest (Figure 2). This approach is different from that reported below taken on commercially available equipment. In that case, the observing frequency is fixed and the pump frequency swept, resulting in a display that is the difference between the perturbed and unperturbed EPR spectra.

The presence of the forbidden transition (termed the "matrix peak") is a phenomenon that has been observed in both ENDOR and ELDOR spectra.[28–31,43–45] In the latter, the matrix transition was found to be associated with the single-line ELDOR. In single-line ELDOR, used as a new technique to study molecular motion in polymeric solids,[28] saturating one

Figure 2. Typical ELDOR spectrum for forbidden transitions of nitroxide spin probe in cyclohexane matrix. (After Kirillov *et al.*[34])

⊢⊣
5 MHz

spin packet within an inhomogeneous envelope and observing another packet within the same envelope provides a means to determine molecular reorientation as saturation transfer de-enhances the EPR transition probability. The presence of matrix ELDOR is evidenced by a secondary peak superimposed on the smooth peak of the single-line spectra. The matrix peak may be taken as additive to the single-line spectral response (Figure 3). In this event, the line shape of the forbidden transition provides information on the nature of the interactions of the electronic spin with neighboring protons in the matrix. Drawing on the ENDOR work,[43–44] the line shape has the following characteristics.

The intensity of the matrix ELDOR is determined by the relative magnitudes of the various relaxation processes involved. This may be described by the line-shape function[30]

$$f(v) = \int_0^r \int_0^\pi \int_0^{2\pi} W(r,\,\phi,\,\theta)g[(v - v_0)(r,\,\theta,\,\phi)]r^2 \sin\theta\,d\phi\,d\theta\,dr \qquad (5)$$

where g is the coordinate-dependent spin packet distribution function, and W is a generalized transition probability based on the relative magnitudes of the electronic spin–lattice relaxation W_e, the nuclear spin–lattice relaxation W_n, the cross-relaxation of nuclear and electron spins W_{END}^N, and the

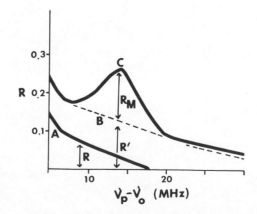

Figure 3. Schematic view of matrix ELDOR response (R_M) superimposed on single-line ELDOR.

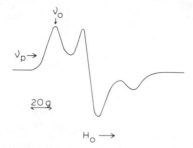

Figure 4. EPR spectrum for nitroxide radical in a solution at $-150°C$. In this solvent, the radical was completely immobilized. In the ELDOR experiments reported herein, v_o was positioned as shown; v_p swept into or away from v_o on the low-field side.[28b]

electronic spin diffusion W_D. W has been derived[34] for simultaneous electron and proton spin flips under an alternating field of frequency ω'.

$$W = \frac{1}{2\hbar\omega_N^2} \int_{-\infty}^{\infty} \exp[-i(\Delta' + \omega_N)\tau - \gamma\tau] \ll B(\tau)B(0) \gg d\tau \qquad (6)$$

where $\Delta' = \omega' - \omega_0'$, the detuning relative to the center of the packet (ω_0'), γ the width of the spin packet; the double angle brackets denote averaging over the random rotations and translation of the molecules. γ is defined by $T_{1e}^{-1} + V_n$.

The detailed procedure for measuring single-line ELDOR has been previously discussed.[28-30] Briefly, the EPR spectrum of the sample is first scanned. Then the observing frequency was positioned as desired (for example, such as that indicated in Figure 4). The pump frequency is then swept, and the ELDOR effect is calculated, expressed by the factor $R(\Delta v)$ given by

$$R(\Delta v) = \frac{\text{ELDOR peak height}}{\text{EPR signal height with pump off}} \qquad (7)$$

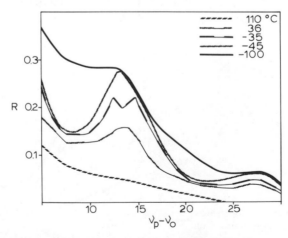

Figure 5. $\Delta M_I = 0$ ELDOR for maleimide in amorphous polystyrene. v_o is at the low-field turning point. Lines represent averages of a number of scans. $v_p > v_o$.[28b]

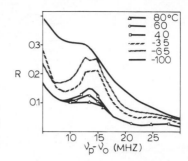

Figure 6. $\Delta M_I = 0$ ELDOR for maleimide in semi-crystalline polystyrene. v_o is at the low-field turning point. Lines represent averages of a number of scans. $v_p > v_o$.[28b]

The ELDOR results given in Figures 5 and 6 are for maleimide in amorphous and semicrystalline polystyrene. In these figures, averages for the line shape at each temperature are plotted for clarity. The spectra were assumed to be made up of the monotically decreasing $R(v_p - v_o)$ function of the single-line ELDOR and the matrix ELDOR as the additional contribution. The latter is shown in Figure 7 to reach a maximum at about $-44°C$. At the high-temperature region, the R values for matrix ELDOR extrapolates to zero at $101° \pm 5°C$. This corresponds to the T_g of polystyrene. We will use the subscript M, e.g., R_M, to indicate the matrix signal, where R without the subscript refers to the single-line ELDOR de-enhancement. The latter increases monotonically with $1/T$.

There is also a second matrix signal at 27 MHz. Because it is weaker than the 13.5-MHz matrix signal, the error involved in estimating its intensity is correspondingly greater. The signal is more prominent at $-45°C$ with $R_M \approx 0.065$, which is about 40% of the R_M at 13.5 MHz.

The data for amorphous and semicrystalline polystyrene are the same within experimental error. At the highest temperatures the values of R in one semicrystalline sample seem to rise. This behavior is sometimes observed

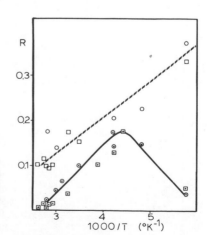

Figure 7. ELDOR magnitude for matrix and single-line contributions to the line shape for maleimide in polystyrene: R at 6 MHz for amorphous polymer ($\square$); R at 6 MHz for semicrystalline polymer ($\bigcirc$); R_M for amorphous polymer ($\boxdot$); R_M for semicrystalline polymer ($\odot$).[28b]

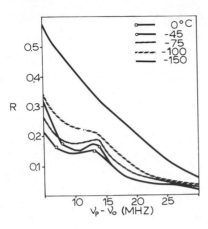

Figure 8. $\Delta M_I = 0$ ELDOR for TANOL in amorphous polystyrene. v_o is at the low-field turning point. Lines are averages of a number of scans.[28b]

also in other amorphous polystyrene samples. This point will be discussed further below.

Two other nitroxides, TANOL and DTBN, both smaller in sizes than maleimide, were incorporated into amorphous polystyrene. The results are shown in Figures 8 and 9. The matrix ELDOR signals in these samples are generally much weaker than those for maleimide. R_M values reach a maximum at about $-64°C$ (Figure 10a), which is about 20°C below the temperature for maximum R_M in maleimide. Aside from the anamolous increase of R value at the highest temperatures, which has already been commented on above, the single-line ELDOR has R values that increases linearly with $1/T$.

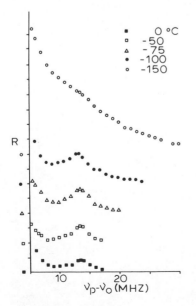

Figure 9. $\Delta M_I = 0$ ELDOR for DTBN in amorphous polystyrene. Divisions on the R scale are 0.05 units each. The zero has been shifted for each curve for clarity; zero point has been marked on the axis with the appropriate symbol. v_o is at the low-field turning point. $v_p > v_o$.[28b]

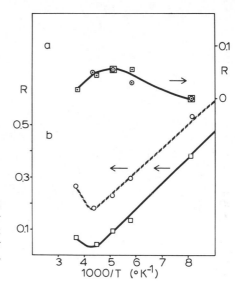

Figure 10. ELDOR magnitudes for separated contributions to line shape: R at 6 MHz for amorphous polymer with DTBN (□); R at 6 MHz for TANOL (○); matrix signal for DTBN (▣); matrix signal for TANOL (◎).[28b]

The slopes for this variation are the same for the three nitroxides (Figures 7 and 10b), but the lines were shifted horizontally with respect to one another. The value of $R = 0.25$ was reached at 63.6°, $-80.7°$, and $-125°$C for maleimide, TANOL, and DTBN, respectively. This order parallels the decreasing molecular size and deviation from a spherical shape.

The ELDOR spectra of maleimide in amorphous and semicrystalline polypropylene are shown in Figures 11 and 12, respectively. The matrix signal begins to emerge at 0°C and is very intense at $-50°$C. The dominant mechanical relaxation at 1 Hz for polypropylene is the β relaxation at 0°C,[46,47] which is believed to be glass–rubber relaxation of the amorphous polypropylene. The dilatometric T_g of polypropylene[48] is, however, at $-35°$C.

The correspondence of matrix ELDOR with some major relaxation process for polypropylene as it did for polystyrene seems to be a valid and probably a general conclusion.

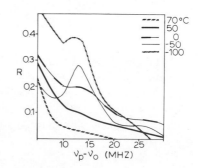

Figure 11. $\Delta M_I = 0$ ELDOR for maleimide in amorphous polypropylene. ν_o is at the low-field turning point. Lines are averages for a number of scans.[28b]

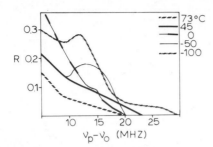

Figure 12. $\Delta M_I = 0$ ELDOR for maleimide in semi-crystalline polypropylene. v_o is at the low-field turning point. Lines are averages of a number of scans.[28b]

These are some noted dissimilarities in the ELDOR spectra of polypropylene and polystyrene. In polypropylene, the matrix signal almost seems to be borrowing intensity from the normal ELDOR. That is, for the spectra at −50°C in Figures 11 and 12, the single-line ELDOR intensities are much lower than expected based on observations on polystyrene and paraffin oil.[29] This "borrowing effect" may in fact account for the increased R values at high T at 6 MHz for the maleimide–polystyrene system above. The "borrowing" in the latter case is of much smaller magnitude than for polypropylene.

In general, the results for amorphous and semicrystalline polypropylenes are the same. The intensities of the matrix signals and v_M are nearly the same. The spectra at elevated temperatures are actually superimposable.

3.1. Temperature Dependence of Single-Line ELDOR

The single-line ELDOR de-enhancement is expected to have an η/T dependence.[30] However, the actual results in many of the above figures show a simple dependence. This is because polymer viscosity cannot be satisfactorily represented by an equation of the Arrhenius type[49] but is sensitive to free volume.[50] Above T_g, η is inversely proportional to the excess free volume due to thermal expansion and kinetic agitation. At temperatures below T_g, any changes in free volume with temperature will be much smaller than above it, as determined by the polymers' thermal expansion coefficients. In our measurements below T_g, the observed $1/T$ dependence of R is reasonable.

Matrix ELDOR

The matrix ELDOR effect described above can be definitely attributed to relaxation processes involving nonbonded protons. For instance, R_M for maleimide in perdeuterated polystyrene is only about 65% as large as R_M in normal polystyrene.[30] It will be interesting to see whether the matrix signal can be completely eliminated for perdeuterated maleimide in perdeuterated polystyrene.

The interaction between the electron spin and the proton spin is probably dipolar in nature. This is supported by the observation that R_M at $2v_H$ is about 40% of the value at $1v_H$. Similar dependence of M_I was observed in hyperfine ELDOR where electron–nuclear dipolar process is the main relaxation mechanism.[22]

In Figure 7, R_M is seen to have a maximum value at $-44°C$. Similar behaviors are observed in Figure 10a. The explanation may be the following: the probability of W_{END}^N is proportional to $\tau_c/(1 + \omega_H^2\tau_c^2)$. At high temperatures, $\omega_H^2\tau_c^2 \ll 1$ and $W_H(END) \propto \tau_c$; at low temperatures $\omega_H^2\tau_c^2 \gg 1$ and $W_H(END) \propto \tau_c^{-1}$. The turning point should occur at $\omega_H^2\tau_c^2 = 1$. For the free proton precession frequency, this corresponds to $\tau_c = 10^{-8}$ sec.

Even though matrix ELDOR is not seen in glassy nonpolymeric systems, it is prominent in amorphous polymers. Hyde and co-workers[43,44] have concluded that only protons less than 6 Å away can interact with the electron to give END relaxation. If the process is reversible, as has been proposed,[30] then it will be facilitated if several protons having nearly the same orientation are present in the vicinity of the electron spin. The proton spin-flipped by the END interaction can, in turn, flip another one, and so on. In this manner, W_{END}^N may be facilitated. According to this argument, the results suggest some short-range order in the polymeric systems.

3.2. Motion of Interstitial Molecules

It has already been noted above that at a given temperature, the values of R for various nitroxides are different. This is a manifestation of the τ_c for the molecule being considered. If this is true, then the R values for all the nitroxides should be comparable for a given τ_c. Using the method devised by McConnell and co-workers,[51] τ_c can be estimated from the EPR spectra. This has been done, and the results of single-line ELDOR at $\Delta v = 6$ MHz have been compared at identical τ_c for the three nitroxides. The data do not fall on a single curve, as might have been expected from our previous work.[29] This may reflect the size differential for these nitroxides, the molecular volumes of DTBN, TANOL, and maleimide are in the approximate ratio of $1 : 1.2 : 1.6$ from molecular models. From the τ_c estimated from the EPR, we find R values of 0.13, 0.19, and 0.32 at $\tau_c = 2 \times 10^{-8}$ sec for maleimide, TANOL, and DTBN, respectively. The ratios of R for the three are $1 : 1.5 : 2.5$, not particularly good agreement with the molecular volume ratios. The trend is correct, but the ratios are functions of τ_c. At $\tau_c = 9 \times 10^{-9}$ sec, for instance, the R values are 0.06, 0.12, and 0.18 for ratios of $1 : 2 : 3$.

It is not surprising that the ELDOR spectra of maleimide in amorphous and semicrystalline polymers are about the same. In the latter cases, the additive is actually found only in the amorphous region of the polymer.

The interstitial additive has lower mobility in polystyrene than in polypropylene at comparable temperatures. For instance, at $-100°C$, τ_c for maleimide in polystyrene is 5×10^{-7} sec; it is an order of magnitude shorter for maleimide in polypropylene.

Matrix ENDOR

Matrix ENDOR has been observed on one polymeric system by Kevan and co-workers.[45] For polyenyl radicals created by irradiation with a γ source, a spectrum of three lines was observed for polyvinyl fluoride. ENDOR were taken for heavily irradiated films and showed free fluorine and free proton resonances at 13.6 and 14.4 MHz, respectively. Using the lineshape model they have developed (see Chapter 6), these workers estimated that six to seven protons at distances of ~ 4–5 Å and seven fluorines approximately 4.7 Å away contribute to the radical environment. From their work, it seems reasonable to hope that matrix ELDOR can be observed under the proper conditions for matrix species other than protons.

4. Spin-Labeled Polymers

Two different types of spin-labeled polystyrene have been explored via ELDOR. One was labeled at the *para* position of the phenyl group (I). The other was terminal-labeled (II).

$$-(CH_2-CH)_n-$$

$$t\text{-Bu}-N-O\cdot$$

(I)

$$t\text{-Bu}-N-(CH-CH_2)_n-CH-N-t\text{-Bu}$$
$$\overset{|}{O}\cdot \qquad\qquad\qquad \overset{|}{O}\cdot$$

(II)

The ELDOR spectra of the two spin-labeled polystyrenes (Figures 13 and 14) showed significant differences between themselves and with other systems discussed above. The (I) sample differs most markedly from all the others. The matrix signal in (I) is well-resolved; R_M is nearly constant between 0.06 and 0.10 from $-100°$ to $+100°C$. The single-line ELDOR intensity also has small temperature dependence. The plot of R (6 MHz) versus $1/T$ in

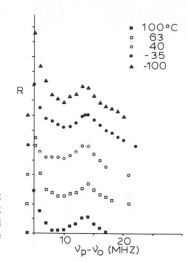

Figure 13. Single-line ELDOR for polystyrene labeled at the para position of the phenyl side chains (I). v_o is at the low-field turning point. The R scale has zero shifted for each curve; zero is indicated by the appropriate symbol. Units of the R scale are 0.05/division.[28b]

Figure 15 is only about one-third of that for maleimide in polystyrene (Figure 7) and for (II).

The matrix signal is also small for (II). However, it does show the kind of temperature dependence shown earlier reaching a maximum value of R_M of about 0.07 and disappears completely at $-150°C$. The slope of R versus $1/T$ plot for this material is slightly smaller than those seen with interstitial nitroxides.

The results in Figure 13 showed that the side-chain spin labels in (I) have short correlation time with only a slight temperature dependence. In fact, τ_c does not become long enough even at $-150°C$ to cause a decrease in the matrix ELDOR intensity. The nitroxide group is single-bonded to a *tert*-butyl group and a phenyl group. It is probably capable of undergoing low activation energy motion such as torsional oscillation. This is not true for

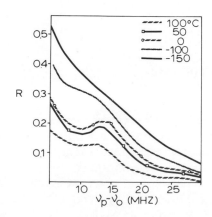

Figure 14. Single-line ELDOR response for terminally labeled polystyrene (II). Lines are averages of a number of scans. v_o is at the low-field turning point.[28b]

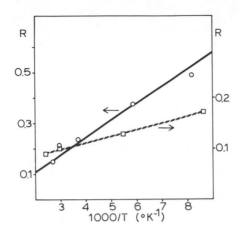

Figure 15. *R* at 6 MHz for (I) (○) and (II) (□) versus T^{-1}. Data are from Figures 13 and 14.[28b]

either maleimide or TANOL, where the nitrogen is in the heterocyclic ring. Even in the case of DTBN, the nearly spherical molecule probably moves as a whole unit.

On the other hand, the terminal spin labels in (II) behave more normally than in (I), both in the magnitude of *R* values and its temperature dependence. The > NO· group is singly bonded to a *tert*-butyl and benzyl group. The nitroxide in (II) apparently has long τ_c and has greater barrier for motion than (I). The restriction may be the result of hyperconjugation

or contribution of

to the structure.

In recent work,[52] Chien and Yang have investigated (II) in tetrahydrofuran solutions of both the protonated and deuterated variety. Using the analysis of linewidths, they calculated correlation times from the EPR spectra. From the ELDOR, using the equations developed by Freed,[22] correlation times were calculated that agree with those of the EPR. It is interesting to note that a significant ELDOR response arising from intermolecular exchange interactions was observed in their work. This manifests itself in Figure 16 by the presence of two pair of ELDOR peaks instead of the usual single pair separated by 44 MHz.

Figure 16. ELDOR response for di-end-labeled polystyrene in tetra-hydrofuran. v_o is at peak of the low field EPR peak. (After Yang and Chien.[52])

25MHz

5. *Concluding Remarks*

In conclusion, though the amount of work done on polymeric systems is not nearly so large as for some of the others in this volume, the methods appear to hold great promise for increased information. Very slow molecular motions, whether in the form of interstitial molecules or spin labels, appear tenable as probes of polymer matrix motions. As calculational capabilities are applied, it seems reasonable to expect quantitative rotational correlations times in the slow-motion region to be derivable. Unfortunately, to date, the utility demonstrated for ELDOR in this area has been for those systems where the EPR linewidths are still changing. It remains to be demonstrated that these correlation times can be extracted from the ELDOR responses without such line-shape changes being available.

On the other hand, the line-shapes analyses of the matrix peaks for both ELDOR and ENDOR appear to be usefully applicable to extract information about the radical environments. In particular, these techniques, coupled with the capabilities of separating overlapped EPR spectra, seem to direct the near-term research to some of those problems of irradiated polymers discussed in Section 2.

References

1. N. G. McCrum, B. E. Read, and G. Williams, *Anelastic and Dielectric Effects in Polymeric Solids*, John Wiley and Sons, New York (1967).
2. I. Ya. Slovim and A. N. Luyubimov, *The NMR of Polymers*, Plenum Press, New York (1970).
3. C. L. Hamilton and H. M. McConnell, in *Structural Chemistry and Molecular Biology* (A. Rich and N. Davidson, eds.), Freeman, San Francisco (1968).
4. A. T. Bullock, G. C. Cameran, and P. M. Smith, *J. Chem. Soc. Faraday Trans. II* **70**, 1202 (1974).
5. P. Tormala, H. Lattila, and J. J. Lindberg, *Polymer* **14**, 481 (1973).
6. V. B. Stryukov, E. G. Rozantsev, A. I. Kashlinskii, N. G. Mat'tseva, and I. F. Tibanov, *Dokl. Acad. Nauk. SSSR* **190**, 895 (1970).
7. Z. Veksli and W. G. Miller, *Macromolecules* **10**, 1245 (1977).
8. A. L. Buchachenro, A. L. Kovarskii, and A. M. Wasserman, in *Advances in Polymer Science* (Z. A. Rogovin, ed.), John Wiley and Sons, New York (1974), pp. 26–57.
9. A. M. Wasserman, T. A. Alexandrova, and A. L. Buchachenico, *Eur. Polym. J.* **12**, 691 (1976).
10. P. L. Kumler and R. F. Boyer, *Macromolecules* **9**, 903 (1976).
11. P. Tormala and J. J. Lindberg, in *Structural Studies of Macromolecules by Spectroscopic Methods* (K. J. Ivin, ed.), John Wiley and Sons, New York (1976).

12. E. Schneider, *Nature* **168**, 645 (1951).
13. D. Libby, *Polymer* **1**, 212 (1960).
14. B. Ranby and P. Cartensen, *Adv. Chem. Ser.* **66**, 256 (1967).
15. M. Dole, *Radiation Chemistry of Macromolecules*, Chapter 14, Vol. I, Academic Press, New York (1972).
16. N. Z. Searle, in *Analytical Photochemistry and Photochemical Analysis* (J. M. Fitzgerald, ed.), Marcel Dekker, New York (1971).
17. P. Ayscough and S. Munari, *J. Polym. Sci.* **B4**, 503 (1966).
18. A. Charlesby, D. Libby, and M. Ormerod, *Proc. Roy. Soc. (London)* **A262**, 207 (1961).
19. T. Fujimara, N. Hayakawa, and N. Tamura *Repts. Prog. Polym. Phys. Japan* **14**, 557 (1971).
20. S. Ohnishi, S. Sugimoto, and I. Nitta, *J. Chem. Phys.* **39**, 2647 (1963).
21. R. W. Fessenden and R. H. Schuler, *J. Chem. Phys.* **39**, 2147 (1963).
22. J. S. Hyde, J. C. W. Chien, and J. H. Freed, *J. Chem. Phys.* **48**, 4211 (1968).
23. J. S. Hyde, "Electron–Electron Double Resonance," Varian Reprint No. 256.
24. J. S. Hyde, L. D. Kispert, R. C. Sneed, and J. C. W. Chien, *J. Chem. Phys.* **48**, 3824 (1968).
25. M. C. R. Symons, *Adv. Phys. Org. Chem.* **1**, 340 (1963); *J. Chem. Soc.*, 1186 (1963).
26. J. C. W. Chien, unpublished results.
27. M. M. Dorio and J. C. W. Chien, unpublished results.
28a. M. M. Dorio and J. C. W. Chien, *Macromolecules* **8**, 734 (1975).
28b. M. M. Dorio, Ph.D. Dissertation, University of Massachusetts (1975).
29. M. M. Dorio and J. C. W. Chien, *J. Mag. Res.* **20**, 114 (1975).
30. M. M. Dorio and J. C. W. Chien, *J. Mag. Res.* **21**, 491 (1976).
31. M. M. Dorio and J. C. W. Chien, *Polym. Preprints* **17**(2), 23 (1976).
32. H. W. H. Yang and J. C. W. Chien, *Polym. Preprints* **18**(2), 149 (1977).
33. P. A. Stunzhas, V. B. Stryukov, V. A. Benderskii, and S. T. Kirillov, *JETP Lett.* **17**(2), 461 (1973).
34. S. T. Kirillov, M. A. Kozhushner, and V. G. Stryukov, *Sov. Phys. JETP* **41**(6), 1124 (1975).
35. S. T. Kirillov, M. A. Kozhushner, and V. B. Stryukov, *Chem. Phys.* **17**, 243 (1976).
36. S. T. Kirillov and A. V. Melnikov, *Mol. Cryst. Liq. Cryst.* **36**, 217 (1976).
37. S. T. Kirillov, V. B. Stryukov, *Russ. J. Phys. Chem.* **50**(11), 1746 (1976).
38. E. Stetter, H. M. Vieth, and K. H. Hausser, *J. Mag. Res.* **23**, 493 (1976).
39. L. R. Dalton, L. A. Dalton, N. Galloway, and J. S. Hyde, in *Fifth Southeastern Magnetic Resonance Conference, Oct. 11–12, 1973, Tuscaloosa, Alabama.*
40. B. H. Robinson, J. L. Monge, L. A. Dalton, L. R. Dalton, and A. L. Kwirom, *Chem. Phys. Lett.* **28**, 169 (1974).
41. J. S. Hyde, M. D. Smigel, L. R. Dalton, and L. A. Dalton, *J. Chem. Phys.* **62**, 1655 (1975).
42. M. D. Smigel, L. R. Dalton, J. S. Hyde, and L. A. Dalton, *Proc. Nat. Acad. Sci. (USA)* **71**, 1925 (1974).
43. J. S. Hyde, G. H. Rist, and L. E. Goren-Eriksson, *J. Phys. Chem.* **72**, 4269 (1968).
44. D. S. Leniart, J. S. Hyde, and J. C. Vedrine, *J. Phys. Chem.* **76**, 2079 (1972).
45. J. N. Helbert, B. E. Wagner, E. H. Poindexter, and L. Kevan, *J. Polym. Sci. Phys.* **13**, 825 (1975).
46. H. A. Flocke, *Kolloid Z.* **180**, 118 (1962).
47. E. Passaglia and G. M. Martin, *J. Res. Nat. Bur. Standards* **68**, 519 (1964).
48. G. Natta, F. Damesso, and G. Moraglio, *J. Polym. Sci.* **25**, 119 (1957).
49. E. N. da C. Andrade, *Nature* **125**, 309, 582 (1930).
50. A. J. Batchinski, *Z. Phys. Chem.* **84**, 643 (1913).
51. R. C. McCalley, E. J. Shimshick, and H. M. McConnell, *Chem. Phys. Lett.* **13**, 115 (1972).
52. H. W. H. Yang and J. C. W. Chien, *Polym. Preprints* **18**(2), 149 (1977).

12

ENDOR of Triplet State Systems in Solids

Marvin D. Kemple

1. Introduction

Physical systems possessing one unit of effective electron spin angular momentum have been the subject of extensive study. Such triplet state systems may be excited- and ground-state molecules, pairs of free radical molecules, transition element ions, pairs of ions, and defects in solids. The molecules may range from relatively simple diatomic species to multiatom biological complexes. Detailed magnetic resonance measurements can yield information concerning wave functions, geometry and structure, lifetime of excited states, interactions with the environment, and other properties of these triplet state entities. We will discuss here the particular usefulness of electron nuclear double resonance[1] (ENDOR)—nuclear magnetic resonance transitions detected by means of electron paramagnetic resonance (EPR). Emphasis will be placed on triplet species in solids.

The advantages that ENDOR has over EPR center on the fact that one can measure directly, with high precision, the interactions of the electron magnetic moment with nuclear magnetic moments. The nuclei in the case of molecules may be those of the paramagnetic molecule itself or nuclei of molecules in the environment. Likewise for ions, the nucleus of the paramagnetic ion or nuclei of the environment may be studied. ENDOR has been very useful in identifying and locating the nuclei in the vicinity of paramagnetic defects in solids. Both the anisotropy and algebraic signs of electron–nuclear

Marvin D. Kemple • National Bureau of Standards, Washington, D.C.; NRC/NBS Postdoctoral Research Associate, 1976–1977. An official contribution of the National Bureau of Standards; not subject to copyright. *Present address*: Department of Physics, Indiana University–Purdue University, Indianapolis, Indiana

magnetic interactions can be determined with ENDOR. From that knowledge, molecular and ionic wave functions can be constructed, molecular spin distribution deduced, structural and geometrical details discerned, and the environment of the molecules or ions can be modeled. Electron–nuclear, nuclear, and nuclear–nuclear magnetic interactions that are not resolved in EPR can often be resolved and measured in ENDOR.

2. Nature of Triplet States of Organic Molecules

Much of the present interest in triplet states of organic molecules stems from the proposal of Lewis and others[2–5] that the long-lived phosphorescence of organic molecules in solids originates from the decay of molecules in excited triplet states to their ground singlet states. The EPR investigations by Hutchison and Mangum[6,7] on naphthalene molecules in durene single crystals and by other workers for molecules in glassy mediums[8–10] showed the proposal to be essentially correct. Since then, both excited- and ground-state triplet molecules have been studied in detail with numerous single and multiple resonance techniques.

In the simplest model, a molecular triplet state results when two electrons on a molecule combine to form a state with total spin angular momentum quantum number $S = 1$. In the molecular orbital approximation, one constructs electron orbitals as linear combinations of atomic orbitals centered on the nuclei of the molecule. For planar aromatic molecules, the appropriate orbitals used for the carbon atoms are sp^2 hybrids and regular p orbitals. In a typical ground state, the molecular orbitals are filled with electrons with paired spins giving a singlet state. The low-lying excited states are formed by elevating an electron to an excited nonbonding or antibonding orbital. The excited electron and the electron left behind in a bonding orbital together form excited molecular singlet or triplet states. The lowest excited triplet will generally be lower in energy than the first excited singlet state.

For planar aromatic molecules, the molecular orbitals relevant for the low-lying triplet states are of the π type composed of carbon $2p_z$ orbitals, with z chosen perpendicular to the plane. The states are referred to as $\pi\pi^*$ triplet states. The distribution of triplet spin angular momentum over the molecular framework is characterized by the spin density on each carbon atom. Spin density, unlike electron density, can be negative as well as positive. The orbital angular momentum is almost completely quenched. Spin-orbit coupling is generally small, but its presence is essential to allow the triplet–singlet transitions responsible for the phosphorescence. The typical wave number separation between the ground state and first excited singlet state is of the order of 30,000 cm^{-1}, while ground-singlet–first-excited-triplet separations are of the order of 25,000 cm^{-1}. Lowest excited triplet state lifetimes can be quite long (of the order of seconds) because of the spin-forbidden nature of the transition.

Molecules with singlet ground states are generally promoted to their lowest-energy excited triplet state by first exciting them to higher singlet states with electromagnetic radiation in the ultraviolet range. Some fraction of the excited singlet molecules undergo a process called "intersystem crossing" from the singlet manifold to the triplet manifold, arriving finally in the lowest triplet state. Molecular vibrations and spin–orbit coupling figure prominently in these considerations. In mixed crystals (crystals containing more than one kind of molecule), the lowest triplet level is generally the one populated, regardless of the molecular species. In crystals of only one molecular species, there often exist localized triplet trap states that are populated. The traps are molecules whose lowest triplet state lies lower in energy than any triplet state of the bulk molecules as a result of crystal irregularities or differences in chemical environment. Also the triplet excitation energy in such crystals may be shared by several molecules of the lattice to form a triplet exciton. All of these types of triplets have been studied by magnetic resonance.

Nonaromatic molecules likewise have excited triplet states. The molecular orbitals may be different, but the general properties of spin distribution and molecular structure are still of interest. Some molecules have a ground triplet state. Typically they have two nearly degenerate orbitals that are to be filled with a total of two electrons. As a result of election correlations, the $S = 1$ state is lower in energy than the $S = 0$ state.

The above discussion in terms of a two-electron model of the triplet state is oversimplified but is sufficient to describe many of the magnetic resonance results that will be discussed below.

3. Spin Hamiltonian

3.1. Form and Determination of Parameters

Magnetic resonance of an $S = 1$ system is described by the spin Hamiltonian

$$\mathscr{H} = |\beta_e| \mathbf{H} \cdot \mathbf{g}_e \cdot \mathbf{S} + \mathbf{S} \cdot \mathbf{T} \cdot \mathbf{S} + \sum_k \mathbf{I}_k \cdot \mathbf{A}_k \cdot \mathbf{S}$$

$$- |\beta_n| \sum_k \mathbf{H} \cdot \mathbf{g}_{nk} \cdot \mathbf{I}_k + \sum_{k,q} \mathbf{I}_k \cdot \mathbf{P}_{kq} \cdot \mathbf{I}_q$$

$$\equiv \mathscr{H}_S + \mathscr{H}_{SS} + \mathscr{H}_{SI} + \mathscr{H}_I + \mathscr{H}_{II} \tag{1}$$

in which β_e and β_n are the Bohr and nuclear magnetons, respectively, $\mathbf{H}$ is the applied magnetic field, $\mathbf{S}$ is the electron spin angular momentum operator, and $\mathbf{I}_k$ is the nuclear spin angular momentum operator of the kth nucleus. The sums are over the magnetic nuclei of the molecule. $S = 1$, of course, and $\mathbf{I}_k$ is unrestricted; higher-order terms in $\mathbf{I}_k$ are neglected. $\mathbf{g}_e$, $\mathbf{T}$, $\mathbf{A}_k$, $\mathbf{g}_{nk}$, and $\mathbf{P}_{kq}$ of the respective terms $\mathscr{H}_S$, $\mathscr{H}_{SS}$, $\mathscr{H}_{SI}$, $\mathscr{H}_I$, and $\mathscr{H}_{II}$ of equation (1) are the spin Hamiltonian parameters in tensorial form. Typical magnitudes of the terms of equation (1) for protons and nitrogen nuclei of organic triplet state molecules in a 9-GHz EPR experiment are listed in Table 1.

$\mathscr{H}_S = |\beta_e| \mathbf{H} \cdot \mathbf{g}_e \cdot \mathbf{S}$ represents the electron Zeeman interaction. $\mathbf{g}_e$ is nearly isotropic for triplet states of aromatic molecules, the principal values being about equal to the free electron g value. This is a further indication of small spin–orbit interactions. For ions in crystals, $\mathbf{g}_e$ may be more anisotropic. In the case of defects, $\mathbf{g}_e$ is usually nearly isotropic.

$\mathscr{H}_{SS} = \mathbf{S} \cdot \mathbf{T} \cdot \mathbf{S}$ is the fine-structure term. The zero magnetic field splittings of the triplet are reflected in this term, which is often written

$$\mathscr{H}_{SS} = D\left(S_z^2 - \tfrac{2}{3}\right) + E\left(S_x^2 - S_y^2\right) \tag{2}$$

where D and E are the fine structure, or zero field, parameters; x, y, and z, the principal axes of $\mathbf{T}$, can normally be associated with molecular and/or crystallographic symmetry axes. For triplet states of organic molecules, $\mathscr{H}_{SS}$ accounts for most of the observed anisotropy of the nonzero static magnetic field EPR spectrum and has been shown to be primarily a result of magnetic dipolar interactions of the unpaired electrons of the triplet state. Spin–orbit coupling usually does not make a significant contribution. It is often customary for planar aromatic molecules to choose the z axis perpendicular to the plane. One then finds $|D| > |E|$, and $D > 0$. The sign of E is based on the

Table 1

Typical Sizes of the Spin Hamiltonian Terms
for Protons and ^{14}N Nuclei

	Value (MHz)	
Term	^{1}H	^{14}N
$\mathscr{H}_{SS}/h$	3000	9000
$\mathscr{H}_S/h$	3000	9000
$\mathscr{H}_I/h$	15	5
$\mathscr{H}_{SI}/h$	0–30	0–25
$\mathscr{H}_{II}/h$	0–0.020	2

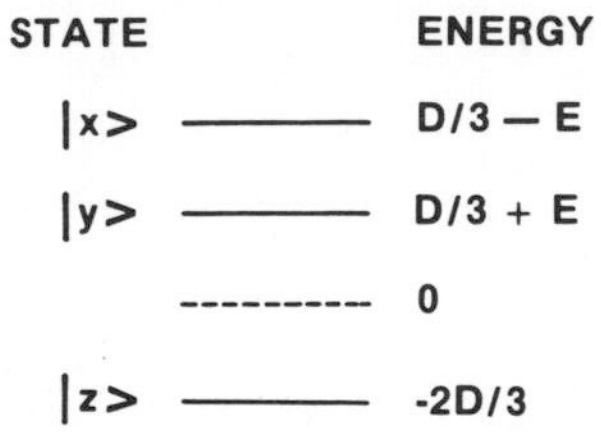

STATE		ENERGY
$\lvert x \rangle$	————	D/3 − E
$\lvert y \rangle$	————	D/3 + E
	- - - - - - - - -	0
$\lvert z \rangle$	————	-2D/3

Figure 1. Triplet zero-magnetic-field energy levels.

assignment of the x and y axes and is not physically significant. For ions and for defects in solids, crystal field interactions may also contribute to $\mathcal{H}_{SS}$.

The energy levels and the states that diagonalize equation (2) alone are indicated in Figure 1. These zero magnetic field states have the property that $S_u \lvert v \rangle = i\varepsilon_{uvw} \lvert w \rangle$ in which u, v, and w can take each of the values x, y, and z. The ε_{uvw} are the components of the completely antisymmetric tensor of rank 3; $\varepsilon_{xyz} = 1$, $\varepsilon_{xzy} = -1$, $\varepsilon_{xxz} = 0$, ..., and $i = \sqrt{-1}$. For example, then $S_x \lvert y \rangle = i \lvert z \rangle$ and $S_x \lvert x \rangle = 0$. The expectation value of $\mathbf{S}$, $\langle \mathbf{S} \rangle$, vanishes in each of the zero field states.

Later, reference to high- and low-magnetic-field experiments will be made. High-field experiments are normally performed at microwave frequencies of 8 GHz and above (which implies magnetic fields $\gtrsim 0.3$ T). In such cases, the matrix elements of $\mathcal{H}_S$ are of the order of or greater than those of $\mathcal{H}_{SS}$. Low-field experiments are those in which $\mathcal{H}_S$ is small compared with $\mathcal{H}_{SS}$. Zero magnetic field experiments are the extreme examples of the latter.

$\mathbf{g}_e$ and $\mathbf{T}$ can be determined from magnetic resonance at nonzero static magnetic field by least-squares fitting the observed magnetic field values of the allowed transitions to the sum of $\mathcal{H}_S$ and $\mathcal{H}_{SS}$ with $\mathbf{g}_e$ and $\mathbf{T}$ as parameters. $\mathcal{H}_S + \mathcal{H}_{SS}$ is generally diagonalized by computer and the parameters adjusted to match the energy level spacings corresponding to resonance for the various magnetic field directions. $\mathbf{g}_e$ and $\mathbf{T}$ can also be found by measuring the values of the magnitude H_0 of the applied static magnetic field $\mathbf{H}_0$ for resonance at which H_0 is stationary with respect to the angle of rotation of $\mathbf{H}_0$ relative to a coordinate system fixed in the sample, and following an iterative procedure such as that described by Hutchison and Mangum.[7] When $\mathbf{g}_e$ is nearly isotropic, these stationary points of the EPR spectrum occur with $\mathbf{H}_0$ parallel to the principal axes of $\mathcal{H}_{SS}$. D and E can also be determined directly, often more precisely than at high field, from zero static magnetic field magnetic resonance experiments. As noted above, the states that diagonalize $\mathcal{H}_{SS}$ alone are ones in which $\langle \mathbf{S} \rangle = 0$; thus, zero field magnetic resonance linewidths are usually narrow compared with high-field linewidths. A typical magnetic field dependence of the energy levels for the applied static field $\mathbf{H}_0$ parallel to the z principal direction of $\mathcal{H}_{SS}$ is given in Figure 2a.

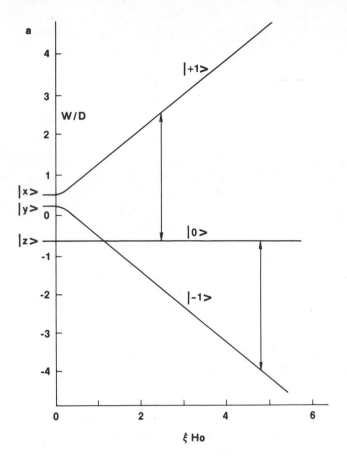

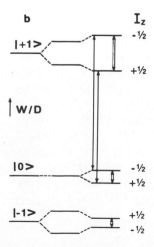

Figure 2. (a) Dependence of the energy W of the triplet levels on magnetic field for H_0 parallel to z of equation (2), $\xi = g_{zz}|\beta_e|/D$. The high-field states are designated by the projection of S along H_0. Except for the state $|0\rangle$, the designations are exact only in the infinite field limit. Arrows indicate allowed EPR transitions. (b) The energy levels at a magnetic field appropriate for the lower-field-allowed EPR transition of (a) at fixed microwave frequency. Effects of successively, left to right, including $\mathscr{H}_{SI}$ and $\mathscr{H}_I$ for interactions with a single $I = \frac{1}{2}$ nucleus are shown (not to scale). Single-line arrows indicate EPR transitions; double-line arrows indicate ENDOR transitions. Often only the upper two ENDOR transitions are observed.

$\mathscr{H}_{SI} = \sum_k \mathbf{S} \cdot \mathbf{A}_k \cdot \mathbf{I}_k$, referred to as the hyperfine interaction term, represents electron spin–nuclear spin interactions. It is dominated by magnetic interactions as expressed in terms of Fermi contact[11] and magnetic dipole-dipole interactions. $\mathbf{A}_k$ is then symmetric when $\mathbf{g}_e$ is isotropic.

$\mathscr{H}_I = -|\beta_n| \sum_k \mathbf{H} \cdot \mathbf{g}_{nk} \cdot \mathbf{I}_k$ takes account of nuclear spin operator-magnetic field couplings. For organic triplets, and ligands of paramagnetic ions and defects, this term is usually isotropic with $\mathbf{g}_{nk} = g_{nk}\mathbf{E}$, where $\mathbf{E}$ is the 3×3 unit matrix and g_{nk} is the g value of the kth nucleus. For a magnetic nucleus of a paramagnetic ion, however, hyperfine-interaction-induced admixtures of excited electronic states into the ground electronic state, coupled with the electronic Zeeman interaction, will contribute terms to the energy which can be cast into the form of $\mathscr{H}_I$.[12] These contributions often dominate the contributions of the direct nuclear Zeeman interaction.

$\mathscr{H}_{II} = \sum_{k,q} \mathbf{I}_k \cdot \mathbf{P}_{kq} \cdot \mathbf{I}_q$, for $k \neq q$, can represent magnetic dipolar interactions between different nuclei. For $k = q$ and $I_k > \frac{1}{2}$, the term can often be associated with nuclear electric quadrupolar interactions. Again for ions, one must watch for excited-state admixtures.

The parameters of $\mathscr{H}_{SI}$, $\mathscr{H}_I$, and $\mathscr{H}_{II}$ are readily determined from ENDOR experiments. The latter two terms generally cannot be investigated by EPR. Components of $\mathbf{A}_k$ can be found from hyperfine structure resolved in EPR. The resonance lines are, however, often complex, and structure due to interactions with only certain of the magnetic nuclei is usually observed. ENDOR transitions occur for the electron spin system in a given state and are normally just single nuclear spin flips. Typical nonzero field ENDOR transitions are indicated in Figure 2b for a single $I = \frac{1}{2}$ nucleus. When several equivalent and inequivalent nuclei are present, the ENDOR spectrum is simpler than the EPR spectrum. A primary advantage of ENDOR is that the ENDOR linewidths are generally characteristic of nuclear dipole–dipole interactions and are much smaller than corresponding EPR linewidths—20 kHz versus 10 MHz, for example, for protons in aromatic triplet states—and thus, nuclear interactions not resolved in EPR may be seen in ENDOR.

There are numerous approaches for determining the parameters $\mathbf{A}_k$, $\mathbf{g}_{nk}$, and $\mathbf{P}_{kq}$ from the data. The ENDOR spectrum is often taken for several different orientations of the magnetic field, and for ions, a least-squares fit of the ENDOR transition frequencies is made to computer-calculated frequencies obtained from the diagonalization of the complete spin Hamiltonian, equation (1).

For triplet states of molecules in a large applied magnetic field $\mathbf{H}_0$, a somewhat different approach is taken. Consider the specific case of hyperfine interactions of the electron spin with the protons of the molecule. Neglecting $\mathscr{H}_{II}$, one approximates the proton terms of the spin Hamiltonian by

$$\mathscr{H}_{SI} + \mathscr{H}_I = \langle \mathbf{S} \rangle \cdot \sum_k \mathbf{A}_k \cdot \mathbf{I}_K - g_p|\beta_n|\mathbf{H}_0 \cdot \sum_k \mathbf{I}_k, \qquad I_k = \tfrac{1}{2}, \text{all } k \quad (3)$$

The proton g_p is assumed isotropic as discussed above; g_p is the free proton g value. S has been replaced by $\langle S \rangle$, the expectation value of S in the electron spin states that diagonalize $\mathcal{H}_S + \mathcal{H}_{SS}$ for the appropriate H_0 for magnetic resonance. Off-diagonal matrix elements of the hyperfine interaction between the electron spin states that diagonalize $\mathcal{H}_S + \mathcal{H}_{SS}$ are thus neglected. This approximation is reasonable since $\mathcal{H}_{SI}$ is small compared with $\mathcal{H}_S + \mathcal{H}_{SS}$. The elements of A_k/h, where h is Planck's constant, are typically on the order of 10^7 Hz; the energy separation of the electronic levels is about equal to the microwave resonance frequency. Thus, from second-order perturbation theory, terms of approximately $(10^7)^2$ Hz$/10^{10} = 10$ kHz are being neglected in a 10-GHz EPR experiment. Proton ENDOR linewidths are usually larger than this.

The ENDOR transition frequency v_k of the kth proton, i.e., the frequency corresponding to a proton spin flip with the electron spin state remaining unchanged, can readily be found from equation (3) as

$$hv_k = |\langle S \rangle \cdot A_k - g_p|\beta_n|H_0|| \tag{4}$$

It is convenient to work with the ENDOR shift Δv_k of the kth proton, defined as the difference between the ENDOR frequency of that proton and the resonance frequency $v_p = |g_p|\beta_n|H_0||/h$ of an isolated proton in the field H_0,

$$\Delta v_k \equiv v_k - v_p \tag{5}$$

If we suppose H_0 has direction cosines l, m, and n with respect to the axes x, y, and z of equation (2), and take A_k to be symmetric, we arrive finally from equations (4) and (5) at the relation

$$
\begin{aligned}
h \, \Delta v_k = \{ & (\langle S \rangle_x A_{kxx} + \langle S \rangle_y A_{kxy} + \langle S \rangle_z A_{kxz} - lhv_p)^2 \\
& + (\langle S \rangle_x A_{kxy} + \langle S \rangle_y A_{kyy} + \langle S \rangle_z A_{kyz} - mhv_p)^2 \\
& + (\langle S \rangle_x A_{kxz} + \langle S \rangle_y A_{kyz} + \langle S \rangle_z A_{kzz} - nhv_p)^2 \}^{1/2} - hv_p
\end{aligned} \tag{6}
$$

At this point, we can notice an interesting facet of triplet state ENDOR. In the very high-field limit, where we can neglect $\mathcal{H}_{SS}$ with respect to $\mathcal{H}_S$, the triplet states that diagonalize $\mathcal{H}_S$ are, of course, those that have, respectively, $\langle S \rangle = \hat{H}_0$, 0, and $-\hat{H}_0$, where $\hat{H}_0$ is a unit vector in the direction of H_0. For an EPR transition between the states with $\langle S \rangle \cdot \hat{H}_0 = 1$ and $\langle S \rangle \cdot \hat{H}_0 = 0$, usually two primary ENDOR transitions are observed from a particular proton (Figure 2b). (Also at times, the ENDOR transition associated with the third state of the triplet is seen, even though that state is not directly involved in the EPR transition.) The transition in the electron states with $\langle S \rangle \cdot \hat{H}_0 = 0$ will have zero ENDOR shift. Likewise for the EPR transition between the states $\langle S \rangle \cdot \hat{H}_0 = 0$ and $\langle S \rangle \cdot \hat{H}_0 = -1$, one ENDOR shift is 0, and the other primary shift is of opposite sign to the nonzero shift observed for the other EPR transition. In the actual experimental situations, D and E are not negligible. Only for H_0 parallel to one of the principle axes of T will one of the electronic states rigorously have $\langle S \rangle = 0$ and correspondingly zero ENDOR shift for one of the ENDOR transitions. For general directions of H_0, there will be a state, however, with $\langle S \rangle$ close to zero; as a result, one ENDOR transition will occur near zero shift. ENDOR in the state with $\langle S \rangle \simeq 0$ is sometimes called zero-level ENDOR.

The sign of the ENDOR shift depends upon the signs of the components of A_k as well as those of $\langle S \rangle$. Thus, if one knows the sign of $\langle S \rangle$ for the EPR transition, the signs of A_k are readily found from the measured ENDOR shifts. Knowing the signs of the components of $\langle S \rangle$ amounts to knowing the sign of D. There are various approaches[13–18] to finding the sign of D; a simple one is to compare intensities of the low- and high-field EPR lines for the field in various orientations. From EPR, except under special circumstances,[15,16] one cannot determine the signs of the components of A_k.

ENDOR lines near zero shift also occur for protons with small A_k values. Protons of molecular neighbors of the triplet are in that category. To interpret this "distant ENDOR," one must sort from it the zero-level ENDOR. That is accomplished by finding the A_k by methods discussed below from the non-zero-level ENDOR and then calculating the zero-level ENDOR frequencies.

To determine A_k, the measured ENDOR shifts can be least-squares fitted to equation (6). $\langle S \rangle$ is calculated by diagonalizing $\mathscr{H}_S + \mathscr{H}_{SS}$ for fixed microwave frequency and specified direction of H_0. Another approach is to measure the ENDOR shifts at their stationary orientations. In many circumstances to a good approximation, those orientations will be the principal directions of A_k, and when one takes the value of $\langle S \rangle$ into consideration, one has a direct measure of those principal A_k values. A problem that can arise if the ENDOR signals from all protons cannot be followed for all orientations of H_0 is one of consistently identifying the signals of a given proton. Hutchison and Pearson[19] outlined an extrapolation approach. Also, calculated estimates of A_k can be used. We defer discussion of the assignment of the ENDOR lines to specific protons to Section 3.2.

Similar procedures for finding A_k values for nuclei other than protons, such as nitrogen, carbon-13, and deuterium, in organic triplet states can be followed. One may need to include nuclear quadrupolar effects for $I_k \geq 1$.

The approximation of using $\langle S \rangle$ in $\mathscr{H}_{SI}$ is usually not a good one for ENDOR of the nucleus of a paramagnetic ion or for any case in low field, but it is likely to be valid for ligands of such an ion and for defects in solids at high field. Normally, a least-squares fitting procedure involving exact diagonalization of the complete Hamiltonian is used to extract the parameters when this approximation does not apply.

For organic triplet states, the parameters P_{kk} of $\mathscr{H}_{II}$ for nuclei with $I_k > \frac{1}{2}$ are usually found by introducing them as perturbations to the nuclear levels in high-field ENDOR. The components of $\mathbf{P}_{kk}$ are generally 10^{-2} to 10^{-1} or so of the $\mathbf{A}_k$ values of local nuclei (nuclei of the triplet molecule). In zero field, elements of $\mathbf{P}_{kk}$ are calculated by diagonalizing the Hamiltonian and matching transitions. We will discuss that in more detail in Section 4.

Terms that can be associated with $\mathscr{H}_{II}$ for $k \neq q$ can appear as splittings of proton ENDOR lines in triplet molecules. Those result from magnetic dipolar interactions between the protons, and the coefficients of $\mathscr{H}_{II}$ can be determined directly from the splittings.

3.2. *Physical Properties Deduced from ENDOR*

For triplet states of organic molecules, the fact that individual nuclei can be resolved in ENDOR leads to immediate consequences with regard to molecular structure, independent of any models for calculations of interactions. Consider the example of benzene-h_6 triplet state molecules studied in benzene-d_6 single crystals (generally the protonated molecule triplet levels lie below those of the deuterated molecule and thus trap the triplet excitation energy). Three distinct proton ENDOR lines are observed,[20] implying that at most the molecule has a center of inversion in the crystal. For diphenyl-h_{10} triplets in diphenyl-d_{10} crystals, more than five distinct proton ENDOR signals are observed[21] with the magnetic field in some orientations, indicating that the molecule lacks inversion symmetry. It was thought from x-ray diffraction measurements[22–24] that the symmetry at the site of the diphenyl molecule should be inversion.

Deviations from planarity of supposed planar molecules can be seen readily in ENDOR. Planar molecules are expected to have a common principal direction for the fine structure and all of the hyperfine interactions. In the benzene example above, such is found not to be the case[20]; the benzene molecule in its lowest triplet state in the crystal is not planar. ENDOR of nuclei on neighboring molecules similarly yields information on the local structure and environment. Likewise for ion and defect ENDOR, site symmetry and local structure follow in a straightforward manner.

Another piece of information of interest determined from organic molecular data is the spin distribution in the molecule. Consider again the case of proton hyperfine interactions measured from ENDOR in approximately planar aromatic molecules where the relevant molecular orbitals are of the π type. The fraction of triplet spin on the ith carbon atom is expressed in terms of the spin density ρ_i, with the condition $\sum_{i=1}^{M} \rho_i = 1$ for the M carbons in the molecule. Complete spin density matrices can be constructed.[25,26] The ρ_i represent only the diagonal elements; we will not concern ourselves with the other elements.

The ρ_i can be determined in a number of ways from the measured $\mathbf{A}_k$ of the protons of the triplet molecule, where, as before, k indexes those protons.

A given A_k can be separated into two parts, an isotropic part, $Tr\ A_k/3$, and an anisotropic part, $A_k - E\ Tr\ A_k/3$, where E is the unit 3×3 tensor. The isotropic contribution is the Fermi contact interaction[11] between the kth proton and the carbon atom to which it is bonded. There is firm theoretical and experimental evidence[26–30] that this isotropic contribution to A_k is proportional to the π electron spin density at the kth carbon, where we have adopted the same index for the carbons and protons directly bonded to each other. The electrons in the carbon π orbitals are thought to spin-polarize the electrons of the carbon σ orbitals, which in turn polarize the spin of the electrons in the σ orbitals centered on the proton. These latter σ orbitals have a nonzero value of their wave function at the proton as required for the contact interaction. The proportionality constant Q_k between the isotropic hyperfine interaction and the spin density has been found to be negative and to be roughly the same from molecule to molecule and from C—H fragment to C—H fragment. The value of $Q_k/h = -70$ MHz lies near the average of those found.

The anistropic contribution to A_k arises from dipole–dipole interactions between the kth proton and the triplet spin angular momentum distributed over the M carbon atoms. One writes

$$A_k = \sum_{i=1}^{M} (A_{ik}^d + Q_i \delta_{ik} E)\rho_i \tag{7}$$

in which A_{ik}^d is the dipolar interaction of the spin on the ith carbon with the kth proton, $Tr\ A_{ik}^d = 0$. Q_i, as noted above, is the isotropic hyperfine interaction per unit spin density, and δ_{ik} is the Kronecker delta. A_{ik}^d for $i \neq k$ cannot be neglected in general.

If a value is assumed for Q_k, ρ_k for carbons with directly bonded protons can be immediately calculated from the isotropic part of A_k. Spin densities on the carbons without protons are then calculated by symmetry considerations coupled with the normalization condition of the ρ_i. An alternate procedure is to calculate the A_{ik}^d from a molecular model, assume a value for Q_k or allow Q_k to vary, and least-squares fit equation (7) to the measured A_k with the ρ_i as parameters. A similar approach is to calculate the A_{ik}^d and least-squares fit the measured anisotropic contribution of A_k to $\sum_i A_{ik}^d \rho_i$ with the ρ_i as parameters. The latter technique allows one to calculate the Q_k and check the notion that Q_k is independent of the C—H fragment. Often there are additional parameters that arise in the calculation of A_{ik}^d that are included as parameters of the fit. Examples will be outlined below. The particular procedure used to extract the ρ_i depends on the extent of the data and the inclination of the workers.

Various models have been used to calculate the A_{ik}^d. One commonly employed in nearly planar aromatic molecules with the triplet spin distributed in π-type molecular orbitals is to suppose the electron spin angular momentum at the *i*th carbon to be distributed as two point spins,[25] each of weight $\frac{1}{2}$ located near the most probable distance of an electron from the carbon atom in each lobe of a *p*-like orbital. The dipole–dipole interaction between the point electron spins and the localized nuclear spin is then calculated with the familiar dipole–dipole formula,

$$(-g_n g_e |\beta_n| \, |\beta_e| / r^3)(\mathbf{E} - 3\hat{\mathbf{r}}\hat{\mathbf{r}})$$

where g_e is the free electron g factor, r is the separation of the dipoles, and $\hat{\mathbf{r}}$ is a unit vector along the line of the dipoles. $\hat{\mathbf{r}}\hat{\mathbf{r}}$ is a dyadic.

The electron point spins have the same coordinates in the molecular (xy) plane as the carbon atoms and have coordinates, which are the negatives of each other, in the direction normal (z) to the plane. One value chosen for the magnitude of the z coordinate is the most probable position of an electron in a Slater $2p_z$ orbital. Often the value of 0.7665 Å has been adopted. With a C—H bond length of 1.084 Å, a point spin with coordinates $(x_k, y_k, \pm 0.7665$ Å$)$ gives a value of 0 for the zz component of $\mathbf{A}_{kk}^d$ in approximate agreement with results of EPR experiments[31] and ENDOR experiments,[32,33] where x_k and y_k are the x and y coordinates of the *k*th carbon atom. In many calculations, the point spin model has been used only for other than nearest-neighbor carbons and hydrogens. Specific values for the nearest-neighbor carbon and hydrogen hyperfine interaction are assumed in those cases or they are introduced as parameters in the fitting of the data along with the spin densities.

Some authors have applied the method of McConnell and Strathdee[34] as opposed to the point spin model. Integrations over assumed molecular wave functions are performed. Lack of knowledge of the wave functions is a limiting factor here. The point spin model has generally proven to be adequate. For other than planar aromatic molecules, a point spin model can be used, the particular details of which depend on the molecular orbitals.

A geometry of the molecule is also necessary for the calculations, of course. Often the maximum possible symmetry is used, but there also exists the possibility of including geometrical parameters in the fitting, depending on the amount of data available.

We can now give an idea of the manner in which the assignments of the measured $\mathbf{A}_k$ are made to the protons of the molecule. First, there is usually some idea of what the ρ_i should be based on molecular orbital calculations. The hyperfine interactions can then be estimated and assignments made. Measurements of resolved hyperfine structure in EPR can be helpful also. In addition, the principal directions of $\mathbf{A}_k$ for the *k*th proton of a planar aromatic molecule would be approximately along and perpendicular to the C—H bond direction if $\mathbf{A}_{ik}^d$ for $i \neq k$ is neglected. The inclusion of $\mathbf{A}_{ik}^d$ with $i \neq k$ alters those principal directions somewhat but generally not enough to negate the assignments. There may still be ambiguities after all the above are considered. Those are resolved by choosing the assignments that yield the best least-squares fit of the measured ENDOR frequencies.

The assignments of lines to protons of neighboring molecules are made by calculating the expected hyperfine interactions. Coordinates of the neighboring protons are obtained from x-ray crystallographic results. The triplet spin distribution deduced from the ENDOR measurements of the local protons is used normally, with the point spin model. ENDOR shifts are calculated and compared with the measured values, and assignments are made. In a slight modification of the procedure, the neighboring proton coordinates can be determined from the measured hyperfine interactions. Rather than using the x-ray coordinates, one adjusts the nuclear coordinates to give a match between the ENDOR data and the calculations. This is similar to ligand ENDOR of ions and to defect ENDOR. The molecular case is somewhat complicated by the fact that the molecular spin angular momentum is distributed.

In the discussion above, we have concentrated on protons as the magnetic nuclei. Among others, deuterons, ^{13}C, and ^{14}N nuclei have also been studied. The calculational approach is similar. It may be possible to determine relative nuclear positions from measurement of the P_{kq} of the spin Hamiltonian for $k \neq q$ when that term in the Hamiltonian arises from nuclear dipole–dipole interactions. If $I_k > \frac{1}{2}$, then the P_{kk} can be related to the nuclear electric quadrupole moment, and to electric field gradients at the nucleus. Knowledge of the latter leads to information regarding electron distribution in the molecule.

3.3. Experimental Approaches

ENDOR of triplets has been studied most extensively by the conventional technique[1] of monitoring changes in the absorption of microwave power by the electron spin system induced by a second alternating (rf) magnetic field applied to the sample. The frequency of the rf field is varied through the range appropriate for nuclear magnetic resonance transitions. Often, the amplitude or frequency of the rf field is modulated and phase-sensitive detection employed. Ground- and excited-triplet-state ENDOR have been observed.

Excited-state triplet species are produced by continuous irradiation of the samples, usually with ultraviolet light. Modulation of the decay envelopes of electron spin echoes due to hyperfine interactions has been seen in an excited molecular triplet state.[35] Rowan *et al.*[36] originally demonstrated the manner in which nuclear transition frequencies could be deduced from the echo modulation frequencies in ground-state ions and thereby showed the usefulness of such an approach as an alternate ENDOR technique. This method might be suitable for short-lived excited triplet states, but it has not been pursued in any detail as yet.

ENDOR has been detected optically as well in excited triplet states in zero and near-zero external static magnetic field and at high fields. The transitions are observed by monitoring changes in phosphorescence of the excited molecules when the appropriate microwave and rf fields are applied. The effects of electron spin–nuclear spin interactions have also been observed in optically detected spin-locking experiments at zero field. One object of the zero-field experiments is to measure nuclear quadrupolar interactions under conditions in which the hyperfine interactions do not dominate the nuclear energy level spacings. The ENDOR measurements described below were generally taken at temperatures 4.2°K or lower. Exceptions will be noted.

4. Examples of Systems Studied

4.1. Ground-State Triplets

The first ENDOR of triplet states of molecules was observed in ground-state triplets of organic molecules. Hutchison and Pearson[19,37] studied fluorenylidene molecules in diazofluorene single crystals. Figure 3 shows a diagram of the fluorenylidene molecule. Discretely oriented ground-triplet-state fluorenylidene molecules were produced by photolysis of diazofluorene molecules of the host crystal. The ground-state nature of the triplet was established by noting that the EPR signals were of arbitarily long lifetime at sufficiently low temperature (77°K and below). Most of the triplet spin is found to reside in the two available orbitals of the central carbon atom,[38] a p_x orbital and a sp_y hybrid orbital with the x axis taken normal to the molecular plane, in this case, and the orthogonal y axis taken to be perpendicular to the bond joining the carbon rings. The distribution of spin between those two orbitals was deduced for molecules in which a ^{13}C nucleus was inserted at the central carbon position by studying the ^{13}C hyperfine structure resolved in EPR.[38] Proton hyperfine interactions were not resolved in EPR, so no information regarding the spin distribution in the

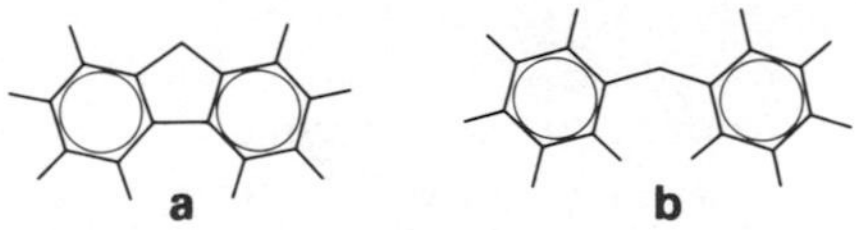

a   **b**

Figure 3. Schematic representations of ground-state triplet molecules [(a) fluorenylidene, (b) diphenylmethylene] examined by ENDOR. Single carbon atoms are understood to be at the vertices of two or more lines, single hydrogen atoms at the open terminations of the lines.

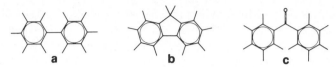

Figure 4. Schematic representations of three molecules [(a) diphenyl, (b) fluorene, and (c) benzophenone] whose excited triplet states have been studied by ENDOR. Single carbon atoms are understood to be at the vertices of two or more lines, single hydrogen atoms at the open terminations of the lines.

two carbon rings was obtained. The proton interactions were resolvable with proton ENDOR, however, and the authors were able to obtain considerable information regarding the ring spin densities.

Methods similar to those outlined in Sections 3.1 and 3.2 were followed. Hyperfine interactions per unit spin density were calculated with a point spin model, and the EPR information on spin distribution at the central carbon atom was essential. Several specific calculations were tried. The model for the molecular structure, along with other quantities, such as Q the isotropic hyperfine interaction per unit π-electron spin density, was varied from calculation to calculation. Although the ENDOR data indicated that the molecules were not planar, only planar geometrical models were considered. By comparing the quality of the least-squares fits of the calculated hyperfine interactions to the ENDOR-measured hyperfine interactions, the authors were able to conclude that the fluorenylidene C—C—C angle is apparently nearly the same as that of the molecule fluorene (Figure 4). The spin densities were the parameters of the fits. With the densities found from the best fit, the zero field parameter D was calculated and compared with the experimental value obtained from EPR. Excellent agreement was found.

Proton ENDOR of the triplet ground-state diphenylmethylene (DPM) (Figure 3) molecule in diphenylethylene (DPE) single crystals has also been investigated.[39] The atomic constituents of DPM are shown in Figure 3. Paramagnetic DPM, oriented in the DPE matrix as noted by EPR, was created by low-temperature photolysis of diphenyldiazomethane that was doped into DPE single crystals in low concentrations. As with fluorenylidene, most of the triplet spin resides at the central carbon atom; proton hyperfine interactions were not resolved in EPR. The triplet spin distribution and the relative orientations of the two phenyl rings of the DPM molecule were determined by comparison of results of least-squares fits of proton anisotropic hyperfine interactions, calculated on a point spin model for different molecular geometries with specific values assumed for the nearest-neighbor C—H hyperfine interactions per unit spin density, to the proton anisotropic interactions measured by ENDOR. The location of the point spins appropriate for the central carbon atom and the apportionment of spin density between the central carbon orbitals were taken in such a manner as to be consistent with ^{13}C hyperfine interactions measured by EPR. The final spin densities, obtained from the best fit of the proton anisotropic interactions, were used to determine Q from the isotropic interactions. Q was seen to vary significantly from one C—H fragment to another. It was indicated that such a variation of Q was not unreasonable for DPM in light of detailed self-consistent field molecular orbital calculations in the intermediate neglect of differential overlap (INDO) scheme, which takes account of details of delocalization of σ-type orbitals.

For DPM in DPE, it was observed[40,41] that at temperatures above 77°K, the triplet EPR signals were not infinitely long-lived but rather decayed at a rate that increased with temperature. It was supposed that the DPM molecules were reacting with DPE host molecules to form tetraphenyl cyclopropane (CP). Proton ENDOR signals were observed from protons on DPE molecules neighboring the ground state triplet DPM molecules. Those signals were near zero ENDOR shift but could be sorted from the signals of protons of the DPM molecules associated with ENDOR in the state with $\langle S \rangle \simeq 0$ because the hyperfine tensors of the DPM protons were already measured from ENDOR data involving the electron spin levels with $|\langle S \rangle| \simeq 1$. Thus, hyperfine tensors for neighboring DPE protons were determined. Those tensors had very small isotropic components, as would be expected. The traceless contributions to the hyperfine interactions were assumed to be dipolar as usual. Since the geometry and spin distribution of the triplet state DPM were determined from the proton ENDOR of DPM molecule protons, the location of the distant molecules relative to the DPM molecules could be calculated from the dipole interaction formula with a point spin model. Taking account of x-ray diffraction measurements of pure DPE crystals, the authors could associate the distant protons with the neighboring DPE molecules. This information led to proposal of a model describing the details of the reaction of the DPM and DPE molecules to form CP. Further detailed comparisons of the x-ray proton coordinates and the ENDOR proton coordinates gave an indication of the distortion of the DPE lattice in the vicinity of the DPM molecule. Of course, it was not possible to distinguish between differences in proton coordinates that would arise from the inadequacies in the calculational models and the actual lattice distortion, but the results give a general idea of the sizes of the distortions.

Proton ENDOR of the ground-state triplet of DPM in benzophenone crystals has also been studied.[42] A data treatment procedure similar to that followed for DPM in DPE was followed. Details of the molecular geometry and spin distribution were again deduced by fitting the anisotropic proton hyperfine interactions, calculated with a point spin model, to the measured anisotropic interactions. The individual sites of the DPM molecules in the benzophenone host at the temperature ($\sim 2°K$) of the ENDOR experiments do not possess twofold symmetry. As a result, the total spin density of one phenyl ring of the molecule differed from that of the other ring by a few percent. Q was found to vary among the C—H fragments essentially as it varied for DPM in DPE. Differences were necessarily found between the DPM conformation and triplet spin distribution in the benzophenone crystal and the corresponding items in the DPE crystal. Those differences were not discussed in detail.

4.2. Excited Triplet States of Organic Molecules

4.2.1. ENDOR in High Magnetic Fields

ENDOR measurements of a number of photoexcited triplet-state molecules in single crystals in high magnetic fields have been accomplished. The first measurements were by Ehret *et al.*[43,44] for naphthalene molecules in durene. Hyperfine interactions of protonated as well as partially deuterated naphthalene molecules were measured. Spin densities were determined from the ENDOR measurements by assuming that the naphthalene molecules had D_{2h} symmetry, assuming that the isotropic hyperfine interaction per unit spin density was the same for different C—H fragments, and using the spin density value for one of the carbon positions as determined from EPR measurements.[31] Calculations of hyperfine interactions were not attempted. The spin densities agreed well with those of the naphthalene negative ion in accordance with considerations of McLachlan.[45] Doublings of ENDOR lines were observed, which the authors suppose indicate that there are four translationally inequivalent sites for naphthalene molecules per unit cell of the durene lattice, as opposed to only two for durene. Deviations of the molecule from planarity were noted. No attempt to discern quadrupolar interactions for the deuterons were made.

ENDOR of photoexcited triplet-state benzene-h_6 in benzene-d_6 crystals was mentioned in Section 3.2. If the benzene molecules had idealized sixfold symmetry in the solid, the zero-field parameter E should vanish, where the z axis has been chosen perpendicular to the molecular plane; proton ENDOR signals from only one type of proton should result. Instead, one finds $E \neq 0$ and three distinguishable ENDOR signals.[20,32] The site of the benzene molecule in the crystal is predicted to be one of inversion symmetry; the observation of three proton ENDOR signals, each signal corresponding to two equivalent protons, is thus expected. Two of the signals were much nearer in frequency to each other than they were to the third. For the subsequent data analysis, those two kinds of protons were assumed equivalent, as were the carbons to which they were bonded. The idea was that there were four equivalent protons bonded to four equivalent carbons, giving rise to one proton ENDOR signal, and two equivalent protons bonded to two equivalent carbons, giving the other ENDOR signal. Spin densities were determined by two approaches. In one, Q was assumed independent of C—H fragment and was calculated along with the two spin densities from the relation $\rho_k = Q \, \mathrm{Tr} \, \mathbf{A}_k/3$, where k indexes the two kinds of carbons, and the spin-density normalization condition. In the second approach, the hyperfine interactions were calculated on a point spin model and least squares fitted to the measured hyperfine interactions with the spin densities and the nearest-neighbor C—H fragment hyperfine interactions per unit spin density, taken the same for each fragment, as parameters. A perfect hexagonal planar molecule geometry was assumed. The spin density and Q values found from the two methods agreed to better than 1%. The spin density for the set of two equivalent carbon positions was about 20% higher than the spin density at the set of four equivalent carbons. Without reference to crystal orientation in the applied magnetic field, it was possible to show that the x axis of the fine structure term of the spin Hamiltonian, equation (2), was along the direction of a line joining the two higher-spin-density carbon atoms with $D/E < 0$. Details of benzene molecular orbitals were inferred from the measured spin densities. In addition, it was estimated that the spin density differences measured could be accounted for by only a 0.005 Å increase of two of the C—C bonds relative to the other four bonds, implying the molecule was slightly elongated. Theoretical calculations for isolated benzene molecules predicted that benzene would be compressed.[46,47] It should be noted that the ENDOR spectrum showed that the molecule deviated from planarity. No estimates of the extent of the deviation were made.

Proton ENDOR of excited triplet-state anthracene molecules in phenazine crystals has been reported.[33] From the number of proton ENDOR signals, it was apparent that the molecule had no more than inversion symmetry. There were some additional splittings of ENDOR lines, indicating that perhaps even inversion symmetry was lost, but those splittings were not analyzed in detail. Deviations of the anthracene triplet molecule from planarity were noted and neglected. Spin densities were found by fitting the anisotropic hyperfine interactions calculated on a point spin model with an ideal molecular geometry assumed. The nearest C—H fragment hyperfine interactions and the spin densities were the parameters of the fit. Q was then determined for individual C—H fragments from the measured isotropic components of $\mathbf{A}$ and was found to vary by about 10% over the different fragments. There were some difficulties in assignment of the ENDOR lines. The assignments adopted were those that gave the best fit to the data. The spin densities agreed reasonably well with values measured for the anthracene negative ion.

Deuterium ENDOR of photoexcited triplet-state phenanthrene molecules in diphenyl single crystals was studied.[48] It was found from EPR that there were four discrete, differently oriented triplet-state phenanthrene-d_{10} molecules at a given set of translationally equivalent substitution sites in the diphenyl crystal. The ENDOR spectrum was accordingly complicated; assignment of the deuteron signals was difficult and ambiguous in some cases. Spin densities were determined by fitting of the calculated hyperfine interactions to the measured values assuming a single fixed value of Q. Instead of the point spin approach, the method of McConnell and Strathdee[34] was used in the calculation. Integrations over p-type Löwdin atomic orbitals[49] were performed. Results of two different fits are given, depending on assignments of the ENDOR lines to deuterons of the molecule. Spin densities were also calculated directly from the isotropic part of the hyperfine interactions. Spin densities thereby obtained were in reasonable agreement with each other and with measured positive and negative ion values. For the calculation, the phenanthrene molecule was assumed to be planar and to have two mutually perpendicular symmetry planes; however, as with other molecules, the ENDOR data clearly showed that deviations from that idealized geometry existed. No attempts to determine deuterium quadrupole interactions were described.

EPR and ENDOR of excited triplet states of diphenyl-h_{10} molecules in dyphenyl-d_{10} single crystals have also been studied of late.[21] There has been considerable interest in the structure of diphenyl crystals. Optical spectroscopic,[50-52] zero-field magnetic resonance,[52] and the proton ENDOR measurements at low temperatures have indicated that the individual molecular sites are not sites of inversion symmetry. Room temperature x-ray diffraction measurements[22-24] and subsequent neutron diffraction[52] measurements at 4.2°K had indicated that the sites should have inversion symmetry. With inversion symmetry, a maximum of five distinct proton ENDOR signals should be observed, as can be seen from Figure 4. Instead, in several orientations of the static magnetic field, ENDOR lines[21,53] associated with three of the five kinds of protons are split, implying a loss of inversion symmetry. The analysis is complicated by the appearance of additional splittings of the proton ENDOR lines, apparently due to proton magnetic dipolar interactions. The contributions from the two sources can be separated, however, based on the angular dependence of the proton magnetic dipolar interactions. It is of interest to note that the ENDOR data of this molecule show no appreciable deviations of the molecule from planarity, as opposed to the majority of cases discussed so far in this chapter. Diphenyl molecules in the gas phase have been shown to have a substantial angle of twist of the two phenyl rings about the central carbon bond,[54] a dramatic example of the importance of crystal field forces in determining molecular geometry. The spin-density analysis is still in progress. Preliminary indications are that the difference in the relative orientations of the two translationally inequivalent molecules of the diphenyl crystal as measured for triplet states by EPR[21] and for ground states by x-ray diffraction[22,24] may be explained in part in terms of the triplet spin distribution as opposed to gross geometrical deviations of the molecule from D_{2h} symmetry. As noted earlier for benzene, there can exist significant differences in spin density distribution with very small changes in molecule geometry.

ENDOR of triplet states of so called X-traps of fluorene molecules, Figure 4, has likewise been reported.[55,56] If crystals of fluorene are doped with dibenzothiophene, phosphorescence from the fluorene molecules adjacent to dibenzothiphene molecules occurs. These phosphorescent molecules are thought to trap the triplet excitation energy; that is, their lowest excited triplet energy is lower than that of the bulk fluorene molecules. Such an impurity-induced trap is called an X-trap.* Proton ENDOR of the X-trap of fluorene showed that the carbon rings of the trap molecule were planar. The CH_2 combination joining the rings apparently was not planar. Spin densities at the carbons with nearest-neighbor protons were found from the isotropic hyperfine interactions assuming a fixed, specific value of Q. Those spin densities agreed reasonably well with spin densities calculated from SCF molecular orbital theory.

ENDOR signals from protons of molecules neighboring the fluorene X-traps were also observed and analyzed in detail.[56] Those signals were identified with specific protons of neighboring fluorene molecules by calculating the expected ENDOR frequencies from proton coordinates deduced from x-ray crystallographic measurements. A point spin model was used along with spin densities of the X-trap molecule determined from the local proton ENDOR for carbons with nearest-neighbor hydrogens and with spin densities consistent with calculated values for the other carbons. Signals from five protons of neighboring molecules were identified. With that information coupled with the observation that the ENDOR results were consistent with the notion that a single molecule acts as the X-trap, the authors were able to determine which fluorene molecule was likely to be the X-trap when it was a nearest neighbor of a dibenzothiophene molecule.

Proton ENDOR of nearest-neighbor pairs of naphthalene-h_8 molecules in naphthalene-d_8 single crystals has been reported.[57] Schwoerer and Wolf[58,59] originally found that, in crystals of naphthalene-d_8 with relatively large concentrations ($>0.05\%$) of naphthalene-h_8, triplet EPR signals in addition to those normally present in more dilute samples were seen. Certain of those additional resonances, called M lines, were shown to arise from the exchange of triplet excitation between the members of isolated nearest-neighbor pairs of naphthalene-h_8 molecules. These pairs are examples of a localized triplet excitation. Clever spin echo and optically detected magnetic resonance observations[60] indicate that the lowest triplet state of the pair is properly considered, at temperatures below 4.2°K, as a coherent superposition of triplet and singlet states of the molecules comprising the pair. Incoherent hopping of the excitation between the molecules is

---

* See Note 1 Added in Proof on p. 436.

not significant at those temperatures. The exchange rate is large compared with the hyperfine interaction, so in the simplest approximation, since the exchange is between two distinct molecular sites, the hyperfine interactions would be expected to be characteristic of the average of that of the two sites. Hyperfine structure should, therefore, be resolved in EPR, contrary to the case of delocalized triplet excitons,[61] and indeed, hyperfine structure on the pair EPR lines was observed, as was proton ENDOR.[57] The ENDOR, which was observed only for the electronic state with $\langle S \rangle \simeq 0$, was consistent with the average hyperfine interaction hypothesis, but the EPR hyperfine structure results were not completely consistent with that idea. The discrepancy has yet to be resolved.

Deuteron ENDOR has been investigated in excited triplet states of quinoline-d_7 in durene and quinoxaline-d_6 in durene.[62] The main interest was in measuring the nuclear quadrupolar terms in the spin Hamiltonian. Those terms were determined with the magnetic field parallel to a principal axis of the fine structure tensor from the ENDOR associated with the $\langle S \rangle = 0$ electronic spin state. Two ENDOR lines were observed for each deuteron; the separation of those lines was dominated by the quadrupolar interaction terms, since the hyperfine interaction vanishes to first order in perturbation theory for the state with $\langle S \rangle = 0$. From ENDOR in the states with $|\langle S \rangle| \simeq 1$, spin densities were estimated. An extensive orientational study was not done, nor were attempts made to interpret the quadrupolar terms.

Also, reports of the study of hyperfine interactions in charge transfer triplet excited states have appeared.[63] Information regarding the charge transfer character of the states can be deduced from the sizes of proton hyperfine interactions. ENDOR was attempted, but the signals observed were unusually broad, and no significant information could be obtained.

Nitrogen-14 ENDOR and proton ENDOR have been optically detected for pyrazine and pyrimidine triplets in low and intermediate magnetic fields.[64] A dramatic difference in the values of the components of the nitrogen quadrupole interaction tensor between the excited-state measurements and ground state measurements on pyrazine were noted. No attempt at interpreting that difference was reported.

4.2.2. ENDOR in Zero and Near-Zero External Static Magnetic Field

ENDOR transitions of ^{14}N, $^{35, 37}$Cl, ^{13}C, and ^{1}H nuclei in excited triplet states of organic molecules have been observed in near-zero applied static magnetic field by optically detected magnetic resonance techniques.

A primary object of studying ^{14}N and $^{35, 37}$Cl in zero field is to determine nuclear quadrupolar coupling terms under circumstances in which the effects of hyperfine interactions are small. The zero-field triplet states necessarily are those with $\langle S \rangle = 0$; thus, the effects of hyperfine interactions on the energy levels will vanish to first order, and quadrupolar interaction terms (with parameters typically 10–100 times smaller than hyperfine parameters) will primarily determine the nuclear energy spacings. It is of some interest to compare nuclear quadrupolar interactions in ground- and excited-state molecules and to determine what implications any observed differences might have on molecular geometry and electron distributions.

The first reports of zero-field ENDOR experiments were by Chan *et al.*[65] for ^{14}N nuclei of quinoxaline and quinoline molecules in durene crystals and by Harris *et al.*[66] for ^{14}N in 2,3-dichloroquinoxaline molecules in durene crystals. For these molecules, so-called forbidden lines are seen in the optically detected magnetic resonance spectrum corresponding to simultaneous nuclear and electron spin flips. ENDOR observations are made on the forbidden EPR lines. Since the ^{14}N nucleus has $I = 1$, the nuclear quadrupolar Hamiltonian is written quite similarly in form to the triplet fine structure Hamiltonian of equation (2). There are two parameters to be determined. For the case of 2,3-dichloroquinoxaline, the parameters could not be determined independently of each other from the ENDOR transitions observed. Likewise, for the initial quinoxaline and quinoline measurements,[65] once the correct assignments[67] of transitions were made, effectively only a single parameter was known. However, from later measurements on quinoline in durene, both parameters were obtained,[68] but no attempt at interpreting them was reported.

Nitrogen-14 as well as ^{35}Cl ENDOR transitions were investigated for 8-chloroquinoline in durene.[69] Spin Hamiltonian parameters were deduced primarily from the EPR spectrum. The transition assignments were verified by the ENDOR measurements. Again, effectively only one quadrupolar parameter was found for each of the nuclei. The ^{35}Cl quadrupolar parameter value was not much different from values found in the ground state of similar molecules.

Chlorine-35 and -37 quadrupole interaction parameters were also determined for *p*-dichlorobenzene traps in neat crystals.[70] ENDOR was observed, but again, parameter determination was from the EPR measurements. Interpretation of the parameters was given in terms of electric field gradients at the molecule and differences in excited and ground state measurements. Conclusions were drawn with regard to molecular distortions. It is worth noting here that quadrupolar interaction parameters have also been moderately successfully obtained from high-field EPR measurements.[71]

In the above examples of this section, interactions involving the protons and deuterons of the triplet-state molecules were not considered in the interpretations of the magnetic resonance spectra. Dennis and Tinti[72] have shown for 1,8-diazanaphthalene and quinoline in durene that proton hyperfine interactions must be taken into account to explain the details of the zero-field EPR transitions observed. It was noted that even in perdeuterated molecules a significant effect on the spectrum can arise due to the deuteron quadrupolar interactions. From separations of ^{14}N ENDOR transitions detected optically in 1,8-diazanaphthalene molecules, the authors were able to deduce the value of one component of the proton hyperfine interaction of a particular type of proton in the molecule. As with the previous work, only a single ^{14}N quadrupole parameter was obtained.

Hyperfine interactions of protons with the electron spin in triplet states of organic molecules in zero applied static magnetic field have in general only been partially resolved in zero-field EPR.[72-74] Their contribution to the zero-field EPR linewidths have been calculated,[72,73] and it has been of some interest to attempt direct measurement of the interactions with proton ENDOR in zero field. Hochstrasser *et al.*[75] have optically detected proton ENDOR transitions in excited triplet states of benzophenone molecules in dibromodiphenylether crystals at low field (0.01 T). ENDOR of ^{13}C substituted for the central carbon in benzophenone was also observed, and the dependence of the ENDOR frequencies on magnetic field orientation was measured.* The proton ENDOR was not analyzed in detail but hyperfine tensor elements determined for ^{13}C were useful in drawing conclusions regarding spin distribution in benzophenone. The high field approximations outlined earlier were not applicable in this case so an alternate procedure was followed.

Effects of proton and deuteron interactions with the triplet electron system of protonated and deuterated 1,2,4,5-tetrachlorobenzene molecules in protonated and deuterated durene crystals were observed in zero magnetic field optically detected magnetic resonance experiments designed to monitor coherence in the triplet spin system.[76,77] With an appropriate microwave pulse sequence, a spin-locked magnetization of the two states in resonance can be produced.[78] Interaction with the surrounding nuclear moments is one cause of loss of the spin-locked magnetization. If the loss of spin-locked magnetization is examined as a function of the spin-locking magnetic field, which is directly related to the applied microwave power, the relative importance of the various nuclear interactions can be determined. The technique is similar to the Hartman–Hahn method[79] of detection of nuclear magnetic resonance transitions of so-called rare spins in solids. So far, only rough estimates of the strength of the nuclear interactions have been given for triplet states.

4.3. Defects in Solids

ENDOR has figured prominately in the identification of the nature of defects in solids. Isolated F centers in alkali halide crystals, single electrons bound to negative ion vacancies, were the first defect centers studied with ENDOR.[80] These defects have $S = \frac{1}{2}$, but there are $S = 1$ defects, as we will now relate.

---

* See Note 2 Added in Proof on page 436.

The so-called M centers of alkali halide crystals were clearly demonstrated to be nearest-neighbor F-center pairs by the ENDOR and ENDOR work of Seidel[81] in KCl at 90°K. Excited triplet states of the M center were produced in crystals of relatively large F-center concentrations by irradiation with ultraviolet light. Anisotropic EPR signals characteristic of the $\mathcal{H}_{ss}$ term of equation (1) were observed while the sample was illuminated. The fine structure tensor principal directions were consistent with the F-center pair model. Potassium hyperfine interactions were measured with ENDOR and compared with those of the isolated F center. The M-center hyperfine interaction tensors for a given potassium nucleus were nearly equal to the average of the corresponding tensors of the isolated F centers that would compose the M center when they were nearest neighbors. Not only was that observation consistent with the model, but also it led the author to conclude that the M-center wave function was essentially a superposition of the two F-center wave functions. ENDOR of more than two nearest-neighbor F centers has also been observed[82] at 90°K.

ENDOR has likewise been reported for the V^0 center of BeO.[83] The V^0 center is a positive ion vacancy with two trapped holes. The model of the defect is that the positive ion vacancy has two near-neighbor O^- ions. Each O^- has an unpaired electron, with the two combining to form a triplet state, which apparently lies lower in energy than their lowest singlet state. These vacancies are created after the BeO crystals are irradiated with electrons or neutrons. From the principal values and directions of the g and T tensors, the authors could determine the lattice positions of the O^- ions. The size and the algebraic signs of the elements of the hyperfine interaction tensors measured by ENDOR for the surrounding beryllium nuclei were consistent with the model for the defect.

Another defect center of current concern is the self-trapped exciton of alkali halide crystals. These defects are responsible for intrinsic luminescence observed from these crystals when they are excited by x-rays. There is no vacancy involved here, but rather an excited state of the molecular ion X_2^{2-}, where X is the halogen atom. A hole is thought to be localized on two adjacent covalently bonded halide ions. This excited state is a triplet state. Optically detected magnetic resonance has been observed in NaCl, RbBr, KBr, and CsBr.[84,85] The possibility of ENDOR is being pursued. There are, of course, other defect candidates for triplet ENDOR, such as triplet states of F-type centers in CaO, where two electrons are bound at a negative ion vacancy.

4.4. *Other $S = 1$ Systems*

ENDOR of the nuclei of transition element ions with an effective spin of 1 has been observed. For example, Ni^{2+} ions[86,87] and Cu^{3+} ions[88] in

Al_2O_3 have been studied by ENDOR techniques. Nuclear quadrupole and Zeeman interactions and ionic wave function information can often be obtained. The reader is referred to Abragam and Bleaney[12] for further details in the case of ENDOR of paramagnetic ions.

ENDOR of triplet radical pairs of molecules in solids has been observed.[89] $S = \frac{1}{2}$ molecular radicals in solids have been investigated in some detail with ENDOR. (See the review by Iwasaki[90] for further details.) Neighboring radicals can pair to form an $S = 1$ system. Those have been studied extensively in EPR,[15,16,91] and one might expect ENDOR to be applied more often in the future.

ENDOR of triplet states of molecules of biological interest is another area of possible future study. Triplet states of porphyrin molecules have been examined in EPR,[92] but not by ENDOR as yet. Triplet states of chlorophyll have also been investigated by EPR.[93] Biological molecules may be particularly amenable to zero-field studies because of difficulties in synthesizing single crystal samples.

5. Summary

We have attempted to give a representative outline of the sorts of problems attacked and solved by ENDOR of paramagnetic triplet-state systems. As noted for molecules, the utility of ENDOR is its ability to resolve structural details of the molecule and the environment of the molecule. Slight losses of symmetry, which are masked by the broad lines of EPR, are quite apparent in ENDOR. Defects in solids, traps in molecular crystals, and chemical reactions have been modeled based on environmental information obtained from ENDOR. Electron spin–nuclear spin interactions, as well as nuclear Zeeman, quadrupolar, and dipole–dipole interactions, can be measured with high precision. The overall impression is that the interpretation of the results is limited by uncertainties in the calculation of the interactions and not by inadequacies of the experiments. We expect ENDOR to continue to be applied as an important technique for understanding basic interactions in solids.

References

1. G. Feher, *Phys. Rev.* **103**, 834–835 (1956).
2. G. N. Lewis, D. Lipkin, and T. T. Magel, *J. Am. Chem. Soc.* **63**, 3005–3018 (1941).
3. G. N. Lewis and M. Kasha, *J. Am. Chem. Soc.* **66**, 2100–2116 (1944).
4. G. N. Lewis and M. Kasha, *J. Am. Chem. Soc.* **67**, 994–1003 (1945).
5. A. Terenin, *Acta Physiochim. URSS* **18**, 210–241 (1943).
6. C. A. Hutchison, Jr., and B. W. Mangum, *J. Chem. Phys.* **29**, 952–953 (1958).

7. C. A. Hutchison, Jr., and B. W. Mangum, *J. Chem. Phys.* **34**, 908–922 (1961).
8. J. H. van der Waals and M. S. de Groot, *Molec. Phys.* **2**, 333–340 (1959).
9. J. H. van der Waals and M. S. de Groot, *Molec. Phys.* **3**, 190–200 (1960).
10. W. A. Yager, E. Wasserman, and R. M. R. Cramer, *J. Chem. Phys.* **37**, 1148–1149 (1962).
11. E. Fermi, *Z. Phys.* **60**, 320–333 (1930).
12. A. Abragam and B. Bleaney, *Electron Paramagnetic Resonance of Transition Ions*, Clarendon Press, Oxford (1970).
13. A. W. Hornig and J. S. Hyde, *Molec. Phys.* **6**, 33–41 (1963).
14. R. M. Hochstrasser and T.-S. Lin, *J. Chem. Phys.* **49**, 4929–4945 (1968).
15. M. Iwasaki, K. Minakata, and K. Toriyama, *J. Chem. Phys.* **54**, 3225–3226 (1971).
16. K. Minakata and M. Iwasaki, *Molec. Phys.* **23**, 1115–1131 (1972).
17. D. S. Tinti, G. Kothandaraman, and C. B. Harris, *J. Chem. Phys.* **59**, 190–193 (1973).
18. G. Kothandaraman, H. J. Yue, and D. W. Pratt, *J. Chem. Phys.* **61**, 2102–2111 (1974).
19. C. A. Hutchison, Jr., and G. A. Pearson, *J. Chem. Phys.* **47**, 520–533 (1967).
20. A. M. Ponte Goncalves and C. A. Hutchison, Jr., *J. Chem. Phys.* **49**, 4235–4236 (1968).
21. H. C. Brenner, C. A. Hutchison, Jr., and M. D. Kemple, *J. Chem. Phys.* **60**, 2180–2181 (1974).
22. G. B. Robertson, *Nature* (*London*) **191**, 593–594 (1961).
23. J. Trotter, *Acta Crystallog.* **14**, 1135–1140 (1961).
24. A. Hargreaves and S. H. Rizvi, *Acta Crystallog.* **15**, 365–373 (1962).
25. R. McWeeny, *J. Chem. Phys.* **34**, 399–401 (1961).
26. H. M. McConnell, *J. Chem. Phys.* **28**, 1188–1192 (1958).
27. S. I. Weissman, *J. Chem. Phys.* **22**, 1378–1379 (1954).
28. H. M. McConnell, *J. Chem. Phys.* **24**, 764–766 (1956).
29. H. M. McConnell and D. B. Chesnut, *J. Chem. Phys.* **28**, 107–117 (1958).
30. H. M. McConnell, C. Heller, T. Cole, and R. M. Fessenden, *J. Am. Chem. Soc.* **82**, 766–775 (1960).
31. N. Hirota, C. A. Hutchison, Jr., and P. Palmer, *J. Chem. Phys.* **40**, 3717–3725 (1964).
32. A. M. Ponte Goncalves, Ph.D. thesis, University of Chicago, Chicago (1969).
33. R. H. Clarke and C. A. Hutchison, Jr., *J. Chem. Phys.* **54**, 2962–2968 (1971).
34. H. M. McConnell and J. Strathdee, *Molec. Phys.* **2**, 129–138 (1959).
35. B. J. Botter, D. C. Doetschman, J. Schmidt, and J. H. van der Waals, *Molec. Phys.* **30**, 609–620 (1975).
36. L. G. Rowan, E. L. Hahn, and W. B. Mims, *Phys. Rev.* **137**, A61–A71 (1965).
37. C. A. Hutchison, Jr., and G. A. Pearson, *J. Chem. Phys.* **43**, 2545–2546 (1965).
38. R. W. Brandon, G. L. Closs, C. E. Davoust, C. A. Hutchison, Jr., B. E. Kohler, and R. Silbey, *J. Chem. Phys.* **43**, 2006–2016 (1965).
39. C. A. Hutchison, Jr., and B. E. Kohler, *J. Chem. Phys.* **51**, 3327–3335 (1969).
40. C. A. Hutchison, Jr., *Pure Appl. Chem.* **27**, 327–360 (1971).
41. D. C. Doetschman and C. A. Hutchison, Jr., *J. Chem. Phys.* **56**, 3964–3982 (1972).
42. R. J. M. Anderson and B. E. Kohler, *J. Chem. Phys.* **65**, 2451–2459 (1976).
43. P. Ehret, G. Jesse, and H. C. Wolf, *Z. Naturforsch.* **23a**, 195–196 (1968).
44. P. Ehret and H. C. Wolf, *Z. Naturforsch.* **23a**, 1740–1746 (1968).
45. A. D. McLachlan, *Molec. Phys.* **5**, 51–62 (1962).
46. M. S. de Groot and J. H. van der Waals, *Molec. Phys.* **6**, 545–562 (1963).
47. J. H. van der Waals, A. M. D. Berghuis, and M. S. de Groot, *Molec. Phys.* **13**, 301–321 (1967).
48. C. A. Hutchison, Jr., and V. H. McCann, *J. Chem. Phys.* **61**, 820–832 (1974).
49. P.-O. Löwdin, *Phys. Rev.* **90**, 120–125 (1953).
50. R. M. Hochstrasser, R. D. McAlpine, and J. D. Whiteman, *J. Chem. Phys.* **58**, 5078–5088 (1973).

51. P. S. Friedman, R. Kopelman, and P. N. Prasad, *Chem. Phys. Lett.* **24**, 15–17 (1974).
52. R. M. Hochstrasser, G. W. Scott, A. H. Zewail, and H. Fuess, *Chem. Phys.* **11**, 273–279 (1975).
53. C. A. Hutchison, Jr., and M. D. Kemple, to be published.
54. H. Suzuki, *Electronic Absorption Spectra and Geometry of Organic Molecules*, Chapter 12, Academic Press, New York (1967).
55. V. Zimmerman, M. Schwoerer, and H. C. Wolf, *Chem. Phys. Lett.* **31**, 401–405 (1975).
56. V. Zimmerman, H. C. Wolf, and M. Schwoerer, *Chem. Phys. Lett.* **31**, 406–409 (1975).
57. C. A. Hutchison, Jr., and J. S. King, Jr., *J. Chem. Phys.* **58**, 392–393 (1973).
58. M. Schwoerer and H. C. Wolf, in *The Triplet State* (A. B. Zahlen, ed.), pp. 133–140, Cambridge University Press, Cambridge, England (1967).
59. M. Schwoerer and H. C. Wolf, *Molec. Cryst.* **3**, 177–213 (1967).
60. B. J. Botter, C. J. Nonhof, J. Schmidt, and J. H. van der Waals, *Chem. Phys. Lett.* **43**, 210–216 (1976).
61. H. Sternlicht and H. M. McConnell, *J. Chem. Phys.* **35**, 1793–1800 (1961).
62. H. Blok, J. A. Kooter, and J. Schmidt, *Chem. Phys. Lett.* **30**, 160–164 (1975).
63. N. S. Dalal, D. Haarer, J. Bargon, and H. Möhwald, *Chem. Phys. Lett.* **40**, 326–330 (1976).
64. A. L. Kwiram, in *Magnetic Resonance, MTP International Review of Science, Physical Chemistry Series 1*, Vol. 4 (C. A. McDowell, ed.), pp. 271–316, Butterworths, London (1972).
65. I. Y. Chan, J. Schmidt, and J. H. van der Waals, *Chem. Phys. Lett.* **4**, 269–273 (1969).
66. C. B. Harris, D. S. Tinti, M. A. El-Sayed, and A. H. Maki, *Chem. Phys. Lett.* **4**, 409–412 (1969).
67. M. J. Buckley, C. B. Harris, and A. H. Maki, *Chem. Phys. Lett.* **4**, 591–595 (1970).
68. I. Y. Chan and J. H. van der Waals, *Chem. Phys. Lett.* **20**, 157–162 (1973).
69. M. J. Buckley and C. B. Harris, *Chem. Phys. Lett.* **5**, 205–213 (1970).
70. M. J. Buckley and C. B. Harris, *J. Chem. Phys.* **56**, 137–146 (1972).
71. G. Kothandaraman, D. W. Pratt, and D. S. Tinti, *J. Chem. Phys.* **63**, 3337–3348 (1975).
72. L. W. Dennis and D. S. Tinti, *J. Chem. Phys.* **62**, 2015–2025 (1975).
73. C. A. Hutchison, Jr., J. V. Nicholas, and G. W. Scott, *J. Chem. Phys.* **53**, 1906–1917 (1970).
74. C. A. Hutchison, Jr., and G. W. Scott, *J. Chem. Phys.* **61**, 2240–2249 (1974).
75. R. M. Hochstrasser, G. W. Scott, and A. H. Zewail, *Molec. Phys.* **36**, 475–499 (1978).
76. H. Schuch and C. B. Harris, in *Magnetic Resonance and Related Phenomena, Proc. 18 Congr. Ampere* (P. S. Allen, E. R. Andrew, and C. A. Bates, eds.), Vol. 1, pp. 223–224, North-Holland, Amsterdam (1975).
77. H. Schuch and C. B. Harris, *Z. Naturforsch* **30a**, 361–371 (1975).
78. C. B. Harris, R. L. Schlupp, and H. Schuch, *Phys. Rev. Lett.* **30**, 1019–1022 (1973).
79. S. R. Hartman and E. L. Hahn, *Phys. Rev.* **128**, 2042–2053 (1968).
80. G. Feher, *Phys. Rev.* **105**, 1122–1123 (1957).
81. H. Seidel, *Phys. Lett.* **7**, 27–29 (1963).
82. H. Seidel, M. Schwoerer, and M. Schmid, *Z. Phys.* **182**, 398–426 (1965).
83. H. Maffeo and A. Hervé, *Phys. Rev. B* **13**, 1940–1959 (1976).
84. M. J. Marrone, F. W. Patten, and M. N. Kabler, *Phys. Rev. Lett.* **31**, 467–471 (1973).
85. A. Wasiela, G. Ascarelli, and Y. Merle d'Aubigne, *Phys. Rev. Lett.* **31**, 993–996 (1973).
86. P. R. Locher and S. Geschwind, *Phys. Rev. Lett.* **11**, 333–336 (1963).
87. S. Geschwind, in *Hyperfine Interactions* (A. J. Freeman and R. B. Frankel, eds.), Chapter 6, Academic Press, New York (1967).
88. W. E. Blumberg, J. Eisinger, and S. Geschwind, *Phys. Rev.* **130**, 900–909 (1963).
89. R. J. Cook, in *Magnetic Resonance and Radiofrequency Spectroscopy, Proceedings of the 15th Colloque AMPERE* (P. (Averbuch, ed.), pp. 269–273, North-Holland, Amsterdam (1969).

90. M. Iwasaki, in *Magnetic Resonance, MTP International Review of Science, Physical Chemistry Series 1*, Vol. 4 (C. A. McDowell, ed.), pp. 317–346, Butterworths, London (1972).
91. K. Nunome, K. Toriyama, and M. Iwasaki, *J. Chem. Phys.* **62**, 2927–2933 (1975).
92. W. G. van Dorp, M. Soma, J. A. Kooter, and J. H. van der Waals, *Molec. Phys.* **28**, 1551–1568 (1974).
93. R. H. Clarke and R. H. Hofeldt, *J. Chem. Phys.* **61**, 4582–4587 (1974).

Note 1 Added in Proof

ENDOR of X-traps in naphthalene crystals doped with 2-fluoronaphthalene and 2-chloronaphthalene has also been measured [H. Dörner and D. Schmid, *Chem. Phys. Lett.* **61**, 171–174 (1979)]. In the first-mentioned case the observation of fluorine ENDOR aided in the determination of the nature of the trap.

Note 2 Added in Proof

^{13}C ENDOR of benzophenone has also been optically detected in high field [P. F. Brode III and D. W. Pratt, *Chem. Phys. Lett.* **54**, 591–593 (1978); *Semicond. Insul.* **4**, 277–293 (1978)]. In the first-mentioned article a method for determining the algebraic signs of the elements of the hyperfine tensor when the standard methods fail is discussed. Other molecules with excited $n\pi^*$ triplet states have been studied by high-field optical detection of ENDOR as well [See K. Itoh, T. Zentoh, and H. Kawakami, *J. Lumin.* **12–13**, 397–402 (1976); J. H. Lichtenbelt and D. A. Wiersma, *Chem. Phys.* **34**, 47–54 (1978); and C. von Borczyskowski, M. Plato, K.-P. Dinse, and K. Möbius, *Chem. Phys.* **35**, 355–366 (1978).] Detailed SCF calculations of hyperfine interactions are given for the *p*-benzoquinone triplet in a paper by J. H. Lichtenbelt, D. A. Wiersma, H. T. Jonkman, and G. A. van der Velde [*Chem. Phys.* **22**, 297–305 (1977)].

13

Principles and Applications of Optical Perturbation–Electron Paramagnetic Resonance (OPEPR)[†]

Haim Levanon

1. Introduction

The study of the dynamics associated with the production of paramagnetic species or states upon optical pumping has been gaining interest in recent years in a large variety of fields. In this chapter, we confine ourselves to describing the direct EPR detection of two groups of transient paramagnetic systems that are generated under light excitation: (a) molecules promoted to their photoexcited triplets and (b) free radicals.

The search for the dynamics associated with the photoexcited triplet manifold resulted in several methods of optical detection of magnetic resonance (ODMR). The important ones that are currently used are phosphorescence and fluorescence microwave double resonance (PMDR) and (FDMR), respectively. A complementary method to FDMR is the so-called technique of triplet absorption detection magnetic resonance (TADMR). A common feature in all these techniques is that the luminescence properties such as phosphorescence, fluorescence, or triplet–triplet absorption are optically monitored when the microwave frequency is swept through resonance.

[†] The Radiation Laboratory of the University of Notre Dame is operated under contract with the U.S. Energy Research and Development Administration. This is Document No. S.R.-27.

Haim Levanon • Radiation Laboratory and Department of Chemistry, University of Notre Dame, Notre Dame, Indiana (1976–1977). *Permanent address*: Department of Physical Chemistry, The Hebrew University, Jerusalem, Israel

These methods and some others have been discussed recently in detail by Olmsted and El-Sayed,[1] El-Sayed,[2] and Hausser and Wolf.[3] In most cases, such experiments are performed at zero magnetic field on oriented single crystals at liquid helium temperatures in order to retard of the spin-lattice relaxation (SLR) processes among the spin sublevels.

The basic method for the observation of free radicals that are generated upon light excitation still remains EPR spectroscopy. Nevertheless, improvement of the response time of detection is the main goal in many laboratories where, indeed, significant progress has been made on fast EPR detection of free radicals in solution.[4–6]

Conventional EPR–optical resonance (EOR) spectroscopy is employed to detect paramagnetic species that are formed under steady-state conditions, i.e., upon continuous light excitation. To observe a reasonable signal-to-noise ratio in the EPR spectrum, it is essential that the steady state concentration exceed $\sim 10^{-6}$ M. This concentration requirement implies that the lifetime of the detectable species should be relatively high, i.e., of the order of 0.1 sec. Hence, in the early EOR experiments, only the magnetic parameters, namely, the zero-field splitting (ZFS) components, the spectroscopic splitting factor (g), the hyperfine components (A), and the total lifetime of the detectable paramagnetic species could be evaluated. Conventional EOR is not sensitive enough to give a detailed description of the dynamics of the triplet state or the radical formation.

In principle, the EOR technique may be considered as consisting of an optical excitation of a diamagnetic system giving rise to paramagnetic species that may be detected by the EPR method. Figure 1 shows schematically a typical experiment. The compound A in Figure 1a represents an organic molecule doped into a single crystal, which is immersed at liquid nitrogen (or helium). Alternatively, the compound A may be dissolved in a solvent that remains transparent to the exciting light on freezing to low temperatures. When either of these systems is irradiated with photons of energy $h\nu$, which correspond to $S_0 \rightarrow S_i$ (ground singlet $\rightarrow$ excited singlets) transitions, the photoexcited triplet state T_1 is formed via an intersystem crossing (ISC) process $S_i \rightsquigarrow T_i$.[3] On continuous irradiation, a steady state of the photoexcited triplets is maintained, and two EPR transitions $h\nu'$ and $h\nu''$ in the microwave region should be observed when one of the principal axes of the ZFS tensor is oriented along the external magnetic field. In a randomly oriented sample, these orientations are often referred to as the canonical orientations. In Figure 1b, the compound R'—R", which is dissolved in a solution, is being irradiated by photons of energy $h\nu$. This results in a bond cleavage, giving rise to a pair of radicals (doublets), each of which may be EPR detected in the microwave region with energies $h\nu'$ and $h\nu''$, for R'· and R"· radicals, respectively. In both examples given in Figure 1, EPR

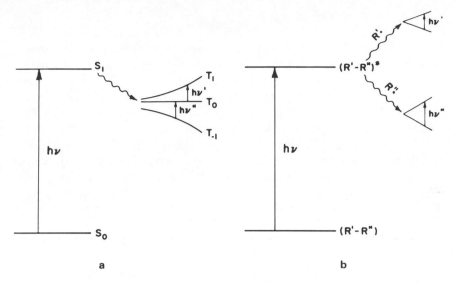

Figure 1. A schematic representation of an EOR experiment. (a) Two EPR transitions in the microwave region for a molecule promoted to its photoexcited triplet by light excitation. (b) Two EPR transitions in the microwave region for two radicals $R'\cdot$ and $R''\cdot$ formed during light photolysis.

transitions are normally observed, provided that a reasonable steady-state concentration of either species is maintained.

Recent experiments employing periodic light modulation excitation in conjunction with direct EPR detection of transient signals due to triplets or radicals in solution, were found to be successful in elucidating the dynamics associated with the paramagnetic species.[7–10] In most cases, transient EPR spectra are accompanied by anomalous line intensities, indicating that the spin sublevels have not reached a thermal equilibrium population. This phenomenon is due to a selective mechanism for population and depopulation of the spin states giving rise to electron spin polarization (ESP) effect.[3] For polarized spins in radicals, this effect is normally referred to as chemically induced dynamic electron polarization (CIDEP).[5,11] The fact that ESP in triplets can be observed also at relatively high temperatures (above 77°K) where SLR is not negligible, provides an independent experimental approach that is complementary to the ODMR methods. This has been demonstrated recently by Kim and Weissman[12] on the observation of transient kinetics in the photoexcited triplet state of phenazine at room temperature. This chapter covers mainly this experimental technique, which will be referred to hereafter as optical perturbation–electron paramagnetic resonance (OPEPR).

2. Experimental Technique

Complementary types of OPEPR experiments are performed using periodic modulation of the exciting light: (a) single-light-pulse (SLP) experiment and (b) continuous-wave (CW) experiment. Figure 2 shows schematically the experimental setup that is applied for both types of experiments.

Modulation of the light intensity produces a variation in time of the susceptibility χ'', which is a function of the field and time. We write $\chi'' = \chi''(H, t)$, where H is the external magnetic field. The explicit time dependence results from the modulated photoexcitation; there may be, in addition, an implicit time dependence resulting from field modulation. We may write the effect of field modulation with small amplitude δ at frequency ω_M around H as producing a susceptibility[7]

$$\chi'' = \chi''(H, t) + \delta \cos \omega_M t \, \frac{\partial \chi''(H, t)}{\partial H} \tag{1}$$

Phase-sensitive detection at ω_M with integrating time τ produces a detected signal proportional to $\delta \langle \partial \chi''(H, t)/\partial H \rangle_\tau$, where the symbol $\langle \ \rangle_\tau$ means integration over time τ.

In an SLP experiment in which the kinetic behavior is displayed directly, a signal proportional to $\langle \partial \chi''(H, t)/\partial H \rangle_\tau$ at fixed field H is recorded.

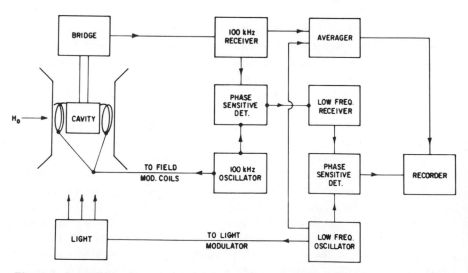

Figure 2. A schematic representation of the experimental setup used in OPEPR experiment. The light source may produce square wave light pulses or sine wave modulation. For detection of higher harmonics, e.g., $\partial \chi_2''(H)/\partial H$, the phase-sensitive detector should be tuned by the corresponding frequency, $2\omega_L$.

In order to analyze the observations under periodic light modulation at frequency ω_L with phase-sensitive detection referenced to the light intensity, we expand the susceptibility in a Fourier series[7]:

$$\chi''(H, t) = \sum_{n=0}^{\infty} \chi_n''(H)\cos(n\omega_L t + \phi_n) \tag{2}$$

The detected signal emerging from the first, i.e., field modulation, phase-sensitive detector is proportional to

$$\delta \sum_{n=0}^{\infty} \left\langle \frac{\partial\chi_n''(H)}{\partial H} \cos(n\omega_L t + \phi_n) \right\rangle_\tau \tag{3}$$

Thus, for high harmonics, $n > 1/\omega_L \tau$, the integration over τ loses the kinetic information. The kinetic information contained in the harmonics for which $n > 1/\omega_L \tau$ passes through the first phase-sensitive detection.

A second phase-sensitive detection at frequency $n\omega_L$ yields $\partial\chi_n''(H)/\partial H$ and ϕ_n, i.e., the Fourier component at the fundamental frequency of the light modulation of the derivative of susceptibility with respect to the field and the phase angle, respectively (CW experiment). Thus, $\partial\chi_1(H)/\partial H$ is the first harmonic representation of the EPR signal intensity, etc. The relationship between ϕ_n and the instrumental phase angle ψ' will be discussed in Section 3.

2.1. Single-Light-Pulse (SLP) Experiment

In this type of experiment, the variation with time of the EPR signal intensity during and after the pulse is recorded directly, either from the unfiltered output (after phase-sensitive detection at the field modulation frequency ω_M), or directly from the detection diodes in the spectrometer. The typical band width of EPR spectrometers employing 100-KHz field modulation frequency is about 10 KHz, which corresponds to a response time of about 100 μsec. Square wave light pulses are obtained either by electronically modulated light sources or by chopping the continuous light beam using a mechanical sector. The advantage of the former version is mainly the relatively short rise and decay times of the light pulses, which are of the order of 10–15 μsec.

Limitations imposed by the feebleness of the signals requires for most experiments the use of accumulation or averaging of the signal intensity over many repetitions of the exciting light pulses. This amplification of the signal intensity prior to its display is accomplished under various experimental methods.

2.1.1. *CAT Method*

The EPR signal output is fed into a computer of averaged transients (CAT). It transforms the signal into a digital form, which is stored in the CAT memory. The time resolution of a conventional CAT is sufficient for EPR signals driven out of the phase-sensitive detection at 100 KHz field modulation [equation (3)]. This type of averaging was discussed in detail by Bolton and Warden.[13]

2.1.2. *Wave-Form Eductor Method*

A somewhat different method of averaging is the employment of a wave-form eductor.[14,15] This type of averager divides repetitive input wave forms into a set of capacitors (100 or 50 in commercial eductors). The EPR signal output is averaged and stored in the eductor's memory in an analog form. The wide-band amplification and the available long time constant for storage and averaging permit reasonable signal-to-noise ratio (S/N).

Figure 3 shows a typical time-dependent curve during and on termination of a light pulse. This curve describes the formation and the decay process of photoelectrons which are generated from rubidium anion present in rubidium–tetrahydrofuran solutions.[16] The experimental setup for this experiment is as described in Figure 2.

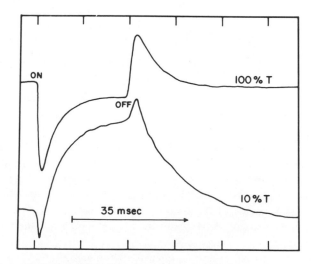

Figure 3. Rise and decay curves of the EPR signal intensity in an SLP experiment of the photoelectron in Rb–THF solution at 212°K using the experimental setup as described in Figure 2. The time between *on* and *off* is the duration of the light pulse. The percentage transmittance of the exciting light is indicated on each curve. (Reproduced with permission from Friedenberg and Levanon.[16])

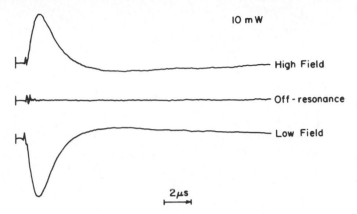

Figure 4. Time-resolved EPR signal intensities in an SLP experiment of the low(emission)- and high(enhanced absorption)-field lines of triplet phenazine at room temperature. (Reproduced with permission from Kim and Weissman.[12])

2.1.3. Pulsed Laser Method

The limitations of the EPR time resolution resulting from relatively long response time of the spectrometers (mainly due to field modulation) and also from long rise and decay times of the light pulses brought about many attempts to improve the EPR time resolution. This is currently being accomplished by the use of pulsed lasers producing light pulses with rise and decay times of about 10 nsec in conjunction with the improvement of the response time of EPR spectrometers. In these experiments, time-resolved EPR spectra in the submicrosecond region have been reported.[5,6,12]

Typical SLP curves for the low- and high-field resonance lines of triplet phenazine at room temperature are shown in Figure 4.

2.2. Continuous-Wave (CW) Experiment

In this type of experiment, the time-dependent EPR spectrum with respect to the light modulation frequency is obtained using phase-sensitive detection at the fundamental frequency of the light modulation. A particular advantage of this arrangement is that the EPR signals from radiation-damaged centers are suppressed. Moreover, continuous illumination in conjunction with conventional EPR detection may not reveal any transient effects associated with the spectrum. However, when a CW OPEPR experiment is performed, transient effects may clearly appear. These are illustrated in Figure 5, where EPR spectra at 100-KHz field modulation of phenazine dissolved in toluene glass at 77°K were recorded at different light modulation frequencies.[7] The top trace, which was recorded under continuous

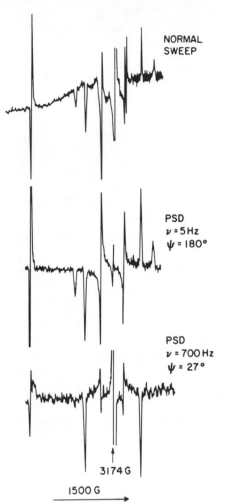

Figure 5. $\partial\chi_1''(H)/\partial H$ versus H in a CW experiment performed on phenazine dissolved in toluene glass at 77°K under continuous illumination (*top trace*) and under pulsed illumination (*middle and bottom trace*). The experimental setup is as described in Figure 2. In this experiment, 100-kHz field modulation was employed. (Reproduced with permission from Levanon and Weissman.[7])

light excitation, contains the spectrum of triplet phenazine, a broad background resonance of an impurity in the cavity of the spectrometer, two sharp lines from hydrogen atoms, and a resonance at $g = 2$ of unknown origin. The lower two trace spectra at different light modulation frequencies ω_L with phase-sensitive detection at ω_L and phase ψ are due to triplet phenazine only. The bottom trace spectrum is a typical transient triplet, which is reflected by its inverted lines (see Section 3). Notice that the broad background and hydrogen atoms and most of the $g = 2$ resonances have been eliminated.[7]

For short-lived paramagnetic species, it is possible to employ diode detection without employing field modulation. This is an important in-

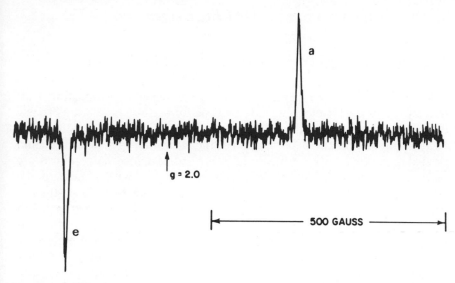

Figure 6. $\partial\chi_1''(H)/\partial H$ versus H in a CW experiment performed on anthracene doped in phanazine crystal at 77°K using the experimental setup as described in Figure 2. (Modulation frequency 2050 Hz.) In this experiment, no field modulation was employed, e and a represent emission and absorption lines, respectively. (Reproduced from Levanon.[17])

strumental approach in that pure absorption (or emission) lines are seen and the usual EPR field modulation compromise between line shape and line intensity is avoided. Thus, the signal output from the preamplifier at the detection diodes is obtained using phase-sensitive detection. A mixed absorption emission photoexcited triplet spectrum of 1% anthracene doped in phenazine single crystal at 77°K, using a light modulation frequency of 2050 Hz, is shown in Figure 6.[17]

An alternative experimental approach to diode detection is given by Verma and Fessenden.[6] A two-channel sampler (boxcar integrator) simultaneously subtract the transient signal intensity on resonance from the background while the magnetic field is being swept. This technique is applicable in OPEPR experiments where repetitive light pulses of different duration and intervals are used.

A somewhat different CW experiment has been described recently by Paul[9] as a harmonic effect modulation. In his experiment, instead of a square wave train of pulses, a sine (or cosine) modulation excitation is applied to the sample in the EPR cavity. The EPR signal output is phase sensitive detected at the fundamental and first harmonic. The kinetics associated with the paramagnetic species produced by modulated light excitation are analyzed in terms of the frequency dependence of the signal amplitude and phase.

3. Theoretical Treatment of OPEPR Experiments

3.1. Kinetic Theory of the Transient Triplet EPR Spectrum

Most of the OPEPR experiments that were designed to demonstrate ESP in the photoexcited triplet state have been performed at temperature of a few degrees Kelvin on oriented single crystals.[18,19] From the results, the selectivity in population and depopulation of the individual spin sublevels could be deduced. Apparently, it has been presumed that low temperatures are necessary for the detection of the effects in order to retard sufficiently the rates of equilibrium between states (SLR) so as to make them slower than or comparable to the rates of dissipation of the states. OPEPR experiments in the range of $100°K$ demonstrated unambiguously a variety of transient effects in the EPR spectra of nonoriented organic and biological molecules. Thus, the requirements for observation of transients are simply that their amplitudes and duration, respectively, lie within the sensitivity and frequency response of the detecting instruments. Experimentally, it is simpler to prepare nonoriented systems than to grow the corresponding oriented systems. Moreover, a substantial number of compounds cannot be doped into single crystals. This, however, does not imply that oriented and nonoriented systems will exhibit the same transient effect. This was demonstrated by Lin[10] in his study of 1,3-diazaazulene using OPEPR method. The theory described below has been originally applied to random triplets that show ESP in OPEPR experiments performed at relatively high temperatures.[20] From the experimental results, one can determine the relative population, the depopulation, and SLR rates.

To designate the different lines measured in an EPR experiment, we represent in Figure 7 schematically a typical derivative of a nonaxial random

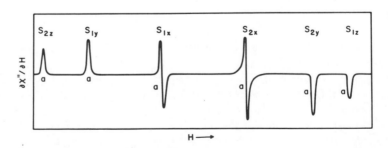

Figure 7. First-derivative $\partial\chi''/\partial H$ of the EPR absorption spectrum for a randomly oriented triplet system in thermal equilibrium. S_{ip} ($i = 1, 2$; $p = x, y, z$) is the EPR intensity for a particular canonical orientation. (Reproduced with permission from Levanon and Vega.[20])

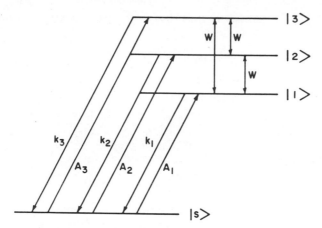

Figure 8. Four-energy-level system representing the ground-state singlet and the spin levels of the excited triplet state. W is the SLR rate constant; A_i and k_i $(i = 1, 2, 3)$ are the population and depopulation rate constants, respectively. (Reproduced with permission from Levanon and Vega.[20])

triplet absorption spectrum.[21] Each peak in this spectrum is connected with a canonical orientation[22] and is labeled S_{1p} or $S_{2p}(p = x, y, z)$ as shown in the figure. The indices 1 and 2 refer to the transitions (1–2) and (2–3), respectively (cf. Figure 8), and the indices x, y, z correspond to the principal directions of the ZFS tensor. From Figure 7, it is clear that the absorption signals S_{2y} and S_{1z} are obtained in the opposite direction to their counterparts.

The intensities of the EPR spectral lines of a photoexcited triplet state are proportional to the differences between the spin populations of the energy levels. We consider a model that consists of four energy levels—the singlet ground state and the first excited triplet state. The latter are labeled 1, 2, and 3 with increasing energy, as shown in Figure 8. Excitation to the triplet is represented by the population rates A_i, the depopulation rates to the ground singlet by k_i, and the spin–lattice relaxation between the three triplet sublevels W_{ij}. The rate equations are expressed in terms of the deviation of the populations from their thermal equilibrium values n_i. The total rate equations in the high temperature approximation for the populations of the four energy levels are:

$$\frac{d}{dt}N_T = A_T - B_T N_T - k_1 n_1 - k_2 n_2 - k_3 n_3$$

$$\frac{d}{dt}N_i = A_i - B_i N_T - (k_i + W_{ij} + W_{ik})N_i + W_{ij}n_j + W_{ik}n_k$$

(4)

with $i, j, k = 1, 2, 3$ or cyclic permutations. The parameters in equations (4) are defined and summarized as

$$n_i = N_i - \varepsilon_i N_T, \qquad \varepsilon_i = \tfrac{1}{3}(1 + f_i)$$
$$f_i = [(E_j - E_i) + (E_k - E_i)]/3k_B T \tag{5}$$

$$N_1 + N_2 + N_3 = N_T, \qquad A_T = A_1 + A_2 + A_3$$
$$B_i = A_i + \varepsilon_i k_i, \qquad\qquad B_T = B_1 + B_2 + B_3 \tag{6}$$

From equations (5), it follows that

$$n_i + n_2 + n_3 = 0 \tag{7}$$

In deriving equations (4), we have used the normalization condition $N_S + N_T = 1$. For a magnetic field in the direction of the principal p axis, the population and depopulation rate constants are

$$k_1 = k_3 = \tfrac{1}{2}(k_q + k_r), \qquad k_2 = k_p$$
$$A_1 = A_3 = \tfrac{1}{2}(A_q + A_r), \qquad A_2 = A_p \tag{8}$$

where $p, q, r = x, y, z$ or a cyclic permutation. A_x, A_y, A_z and k_x, k_y, k_z are the rate constants at zero field.

The four-dimensional equations (4) can be represented as a master equation in a vector form:

$$\frac{d}{dt}\mathbf{n} = \mathbf{A} - R\mathbf{n} \tag{9}$$

where $\mathbf{A}$ and R are considered as the driving and relaxation terms, respectively. Equation (9) is further reduced into a two-dimensional expression[20]:

$$\frac{d}{dt}\mathbf{v} = \mathbf{a} - \rho\mathbf{v} \tag{10}$$

where

$$\mathbf{v} = \begin{pmatrix} N_T \\ \sqrt{6}\,n_1 \end{pmatrix}, \qquad \mathbf{a} = \begin{pmatrix} A_T \\ \tfrac{2}{3}\sqrt{6}\,\alpha A_T \end{pmatrix} \tag{11}$$

$$\rho = \begin{pmatrix} K_T & 2/\sqrt{6}\,\kappa K_T \\ \tfrac{2}{3}\sqrt{6}\,\kappa K_T & K_T[(1 - \tfrac{1}{3}\kappa) + 3W] \end{pmatrix}$$

In these equations, the following definitions were used:

$$K_T = \tfrac{1}{3}(2k_1 + k_2), \qquad \kappa K_T = k_1 - k_2, \qquad \alpha A_T = A_1 - A_0 \tag{12}$$

The values of κ and α lie in the ranges

$$-1 \leq \alpha \leq \tfrac{1}{2}, \qquad -3 \leq \kappa \leq \tfrac{3}{2} \tag{13}$$

From the definitions in equations (6), (7), and (8), it follows that A_T and K_T are fixed for all signal intensities for a particular compound, and κ_p and α_p obey

$$\kappa_x + \kappa_y + \kappa_z = 0, \qquad \alpha_x + \alpha_y + \alpha_z = 0 \tag{14}$$

In deriving the two-dimensional expression (10), we have considered the case for nonsaturation conditions, where the population rates A_i are much smaller than the depopulation rates k_i. Thus, the signal intensities for two transitions in a particular canonical orientation are proportional to the population differences $N_i - N_j$, or in terms of the components of $\mathbf{v}$,

$$S_1 = c(N_1 - N_2) = c[\tfrac{3}{2}v_2 + \tfrac{1}{3}(f_1 - f_2)v_1]$$
$$S_2 = c(N_2 - N_3) = c[-\tfrac{3}{2}v_2 + \tfrac{1}{3}(f_2 - f_3)v_1] \tag{15}$$

where c is a proportionality factor that depends on the experimental conditions; i.e., one should take into account that the ratio between the signal intensities at the canonical orientations in thermal equilibrium are as shown in Figure 7.

3.2. Expressions for the Transient Triplet EPR Signal Intensities in an SLP Experiment

The components v_1 and v_2 can be derived explicitly by diagonalization of ρ in equation (10).[20] The final expressions for the signal intensities in an SLP experiment during a light-pulse period are

$$S_{1p} = S_{11}^p(1 - e^{-\lambda_- t}) + S_{12}^p(1 - e^{-\lambda_+ t})$$
$$S_{2p} = S_{21}^p(1 - e^{-\lambda_- t}) + S_{22}^p(1 - e^{-\lambda_+ t}) \tag{16}$$

where

$$S_{11}^p = \frac{c_p A_T}{\lambda_-}\left[\tfrac{1}{3}(f_1 - f_2) - \frac{\kappa_p K_T}{9W}\right] = \frac{c_p A_T}{\lambda_-}\left[\frac{h\nu_E}{3k_B T} - \frac{\kappa_p K_T}{9W}\right]$$
$$S_{21}^p = \frac{c_p A_T}{\lambda_-}\left[\tfrac{1}{3}(f_2 - f_3) + \frac{\kappa_p K_T}{9W}\right] = \frac{c_p A_T}{\lambda_-}\left[\frac{h\nu_E}{3k_B T} + \frac{\kappa_p K_T}{9W}\right] \tag{17}$$

$$S_{12}^p = -S_{22}^p = \frac{c_p A_T}{\lambda_+}\left[\alpha_p + \frac{\kappa_p K_T}{9W}\right] \tag{18}$$

The right-hand sides of equations (17) are obtained using equations (5) and from the fact that the microwave frequency ν_E is constant at a particular experiment. The coefficients S_{11}^p and S_{21}^p represent the thermal signals that build up with a small characteristic rate constant λ_-, whereas $S_{12}^p = -S_{22}^p$

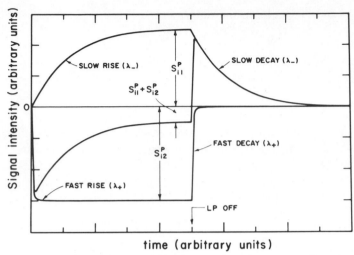

Figure 9. Resolution of an experimental kinetic curve into the corresponding slow and fast components for a particular transition. The rate constants λ_- and λ_+ and the signal amplitudes were chosen arbitrarily.

represent the transient behavior of the signals with a large characteristic rate constant λ_+.

The expressions for the signal intensities upon cessation of the light pulse are derived similarly, except that the steady-state conditions in equations (16) serve in that case as the initial conditions.

$$S_{1p} = S_{11}^p e^{-\lambda_- t} + S_{12}^p e^{-\lambda_+ t}$$
$$S_{2p} = S_{21}^p e^{-\lambda_- t} + S_{22} e^{-\lambda_+ t} \tag{19}$$

As an illustration, we have constructed a typical kinetic curve using equations (16) and (19) as shown in Figure 9.

For the derivation of k_p and A_p, we deduce from equations (17) and (18) the following expressions, which connect the kinetic parameters with the measured signal intensities:

$$\frac{\kappa_p K_T}{9W} = \frac{S_{21}^p - S_{11}^p}{S_{11}^p + S_{21}^p} \frac{h\nu_E}{3k_B T} \tag{20}$$

$$c_p A_T = (S_{11}^p + S_{21}^p) \frac{3k_B T}{2h\nu_E} \lambda_-$$
$$\tag{21}$$
$$\alpha_p = S_{12}^p \frac{\lambda_+}{c_p A_T} - \frac{\kappa_p K_T}{9W}$$

$$\lambda_- = K_T, \qquad \lambda_+ = K_T + 3W \tag{22}$$

Knowing the values of $\lambda_\pm$, S_{ij}^p $(i, j = 1, 2, 3; p = x, y, z)$, the values of κ_p, α_p, K_T, and $c_p A_T$ can now be evaluated. From equations (12) we obtain

$$k_p = \tfrac{1}{3}[3 - 2\kappa_p]K_T, \qquad A_p = \tfrac{1}{3}(1 - 2\alpha_p)A_T \tag{23}$$

from which k_p, A_p, and W can be determined. It should be noted that in the above derivations, the spin–lattice relaxation rate W is assumed to be isotropic. In this approximation, the following useful relations are derived:

$$(S_{11}^x - S_{21}^x) + (S_{11}^y - S_{21}^y) + (S_{11}^z - S_{21}^z) = 0$$
$$S_{12}^x + S_{12}^y + S_{12}^z = 0 \tag{24}$$

From the experimental results for S_{ij}^p, one can check on the validity of an isotropic spin–lattice relaxation assumption.

3.3. Expressions for the EPR Signal Intensities in a CW Experiment

In the CW experiment, we apply a periodic modulation of square light pulses with a frequency ω_L. The treatment is basically the same as was discussed in the SLP experiment, except that the **a** vector in equation (10) is replaced by a time-dependent vector.

$$\mathbf{a}(t) = \tfrac{1}{2}a_0 + \sum_{n=1}^{\infty} \mathbf{a}_n \sin(n\omega_L t) \tag{25}$$

Considering only the first harmonic component, the expression for the master equation is

$$\frac{d}{dt}\mathbf{v} = \tfrac{1}{2}a_0(1 + \sin \omega_L t) - \rho\mathbf{v} \tag{26}$$

After solving for the components of **v** and recalling that the observed EPR lines are measured via a second phase-sensitive detector, the signal intensities T_{1p} for the (1–2) transition and T_{2p} for the (2–3) transition after integration become

$$T_{1p} = S_{11}^p \frac{\cos(\phi_- - \psi)}{4(\omega_L^2 + \lambda_-^2)^{1/2}} + S_{12}^p \frac{\cos(\phi_+ - \psi)}{4(\omega_L^2 + \lambda_+^2)^{1/2}}$$
$$T_{2p} = S_{21}^p \frac{\cos(\phi_- - \psi)}{4(\omega_L^2 + \lambda_-^2)^{1/2}} + S_{22}^p \frac{\cos(\phi_+ - \psi)}{4(\omega_L^2 + \lambda_+^2)^{1/2}} \tag{27}$$

where $\phi_\pm$ are the intrinsic phase angles of the fast and slow components obeying

$$\tan \phi_\pm = -\omega_L/\lambda_\pm \tag{28}$$

ψ is the detector's phase angle and S_{ij}^p are given in equations (17) and (18). The resulting line intensity and its detection should vary with respect to the slow and fast components and are a function of the phase angle ψ. Inspection of Figure 9 and equations (27) shows that by controlling the light modulation frequency ω_L to satisfy the conditions $\omega_L < \lambda_\pm$, one should observe both the slow and fast components. By changing the experimental conditions so that $\lambda_- < \omega_L < \lambda_+$, only the fast component should be observed.

3.4. Analysis of the Transient Kinetics in a Harmonic Modulation CW Experiment

The evaluation of the intermediate lifetimes from the frequency dependence of the amplitude and phase shift was first introduced experimentally by Gaviola[23] in 1927 and later by Baily and Rolefson.[23a] This method was extended by Günthard and co-workers[24,25] applying it in EPR experiments to measure the lifetimes of paramagnetic intermediates that are decaying by following a first-order rate law.

Recently, Paul[9] has extended this work to analyze the kinetics of free radicals, which are generated in solutions in an OPEPR experiment.† In his treatment, the radicals recombination process may follow either a first-, or a second-, or a mixed-order decay rate law.

$$S \xrightarrow{\text{uv}} 2R \cdot \begin{array}{c} \xrightarrow{k_1} P_1 \\ \xrightarrow{k_2} P_2 \end{array} \qquad (29)$$

where S is a substrate; k_1 and k_2 are the first- and second-order rate constants, respectively; and P_1 and P_2 are the corresponding products.

Applying a pure sine wave modulation excitation, $I = \frac{1}{2}I_0(1 - \cos\omega_L t)$, the rate equation for $[R\cdot]$ is

$$\frac{d}{dt}[R\cdot] = \frac{1}{2}I_0(1 - \cos\omega_L t) - k_1[R\cdot] - 2k_2[R\cdot]^2 \qquad (30)$$

Equation (30) can readily be solved for two limiting cases:

1. $k_1 \neq 0$, $k_2 = 0$. The solution is the same, as was already given in equations (27) and (28) for the amplitude and phase, respectively. In that case, no higher harmonics than the fundamental are generated.

2. $k_1 = 0$, $k_2 \neq 0$. Equation (30) can be solved under the condition of slow light modulation frequency, i.e., $\omega_L \tau \ll 1$ (τ is the radical lifetime). For

---

† A similar approach applied for gas-phase kinetics has been proposed by E. H. Hunziker, *I.B.M. J. Res. Develop.* **15**, 10 (1971).

such conditions, the radical concentration follows the light excitation, so that $d[R \cdot]/dt \approx 0$, yielding

$$[R \cdot] = \tfrac{1}{2}\left[\frac{I_0}{k_2}(1 - \cos \omega_L t)\right]^{1/2} \tag{31}$$

For that case, higher harmonics are generated in addition to the fundamental. A Fourier expansion of equation (31) yields the coefficients and the phase angles for the free term and for the fundamental and first harmonic terms:

$$A_0 = \frac{2}{\pi}\left(\frac{I_0}{2k_2}\right)^{1/2}, \qquad A_1 = \tfrac{2}{3}A_0, \qquad A_2 = \tfrac{2}{15}A_0$$

$$\phi_1 = \phi_2 = 0 \tag{32}$$

Thus, both the fundamental $\partial\chi_1''(H)/\partial H$ and the first harmonic $\partial\chi_2''(H)/\partial H$ should be phase-sensitive detected at ω_L and $2\omega_L$ with an amplitude ratio of $5:1$, respectively.

For $k_1 \neq 0$, $k_2 \neq 0$ equation (30) cannot be solved analytically. However, there exists a linear approximation under the condition $\omega_L \gg 1$ for which

$$A_0 = \frac{1}{4k_2}[(k_1^2 + 4k_2 I_0)^{1/2} - k_1]$$

$$A_1 = \frac{I_0}{2}(\omega_L^2 + k_1^2 + 4k_2 I_0)^{-1/2} \tag{33}$$

$$\tan \phi_1 = \frac{\omega_L}{(k_1^2 + 4k_2 I_0)^{1/2}}$$

To check upon the validity of that approximation [equations (33)], one should compare the coefficient A_1 and phase ϕ_1 to the corresponding exact coefficient A_1^{ext} and phase ϕ_1^{ext}, which are obtained by solving numerically the exact rate equation (30).[9] Figure 10 shows a plot of A_1^{ext}/A_1 and ϕ_1^{ext}/ϕ_1 versus $\omega_L \tau_2$ for the conditions where $k_1 = 0$ (notice that the maximum deviation is expected when no first-order contribution exists). It is evident from Figure 10 that ϕ_1 is almost unaffected over the entire range of $\omega_L \tau_2$ (maximum deviation at $\omega_L \tau_2 \approx 1$). Regarding A_1, it is almost unaffected over the range $\omega_L \tau_2 \gtrsim 1$, and the maximum deviation of 20% is at $\omega_L \tau_2$ approaching zero.

In practice, one should plot the amplitude of the fundamental versus ω_L [equation (33)], where a straight line should be obtained. From the intercept and slope, the apparent lifetime is evaluated, $\tau_L^{-2} = k_1^2 + 4k_2 I_0$. To separate the first- and second-order lifetimes, similar experiments at different intensities of light excitation should give separately k_1 and $k_2 I_0$. The absolute

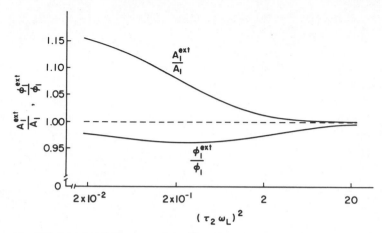

Figure 10. A_1^{ext}/A_1 and $\phi_1^{\text{ext}}/\phi_1$ versus $(\omega_L \tau_2)^2$. A_1^{ext} is the exact value for the fundamental signal amplitude, and A_1 is the corresponding value using the linear approximation. Similar definitions are for the phase angles ϕ_1^{ext} and ϕ_1; ω_L is the light modulation frequency and τ_2 is the chemical lifetime. These curves were plotted for the case $k_1 = 0$ and $k_2 \neq 0$. (Reproduced with permission from Paul.[9])

radical concentration, obtained using a standard calibration of the EPR signal intensity, permits one to calculate the second-order rate constant k_2. An alternative method, which is less sensitive, is to compare the amplitudes of the fundamental and first harmonic.[9] The same results should be obtained by measuring the frequency shift versus the phase shift. It is important to note that the derivations above are meaningful under the condition where $\omega_L \ll 1/T_1$, implying that the chemical lifetime is longer than SLR times.

4. *Examples and Applications*

In the subsequent paragraphs, we shall discuss some specific examples on which the OPEPR method as described above has been applied.

4.1. *Porphyrins*

Porphyrins have been subjected to extensive theoretical and experimental studies employing optical and magnetic resonance methods.[26,27] As the photoexcited triplet state of free-base porphyrin is apparently nonradiative, it is of interest to apply the OPEPR method to molecules of that type. Free-base and metal-substituted porphyrins are known to dissolve in *n*-octane, known as Shpolski'i-type solvent,[28] and form polycrystalline matrices on quick sample freezing. In fact, it is possible on very slow cooling to grow mixed single crystals, as was demonstrated first by van der Waals and co-workers.[29]

Let us consider a glass matrix of randomly oriented triplets of tetra-phenyl porphyrin (H_2TPP) as a typical example.[20] First harmonic representations of triplet EPR spectra taken at 100°K are shown in Figure 11. These spectra were taken under different light modulation frequencies and

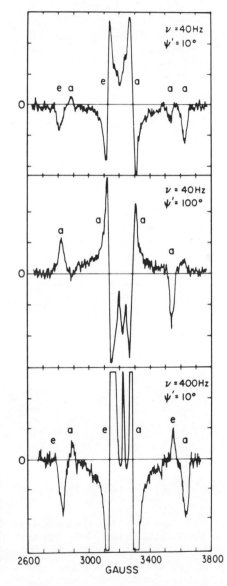

Figure 11. $\partial\chi_1''(H)/\partial H$ versus H in a CW experiment performed on H_2TPP dissolved in n-octane at 100°K. The light modulation frequency ν and the phase angle ψ' are indicated on each trace, e and a indicate emission and absorption lines, respectively. The experimental setup is as described in Figure 2. (Reproduced with permission from Levanon and Vega.[20])

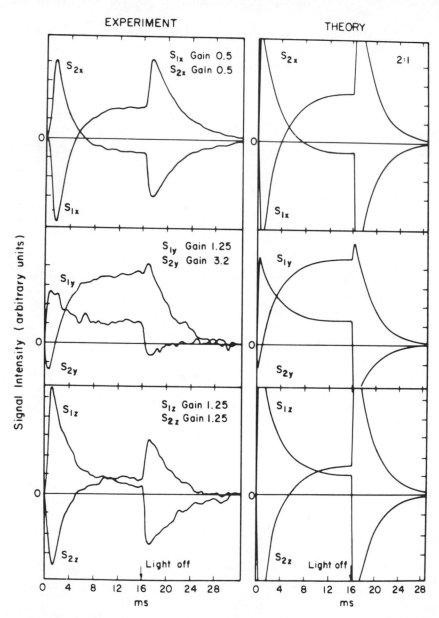

Figure 12. Experimental and calculated rise and decay curves of the EPR signal intensity in an SLP experiment of H_2TPP. The experiment is the same as in Figure 11. The light pulse duration is 15.8 ms. (Reproduced with permission from Levanon and Vega.[20])

different instrumental phase angles.† The anomalous line intensities and their directions can be well interpreted in terms of equations (27). In this experiment, two light modulation frequencies, 40 and 400 Hz, were used, and the phase detector was set at two different phase angles ψ', 10° and 100°. One may notice from the spectra taken at 40 Hz that for different phase angles the relative intensities are different, and the polarization direction (e for emission and a for absorption) for some lines changes sign; e.g., an absorption changes into emission (compare also to a typical absorption spectrum of a triplet, shown in Figure 7). At 400 Hz, the coefficients of $\cos(\phi_- - \psi)$ in equation (27) become small and the measured intensities are proportional to $\pm S^p_{12} \lambda_+ (\lambda_+^2 + \omega_L^2)^{-1/2} \cos(\phi_+ - \psi)$.

The coefficients S^p_{ij} can be extracted either from an SLP or from a CW experiment. We adopt the former, as it has been worked out in the literature in more detail. The experimental signal intensities for H_2TPP during and after a light pulse are shown on the left-hand side of Figure 12. The gain and the canonical orientation at which each of these curves was recorded is indicated on each trace; the duration of the light pulse is labeled by an arrow at the bottom of the figure. The coefficients S^p_{ij} are calculated by extrapolating each of the signal intensities to its initial value at $t = 0$. By combining this result with the steady-state value, the coefficient of the fast component is determined (for illustration, cf. Figure 9). Also, the values for λ_- are evaluated directly from the experimental curves. To complete the analysis, we must know λ_+ [assuming that equations (24) are fulfilled]. It is impossible to obtain λ_+ directly, since it lies within the time constant of the EPR spectrometer ($\sim 100 \ \mu sec$). However, with the restrictions on κ_p and α_p in equations (13) and a few iterations using equation (18), it is possible to determine λ_+ within a narrow range of allowed values. The right-hand side of Figure 12 shows the calculated curves using the coefficients S^p_{ij} and λ_+ that have been determined from the kinetic analysis. Finally, inserting the obtained values for S^p_{ij} and λ_+ in equations (20)–(23), we obtain k_x, k_y, k_z, the ratio between the population rates A_x, A_y, A_z, and the isotropic SLR rate W. Typical values of k_p and W lie in the range of $10^3 \ sec^{-1}$ and 10^4–$10^5 \ sec^{-1}$, respectively. The absolute values for A_p can be derived after performing quantum yield measurements.

In the CW experiment where the sample is subjected to a periodic light modulation of frequency ω_L, the EPR signal intensities should be analyzed in terms of equations (27). The solid lines in Figure 13 represent the calculated relative intensities at each canonical orientation as a function of the phase angle ψ. In calculating these curves, the parameters S^p_{ij} and λ_+ derived in the

---

† ψ' is the actual reading on the phase-sensitive detector. For a particular experiment the difference $\psi - \psi'$ should be constant. This difference in phases is due to an additional instrument phase angle.

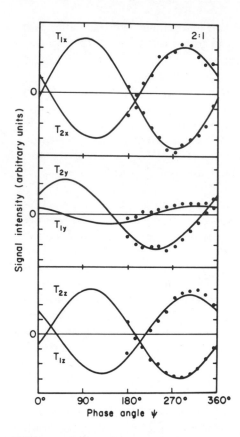

Figure 13. Calculated relative intensities T_{ip} ($i = 1, 2; p = x, y, z$) for each canonical orientation versus phase angle ψ using equations (27) for H_2TPP. The experiment is the same as in Figure 11. The solid circles are the normalized experimental EPR intensities $\partial\chi_1''(H)/\partial H$ in a CW experiment.

SLP experiment were used. The solid circles in Figure 13 are the normalized experimental EPR line intensities measured as a function of the phase angle ψ.

Similar experiments have been performed on a variety of free-base porphyrins,[30] and recently also on tetraphenyl chlorin (TPC) and magnesium tetraphenyl porphyrin (MgTPP).[31] In all compounds except MgTPP, the effect of ESP is remarkable at rather high temperatures. From the transient spectra and kinetics, the dynamic parameters could be extracted. In general, the same ESP patterns were observed in other solvents used. For TPC, for example, a solvent dependence of the rate parameters has been performed.[32] It was found that the triplet decay rate constant K_T does not show a significant dependence on changing the solvent polarity. This implies that the nonradiative decay rates, which are the predominant ones in free-base porphyrins, are not affected on changing the solvent polarity (*n*-octane to ethanol) and temperature (4.2°K to 100°K). Independent measurements on H_2TPP employing the TADMR method[32] give essentially very similar results to those reported by OPEPR.[20]

It is noteworthy that the active spin component of the porphyrins described above lies in the molecular plane, i.e., $k_x \gtrsim k_y > k_z$ and $A_x \gtrsim A_y \gtrsim A_z$. The same trend in the rate parameters has also been reported by van Dorp et al.[33] for free-base porphyrin (H_2P), which was studied at helium temperatures by microwave-induced delayed fluorescence (MIDF) technique. These authors interpreted the predominant selectivity in populating and depopulating the in-plane spin components in terms of the theory of radiationless transitions applied for planar aromatic hydrocarbons.[34] The rate constants for population (and depopulation) were shown to be proportional to the product of the Franck–Condon factor and the first derivatives of the matrix elements of the spin–orbit coupling operator with respect to the nuclear displacement coordinate. Utilizing wave functions for the photoexcited singlet S_i and triplet T_1, which are expressed in a Herzberg–Teller expansion about the equilibrium position, they showed qualitatively that the first-order terms in that expansion contribute to k_x and k_y as well as to A_x and A_y, whereas the second-order terms in that expansion contribute to k_z and A_z. Treatments employing symmetry considerations in conjunction with spin–orbit mixing have been worked out to interpret the selectivity in the intersystem crossing processes between the singlet and triplet manifolds in a considerable number of aromatic and N-heterocyclic molecules,[3,34–36] and to a lesser extent in nonplanar molecules.[10]

4.2. Chlorophylls in Vitro and in Vivo

The photosynthetic system serves as a natural model for demonstrating the efficient conversion of light energy into electromotive force for chemical energy consumption. Numerous publications on the photosynthetic process employing a variety of experimental and theoretical approaches have been reported in the literature.[37–42]

The possibility of triplet-state participation in photosynthesis[37] intrigued many investigators to search for the triplet state and its dynamics in chlorophylls in vitro and also in the photosynthetic apparatus constituents.[43–45] The successful observations of transient EPR triplet spectra in OPEPR experiments performed on chlorophylls in vitro and in vivo indicated unambiguously the existence of selective routes for population and depopulation of the triplet manifold. As was first pointed out by Norris and co-workers[46] and recently by Clarke and co-workers,[47] the small variation in the ZFS parameters on changing the photosynthetic constituents does not serve as a conclusive evidence for distinguishing between the various components involved in the photosynthetic process.

ESP patterns in the EPR triplet spectra were found to be dependent upon the aggregation state of the chlorophylls, as well as on the experimental conditions, i.e., whether the experiment is being performed in an in vitro

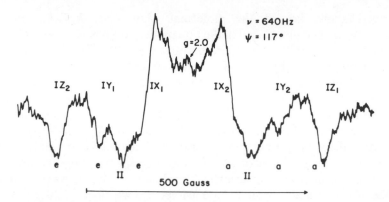

Figure 14. $\partial\chi_1''(H)/\partial H$ versus H in a CW experiment performed on *Chlamydomonas reinhardi* whole cells at 35°K. The labels I and II indicate two different spectra. In the present discussion, we interpret only spectrum I. (Reproduced with permission from Nissani *et al.*[31])

or *in vivo* system.[46] It is evident that a detailed kinetic analysis is essential for establishing the details of the polarization.

Such an analysis has been performed recently[31] on the photoexcited triplet state of chloroplasts in whole cells of *Chlamydomonas reinhardi* using the OPEPR method. The triplet spectrum at 35°K at light modulation frequency of 640 Hz is shown in Figure 14. This anomalous spectrum, reflected by the mixed enhanced absorption and emission line intensities, results from ESP in the triplet spin sublevels. The EPR line intensities (labeled I) are assigned to a nonaxial ($|E| \neq 0$) triplet spectrum with ZFS parameters very close to those found for spinach chloroplasts.[48] Nevertheless, this spectrum may still be a superposition of two spectra attributed to chlorophyll *a* and chlorophyll *b* species, which exhibit almost identical ZFS parameters.[44,47]

The experimental kinetic curves in an SLP experiment are shown on the left-hand side of Figure 15. Notice that poor S/N at the canonical orientation S_{1y} prevented the recording of this line intensity versus time. In the analysis,[31] the slow components (during and after cessation of the light pulse) could be resolved into two first-order decay curves, which were attributed to chlorophyll *a* and chlorophyll *b*. A complete analysis of the transient kinetics resulted in the dynamic parameters for both chlorophylls, namely k_p^a, k_p^b; A_p^a, A_p^b; W^a, W^b. The results were found to be in good agreement with those found separately for chlorophyll *a* and chlorophyll *b* using zero-field and OPEPR experiments.[49–51] In regard to the active spin state, it is found that for both chlorophylls it lies in the molecular plane. However, no change in the sign of polarization is observed when the external field is directed along the active state. In that respect, it contradicts the unique polarization observations that were found in some *in vivo* bacterial

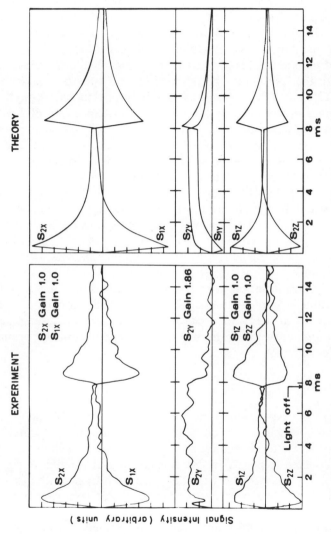

Figure 15. Experimental and calculated rise and decay curves of the EPR signal intensity in an SLP experiment of species I of Figure 14 at 35°K. The calculated curves were constructed by inserting the separate parameters for chlorophyll *a* and chlorophyll *b* into the corresponding equations. (Reproduced with permission from van Dorp *et al.*[33])

systems exhibiting only one photosystem.[48] In those experiments, a change in sign of the polarization was observed when the external field is directed along the active axis and could be interpreted in terms of the "special pair" model proposed by Norris *et al.*[40]

Apparently, chloroplasts that exhibit two active photosystems show triplet state dynamics that are due to oligomers present in antenna chlorophyll rather than the reactive center. In fact, this has been proposed previously on a qualitative basis by Uphaus and co-workers.[48]

One should inquire what is the part in the chlorophyll molecule that is "responsible" for the ESP in the photoexcited triplet. Apparently, the combination of MgTPP and TPC may serve as a naive model for chlorophyll.[30] It was found, quite surprisingly, that the transient EPR spectrum of TPC is very close to that of chlorophylls *a* and *b in vitro*. MgTPP, on the other hand, does not show any apparent ESP in its photoexcited triplet, and its lifetime is substantially longer than in TPC. Evidently, the dynamics of the photoexcited triplet state in chlorophyll are mainly governed by the chlorin macrocycle. This hypothesis is supported by the study of Clarke and co-workers,[47] where they compare the triplet dynamics of chlorophylls *a* and *b* to those in which zinc cation was incorporated instead of the magnesium cation. They found that in the modified compounds, the out-of-plane spin component is the most active, from which they conclude that magnesium has a minor role in the triplet dynamics of chlorophylls.

Undoubtedly, significant progress has been undertaken in regard to the triplet participation in photosynthesis. However, it still remains an open question to what extent OPEPR or ODMR experiments in condensed phases may provide unique answers to the unknown mechanism of the primary process in photosynthesis.

4.3. *Triplet–Triplet Absorption*

Inspection of the experimental kinetic curves in Figure 14 shows that λ_- (the slow-component rate constant) during the light pulse is somewhat greater than its value after termination of the light pulse. Such a behavior has also been observed in porphyrin systems and was attributed to triplet–triplet absorption.[30] Since triplet–triplet absorption provides an extra path for depopulation,[52,53] we expect that the time constant associated with the formation of the triplet will differ from its decay time constant upon cessation of the excitation. This effect has been demonstrated first in an OPEPR experiment performed on phenanthrene-d_{10} doped in benzophenone host.[54] When phenanthrene-d_{10} at 77°K is excited by ultraviolet light, the magnetic triplet state is formed with a rise time of 10 sec. When the ultraviolet source is turned off, the observed decay time is 9.5 sec. When the ultraviolet source is turned off but the sample is illuminated by an intense source that contains all wavelengths above 475 nm, the observed

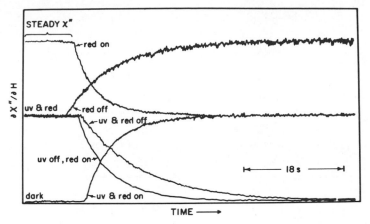

Figure 16. A typical set of SLP experimental measurements at 77°K. The red filter used in that experiment was transparent above 550 nm.

decay time is 4.9 sec. The same rise time is obtained when the sample is illuminated simultaneously by both light sources.

In this OPEPR experiment, two independent light sources, both focused on the same portion of the sample, were used—an ultraviolet source with a filter that transmits up to 380 nm and a xenon lamp with appropriate filters serving as a red source mounted perpendicular to the ultraviolet source. A quartz window mounted at 45° in the ultraviolet light path was used to transmit the exciting light and the red light simultaneously. Typical rise and decay curves, where the red filter used in that experiment was transparent above 550 nm, are shown in Figure 16.

In benzophenone at 77°K, triplet–triplet annihilation[55] or triplet-state energy transfer from host to guest are negligible.[56] The rate equation for the direct photoexcitation into the triplet state can be written as[54]

$$\frac{d}{dt}[N_T] = A_T([N_S] - [N_T]) - (k_{T-T} + K_T)[N_T] \qquad (34)$$

A_T corresponds to the intersystem crossing rate constant, k_{T-T} is the rate constant for the triplet–triplet transition,† and K_T is the depopulation rate constant from the triplet to the ground singlet. Both A_T and k_{T-T} include the light intensity for the $S_0 \rightarrow S_i$ and $T_1 \rightarrow T_i$ transitions, respectively. Under steady-state conditions, the triplet concentration $[N_T]_{SS}$ is given by

$$[N_T]_{SS} = \frac{A_T[N_S]}{A_T + k_{T-T} + K_T} = \frac{A_T[N_S]}{k_{T-T} + K_T} \qquad (35)$$

† In the original paper,[54] this transition was ascribed to an apparent excimer transition that has not been verified since.

The right-hand side of equation (35) is obtained under the condition that $A_T \ll k_{T-T} + K_T = \lambda_-$. Under these conditions, integration of equation (34) gives the rise and decay of the magnetic triplets. Their time constants are

$$\tau_{\text{rise}} = \tau_{\text{decay}} = 1/\lambda_- \tag{36}$$

The observed experimental results can be explained through equations (35) and (36). When the sample is subjected to excitation by unfiltered light, both channels for depopulation of the triplet k_{T-T} and K_T exist. Upon cessation of the light pulse, $k_{T-T} = 0$, and the decay is governed by K_T only. Moreover, the ratio $K_T/(k_{T-T} + K_T) \equiv \Delta_\tau$ is expected to be the same as the ratio of the EPR signal intensities with and without filtration. Plotting Δ_τ versus Δ_χ'', where $\Delta_\chi = x_{\text{red off}}''/x_{\text{red on}}''$ should give a straight line with a slope of 1, which was found experimentally.[54] This type of experiment is to some extent analogous to standard flash or laser photolysis experiments.[57] However, it is not sensitive in those cases where $S_0 \rightarrow S_i$ transitions overlap with $T_1 \rightarrow T_i$ transitions or when the EPR signal intensities are poor, as is the case in some porphyrins and chlorophylls.

To eliminate triplet–triplet contributions in the kinetic analysis for the porphyrins and chlorophylls, one should calculate the kinetic parameters on cessation of the light pulse. This is being performed by calculating first κ_p, α_p, and W from the signal intensities during the light pulse, followed by the analysis of the decay curves on cessation of the light pulse.[31]

4.4. *Photoreduction of Porphyrins to Chlorins by Tertiary Amines in the Visible Spectral Range*

Substances having π, π^* configuration are normally inefficient in photoreduction processes involving hydrocarbons or alcohols. On the other hand, these substances are photoreduced efficiently by tertiary amines.[58,59] This has been demonstrated for π, π^* singlets[60,61] and π, π^* triplets.[59] The absorption spectra of porphyrins in the range of 350–700 nm are believed to be due to π, π^* transitions[62]; thus, it is expected that porphyrins will be photoreduced by tertiary amines in the visible range.

Since the course of such a reaction may involve electron transfer forming the porphyrin free radical, a combined optical and OPEPR study has been performed on H_2TPP.[63] Spectroscopic data indicate that chlorin is generated upon irradiation ($\lambda_{\text{excitation}} > 550$ nm) of solutions of H_2TPP with various tertiary amines. When such solutions are pumped into the EPR cavity and irradiated *in situ*, a single EPR line assigned to the porphyrin free radical ($g = 2.0024 \pm 0.002$) is observed. The linewidth between the inflexion points is 5.2 ± 0.2 G, which is very close to other free-base porphyrin radicals.[64] The observed time-dependent EPR signal intensity when an SLP experiment

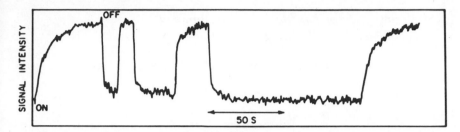

Figure 17. Rise and decay of EPR signal intensity as a function of the dark period in a successive SLP experiment. The curve has been recorded for the same sample solution—H_2TPP $(8 \times 10^{-4}\ M)$, triethylamine $(0.1\ M)$ in benzene at room temperature. (Reproduced with permission from Harel *et al.*[63])

is performed on such a solution is shown in Figure 17. The curve shown in Figure 17 is obtained in the following way: When a light pulse is turned on, a slow rise of the EPR signal is observed. Upon termination of the light pulse, the signal disappears with a second-order decay rate constant ($k_5 = 1.7 \times 10^7\ M^{-1}\ sec^{-1}$). When a second light pulse is turned on, a fast rise followed by a slow rise of the EPR signal is observed. The relative amplitude of the fast and slow components of the signal intensities depend strongly on the length of the dark period. When the sample is subjected to a light pulse after a long dark period, the EPR signal intensity is building up slowly, having a rise time identical to that observed in a fresh solution. These OPEPR observations together with flash photolysis data[63] were explained by slightly modifying the generally accepted set of reactions for photoreduction[64]:

$$PH_2 \xrightarrow{Ik_1} PH_2^*, \quad PH_2^* \xrightarrow{k_2} PH_2,$$

$$PH_2^* + DH \xrightarrow{k_3} [PH_2^- - DH^+] \quad [PH_2^- - DH^+] \xrightarrow{k_4} PH_3 \cdot + D\cdot,$$

$$2PH_3 \underset{Ik_6}{\overset{k_5}{\rightleftharpoons}} [PH_3 PH_3] \xrightarrow{k_7} PH_4 + PH_2 \qquad (37)$$

PH_2, PH_2^*, and PH_4 are the free-base porphyrin, the corresponding photoexcited state, and chlorin, respectively; $[PH_2^- - DH^+]$ is a charge transfer complex that dissociates into porphyrin and amine radicals, $PH_3 \cdot$ and $D\cdot$, respectively; and $[PH_3 PH_3]$ is a radical dimer that disproportionates in the dark period to chlorin and porphyrin. Irradiation of the sample after a dark period, which is short compared with $1/k_7$, results in a fast dissociation of the dimer into the free radicals. This is reflected by the fast rise of the EPR signal intensity. Due to fast recombination, the amine radicals escape EPR detection.[64,65]

4.5. Determination of Kinetic Rate Constants of Recombination of Radicals in Solutions

OPEPR CW experiments have been applied by Paul[9] to investigate the kinetics of two radicals.

4.5.1. Hydroxymethyl Radicals

When methanol solutions are photolyzed by a sine modulation excitation in the presence of di-*tert*-butyl peroxide, hydroxymethyl radicals are formed via the reaction scheme

$$(CH_3)_3C-O-O-C(H_3C)_3 \xrightarrow{\;h\nu\;} 2(CH_3)_3C-O\cdot \qquad (38)$$

$$(CH_3)_3C-O\cdot + CH_3OH \xrightarrow{\;fast\;} (CH_3)_3C-OH + \dot{C}H_2OH \qquad (39)$$

The EPR signal intensity was found to be linear with the square root of the exciting light intensity, indicating that $k_1 = 0$. Such a proportionality implies also that the spin system is in thermal equilibrium, i.e., no CIDEP is observable. Thus, the kinetics of the radicals are mainly governed by a second-order rate law. In that case, equations (33) can be applied by plotting the fundamental EPR amplitude A_1^{-2} versus ω_L^2. This is shown in Figure 18, where such a plot has been drawn for the central lines of the spectrum. The experimental results (open circles) are compared to A_1^{ext} (solid line) and also to the linear approximation [equation (31)], which show a quite good agreement. From the plot, the radical lifetime could be evaluated ($\tau_2 = 3.10 \times 10^{-4}$ sec at 193°K), which, together with the radical concentration, gave $k_2(=0.42 \times 10^9\ M^{-1}\ \text{sec}^{-1}$ at 200°K) and the apparent activation energy $E_K(=10.96\ \text{kJ}\ M^{-1})$. Analysis of the results in terms of the diffusion coefficient and also by applying the Smoluchowski equation leads one to conclude that the recombination rate constant k_2 of hydroxymethyl radicals is a diffusion-controlled reaction.

4.5.2. α-Tetrahydrofuryl Radicals

When methanol is replaced by tetrahydrofuran in reaction (39), the α-tetrahydrofuryl radicals $CH_2CH_2CH_2\dot{C}HO$ are produced. Contrary to the previous example, the EPR spectrum exhibits an anomalous line intensity, indicating that the spin system is polarized, i.e., a CIDEP effect. From the experiments, it is found that A_1^{-2} is linear with ω_L^2. This observation indicates that no initial polarization exists in that case and the polarization observed is generated via the collision of two radicals. This is expected, since the deactivation of the tertiary butoxy radical proceeds via the excited singlets and also because of the fast SLR times in the butoxy radicals.

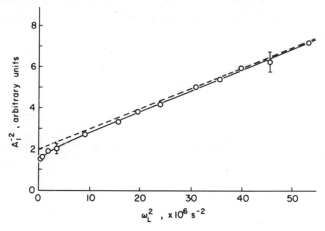

Figure 18. The inverse square of the Fourier component at the fundamental frequency $\partial\chi_1''(H)/\partial H$ versus the light modulation frequency squared ω_L^2 in a harmonic modulation CW experiment of $\dot{C}H_2OH$ radical at 193°K. The open circles are the experimental points, the solid line is the ω_L^2 dependence of A_1^{ext}, and the dashed line is the ω_L^2 dependence of A_1. (Reproduced with permission from Paul.[9])

Performing the same analysis as described above shows that the recombination process is almost second order, yielding the chemical lifetime $\tau_2(=1.1 \times 10^{-3}$ sec) and $k_2(=3.5 \times 10^8 \ M^{-1} \ \text{sec}^{-1})$ at 180°K; SLR time was found to be 1.7×10^{-6} sec. These calculated values together with the experimental observation of the spin polarization effect enables one to evaluate the enhancement parameters (this will not be discussed here).

4.6. *Photoionization of Alkali Metals in Solutions in the Presence of Crown Ethers*

The introduction by Dye and co-workers of the organic macrocyclic molecules such as crown ethers $(CR)^{[66]}$ as complexing agents of alkali metal cations opened many interesting possibilities in the study of blue solutions of ethers and amines.[67–69] By monitoring the CR concentrations, it is possible to control the elementary process

$$2M_s \rightleftharpoons M^+ + M^-$$
$$M^+ + CR \rightleftharpoons MCR^+ \tag{40}$$

Solutions of alkali metals in amines have been extensively investigated by optical and magnetic resonance methods.[69–71] Regarding ethers as solvents, these have mainly been subjected to optical[68,72] and, to a lesser extent, magnetic resonance experiments.[73] The main reason for this is the relatively low solubility of alkali metals in ethers, resulting in too low a concentration of paramagnetic species to be detected.

Blue solutions of amines and ethers in the presence of alkali metals were shown by Glarum and Marshall[73] to be due to the diamagnetic species M^-. They also showed, using optical and EPR measurements, that the long-wavelength absorption bands at approximation 600 nm, 850 nm, 900 nm, and 1100 nm can be assigned to Na^-, K^-, Rb^-, and Cs^- anions, respectively.

It is well accepted that upon illumination of a blue solution at wavelengths corresponding to the absorption band of the anion M^-, photoelectrons are produced and are easily EPR detected according to

$$M^- \rightleftharpoons M\cdot + e \tag{41}$$

On the other hand, the monomer radical is scarcely detectable by EPR, particularly in ether solutions. However, in the presence of small amounts of CR in potassium–tetrahydrofuran (THF) solutions, the EPR signal of the monomer is invariably observed both in the dark and in the light.[16,73] Since the monomer radicals are detected over a very narrow range of CR $(0 \leq [CR] \leq 10^{-4} M)$, we shall describe here only the behavior of the photoelectrons.[74]

The dominant equilibria in nonpolar solvents are[70,74]

$$2M_s \underset{k_2}{\overset{k_1}{\rightleftharpoons}} M^+ + M^-, \qquad k_1/k_2 = K_1 \tag{42}$$

$$M^- \underset{k_4}{\overset{k_3}{\rightleftharpoons}} M\cdot + e, \qquad k_3/k_4 = K_2 \tag{43}$$

$$M\cdot \underset{k_6}{\overset{k_5}{\rightleftharpoons}} M^+ + e, \qquad k_5/k_6 = K_3 \tag{44}$$

$$M^+ + CR \underset{k_8}{\overset{k_7}{\rightleftharpoons}} MCR^+, \qquad k_7/k_8 = K_4 \tag{45}$$

The overall reaction may be summed up as

$$2M_s + CR \rightleftharpoons 2e + M^+ + MCR^+ \rightleftharpoons M^- + MCR^+ \tag{46}$$

Thus, excess of CR will shift equation (46) to the right. This is reflected experimentally by increasing the solvated electron concentration and the intensification of the blue color of the solution.

Figure 19 shows the temperature dependence of the EPR signal intensity of the solvated electron at two CR concentrations. For the low CR concentration, the signal intensity in the dark is too small to be drawn to scale.

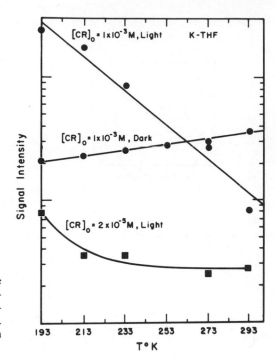

Figure 19. EPR signal intensity of the solvated electron at two CR concentrations in light and dark conditions as a function of the temperature. (Reproduced with permission from Friedenberg and Levanon.[74])

Inspection of Figure 19 shows that strong photo EPR signals are observed on decreasing the temperature and increasing the CR concentration. Employing an SLP experiment, one can follow the temporal behavior of the EPR signal intensity on varying the CR concentration and temperature. Typical curves are shown in Figure 20. Except for the high-temperature region, all the kinetic curves follow a pseudo-first-order recombination process. The effect of increasing the CR concentration on the observed first-order rate constant is analogous to that of decreasing the temperature.

To account for these observations, the following process is proposed:

$$e + KCR^+ \xrightarrow{k_9} e, KCR^+ \qquad (47)$$

in addition to

$$e + K^+ \xrightarrow{k_6} K\cdot \quad (\text{or } e, K^+) \qquad (48)$$

These two processes lead to a pseudo-first-order decay rate k^{obs} given by

$$k^{obs} = [M^+]_0 \frac{k_6 + k_9 k_4 [CR]}{1 + K_4 [CR]} \qquad (49)$$

It follows from equation (49) that when $[CR]_0 = 0$, it is possible to determine k_6, provided that $[M^+]_0$ is known; $[M^+]_0$ was estimated to be

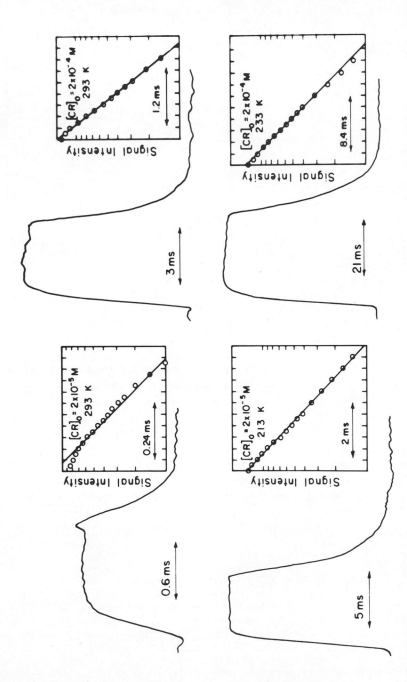

Figure 20. EPR rise and decay curves in an SLP experiment of the photoelectron at two CR concentrations at different temperatures. Inserted are first-order plots. (Reproduced with permission from Friedenberg and Levanon.[74])

5×10^{-6} M at room temperature,[74] which in turn gives $k_6 \approx 6 \times 10^8$ M^{-1} sec^{-1}, in agreement with previous reports.[73] The other limit of equation (49) enables one to estimate k_9. When $[CR]_0$ is in excess and assuming that $[CR]_0 \approx [M^+]_0$, k_9 is calculated to be $\sim 10 \times 10^6$ M^{-1} sec^{-1} at room temperature.

In some of these systems, the photoelectrons are found to be polarized. This effect was reported for the first time by Glarum and Marshall on rubidium–dimethoxyethane solutions and was explained by combination of spin–orbit and exchange interactions with an adiabatic dissociation.[6,73] This conspicuous polarization effect of photoelectrons has been neglected for several years. Recently, it has been reported in the observation of enormous spin polarization of photoelectrons in M–THF (M = Rb, Cs) solutions in the presence of CR.[16] Also, it has been demonstrated that potassium radicals and photoelectrons in K–THF solutions exhibit spin promising future prospects for OPEPR and probably OPNMR methods.[77,78] kinetic curve at low CR concentrations in Figure 20.)

5. Concluding Remarks

In this chapter, we have demonstrated the versatile experimental features of the OPEPR method and have presented some applications and typical examples in fields that are of interest to photochemists and photobiologists. We have not included in this chapter the relevant theories and applications of CIDEP in SLP or CW experiments,[5,8,11,75] nor have we included the applications of OPEPR to paramagnetic centers in crystals.[76]

We hope that we have shown in this survey that the OPEPR method is compatible with other methods for gaining a closer insight into the dynamic processes of transient spin systems. In particular, the recent achievements in increasing the sensitivity of the magnetic resonance detection indicate very promising future prospects for OPEPR and probably OPNMR methods.[77,78]

Note Added in Proof

Since this manuscript was submitted for publication the field of OPEPR has developed quite dramatically and, naturally, many of the subjects which were covered here have been much advanced in many laboratories.[79–84] In particular, the experimental approach has been developed to a level where OPEPR experiments are performed in a time scale of a few tens of nanoseconds.[85] This reduction in time scale has an important implication for the study of fast chemical, photochemical, and photophysical processes that involve paramagnetic transients, and complements, to a large extent, fast methods in optical spectroscopy.[86] Some other substantial progress was made in the application of OPEPR to problems related to primary events in photosynthesis. These experimental improvements have significantly contributed to the up-to-date conclusion that the photoexcited triplet state is not on the main pathway of primary bacterial photosynthesis.[87]

ACKNOWLEDGMENTS

I wish to thank my colleagues at the Radiation Research Laboratory at the University of Notre Dame for their very kind hospitality. I am indebted to Professor A. M. Trozzolo, Dr. P. Neta, and Dr. H. Paul for reading the manuscript and for helpful suggestions. The technical assistance of Dr. A. B. Ross of the Radiation Chemistry Data Center and of Mr. J. Lute for redrawing all of the figures in the manuscript, are highly acknowledged. Finally, I would like to express my deep appreciation to Professor S. I. Weissman and Professor Z. Luz for their inspiration and encouragement.

This work was supported in part by grants from the United States–Israel Binational Science Foundation (BSF), Jerusalem, Israel, and from the Israel Commission for Basic Research.

Abbreviations

CIDEP	Chemically induced dynamic electron polarization
CR	Crown
CW	Continuous wave
EOR	EPR optical resonance
ESP	Electron spin polarization
FMDR	Fluorescence microwave double resonance
H_2TPP	Tetraphenyl porphyrn
ISC	Intersystem crossing
MgTPP	Magnesium-tetraphenyl porphyrin
ODMR	Optical detection magnetic resonance
OPEPR	Optical perturbation electron paramagnetic resonance
OPNMR	Optical perturbation nuclear magnetic resonance
PMDR	Phosphorescence microwave double resonance
SLP	Single light pulse
SLR	Spin–lattice relaxation
TADMR	Triplet absorption detection magnetic resonance
TPC	Tetraphenylchlorin
ZFS	Zero-field splitting

References

1. J. Olmsted and M. A. El-Sayed, in *Creation and Detection of the Excited State* (W. R. Ware, ed.), Vol. 2, pp. 1–62, Marcel Dekker Inc., New York (1974).
2. M. A. El-Sayed, *Ann. Rev. Phys. Chem.* **26**, 235–258 (1975).
3. K. H. Hausser and H. C. Wolf, in *Advances in Magnetic Resonance* (J. S. Waugh, ed.), Vol. 8, pp. 85–121, Academic Press, New York (1976), and references therein.
4. B. Smaller, J. R. Remko, and E. C. Avery, *J. Chem. Phys.* **48**, 5174–5181 (1968).
5. P. W. Atkins and K. A. McLauchlan, in *Chemically Induced Magnetic Polarization* (A. R. Lepley and G. L. Closs, eds.), Chapter 2, pp. 42–93, Wiley-Interscience New York (1973).
6. N. C. Verma and R. W. Fessenden, *J. Chem. Phys.* **65**, 2139–2155 (1976).
7. H. Levanon and S. I. Weissman, *Israel J. Chem.* **10**, 1–5 (1972).

8. J. B. Pedersen, C. E. M. Hansen, H. Parbo, and L. T. Muus, *J. Chem. Phys.* **63**, 2398–2405 (1975).
9. H. Paul, *Chem. Phys.* **15**, 115–129 (1976).
10. T. S. Lin, *Chem. Phys. Lett.* **28**, 77–82 (1974).
11. J. H. Freed and J. B. Pedersen, in *Advances in Magnetic Resonance* (J. S. Waugh, ed.), Vol. 8, pp. 1–84, Academic Press, New York (1975).
12. S. S. Kim and S. I. Weissman, *J. Mag. Res.* **24**, 167–169 (1976).
13. J. R. Bolton and J. T. Warden, in *Creation and Detection of the Excited State* (W. R. Ware, ed.), Vol. 2, pp. 63–97, Marcel Dekker, Inc., New York (1974).
14. H. Levanon, *Chem. Phys. Lett.* **9**, 257–259 (1971).
15. H. Levanon and S. I. Weissman, *J. Amer. Chem. Soc.* **93**, 4309–4310 (1971).
16. A. Friedenberg and H. Levanon, *Chem. Phys. Lett.* **41**, 84–86 (1976).
17. H. Levanon, *Varian Instrument Application* **5**, 6–7 (1971).
18. M. Schwoerer and H. Sixl, *Z. Naturforsch.* **A24**, 952–967 (1969).
19. R. H. Clarke, *Chem. Phys. Lett.* **6**, 413–416 (1970).
20. H. Levanon and S. Vega, *J. Chem. Phys.* **41**, 2265–2274 (1975).
21. E. Wasserman, L. C. Snyder, and W. A. Yager, *J. Chem. Phys.* **41**, 1763–1772 (1964).
22. S. P. McGlynn, T. Azumi, and M. Kinoshita, *Molecular Spectroscopy of the Triplet State*, Prentice-Hall, Englewood Cliffs, New Jersey (1969).
23. Von E. Gaviola, *Z. Phys.* **42**, 853–861 (1927).
23a. E. A. Baily, Jr., and G. K. Rolefson, *J. Chem. Phys.* **21**, 1315–1322 (1953).
24. M. Forster, U. P. Fringeli, and Hs. H. Günthard, *Helv. Chim. Acta* **56**, 389–407 (1973).
25. K. Loth, M. Andrist, F. Grat, and Hs. H. Günthard, *Chem. Phys. Lett.* **29**, 163–168 (1974).
26. S. R. Langhoff, E. R. Davidson, M. Gouterman, W. R. Leenstra, and A. Kwiram, *J. Chem. Phys.* **62**, 169–176 (1975), and references therein.
27. J. Subramanian, in *Porphyrins and Metalloporphyrins* (K. M. Smith, ed.), Chapter 13, pp. 555–586, Elsevier Scientific Publishing Co., Amsterdam (1975).
28. E. V. Shpolskii, *Sov. Phys. Usp.* **3**, 372–389 (1960); *Sov. Phys. Usp.* **5**, 522–531 (1962); *Sov. Phys. Usp.* **6**, 411–427 (1963) (English translation).
29. I. Y. Chan, W. G. van Dorp, T. J. Schaafsma, and T. H. van der Waals, *Molec. Phys.* **22**, 741–751 (1971); *Molec. Phys.* **22**, 753–760 (1971).
30. A. Scherz, N. Orbach, and H. Levanon, *Israel J. Chem.* **12**, 1037–1048 (1974).
31. E. Nissani, A. Scherz, and H. Levanon, *Photochem. Photobiol.* **25**, 93–101 (1977).
32. R. H. Clarke and R. E. Connors, *J. Chem. Phys.* **62**, 1600–1601 (1975).
33. W. G. van Dorp, W. H. Schoemaker, M. Soma, and J. H. van der Waals, *Molec. Phys.* **30**, 1701–1721 (1975).
34. F. Metz, S. Friedrich, and G. Hohlneicher, *Chem. Phys. Lett.* **16**, 353–358 (1972).
35. J. H. van der Waals and M. S. de Groot, in *The Triplet State* (A. Zahlan, ed.) pp. 101–132, Cambridge Univ. Press, London (1967).
36. U. Eliav and H. Levanon, *Chem. Phys. Lett.* **36**, 377–381 (1975).
37. J. Franck and J. L. Rosenberg, *J. Theoret. Biol.* **36**, 377–381 (1964).
38. R. K. Clayton, *Ann. Rev. Biophys. Bioeng.* **2**, 131–156 (1973).
39. W. W. Parson and R. J. Codgell, *Biochim. Biophys. Acta* **416**, 105–149 (1975).
40. J. R. Norris, H. Scheer, and J. J. Katz, *Ann. N.Y. Acad. Sci.* **244**, 260–280 (1975), and references therein.
41. F. K. Fong and N. Winograd, *J. Amer. Chem. Soc.* **98**, 2287–2289 (1976).
42. R. H. Clarke, R. E. Connors, and H. A. Frank, *Biochem. Biophys. Res. Commun.* **71**, 671–675 (1976).
43. J. S. Leigh and P. L. Dutton, *Biochim. Biophys. Acta* **357**, 67–77 (1974).
44. M. C. Thurenauer, J. J. Katz, and J. R. Norris, *Proc. Nat. Acad. Sci. USA* **72**, 3270–3274 (1975).
45. R. H. Clarke, R. E. Connors, H. A. Frank, and J. C. Hock, *Chem. Phys. Lett.* **45**, 523–528 (1977).

46. J. R. Norris, R. A. Uphaus, and J. J. Katz, *Chem. Phys. Lett.* **31**, 157-161 (1975).
47. R. H. Clarke, R. E. Connors, T. J. Schaafsma, J. F. Kleibeuker, and R. J. Platenkamp, *J. Amer. Chem. Soc.* **98**, 3674-3677 (1976) and references therein.
48. R. A. Uphaus, J. R. Norris, and J. J. Katz, *Biochem. Biophys. Res. Commun.* **61**, 1057-1063 (1974).
49. R. H. Clarke and R. H. Hofeldt, *J. Chem. Phys.* **61**, 4582-4587 (1974).
50. H. Levanon and A. Scherz, *Chem. Phys. Lett.* **31**, 119-124 (1975).
51. J. F. Kleibeuker and T. J. Schaafsma, *Chem. Phys. Lett.* **29**, 116-122 (1974).
52. B. N. Srinivasen, M. Kineshita, J. W. Rabalais, and S. P. McGlynn, *J. Chem. Phys.* **48**, 1924-1931 (1968).
53. J. S. Brinen, *J. Chem. Phys.* **49**, 586-590 (1968).
54. H. Levanon and S. I. Weissman, *Chem. Phys. Lett.* **10**, 25-28 (1971).
55. M. Kinoshita, T. N. Misra, and S. P. McGlynn, *J. Chem. Phys.* **45**, 817-821 (1966).
56. E. T. Harrigen and N. Hirota, *J. Chem. Phys.* **49**, 2301-2313 (1968).
57. L. Pakkarinen and H. Linschitz, *J. Amer. Chem. Soc.* **82**, 2407-2411 (1960).
58. S. G. Cohen, A. Parola, and G. H. Parsons, Jr., *Chem. Rev.* **73**, 141-161 (1973).
59. J. C. Scaiano, *J. Photochem.* **2**, 81-118 (1973/4).
60. H. D. Burrows, *Photochem. Photobiol.* **19**, 241-243 (1974).
61. A. Matsuzaki, S. Nagakura, and K. Yoshihara, *Bull. Chem. Soc. Japan* **47**, 1152-1157 (1974).
62. M. Zerner, M. Gouterman, and H. Kobayashi, *Theoret. Chim. Acta* **6**, 363-400 (1966).
63. Y. Harel, J. Manassen, and H. Levanon, *Photochem. and Photobiol.* **23**, 337-341 (1976).
64. D. Mauzerall and G. Feher, *Biochim. Biophys. Acta* **79**, 430-432 (1964); D. Mauzerall and G. Feher, *Biochim. Biophys. Acta* **88**, 658-660 (1969).
65. P. W. Atkins, A. J. Dobbs, G. T. Evans, K. A. McLauchlan, and P. W. Percival, *Molec. Phys.* **27**, 769-777 (1974).
66. G. W. Gokel and H. D. Durst, *Synthesis* **1976**, 168-184 (1976), and references therein.
67. M. T. Lok, F. J. Tehan, and J. L. Dye, *J. Phys. Chem.* **76**, 2975 (1972).
68. J. L. Dye, D. W. Andrews, and S. E. Mathews, *J. Phys. Chem.* **79**, 3065-3070 (1975).
69. J. L. Dye, in *Electrons in Fluids* (J. Jortner and C. R. Kestner, eds.), p. 77, Springer-Verlag, West Berlin (1973).
70. R. Catterall and P. Edwards, *J. Phys. Chem.* **79**, 3010-3017 (1975).
71. A. Gaathon and M. Ottolenghi, *Israel J. Chem.* **8**, 165-180 (1970).
72. J. Eloranto and H. Linschitz, *J. Chem. Phys.* **38**, 2214-2219 (1970).
73. S. H. Glarum and J. H. Marshall, *J. Chem. Phys.* **52**, 5555-5565 (1970).
74. A. Friedenberg and H. Levanon, *J. Phys. Chem.* **81**, 766-771 (1977).
75. J. K. S. Wan, S. K. Wong, and D. A. Hutchinson, *Accounts Chem. Res.* **7**, 58-64 (1974), and references therein.
76. S. Geschwind, *Electron Paramagnetic Resonance*, Plenum Press, New York (1972).
77. S. G. Boxer and G. L. Closs, *J. Amer. Chem. Soc.* **97**, 3268-3270 (1975).
78. D. Stehlik in *Excited States* (E. C. Lim ed.), Vol. 3, pp. 204-300, Academic Press, New York (1977).
79. C. J. Winscon, *Z. Naturforsch.* **30**, 571-582 (1975).
80. U. Eliav and H. Levanon, *Chem. Phys. Lett.* **55**, 369-374 (1978).
81. W. Hägele, D. Schmid, and H. C. Wolf, *Z. Naturforsch.* **339**, 94-97 (1978).
82. S. S. Kim and S. I. Weissman, *Chem. Phys. Lett.* **58**, 326-328 (1978).
83. Y. Harel and J. Manassen, *J. Amer. Chem. Soc.* **100**, 6228-6234 (1978).
84. J. F. Kleibeuker, R. J. Platenkamp, and T. J. Schaafsma, *Chem. Phys.* **27**, 51-64 (1978).
85. A. D. Trifunac and J. R. Norris, *Chem. Phys. Lett.* **59**, 140-142 (1979).
86. A. D. Trifunac and M. C. Thurnauer, in *Time Domain Electron Spin Resonance Spectroscopy* (L. Kevan and R. N. Schwartz, eds.), Wiley-Interscience, New York (in press).
87. For a recent review on this subject see, e.g., H. Levanon and J. R. Norris, *Chem. Revs.* **3**, 185-198 (1978).

14

Electron–Nuclear–Nuclear TRIPLE Resonance of Radicals in Solutions

Klaus Möbius and Reinhard Biehl

1. Introduction

In molecular spectroscopy, the introduction of more than one resonant electromagnetic field is a well-established technique applied in the full spectral range between radiofrequency (rf) and ultraviolet.[1] The main motivation for extending single resonance to double or multiple resonance techniques can be twofold: (1) either one wants to increase the *sensitivity* of detection by "quantum transformation" from low-frequency absorbed quanta to high-frequency detected quanta, or (2) one wants to enhance the *resolution* of the spectra, i.e., reduce the number of spectral lines in a given frequency range by introducing specific coherent time-dependent interactions that average out unwanted static interactions in the Hamiltonian,[2] by imposing additional "selection rules,"[2] or by exploiting degeneracies of certain transitions.[3]

Naturally, the ultimate goal is to design a multiple resonance experiment that combines these two advantages. These considerations have probably found their widest realization and application in NMR. This field has been covered by excellent review articles.[2,4]

In a different spectroscopic field, a convincing example for case (1) is the microwave *optical* *double* *resonance* (MODOR) on barium oxide in the $X^1\Sigma$ ground state.[5] In this experiment, an argon-ion laser pumps an optical transition between the $X^1\Sigma$ ground state and the $A^1\Sigma$ excited state, thereby

Klaus Möbius and Reinhard Biehl ● Institut für Molekülphysik, Freie Universität Berlin, Berlin, Germany

producing intensive fluorescence. The microwave-induced rotational transition is then detected via an increase in the fluorescence intensity. The sensitivity of detection in this case was enhanced by two orders of magnitude as compared to the direct microwave absorption.

In the context of this chapter, the adequate example for case (2) is electron–nuclear double resonance (ENDOR). In this method, the resolution enhancement (compared to ESR) is due to the high degree of transition frequency degeneracies for equivalent nuclei in the high-field approximation. If one defines[6] a spectral line density Ω

$$\Omega = \frac{\text{total number of lines}}{\text{frequency extension of the spectrum}}$$

for ESR and ENDOR, one obtains

$$\Omega_{\text{ESR}} = \frac{\prod_{k=1}^{K}(2n_k I_k + 1)}{2\sum_{k=1}^{K} a_k n_k I_k} \tag{1}$$

and

$$\Omega_{\text{ENDOR}} = \frac{2K}{\max(a_k)} \tag{2}$$

where a_k is the hyperfine coupling constant of the kth group of n_k equivalent nuclei with nuclear spin quantum number I_k, and K denotes the total number of groups.

An illustrative example is given in Figure 1 for the phenylcyclazin radical anion in solution.[7] In this case, $\Omega_{\text{ESR}}/\Omega_{\text{ENDOR}} \approx 30$. Since the widths of the ESR and ENDOR lines are similar, this factor is fully gained as a bonus in resolution. Unfortunately, this bonus has to be paid off by the following drawbacks of the ENDOR method: ENDOR signals normally reach only some percent of the ESR signals, which means that one loses sensitivity. In addition, as is obvious from Figure 1, the ENDOR line intensities normally are not proportional to the number of contributing nuclei, while in ESR, the intensity pattern directly reflects the number of nuclei involved. Only in the case of nonsaturating NMR rf fields should ENDOR signal intensities reflect the number of nuclei.[7,8] This alternative, however, often fails experimentally, because the signal-to-noise ratio is further reduced.

It was proposed by Freed[9] in 1969 that by extending ENDOR to electron–nuclear–nuclear triple resonance (TRIPLE) these drawbacks can be overcome. The first successful TRIPLE experiment on radicals in solution was performed by Dinse *et al.*[10] As was demonstrated by Biehl *et al.*,[11] additional information about relative signs of hyperfine couplings can be obtained by applying TRIPLE to different nonequivalent nuclei (general TRIPLE) of radicals in solution. The analog of this experiment in the solid state at low temperature was performed earlier by Cook and Whiffen,[12] who have

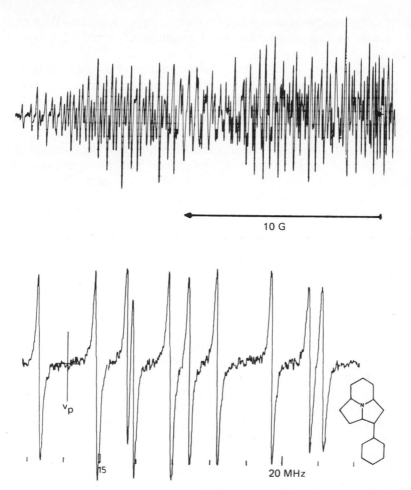

Figure 1. (*Top*) High-field half of the ESR spectrum of the 2-phenyl[3.2.2]azine anion radical (solvent: DME, $T = 210°K$). (*Bottom*) High-frequency half of the "high-power" ENDOR spectrum ($T = 180°K$).

coined the name Double-ENDOR. There have been only very few applications of Double-ENDOR[13-17] in solids, probably owing to experimental difficulties. In addition, it was initially believed[16] that this experiment would only work in the transient limit where all relaxation effects are negligible. In contrast, general TRIPLE resonance, which is performed in the steady state, also gives the sign information.

Like ENDOR, the TRIPLE resonance experiments can be principally considered as NMR experiments, the unpaired electron only serving as a sensitive detector. One observes intensity changes of the saturated ESR because alternative relaxation paths are opened by driving NMR transitions. With respect to the NMR irradiation schemes, we distinguish between the following two cases:

1. Special TRIPLE Resonance. In ENDOR one of the two possible NMR transitions v^+, v^- belonging to a set of equivalent nuclei with resonant frequencies $v^\pm = | -v_0 + am_s |$ is excited to desaturate the ESR saturation. In contrast, in special TRIPLE, both nuclear transitions v^+, v^- are driven simultaneously [m_s denotes the electron spin quantum number, v_0 is the free nuclear Larmor frequency, and a is the hyperfine coupling constant (hfs)].

2. General TRIPLE Resonance. In general TRIPLE, one simultaneously irradiates two nonequivalent nuclei at two of their respective NMR frequencies. Specifically, one observes the intensity change of the ENDOR signal of one nucleus while saturating an NMR transition of the other nucleus.

2. Experimental Aspects

2.1. General Considerations

A TRIPLE-in-solution spectrometer naturally has at least to meet the same requirements as any ENDOR-in-solution spectrometer. Additionally, one has to produce the second pump NMR field with reasonable power levels (up to 10 G in the rotating frame), which, depending on the sample, can differ in frequency from the first NMR field by up to 30 MHz. In the extension of high-power spectrometers described in the literature,[3,18–21] one could make use of two independent tunable resonance circuits. However, because of difficulties in tracking two coupled circuits automatically and independently to different frequencies, the use of a simple " single-coil arrangement " connected to a broad-band rf amplifier is preferable, although the conversion factor between rf power and rf field is thereby decreased. Straightforward calculations show that in principle the actual design of the coil arrangement is immaterial, since the conversion factor mainly depends on the sample volume to be homogeneously irradiated. Additionally, such an arrangement compensates the effect of the "hyperfine enhancement,"[22] since ENDOR signals are a function of induced transition rates δ_n between particular hyperfine levels of an electron–nuclear spin system. These transitions can be classified as pure NMR transitions only to first order in $2\pi a/\gamma_n B_0$, where the hfs constant a is measured in frequency units and B_0 is the static magnetic field. In second order, the transition moment $d_n(d_n^2 \propto \delta_n)$ is given by the well-known expression[22]

$$d_n \propto B_n[1 - m_s(a/v_0)] \tag{3}$$

where B_n is the amplitude of the NMR rf field. On account of the ENDOR resonance condition, which in this context is sufficient to be considered in first order,

$$v^{\pm} = \left| -v_0 + am_s \right| \tag{4}$$

one obtains

$$d_n \propto B_n(v_{ENDOR}/v_0) \tag{5}$$

where v_{ENDOR} now is the resonance frequency of any given ENDOR line. Thus, for d_n to be constant, a requirement that is important when comparing signal intensities at different frequencies, B_n must be kept inversely proportional to the NMR rf frequency. Using the single coil arrangement with reactance L in series with the rf amplifier output and load resistance R, one obtains a frequency response function

$$f(v_{ENDOR}) = [1 + (v_{ENDOR}/v_1)^2]^{-1/2} \tag{6}$$

where $2\pi v_1 = R/L$ is the characteristic frequency for the setup. By choosing v_1 smaller than the free nuclear Larmor frequency, one obtains the desired field-frequency dependence within reasonable frequency ranges. The ideal condition would be $v_1 \ll v_{ENDOR}$, which, however, would further decrease the NMR power-to-field conversion factor. Using 50-Ω rf amplifiers in the first TRIPLE-in-solution experiment, a solenoid with $L = 1.4 \, \mu H$ was therefore placed inside a cylindrical TE_{112} cavity.[10]

Since ENDOR and TRIPLE signal intensities depend strongly on the ESR saturation, one should provide for a constant microwave field over the sample volume. The TE_{112} cavity, however, as well as the commonly used TE_{011} ENDOR cavity,[3,18-21] is not suitable in this respect if long samples are used. Therefore, for subsequent TRIPLE experiments, a cylindrical TM_{110} cavity was chosen.[11] For this mode, the microwave field B_e is constant along the cylinder axis. An experimental comparison of microwave field homogeneity and filling factor for both cavity arrangements clearly favors the TM_{110} cavity. The electric data for the NMR coil are similar to those given above.

Double coding of ENDOR signals by Zeeman field and NMR FM modulation is commonly used[3,18-21] in order to discriminate against spurious signals. These spurious signals result from improper rf shielding and thermal and mechanical instabilities of the microwave cavity. This double coding is a somewhat questionable procedure: First, sinusoidal Zeeman modulation diminishes the sensitivity of the spectrometer by roughly a factor of three. Second, overmodulation of unresolved ESR spectra with amplitudes of typically 5 G contributes 40 kHz to the ENDOR linewidths and thereby obscures the line profiles.

2.2. *Apparatus*

The TRIPLE spectrometer consists basically of a Varian E 12 ESR spectrometer. For general TRIPLE, the block diagram of the additional NMR equipment is shown in Figure 2. For special TRIPLE, the power combiner (ENI PM 400-4) is replaced by a double balanced mixer (ANZAC MD141). The pump rf source (Wavetek 7000) is set to the free nuclear frequency and mixed with the "scan" oscillator frequency, thus producing two side-bands with the right modulation behavior (see Section 2.3).

The cavity–helix arrangement is shown in Figure 3. It differs slightly from the arrangement described earlier.[11] To take advantage of the full homogeneity region of the 12-inch magnet, the cavity length was increased to 38 mm, thereby gaining 40% in active sample volume. The measured microwave and NMR field distributions are shown in Figure 4. The calculated conversion factor for the middle of the NMR helix is 4.1 G_{rot}/A_{eff}. Even when

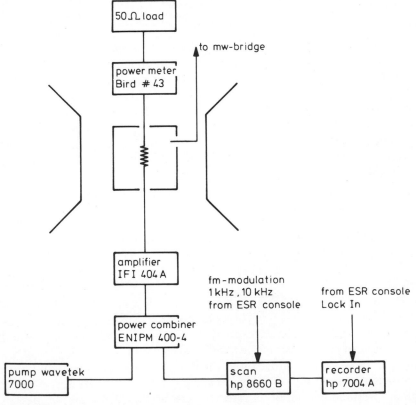

Figure 2. Block diagram of the NMR equipment used for TRIPLE experiments. The basic unit is a Varian E 12 ESR spectrometer.

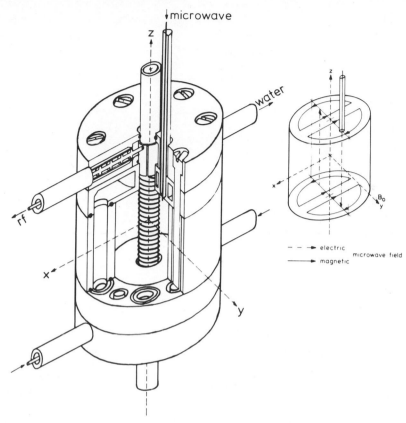

Figure 3. Cutaway view of the microwave cavity and NMR helix of the TRIPLE spectrometer. The field configuration of the TM_{110} mode is also shown.

strongly varying the NMR power, excellent frequency stability of the cavity is maintained due to efficient water cooling. The large water channels are clearly seen in Figure 3. Because of space problems, a coaxial microwave coupling was chosen. To prevent microwave leakage into the NMR channel, multiple $\lambda/4$ rejection filters are connected to both ends of the NMR helix. Compared to the high-power ENDOR spectrometer described earlier,[17] the sensitivity of this spectrometer is increased by about one order of magnitude.

2.3. Modulation Schemes for TRIPLE Resonance

After having discarded the Zeeman modulation on account of the previous general considerations, for general TRIPLE resonance there are still various choices for modulating the rf fields. This can be done by either modulating both NMR rf fields or by modulating the pumping or scanning rf

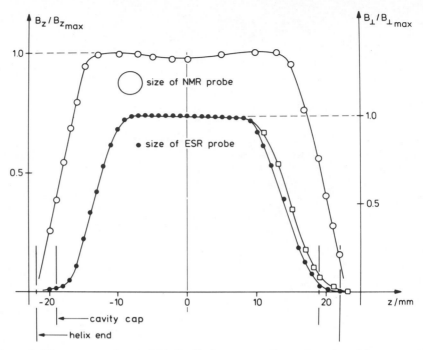

Figure 4. Microwave- and NMR-field distributions versus the cavity axis z. Measurements were performed by ESR and NMR using DPPH and doped water as samples. The size of the samples is depicted. By the ESR measurement, only the effective field components perpendicular to the z axis could be determined. ($\bigcirc$) and ($\bullet$) correspond to $x = y = 0$; ($\square$) corresponds to $x = 0$, $y = 2$ mm.

field. AM modulation, which would be the best choice for effective modulation,[3] causes severe problems; switching the rf power results in mechanical vibrations of the NMR helix and the quartz Dewar to which it is attached, thereby modulating the microwave cavity resonance frequency. Furthermore, the ESR resonance condition is modulated, since the extreme high-field limit is not realized.[3] These difficulties are avoided by FM-modulation. In the case of general TRIPLE, modulation of just one rf field is again preferable because of sensitivity considerations. The following alternatives have been used:

1. FM modulation of the pump field with rate ω_0 and no modulation of the scanning field. The lock-in detector is tuned to $2\omega_0$, thus recording the second derivative of the ENDOR signal. In this modulation scheme, the pump is centered on an ENDOR line, and the TRIPLE spectrum is recorded in the form of absorption lines representing the intensity changes of the pumped ENDOR line versus the frequency of the unmodulated scanning field.[12]

2. FM modulation of the scanning field with rate ω_0 and no modulation of the pump field, the lock-in being tuned to ω_0. The pump again is centered on an ENDOR line. The TRIPLE spectrum in this case looks like a first-derivate ENDOR spectrum, the line intensities, however, being changed by the pump field.

Both modulation schemes yield the same information concerning the intensity changes. Scheme 2 has the advantage of higher resolution in the sense that badly separated ENDOR lines can still be individually pumped. This might be also useful when studying coherence splittings due to large pumping fields.

For special TRIPLE resonance, where pumping and scanning fields are equivalent, modulation of both rf fields gives the maximum sensitivity. The modulation of both rf fields gives the maximum sensitivity. The modulation of the two rf sidebands (see Section 2.2) has to be out of phase by π in order to hit both NMR transitions simultaneously to the same extent. Otherwise, a signal is observed only when the free nuclear frequency oscillator is not properly adjusted.[13]

3. Theoretical Aspects

The theoretical description of a coupled multispin system in solution is generally not an easy task. In recent years, however, there has been considerable progress in this field. This is particularly true for electron–electron and electron–nuclear double resonance, for which a sophisticated theory was developed by Freed.[23–26] Both in NMR and ESR double resonance, one starts with the spin Hamiltonian

$$\mathscr{H}(t) = \mathscr{H}_0 + \mathscr{H}_1(t) + \varepsilon(t) \tag{7}$$

where the static part $\mathscr{H}_0$ contains all the time-independent intramolecular interactions and the Zeeman energies of the various spins. From this operator, all transition frequencies are calculated in first order. $\mathscr{H}_1(t)$ describes all possible relaxation interactions, and $\varepsilon(t)$ represents the interactions of the spins with the coherent radiation fields. Within the limits of certain approximations concerning the time dependence and magnitude of $\mathscr{H}_1(t)$, manageable theoretical expressions were obtained in the frame of the density matrix formalism.[23–26] Whether it is necessary or not to apply the density matrix formalism to the full Hamiltonian [equation (7)], which can be rather elaborate for multiple resonance experiments, depends on the amount of information one wants to obtain from a specific experiment. Multiple resonance experiments can be classified with respect to two main aspects— linewidth and saturation.

As long as $T_{2i}^{-1} \gg \gamma_i B_1, \gamma_i B_2$ (where T_{2i}^{-1} is the apparent homogeneous linewidth of any transition in the spin system), the coherence of the radiation fields (with amplitudes B_1, B_2) will have negligible effects on the observed line shapes. In this limit, low-saturation experiments, where $\sigma_i \approx 1$ ($\sigma_i = \gamma_i^2 B_1^2 T_{1i} T_{2i}$, $\gamma_i^2 B_2^2 T_{1i} T_{2i}$), can be understood phenomenologically in terms of first-order rate equations. They couple the population deviations from the Boltzmann distribution with lattice-induced and rf-induced transition rates[27] (generalized Overhauser effect). This leads to the well-known electric circuit analogy,[28] where the population deviations are identified with voltage drops, which are a linear function of the induced currents, representing the rf-induced transition rates. The conductances in the electric network are proportional to the lattice-induced transition rates.

INDOR (internuclear double resonance) and ELDOR (electron–electron double resonance) can be understood in this limit with respect to T_{2i}^{-1} as low-saturation experiments. Still assuming $T_{2i}^{-1} \gg \gamma_i B_1$, $\gamma_i B_2$ and $T_{2i}^{-1} \gg T_{1i}^{-1}$, ENDOR on the other hand must be considered as a typical nonlinear saturation experiment,[27] where $\sigma \gg 1$. For ENDOR, therefore, the rf-induced nuclear transition rates must be treated to effectively increase the lattice-induced nuclear transition rates, thereby opening a relaxation bypass for the pumped electron spins. In the electric circuit analogy, this is accounted for by inserting resistances parallel to the nuclear relaxation resistances.

In the opposite case, where $T_{2i}^{-1} \ll \gamma_i B_1, \gamma_i B_2$, the coherent nature causes subtle line-shape effects and line splittings. Examples are NMR tickling and decoupling,[4,29,30] as well as coherence effects observable in ENDOR.[19–25] Tickling and decoupling effects in NMR are often visualized using the rotating frame approach, which is equivalent to solving the density matrix neglecting any contribution from $\mathcal{H}_1(t)$ in equation (7). In ENDOR, this is not possible, but the density matrix formalism with the full Hamiltonian [equation (7)] has to be applied, since the limiting case $T_{2i}^{-1} \ll \gamma_i B_1, \gamma_i B_2$ is never fully realized. Principally, the same holds for TRIPLE resonance.

Even for the simple special TRIPLE case, which can be treated in the four-level transition scheme, coherence effects have been neglected in density matrix calculations published so far.[9,10] General TRIPLE will be much more complicated, since the transitions in an eight-level scheme have to be handled. Additional complications arise from degeneracies of certain transition frequencies and from the large number of multiple quantum transitions. Only recently, full density matrix calculations for TRIPLE have been performed by Plato[31] in this laboratory. Examples given in the following section are first results using this program.

From an application point of view, in ENDOR one is primarily interested in obtaining large line intensities, the line-shape information being used at a second stage. In TRIPLE resonance, the main interest is focused on intensity changes of the ENDOR spectrum due to the second NMR field. In the first instance, the following theoretical description of TRIPLE is restricted to the limiting case, where coherence effects can still be neglected, so that the electric circuit analogy can be used. In a second step, coherence effects will be briefly discussed (see Section 3.2.3).

3.1. *Special TRIPLE in the Four-Level Transition Scheme*

The four-level scheme, where the electronic and nuclear lattice-induced transition rates are denoted by W_e and W_n and cross-relaxation has been neglected, is shown in Figure 5. If one defines in the network analog the ENDOR enhancement E of the ESR absorption due to the rf-induced (rates δ_n) change of the ESR input conductance W by

$$E = \frac{W(\delta_n \text{ on}) - W(\delta_n \text{ off})}{W(\delta_n \text{ off})} \tag{8}$$

one obtains for the two limiting cases

$$\text{(i)} \quad \delta_n \to 0, \qquad E_{\text{ENDOR}} = \frac{1}{2} \frac{\delta_n}{2W_e + W_n} \tag{9}$$

$$\text{(ii)} \quad \delta_n \gg W_e, W_n, \qquad E_{\text{ENDOR}} = \left[2\left(2 + \frac{W_n}{W_e} + \frac{W_e}{W_n} \right) \right]^{-1} \tag{10}$$

In low-power ENDOR, therefore, the signal is proportional to δ_n which competes with the sum of the relaxation rates $2W_e + W_n$. In high-power ENDOR the signal is independent of δ_n. The magnitude of E then depends on the relaxation bottleneck formed by W_n or W_e, whichever is smaller. When the remaining W_n in (ii) is shorted out by pumping the second NMR transition with δ_n, i.e., when special TRIPLE is performed, E will always increase. For $W_n \ll W_e$ the limiting value is given by

$$E_{\text{TRIPLE}} = \frac{\delta_n}{\delta_n + 2W_e} \tag{11}$$

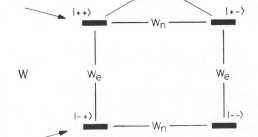

Figure 5. Four-level $S, I = \frac{1}{2}$ spin system classified by $|m_S, m_I\rangle$. Electronic and nuclear relaxation rates are designated by W_e, W_n, the rf-induced rate by δ. W denotes the total input conductance for the observing ESR transition.

which approaches 1 for $\delta_n \gg W_e$. Roughly a factor of 8 should be obtainable as compared to the optimum ENDOR case ($E_{\text{ENDOR}} = \frac{1}{8}$ for $W_e = W_n$). For $W_e > W_n$, a situation frequently met in the experiment, the gain in signal enhancement should be even better. It is therefore always advisable to distribute a given NMR power among both NMR transitions instead of driving just one NMR transition and being stuck in a relaxation bottleneck.

For $W_e > W_n$, the low-power ENDOR enhancement is rather independent of W_n. Since $\delta_n \propto |\langle m_{I_i} + 1 | I_i^+ | m_{I_i} \rangle|^2$, the signal intensities for different nuclei with different spin I_i are proportional to $[I_i(I_i + 1) - m_{I_i}(m_{I_i} + 1)]$, no matter whether they have different W_{n_i}, but depending on the ESR hyperfine component, which is chosen to observe ENDOR. From this consideration, it can be shown that for a system consisting solely of nuclei with $I = \frac{1}{2}$, the *relative* signal intensities of different ENDOR lines are proportional to the number of nuclei contributing to the lines— independent of the hyperfine component chosen.[26] This statement, therefore, is also valid for experiments where Zeeman modulation is applied, thereby averaging the ENDOR effect over several hfs components.[10] The same conclusion essentially holds for high-power TRIPLE.

Besides its increased sensitivity, special TRIPLE has also the advantage of higher resolution. Simplified density matrix calculations have shown[9,10] that at a given power level, the effective NMR saturation, which determines the observed linewidth, is smaller in TRIPLE than in ENDOR. This is due to the coupled effects of both induced NMR transitions.[9] To further clarify this point, detailed calculations on the basis of the full density matrix formalism have been performed.[31] For these calculations, a model radical was chosen, which is representative with respect to relaxation times and hfs coupling constant. It turned out that for NMR saturation parameters larger than 10, TRIPLE linewidths are reduced by 30–50% as compared to the ENDOR linewidths, while the TRIPLE line intensities are still larger by a factor of 4. Details of the model system are given in Section 3.2.2.

3.2. *General TRIPLE in the Eight-Level Transition Scheme*

The solution of the time-independent part of the spin Hamiltonian for the simplest three-spin systems $S = I_1 = I_2 = \frac{1}{2}$ in a magnetic field with isotropic hfs coupling constants leads to first-order energy levels in the basis $|m_S m_{I_1} m_{I_2}\rangle$, which can be clearly arranged to form the eight corners of a cube. Every two corners i, j are then connected by the various relaxation paths W_{ij}. In Figure 6, this energy level arrangement is depicted for the two different cases $a_1, a_2 > 0$ and $a_1 > 0, a_2 < 0$.

There are four ESR transition frequencies, $\nu_{\text{ESR} 1, 2, 3, 4} = |\nu_{e0} - m_{I_1} a_1 - m_{I_2} a_2|$, which are the same for both sign combinations of a_1 and a_2, so that the signs cannot be distinguished from the ESR spectrum. There

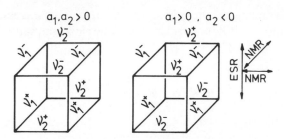

Figure 6. Energy level arrangements for a three-spin S, I_1, $I_2 = \frac{1}{2}$ system with equal and opposite signs of the hyperfine couplings, showing the first-order NMR transitions. The ESR transitions are not distinguished.

are also four different NMR transitions, each of them doubly degenerate in frequency:

$$v_1^\pm = |-v_{n0} + a_1 m_S| \qquad \text{for nucleus 1} \qquad (12)$$

$$v_2^\pm = |-v_{n0} + a_2 m_S| \qquad \text{for nucleus 2} \qquad (13)$$

Again, both sign combinations result in the same ENDOR line frequencies, while the ENDOR line intensities might be quite different, depending on which ESR line was chosen to observe the NMR transitions. With the second NMR field, one can perform three different types of TRIPLE resonance experiments. They can be visualized by three different geometrical figures—a pyramid, a tetrahedron, and a square (Figure 7). They are derived from the cubes by contracting corners that are connected by induced NMR transitions. This represents the limiting case of highly saturated NMR transitions, where the populations of the connected states are equalized. In this representation, an ENDOR experiment forms a prism (Figure 7). Because of short-circuiting of

ENDOR

TRIPLE (general)

Figure 7. Geometric representation for the different multiple resonance experiments in the three-spin system. The observing ESR transitions are marked by circles.

TRIPLE (general)

TRIPLE (special)

several NMR conductances, there is obviously *always* an increase of the ESR input conductance in all four experiments. Therefore, the saturated ESR line will always be enhanced.

3.2.1. Neglect of Cross-Relaxation and Coherence Effects

To handle the eight-level spin system, we assume the nuclear relaxation rates to originate solely from electron-nuclear dipolar interaction (END). The nuclear relaxation rates W_{ni} are assumed to be all equal to W_n. To begin with, we furthermore exclude nonsecular cross-relaxation processes. These assumptions are a realistic approach for highly diluted samples at sufficiently low temperatures containing only protons as nuclear spins with not too different isotropic and anisotropic hfs couplings. All the electron relaxation processes are assumed to be mainly determined by g-factor anisotropy and spin rotational interaction. They are collected in a common W_e, which then is independent of the nuclear spin quantum numbers m_{I_i}.

With the additional assumption $W_e \gg W_n, \delta_n$, this simple system can be treated analytically.[11] It was found that the increase of any ESR input conductance W, caused by adding conductances δ_n to W_n, differs specifically for each of the four experiments. It is convenient to define a quantity

$$V = S_{\text{TRIPLE}}/S_{\text{ENDOR}} \tag{14}$$

where

$$S_{\text{TRIPLE}} \propto \left| W\begin{pmatrix} \text{pump on} \\ \text{scan \ \ on} \end{pmatrix} - W\begin{pmatrix} \text{pump on} \\ \text{scan \ off} \end{pmatrix} \right| \tag{15}$$

$$S_{\text{ENDOR}} \propto \left| W\begin{pmatrix} \text{pump off} \\ \text{scan \ on} \end{pmatrix} - W\begin{pmatrix} \text{pump off} \\ \text{scan \ off} \end{pmatrix} \right| \tag{16}$$

are proportional to TRIPLE and ENDOR signal intensities obtained by modulation scheme (2) (Section 2.3).

For the general case $W_e \lesssim W_n$, the equivalent network problem was solved by numerical calculations.[11] The inequalities

$$V_{\text{pyr}} \leq 1, \qquad V_{\text{tetra}} \geq 1$$

$$V_{\text{spec TRIPLE}} > V_{\text{tetra}} \tag{17}$$

were found valid for all values of $10^{-3} \leq W_e/W_n \leq 10^4$ and $10^{-2} \leq \delta_n/W_n \leq 10^4$. The dependencies of the V on W_e/W_n and δ_n/W_n are shown in Figure 8.

In a next step, the restrictions $W_{n_1} = W_{n_2}$ and $\delta_{n_1} = \delta_{n_2}$ have been dropped. Numerical calculations again proved the validity of expressions (17)

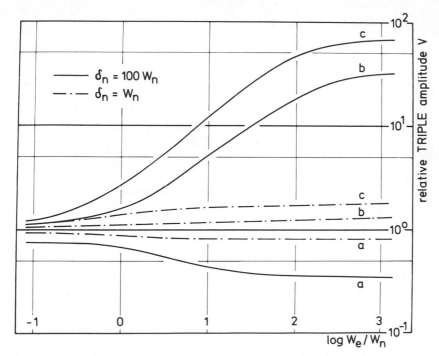

Figure 8. Calculated dependences of TRIPLE amplitude ratios V on W_e/W_n for the eight-level spin system using the network analog. The results for V_{pyr}, V_{tetra}, and $V_{spec\ TRIPLE}$ are designated by a, b, c, respectively. (Reproduced from Biehl *et al.*[11] with permission. Copyright by the American Physical Society.)

to hold for all combinations of the parameters within $10^{-3} \le W_e/W_{n_i} \le 10^4$; $10^{-2} \le \delta_{n_i}/W_{n_i} \le 10^4$; $10^{-2} \le W_{n_1}/W_{n_2|} \le 10$; $10^{-2} \le \delta_{n_1}/\delta_{n_2} \le 10^2$. Therefore, within the scope of this model, from experimentally determined values V, it is easily deduced whether a pyramid or tetrahedron experiment was performed. Using this approach, one directly obtains the sign of a_2 relative to a_1 when pumping an ENDOR transition of nucleus 1.

The numerical results for V_{pyr} and V_{tetra} can be visualized in the following way (Figure 6): Starting from the level arrangement that represents the ENDOR experiment (prism), it can be seen that for the pyramid experiment a relaxation path parallel to ENDOR is opened, thus diminishing the observable ENDOR effect, because this parallel path is not modulated. For the tetrahedron experiment, on the other hand, an additional relaxation path in series of ENDOR is opened and therefore will enhance the observable effect. From this picture, it is obvious that in the case of special TRIPLE the modulation of both rf fields will yield the maximum effect.

3.2.2. Cross-Relaxation Effects

Still restricting ourselves to END relaxation, at elevated temperatures, the neglect of cross-relaxation (rates $W_{x_{1,2}}$) is no longer justified, since $W_n/W_{x_{1,2}} \propto (\eta/T)^2$ (η is the viscosity of the solvent). The ratio of rates W_{x_1} ($|+-\rangle \leftrightarrow |-+\rangle$) to rates W_{x_2} ($|++\rangle \leftrightarrow |--\rangle$) is equal to $\frac{1}{6}$. As long as W_{x_2} is still comparable with W_n, network model calculations yield qualitatively the same result as before. The quantities V now depend on the ESR line chosen for detection, but for any particular ESR line, expressions (17) hold. This is no longer true for $W_{x_2} \gg W_n$, because then the pumping rf field might not have any effect on the ENDOR at all.

This result can be easily visualized by again resorting to the geometric figures in Figure 6. If, for instance, W_{x_1} is the dominant relaxation rate for one nucleus, the eight-level cube can be cut along W_{x_1} into two prisms (Figure 9), because certain transitions are already "shorted out" by W_{x_1}. In this case, one can see that by just one rf field, a pyramid or a tetrahedron experiment is performed, depending on the relative signs of a_1 and a_2. Therefore, the sign information is obtained by doing an ENDOR experiment. This was demonstrated earlier[32] for a radical–ion pair, where a dominant W_{x_1} relaxation was present due to modulation of the isotropic alkali metal hfs. Effects similar to those produced by W_{x_1} relaxation can, however, also result from a correlated modulation of the quadrupole and anisotropic hyperfine couplings.

Concluding, one can say that cross-relaxation does not hamper the applicability of TRIPLE resonance.

3.2.3. Coherence Effects

From Figure 8, it is obvious that the differences of the calculated ratios V_{tetra} and V_{pyr} become quite small when $W_e \approx W_n$, even when large induced nuclear transition rates δ_n are used. For very large δ_n the network analog will not give an adequate description, because the condition $T_{2ni}^{-1} < \gamma_i B_1, \gamma_i B_2$ is no longer fulfilled in an actual experiment on a dilute radical sample. In this

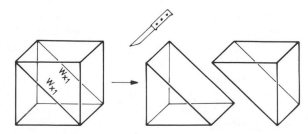

Figure 9. Eight-level spin system with dominant W_{x_1} cross-relaxation for one nucleus. The ENDOR response of this system is not changed in first order when cutting the cube along W_{x_1}.

situation, distinct coherence effects in form of line splittings do occur in ENDOR,[19,25] making the definition of the quantities V meaningless. Therefore, density matrix calculations for the full Hamiltonian [equation (7)] of a hypothetical two-proton radical were performed considering all possible multiple quantum transitions.[31] The relaxation part of $\mathscr{H}_1(t)$ was treated in the approximation given by Freed.[25]

For this radical, a rotational correlation time $\tau_{rot} = 0.5 \times 10^{-9}$ sec was assumed, and the following relaxation processes have been considered (using standard notation[25]):

1. An effective g-factor anisotropy of $Tr(\Delta g^2) = 1 \times 10^{-4}$. This large value was chosen to account phenomenologically for the combined effects of spin rotational interaction and "true" g-factor anisotropy, because both processes contribute solely to electron spin relaxation.[25]

2. A hyperfine anisotropy of $Tr(A^2) = 4 \times 10^{14}$ sec^{-2}. This value of the electron–nuclear dipolar interaction is adapted to the anisotropies frequently found for protons in aromatic hydrocarbon radicals. For simplicity, the same value was assigned to both nuclei.

Terms that mix g-factor and hyperfine anisotropy have been neglected in $\mathscr{H}_1(t)$. For the static part $\mathscr{H}_0$, $v_e = 9.3$ GHz and isotropic hfs couplings $a_1 = 1$ MHz, $a_2 = 3$ MHz have been assumed. The line separations, therefore, are much larger than $\gamma_n B_1$, $\gamma_n B_2$. This set of parameters leads to $W_e = 5.2 \times 10^3$ sec$^{-1} \approx W_n = 4.9 \times 10^3$ sec^{-1} and $T_{2e} = 1.7 \times 10^{-7}$ sec, $T_{2n} \approx 5 \times 10^{-5}$ sec. (The NMR lines are not given by a single Lorentzian with a unique T_{2n} but, because of the degeneracies in the NMR transitions, by a superposition of two Lorentzians with slightly different values of T_{2n}). With reference to the experimental conditions for optimum ESR enhancement, the ESR saturation parameter was set to $\sigma_e = 1$. Variation of the nuclear saturation parameters between $1 < \sigma_n < 9$ resulted in TRIPLE lines with Lorentzian shape, the amplitude ratios V_{pyr} and V_{tetra} being in accordance with the results obtained from the network analog. By increasing σ_n to 15 (B_1, $B_2 = 2.9 G_{rot}$), the line shapes for the ENDOR, special TRIPLE, and tetrahedron experiments still are not distorted. An ENDOR enhancement of 5.6% and $V_{spec\ TRIPLE} = 2$, $V_{tetra} = 1.38$ were obtained, which still agree quite well with the network results. The value of $V_{pyr} = 0.48$, however, is by far too small. As can be seen from Figure 8, this value already reaches the limiting value of V_{pyr} in the network calculation for $W_e/W_n \gg 1$ and $\sigma_n \to \infty$. The reason for this discrepancy is obvious from Figure 10, where the differentiated ENDOR and TRIPLE lines for $\sigma_n = 15$ are depicted. The line shape for the pyramid experiment is already distorted; a further increase of σ_n will tend to split this line completely.

The different behavior of this "pyramid" line can be understood from the energy-level cubes in Figure 6. The pyramid represents the only TRIPLE experiment, where all three radiation fields have one energy level in common. Because of the small ESR saturation, electron–nuclear coherence

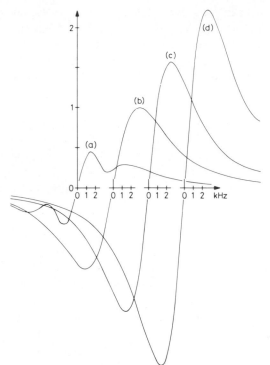

Figure 10. Calculated ENDOR and TRIPLE signals using the full density matrix. Parameters are given in the text. The signals correspond to: (a) a pyramid experiment, (b) an ENDOR experiment, (c) a tetrahedron experiment, (d) a special TRIPLE experiment. The representation is according to modulation scheme 2 in Section 2.3.

effects are suppressed.[19,25] Therefore, for the pyramid configuration—assuming the pump is centered on the transitions v_1^- of nucleus 1—nuclear spin tickling is observed via the ESR enhancement when the second rf field is scanned through the transitions v_2^- for nucleus 2. Principally, this can also be used for sign determination purposes, as is well known from NMR.[33]

Actually, the linewidths observed in low-power ENDOR experiments are larger by a factor of about 4 as compared to the above calculations. This is due to the fact that not all relaxation terms in $\mathcal{H}_1(t)$ have been accounted for. Therefore, these tickling effects often will be masked in the experiment. Furthermore, if there are several equivalent protons, similar effects will show up in the ENDOR spectrum. Nevertheless, this internuclear coherence effect will always tend to give a more pronounced distinction between V_{pyr} and V_{tetra} than is expected from the simple network calculation.

3.2.4. Extension to the Multilevel Transition Scheme

The extension of the simple spin system so far discussed to more realistic cases, which include more than two nuclei all having $I = \frac{1}{2}$ but different hfs constants, is straightforward. In the limit of $W_e \gg W_{n_i}, \delta_{n_i}$, it is always sufficient to consider only the eight-level scheme belonging to the two nuclei of interest. This is possible because from any hyperfine level one can perform NMR transitions belonging to all nuclei, which is the consequence of the simple spin hamiltonian considering only the hfs interactions between one electron and the nuclei. This is no longer true if one accounts for nuclear quadrupole or nuclear dipole interactions (observable in liquid crystals[34] and in the solid state), which lift degeneracies in particular NMR transition frequencies.

The case of a system containing at least two groups of several equivalent protons is more complicated. Since all the interesting time-dependent interactions for these systems are described in the dipolar approximation, it is convenient to divide the spin system into independent subsystems classified by $|J_1, J_2, \ldots, M_1, M_2, \ldots\rangle$, where $0 \leq J_i \leq n_i I_i$ and $-J_1 \leq M_i \leq J_i$, and n_i equals the number of equivalent nuclei in the ith group. Within the limit $W_e \gg [J_i(J_i + 1) - M_i^2] \times (W_{n_i}, \delta_{n_i})$, the reduction to the eight-level problem can again be performed. The enhancements obtained in the different subgroups for the particular ESR transition chosen are then summed up according to their statistical weight. In this case, the inequalities (14)–(17) still hold; the magnitude of V, however, is now dependent on the particular ESR transition monitored. In the case where this transition contains one or more subsystems $|\ldots, J_i, \ldots, J_k, \ldots\rangle$ with $J_i = 0$, the difference between V_{tetra} and V_{pyr} can become quite small, because for nuclei in group k, such subsystems do not respond to the pumping of nuclei in group i.

4. Comparison between Theory and Experiment

4.1. Special TRIPLE Resonance

In Figure 11, the ENDOR and special TRIPLE spectra are shown for the tetraphenylpyrene anion radical.[10] The TRIPLE lines are recorded versus the frequency of the scanning oscillator (see Section 2.2), the line positions, therefore, give directly $a/2$. The ENDOR spectrum has a bad signal-to-noise ratio due to the rather high temperature ($W_e \gg W_n$). Since the NMR transitions are already saturated at the applied field amplitude of $10G_{\text{rot}}$, the ENDOR lines show equal intensity within the noise limits. The special TRIPLE spectrum is recorded at the same total NMR power level; consequently, the NMR field amplitudes per side band are reduced to $7G_{\text{rot}}$. In qualitative accordance with the theoretical predictions, the signal-to-noise is increased in the TRIPLE spectrum; furthermore, the intensity ratios of the TRIPLE lines reflect approximately the number of protons involved in the NMR transitions.

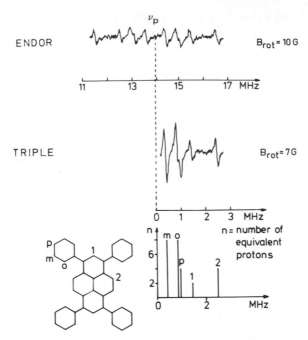

Figure 11. ENDOR and special TRIPLE spectra of the negative tetraphenylpyrene ion. Solvent DME, counter ion Na, $T = 215°K$. (Reproduced from Dinse et al.[10] with permission. Copyright by the American Physical Society.)

A similar behavior is observed for the neutral radical tri-*tert*-butylphenoxyl (TTBP). As is obvious from Figure 12, the ENDOR is saturated at $4G_{rot}$ and therefore does not respond to increasing the NMR power further. The TRIPLE lines, however, continue to increase with increasing NMR power. In the power range given, the proportionality factor is smaller than 1. This does not mean, however, that the induced rate δ_n competes with W_e [equation (11)], since the measured TRIPLE enhancement (normalized to the number of protons) is smaller than 20%. It rather shows that coherence effects are starting to produce additional line broadenings in this power region. Increasing the rf field to $16G_{rot}$ results in pronounced line splittings, so that the information concerning the number of contributing nuclei is lost (Figure 13). From the form of the splitting, however, it can be seen that the line at 2.5 MHz belongs to an even number of protons (splitting into a doublet), whereas the line at 0.5 MHz contains an odd number of protons (splitting into a triplet).[19,25,35] Because of these coherence effects, the theoretically predicted line narrowing, which was calculated for a four-level system, is almost masked in the experimental spectra. Therefore, the expected increase in resolution can hardly be realized for multispin systems.

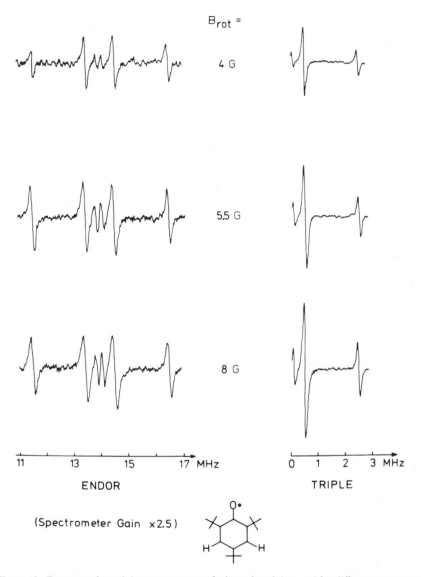

Figure 12. ENDOR and special TRIPLE spectra of tri-*tert*-butylphenoxyl for different NMR power levels. Solvent mineral oil, $T = 300°K$; note the change in spectrometer gain. (Reproduced from Dinse *et al.*[10] with permission. Copyright by the American Physical Society.)

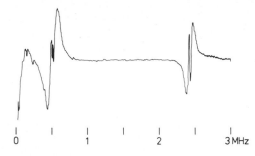

Figure 13. Nuclear coherence splitting in the special TRIPLE spectrum of tri-*tert*-butylphenoxyl at high NMR power levels $\approx (16 G_{rot})^2$.

4.2. General TRIPLE Resonance

In order to test the predictions concerning the intensity changes when pyramid or tetrahedron experiments are performed, perinaphthenyl (PNT), TTBP, and pyrene anion have been selected for which the signs of the hfs couplings are reliably known.

The PNT radical is a very simple molecule that only contains two different groups of protons. The ESR spectrum is completely resolved and therefore allows the dependence of the TRIPLE effect on the M_1, M_2 components to be studied. In Figure 14, ENDOR and TRIPLE spectra are shown that were obtained by saturating the M_1, $M_2 = \frac{1}{2}$, 0 component of the ESR spectrum.[11] Since the ENDOR spectrum was not recorded in the low-power limit, the line intensities do not correspond to the number of protons. (The monotonic increase of the intensities with NMR frequency reflects relation

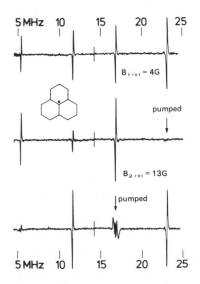

Figure 14. ENDOR and TRIPLE spectra of the perina-phthenyl radical. Solvent mineral oil, $T = 300°$K.

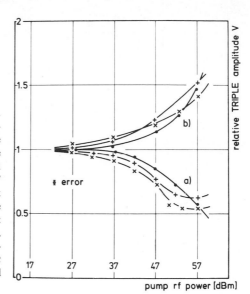

Figure 15. Experimentally determined TRIPLE amplitude ratios of the perinaphthenyl radical versus pump rf power. The pump was held on the 23.1 MHz line. The sets of curves correspond to: (a) a pyramid experiment scanning through the 11.7 MHz line, (b) a tetrahedron experiment scanning through the 16.8 MHz line. The power level of the scan corresponds to: (●) 37 dBm, (×) 47 dBm, (+) 57 dBm. (Conversion factor $0.72 G_{rot}/W^{1/2}$.) (Reproduced with permission from Biehl *et al.*[11] Copyright by the American Physical Society.)

(5), which could not be compensated by the response function (6) over the wide frequency range from 5 to 25 MHz.)

The TRIPLE spectra obtained when pumping the NMR transition at 23.1 MHz or 16.8 MHz show drastic intensity changes; strictly speaking, two lines are enhanced as compared to ENDOR, whereas one line is strongly deenhanced. For PNT, the two hfs couplings have opposite signs.[36] Therefore, when pumping the 23.1 MHz transition with respect to the 16.8 MHz line, a pyramid experiment is performed; with respect to the 11.7 MHz line, a tetrahedron experiment is performed. The measured intensities of these two lines as compared to ENDOR are in accordance with the theoretical predictions $V_{pyr} < 1$, $V_{tetra} > 1$. The same result naturally holds when the 16.8 MHz transition is pumped (Figure 14). It is therefore sufficient to pump any one of the NMR transitions in order to obtain the full sign information.

In order to verify the conclusions drawn in Section 3.2, V_{pyr} and V_{tetra} were measured as functions of the scan and pump power. The result is depicted in Figure 15 and clearly shows that, for any power combination within the available power range, the inequalities $V_{pyr} < 1$, $V_{tetra} > 1$ safely hold. Their validity was also proved by studying the TRIPLE spectra of PNT on different ESR components and at various temperatures.

The radical TTBP was chosen because it consists of three groups of equivalent protons, two of which have large resultant nuclear spins. The hfs coupling constants are $a_1 = +0.2$ MHz (18 protons), $a_2 = +1.0$ MHz (9 protons), and $a_3 = +4.9$ MHz (2 protons).[37,38] The small magnitude of a_2, which is comparable to the unsaturated homogeneous ESR linewidth,

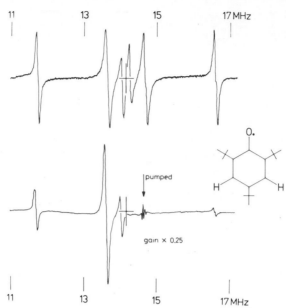

Figure 16. ENDOR and TRIPLE spectra of tri-*tert*-butylphenoxyl. Solvent mineral oil, $T = 300°\text{K}$; note the change in spectrometer gain.

implies that at least two ESR transitions in the eight-level cube are induced simultaneously. While this complication has been discussed in the ENDOR case,[39] its effect on the TRIPLE resonance is hard to predict. The experimental result (Figure 16) shows that even for small couplings, pronounced changes can fortunately be observed in the TRIPLE spectrum. It can also be seen that large resultant nuclear spins do not dilute the TRIPLE effects. The observed intensity changes verify that the signs of the couplings are equal.

The pyrene negative ion can be considered as a typical representative of the large class of hydrocarbon radical ions that have been studied extensively by ESR and ENDOR in solution. The essential difference from the neutral radicals discussed so far lies in a significantly larger ratio W_e/W_n, mainly on account of the smaller viscosity of the solvents. Although this results in smaller absolute ENDOR signals, the TRIPLE effects should become more pronounced with increasing W_e/W_n (Figure 8). The experimental results[11] support these predictions.

The examples discussed so far are distinguished by their completely resolved ENDOR lines. For the applicability of general TRIPLE to low-symmetry radicals, it has to be assured that the previous conclusions are also valid in the case where the NMR transitions overlap within the ENDOR line-widths. Therefore, general TRIPLE was performed on the completely *p*-methyl-substituted pentaphenylcyclopentadienyl radical (MPPCPD). For this radical, the hyperfine couplings, including their signs, are known.[40] Additionally, the ENDOR line of the methyl protons overlaps strongly with the line of the ortho protons (the apparent first- derivative ENDOR linewidth is 90 kHz, the line separation 100 kHz). When pumping successively the high- and low-frequency meta proton transitions, well-pronounced TRIPLE intensity changes were observed that are in accordance with the sign of the couplings.

Even in the rare case of two degenerate NMR transitions belonging to nuclei with opposite signs of their hfs couplings, TRIPLE effects can be detected, provided the number of nuclei per coupling constant is different. This has been proven for the cation radical of diphenylanthracene by means of specific deuteration.

5. Applications of TRIPLE-in-Solution

One should keep in mind that TRIPLE must always be regarded as an extension of ENDOR that yields additional pieces of information. TRIPLE resonance, therefore, supplements ENDOR in the same way as spin tickling or INDOR supplement NMR. The results presented in this section are preliminary data from the authors' laboratory and have not been published elsewhere. To our knowledge, there are not yet TRIPLE-in-solution applications from other groups.

5.1. Fluorenon–Alkali Metal Ion Pairs

The study of aromatic ion pairs in solution is a field of wide interest in physical chemistry. The standard spectroscopic methods in this area comprise NMR,[41] ESR,[42] and ENDOR.[32,43] ESR often is impaired by bad resolution. While ENDOR has been shown to yield a large improvement in resolution, both methods generally suffer from their difficulties in determining the signs of the hfs couplings. In this respect, NMR is superior, allowing the determination of coupling constants, including their absolute signs.[44] The necessary prerequisite, however, for this method is an extremely short T_1 of the electron spins, which is normally achieved by using radical concentrations as high as 1 mole/liter. This is about four orders of magnitude higher than a typical ESR or ENDOR concentration. From this, it is obvious that both methods need not necessarily give the same result, because at large radical concentrations, the probability of dimer and polymer formation is very high. An example for this aspect is the fluorenon–sodium ion pair in tetrahydrofuran.[45,46] In Figure 17, the temperature dependences of the sodium hfs coupling obtained by NMR and TRIPLE are compared. The NMR

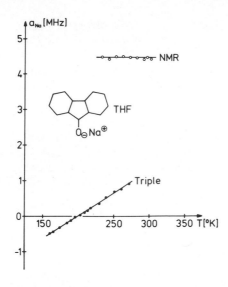

Figure 17. Temperature dependence of the sodium hyperfine splitting in sodium–fluorenon as measured by NMR and TRIPLE. The concentrations are 0.5 moles/liter and 10^{-4} moles/liter, respectively. Solvent THF.

data[46] obtained for a sample of 0.5 mole/liter show that the sodium coupling is essentially temperature independent between 200°K and 320°K and has the value of +4.3 MHz. The ENDOR and TRIPLE data obtained from a sample of approximately 5×10^{-4} mole/liter give a completely different result. There is a linear temperature dependence of the sodium coupling between 150°K and 280°K, with $\partial a/\partial T = 12.5$ kHz/°K, the coupling crossing zero at 190°K. The sign information was easily obtained (Figure 18) by observing the TRIPLE intensities of the alkali and proton lines when a proton transition was pumped. The sign of this proton coupling can be safely assumed to be negative. The physical interpretation for the discrepancy

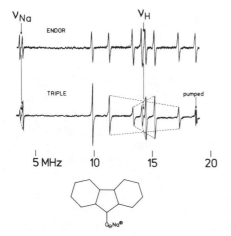

Figure 18. TRIPLE spectrum of the sodium–fluorenon ion pair. Solvent THF, $T = 210$°K.

between NMR and TRIPLE results, although very interesting from the molecular theory point of view, will not be given here, since it is beyond the scope of this chapter.

5.2. *Diphenylpicrylhydrazyl (DPPH)*

This example has been chosen primarily for the purpose of demonstrating the power of the TRIPLE method for analyzing complicated hyperfine structures. A large number of more or less successful attempts to analyze the DPPH hyperfine structure is described in the literature.[21,47] A variety of spectroscopic methods has been employed, such as ESR,[48] NMR,[49] ELDOR,[47] and ENDOR,[21,50,51] both in the solid state and in liquid solution. There is a maximum of seventeen hfs couplings (12 protons, 5 nitrogens) to be expected. The ESR spectrum, therefore, is inhomogeneously broadened to a very large extent. Even by applying elaborate computer simulation techniques, only the large nitrogen couplings $a(N_\alpha) = 22.0$ MHz, $a(N_\beta) = 26.2$ MHz could be reliably determined.[48] The assignment has been established by ^{15}N substitution of the hydrazine nitrogen. With the aid of ELDOR,[47] these results have been finally assured. The other hfs couplings cannot be determined by ESR. NMR results in the solid state show that there are positive and negative proton couplings. Due to hfs anisotropies, the resolution was not sufficient to allow for a complete analysis.[49] The ENDOR-in-solution results published so far[21,50,51] are not thoroughly analyzed and are inconsistently interpreted.

In Figure 19, the low-power ENDOR spectra of DPPH recorded in this laboratory are shown. By using inert solvent mixtures, ENDOR could be detected over a wide temperature range (160–350°K). The ENDOR spectrum at 180°K yields eleven different proton couplings. The doublet around 16 MHz in Figure 19—as well as its mirror image on the low-frequency side—could be further resolved into a triplet under high-resolution conditions, as is shown in Figure 20. From the least-squares fit, carried out with three Lorentzians, couplings, intensities, and linewidths were evaluated. The twelfth proton line is either not resolved or is too weak to be detected. This could be due to a short T_{2e} combined with a very small coupling[29] or due to additional line broadening because of a thermal jump process.[52] With the aid of high-power special TRIPLE, the signal-to-noise ratio could be further improved by about a factor of 4 without any significant change in the appearance of the spectrum.

We have tried to correct the observed intensities for both effects, but an intensity ratio of about 2 : 1 could not be obtained. From studying the spectra over a large temperature range, the presence of at least one jump process affecting several hfs couplings could be safely established. In Figure 19, this is indicated by arrows connecting the corresponding lines. From this,

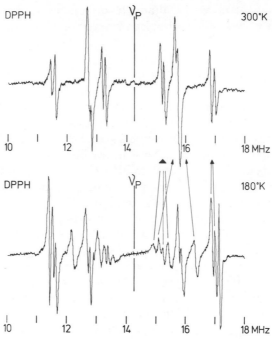

Figure 19. Low-power ENDOR spectra of DPPH at different temperatures. Solvent toluene–mineral oil). Temperature-dependent line positions are indicated by arrows.

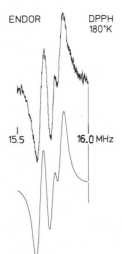

Figure 20. 16-MHz region of the experimental DPPH ENDOR spectrum under high-resolution conditions with simulation. FM deviation 5 kHz, $B_{scan} = 3G_{rot}$.

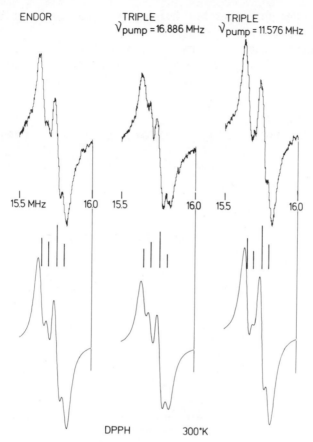

Figure 21. 16-MHz region of the ENDOR and TRIPLE spectra of DPPH. (*Top*) Experimental spectra ($B_{pump} = 10G_{rot}$). (*Bottom*) Simulated spectra using four Lorentzians of 50-kHz width. The stick diagrams show line positions and line amplitudes.

the coupling of the twelfth proton is deduced. This conclusion is further supported by the general TRIPLE experiments. For the spectrum at 180°K, assuming the largest coupling to be negative, the following sign sequence was obtained: $(-), (-), (-), (+), (-), (-), (-), (++), (+), (+), (+)$. For 300°K the sequence $(-), (--), (-), (-), (+), (--), (++), (++)$ was found. In these sequences, the number of protons per resolved line is indicated. In Figure 21, the group of lines around 16 MHz at 300°K is shown under high resolution ENDOR and TRIPLE conditions. For the ENDOR simulation, four Lorentzian lines of equal width with different intensities were required. The TRIPLE spectra (pumping the largest proton coupling at the high- or low-frequency side) then could be fitted consistently solely by varying the line intensities but keeping line positions and linewidths constant (Figure 20).

From the intensity variations, the above sign sequence was concluded. The complete set of coupling constants at 180°K ordered according to their absolute values are (experimental error ± 5 kHz for protons, ± 10 kHz for nitrogens):

Protons: $-5.819, -5.542, -5.328, +4.180, -3.334, -3.234,$
            $-3.059, +2.378, +2.378, +2.000, +1.773, +1.395$ MHz

$^{15}N_\beta$: $+38.32$ MHz

$^{14}N_\alpha$: $+22.30$ MHz

On theoretical grounds, the six positive couplings are assigned to the six meta protons. We refrain from further assignments, because the appropriate discussion is beyond the scope of this chapter but will be given elsewhere.[53]

The sign discussion was based on the assumption of a negative sign of the largest proton coupling. To assure this, a general TRIPLE experiment was performed where this sign was checked against the sign of the large nitrogen couplings. To avoid overlap between nitrogen and proton transitions, a labeled DPPH was used, where the β position was 90% enriched with ^{15}N. In Figure 22, the nitrogen ENDOR is shown. In comparison to the proton ENDOR, temperature, NMR power, and FM-modulation index had to be increased to obtain reasonable signals. In the TRIPLE spectrum, the ^{15}N high-frequency line was pumped. This experiment delivered an opposite sign for the three largest proton couplings with respect to the nitrogen coupling. Since from the interproton high-resolution TRIPLE the same sign for the three largest proton couplings was established, the given sign sequences are confirmed, as there is no doubt about the positive sign of the nitrogen coupling.

6. Conclusion

There are mainly two aspects for extending ENDOR to TRIPLE resonance—signal enhancement and sign determination. Signal enhancement is most favorably achieved by special TRIPLE resonance, i.e., by pumping both NMR transitions for a group of equivalent nuclei by two modulated rf fields. Signal enhancements by a factor of 5 are easily achievable. Attempts to further increase this factor by increasing the NMR power will probably fail, since severe line splittings then occur due to coherence effects.

Couplings as small as a few kilohertz cannot be detected by ENDOR at all, leaving only NMR as the adequate method. The sign determination by general TRIPLE compares favorably with other commonly used methods

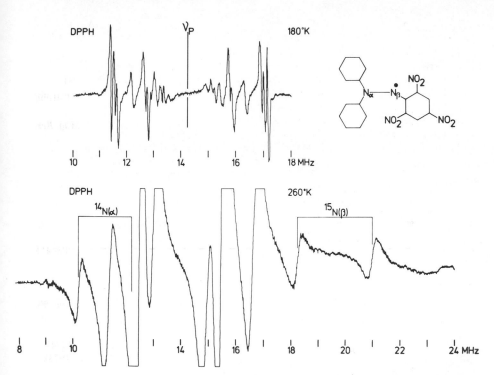

Figure 22. Part of the proton, ^{15}N, and ^{14}N ENDOR spectra of DPPH observed at the center of the ESR spectrum. The picryl–nitrogen transitions could not be detected. FM deviation 100 kHz, $B_{scan} = 7G_{rot}$.

since (1) the NMR contact shift method is restricted to highly concentrated solutions and therefore is often hampered by solubility and stability problems of the radicals; (2) ESR and ENDOR in liquid crystals are restricted to systems for which the hfs anisotropies and ordering parameters have to be known; (3) Analysis of ESR linewidth variations requires a highly resolved hyperfine spectrum. Compared to the NMR method, general TRIPLE has the disadvantage of giving only the relative sign information.

There still remains the problem of unambigiously determining the number of protons contributing to a particular ENDOR line, although TRIPLE resonance is a first step toward a solution.

ACKNOWLEDGMENT

We thank Dr. M. Plato and Dr. K. P. Dinse for many helpful discussions. In particular, we thank Dr. Plato for calculating the ENDOR and TRIPLE spectra and H. Zimmermann, Heidelberg, for synthesizing the ^{15}N-labeled DPPH.

References

1. K. Möbius, *Ber. Buns. Ges.* **78**, 1116–1125 (1974).
2. U. Haeberlen, *Adv. Mag. Res.* suppl. 1 (1976).
3. J. S. Hyde, *J. Chem. Phys.* **43**, 1806–1818 (1965).
4. W. von Philipsborn, *Angew. Chem.* **83**, 470–489 (1971).
5. R. W. Field, G. A. Capelle, and M. A. Revelli, *J. Chem. Phys.* **63**, 3228–3237 (1975).
6. J. S. Hyde, in *Magnetic Resonance in Biological Systems* (A. Ehrenberg, B. G. Malmström, and T. Vänngard, eds.), p. 63, Pergamon Press, Oxford (1967).
7. F. Gerson, J. Jachimowicz, K. Möbius, R. Biehl, J. S. Hyde, and D. S. Leniart, *J. Mag. Res.* **18**, 471–484 (1975).
8. D. S. Leniart, H. D. Connor, and J. H. Freed, *J. Chem. Phys.* **63**, 165–198 (1975).
9. J. H. Freed, *J. Chem. Phys.* **50**, 2271–2272 (1969).
10. K. P. Dinse, R. Biehl, and K. Möbius, *J. Chem. Phys.* **61**, 4335–4341 (1974).
11. R. Biehl, M. Plato, and K. Möbius, *J. Chem. Phys.* **63**, 3515–3521 (1975).
12. R. J. Cook and D. H. Whiffen, *Proc. Phys. Soc. London* **84**, 845–898 (1964).
13. N. S. Dalal and C. A. McDowell, *Chem. Phys. Lett.* **6**, 617–619 (1970).
14. W. Kolbe and N. Edelstein, *Phys. Rev.* **B4**, 2869–2875 (1971).
15. J. A. R. Coope, N. S. Dalal, C. A. McDowell, and R. Srinivasan, *Molec. Phys.* **24**, 403–415 (1972).
16. J. M. Baker and W. B. J. Blake, *Phys. Lett.* **A31**, 61–62 (1970).
17. D. A. Hampton and Grace C. Moulton, *J. Chem. Phys.* **63**, 1078–1082 (1975).
18. A. H. Maki, R. D. Allendoerfer, J. C. Danner, and R. T. Keys, *J. Amer. Chem. Soc.* **90**, 4225–4231 (1968).
19. K. P. Dinse, R. Biehl, and K. Möbius, *Z. Naturforsch.* **A28**, 1069–1080 (1973).
20. K. Gruber, J. Forrer, A. Schweiger, and Hs. H. Günthard, *J. Phys.* **E7**, 569–574 (1974).
21. N. S. Dalal, D. E. Kennedy, and C. A. McDowell, *J. Chem. Phys.* **59**, 3403–3410 (1973).
22. D. H. Whiffen, *Molec. Phys.* **10**, 595–596 (1966).
23. J. H. Freed, *J. Chem. Phys.* **43**, 2312–2332 (1965).
24. J. H. Freed, *J. Phys. Chem.* **71**, 38–51 (1967).
25. J. H. Freed, D. S. Leniart, and J. S. Hyde, *J. Chem. Phys.* **47**, 2762–2773 (1967).
26. J. H. Freed, D. S. Leniart, and H. D. Connor, *J. Chem. Phys.* **58**, 3089–3105 (1973).
27. J. H. Freed, in *Electron Spin Relxation in Liquids* (L. T. Muus and P. W. Atkins, eds.), pp. 503–530, Plenum Press, New York (1972).
28. F. Bloch, *Phys. Rev.* **102**, 104 (1956).
29. W. A. Anderson and R. Freeman, *J. Chem. Phys.* **37**, 85–103 (1962).
30. R. Freeman and W. A. Anderson, *J. Chem. Phys.* **37**, 2053–2073 (1962).
31. M. Plato, unpublished results (1976).
32. H. von Willigen, M. Plato, R. Biehl, K. P. Dinse, and K. Möbius, *Molec. Phys.* **26**, 793–809 (1973).
33. R. Freeman and D. H. Whiffen, *Molec. Phys.* **4**, 321–325 (1961).
34. K. P. Dinse, K. Möbius, M. Plato, and R. Biehl, *Chem. Phys. Lett.* **14**, 196–200 (1972).
35. K. P. Dinse, R. Biehl, K. Möbius, and M. Plato, *J. Mag. Res.* **6**, 444–452 (1972).
36. K. Möbius, H. Haustein, and M. Plato, *Z. Naturforsch.* **A23**, 1626–1638 (1968).
37. K. H. Hausser, *Proc. Colloq. Ampere* **14**, 25–39 (1967).
38. K. P. Dinse, R. Biehl, K. Möbius, and H. Haustein, *Chem. Phys. Lett.* **12**, 399–402 (1971).
39. R. D. Allendoerfer and A. H. Maki, *J. Mag. Res.* **3**, 396–410 (1970).
40. K. Möbius, H. von Willigen, and A. H. Maki, *Molec. Phys.* **20**, 289–304 (1971).
41. E. de Boer and J. L. Sommerdijk, in *Ions and Ion Pairs in Organic Reactions* (M. Szwarc, ed.), Vol. 1, Chap. 7, John Wiley and Sons, New York (1974).
42. N. M. Atherton, in *IUPAC 23rd International Congress*, Vol. 4, pp. 469–483, Butterworth's, London (1971).

43. N. M. Atherton and B. Day, *J. Chem. Soc.* **69**, 1801–1807 (1973).
44. E. de Boer and H. von Willigen, in *Progress in NMR Spectroscopy* (J. W. Emsley, J. Feeney, and L. H. Suttcliffe, eds.), Vol. 2, pp. 111–161, Pergamon Press, Oxford (1967).
45. N. Hirota, *J. Amer. Chem. Soc.* **89**, 32–41 (1967).
46. G. W. Canters and E. de Boer, *Molec. Phys.* **26**, 1185–1198 (1973).
47. J. S. Hyde, R. C. Sneed, Jr., and G. H. Rist, *J. Chem. Phys.* **51**, 1404–1416 (1969).
48. N. W. Lord and S. M. Blinder, *J. Chem. Phys.* **34**, 1699–1708 (1961).
49. J. Heidberg, J. A. Weil, G. A. Janusonis, and J. K. Anderson, *J. Chem. Phys.* **41**, 1033–1044 (1964).
50. N. S. Dalal, D. E. Kennedy, and C. A. McDowell, *J. Chem. Phys.* **61**, 1689–1697 (1974).
51. N. S. Dalal, D. E. Kennedy, and C. A. McDowell, *Chem. Phys. Lett.* **30**, 186–189 (1975).
52. C. von Borczyskowski, K. Möbius, and M. Plato, *J. Mag. Res.* **17**, 202–211 (1975).
53. S. E. O'Connor, R. I. Walter, R. Biehl, and K. Möbius, *J. Phys. Chem.* (to be published).

Index